Deformation of Three-layer Structural Elements in Thermal Radiation Fields

Starovoitov Eduard · Michael Zhuravkov ·
Denis Leonenko · Yongtao Lyu

Deformation of Three-layer Structural Elements in Thermal Radiation Fields

Springer

Starovoitov Eduard
Department of Structural Mechanics,
Geotechnics and Building Structures
Belarusian State University of Transport
Gomel, Belarus

Michael Zhuravkov
Department of Theoretical and Applied
Mechanics
Belarusian State University
Minsk, Belarus

Denis Leonenko
Department of Structural Mechanics,
Geotechnics and Building Structures
Belarusian State University of Transport
Gomel, Belarus

Yongtao Lyu
Department of Engineering Mechanics
Dalian University of Technology
Dalian, Liaoning, China

ISBN 978-981-97-7216-2 ISBN 978-981-97-7217-9 (eBook)
https://doi.org/10.1007/978-981-97-7217-9

This Springer imprint is published by the registered company Springer Nature Singapore Pte Ltd.
The registered company address is: 152 Beach Road, #21-01/04 Gateway East, Singapore 189721, Singapore

Contents

Chapter 1
Introduction

1.1 Introduction

The demand for the use of laminated thin-walled structural elements in aircraft, rocket, machine, instrumentation and shipbuilding, energy production and transport is growing rapidly. This necessitates the development of mathematical models and methods of their calculation for different types and kinds of loads. Rods, plates and shells having a layered structure are usually recruited from materials with significantly different physical and mechanical properties. Bearing layers made of high strength and stiffness materials are designed to take up most of the mechanical load. Bonding layers, which serve to form a monolithic structure, ensure the redistribution of efforts between the load-bearing layers. Another group of layers is designed to protect against thermal chemical, radiation and other undesirable effects. Such combination of layers allows to ensure reliable operation of systems in unfavourable environmental conditions, to create structures combining high strength and rigidity with relatively low weight.

The birth of research on three-layer structural elements was at the end of the first half of the twentieth century. At this time, the first works on the mechanics of multilayer media appear Lekhnitsky [1–3], Chentsov [4, 5].

A significant contribution to the development of the theory of laminated and composite materials and structures in the twentieth century was made by Aleksandrov [6–8], Ambartsumyan [9], Babeshko [10, 11], Bieniek [12], Bolotin [13], Chulkov [14], Ebsioglu [15], Eringen [16], Foss [17], Freudenthal [12], Galimov [18], Gorshkov [19–22], Grigolyuk [14], Grover [23], Habip [24], Kobelev [25], Korolev [26], Mozharovsky [27], K Mushtari [28], Y Novichkov [29], Paimushin [18], Pikul [30], Plantemma [31], Pobedrya [32, 33], Rasskazov [34], Reddy [35], Reissner [36], Shevchenko [11], Solvey [37], Stavsky [38], Stein [39], Teregulov [18], Timoshenko [40], Wang [41].

In recent decades, monographs in this direction have been published: Aghalovyan [42], Altenbach [43], Andreev [44], Carrera et al. [45], Головко et al. [46], Golushko

S. Eduard et al., *Deformation of Three-layer Structural Elements in Thermal Radiation Fields*, https://doi.org/10.1007/978-981-97-7217-9_1

and Nemirovsky [47], Mikhasev and Tovstik [48], Pleskatshevsky et al. [49–51], Yarovaya [52].

In addition, it is necessary to note a number of publications on the study of the stress–strain state of heterogeneous structural elements under quasi-static [53–107] and dynamic [48, 91, 108–146, 147–156] loads.

The first works on the study of plastic properties of materials under cyclic loading were published in the late 19th and early twentieth century. The founders are Bauschinger [156, 157], Huber [158], Masing [159, 160], Mises [161, 162]. In the following years, Ilyushin [163, 164], Moskvitin [165–170], Bell [171], Makhutov [172], Vasin [173], Goldman [46, 174], Gusenkov [175, 176], Rabotnov [177] made a significant contribution to the development of the theory of plasticity under single and variable loading. Publications [178–185] should also be pointed out.

To date, the behaviour of elastic and physically nonlinear three-layer plates in thermal radiation fields under single and cyclic quasi-static external loads has not been sufficiently studied. The proposed monograph is devoted to this topic. In preparing the monograph, the authors have based themselves mainly on their own results [1-A–106-A], published over the last 10 years. The monograph partially uses material from previous monographs, which has been revised and supplemented, more detailed numerical analyses and new calculations have been performed.

References

1. Lekhnitsky, S.G. 1941. Bending of inhomogeneous anisotropic thin slabs of symmetrical structure. *Applied Mathematics and Mechanics* 5 (1): 180–209 (in Russian).
2. Lekhnitsky, S.G. 1938. On some issues related to the theories of bending of thin plates. *Prikladnaia Matematika i Mekhanika. Novaia Seriia* 2 (2): 71–92.
3. Lekhnitsky, S.G. 1977. *Theory of elasticity of an anisotropic body.* Moscow, Nauka, 416 p. (in Russian).
4. Chentsov, N.G. 1936. *Lightest weight racks.* Moscow, Izd-vo i Tip. Tsentr. Aero-Gidrodinamich. In-ta im. Prof. N.E. Zhukovskogo, 48 p. (in Russian).
5. Chentsov, N.G. 1936. *Study of plywood as an orthotropic plate.* Moscow, Izd-vo i Tip. Tsentr. Aero-Gidrodinamich. In-ta im. Prof. N.E. Zhukovskogo, 27 p. (in Russian).
6. Aleksandrov, A.Ya., Borodin, M.Ya., and V.V. Pavlov. 1962. *Structures with foam cores.* Moscow, Oborongiz, 186 p. (in Russian).
7. Aleksandrov, A.Ya., Brucker, L.M., and A.P. Prusakov. 1960. *Calculation of three-layer panels.* Moscow, Oborongiz, 264 p. (in Russian).
8. Alexandrov, A.Ya. et al. 1960. *Calculation of three-layer panels.* Moscow, Oborongiz, 272 p. (in Russian).
9. Ambartsumyan, S.A. 1987. *Theory of anisotropic plates. Strength, stability and vibrations.* Moscow, Nauka, 360 p. (in Russian).
10. Babeshko, M.E., and V.G. Savchenko. 2017. Axisymmetric elastoplastic state of composite shells under thermal force loading and radiation irradiation. *Applied Mechanics* 53 (4): 14–21. (in Russian).
11. Shevchenko, Yu.N., Babeshko, M.E., and R.G. Terekhov. 1992. *Thermovyazko-elastoplastic processes of complex deformation of structural elements.* Kiev, Naukova dumka, 328 p. (in Russian).
12. Bieniek, M.P., and A.M. Freundenthal. 1962. Forced vibration of cylindrical sandwich shells. *Journal of the Aerospace Sciences* 29 (2): 180–184. https://doi.org/10.2514/8.9361.

13. Bolotin, V.V., and Yu.N. Freundenthal. 1980. *Mechanics of multilayer structures.* Moscow, Mashinostroenie, 375 p. (in Russian).
14. Grigolyuk, E.I., and P.P. Chulkov. *Stability and vibrations of three-layer shells.* Moscow, Mechanical Engineering, 172 p. (in Russian).
15. Ebsioglu, J.K. 1965. On the theory on sandwich panels in the reference state. *International Journal Engineering Science* 2 (6): 549–564.
16. Eringen, A.C. 1951. Bending and buckling of rectangular sandwich plates. *Proceedings of First U. S. National Congress Applied Mechanics*, 381–390.
17. Foss, J.I. 1962. *For the Space age, a bibliography of sandwich plates and shells.* Douglas Aircr. Co, Missile & Space System Division, Santa Monica, Calif., 98 p.
18. Galimov, K.Z., Paimushin, V.N., and I.G. Teregulov. 1996. *Foundations of the nonlinear theory of shells.* Kazan, Feng, 216 p. (in Russian).
19. Gorshkov, A.G., and V.I. Pozhuev. 1992. *Stationary problems of dynamics of multilayer structures.* Moscow, Mechanical Engineering, 224 p. (in Russian).
20. Gorshkov, A.G., Starovoitov, E.I., and D.V. Tarlakovsky. 2002. *The theory of elasticity and plasticity.* Moskow, FIZMATLIT, 416 p. (in Russian).
21. Gorshkov, A.G., Starovoitov, E.I., and A.V. Yarovaya. 2005. *Mechanics of laminated viscous elastic-plastic constructional elements.* Moscow, FIZMATLIT, 576 p. (in Russian).
22. Gorshkov, A.G., and D.V. Tarlakovsky. 2001. *Transient aerohydroelasticity of spherical bodies.* Springer, 289 p.
23. Grover, N., Singh, B.N., and D.K. Maiti. An inverse trigonometric shear deformation theory for supersonic flutter characteristics of multilayered composite plates. *Aerospace Science and Technology* (52): 41–51.
24. Habip, L.M. 1966. A curvey of modern developments in the analysis of sandwich structures. *Applied Mechanics Reviews* 18 (2): 93–98.
25. Kobelev, V.N., Kovarsky, L.M., and S.I. Timofeev. 1984. *Calculation of three-layer structures.* Moscow, Mashinostroenie, 303 p. (in Russian).
26. Korolev, V.I. 1971. *Elastic-plastic deformations of shells.* Moscow, Mashinostroenie, 303 p. (in Russian).
27. Mozharovsky, V.V., and V.E. Starzhinsky. 1988. *Applied mechanics of layered bodies made of composites.* Minsk, Navuka and technology, 271 p. (in Russian).
28. Mushtari, Kh.M. 1961. Toward a general theory of shallow shells with filler. *Inzhener. Sb. Izv. AN SSSR. Otd-e Tekhn. Nauk* (2): 24–29 (in Russian).
29. Novichkov, Yu.N. 1980. On various models for describing the deformation of multilayer structures. *Tr. MEI* (459): 40–47 (in Russian).
30. Pikul, B.V. 1985. *Theory and calculation of layered structures.* Moscow, Nauka, 182 p. (in Russian).
31. Plantemma, F.J. 1966. *Sandwich construction.* New York, John Willey and Sons, 246 p.
32. Pobedrya, B.E., and S.V. Sheshenin. 1987. On methods of elastic solutions. *Izv. AN SSSR. Mekhanika Tverdogo Tela* (5): 59–72. (in Russian).
33. Pobedrya, B.E. 1984. *Mechanics of composite materials.* Moscow, Izd-vo MGU, 336 p. (in Russian).
34. Rasskazov, A.O., Sokolovskaya, I.I., and N.A. Shulga. 1986. *Theory and calculation of layered orthotropic plates and shells.* Kiev, Vishcha shkola, 191 p. (in Russian).
35. Reddy, J.N. 2003. *Mechanics of laminated composite plates and shells: Theory and analysis.* Boca Raton: CRC Press, 835 p. https://doi.org/10.1201/b12409.
36. Reissner, E. 1948. Finite deflection of sandwich plates. *Journal Aeronautical Science* 15 (7): 435–440.
37. Solvey, J. 1955. *Bibliography and summaries of sandwich constructions (1939–1954).* Australia, Aeronaut. Res. Lab. Melbourne, 86 p.
38. Stavsky, A., and S. Friedland. 1971. Bucling of composite circular plates under radial compression. *Acta Mechanika* 11: 87–98.
39. Stein, M., and J. Mayers. 1951. *A Small Deflection Theory for Curved Sandwich Plates.* National Adviser Complementry Aeronautic Technology. 11 p.

40. Timoshenko, S.P., and S. Woinowsky-Krieger. 1966. *Theory of plates and shells*. Moscow, Nauka, 635 p. (in Russian).
41. Wang, Ch.T. 1952. *Principle and Application of Complementary Energy Method for Thin Homogeneous and Sandwich Plates and Shells with Finite Deflections*. NACA, Techn. Note, iss. 2620, 33 p.
42. Aghalovyan, L. 2015. *Asymptotic theory of anisotropic plates and shells*, 376. Singapore-London: World Scientific Publishing.
43. Altenbach, H. et al. 2012. Bending of a three-layer plate with close to zero transverse shear rigidity. *Fizicheskaia Mezomekhanika* (15) (6): 15–19 (in Russian).
44. Andreev, A.N. 2013. *Elasticity and thermoelasticity of layered composite shells*, 93. Mathematical Model and Some Aspects of Numerical Analysis. Saarbrucken: Deutschland, Palmarium Academic Publishing.
45. Carrera, E., Fazzolari, F.A., and M. Cinefra. 2016. *Thermal stress analysis of composite beams, plates and shells*. Computational Modelling and Applications, Academic Press, 410 p.
46. Golovko, K.G., Lugovoy, P.Z., and V.F. Meish. *Dynamics of inhomogeneous shells under unsteady loads*. Kyiv, Kiev University, 541 p. (in Russian).
47. Golushko, S.K., and Yu.V. Nemirovsky. 2008. *Direct and inverse problems of mechanics of elastic composite plates and shells of revolution*. Moscow, FIZMATLIT, 432 p. (in Russian).
48. Mikhasev, G.I., Botogova, M.G., and A.P. Mikhievich. 2020. Analysis of free vibrations of a sandwich panel with an electrorheological layer based on two models of layered shells. *Zhurnal BGU. Matematika. Informatika* 3: 51–59 (in Russian).
49. Pleskatshevsky, Yu.M., Starovoitov, E.I., and A.V. Yarovaya. 2004. *Deformation of metal-polymer systems*. Minsk, Belaruskaya navuka, 342 p. (in Russian).
50. Pleskchevsky, Yu.M., Starovoitov, E.I., and D.V. Leonenko. 2011. *Mechanics of three-layered bars and plates connected to an elastic foundation*. Moscow, FIZMATLIT, 560 p. (in Russian).
51. Pleskchevsky, Yu.M., Starovoitov, E.I., and A.V. Yarovaya. 2004. *Dynamics of metal-polymer systems*. Minsk, Belaruskaya navuka, 386 p. (in Russian).
52. Yarovaya, A.V. 2013. *Structural mechanics. Statics of rod systems*. Gomel, BelGUT, 447 p. (in Russian).
53. Abdelaziz, H.H., et al. 2011. Static analysis of functionally graded sandwich plates using an efficient and simple refined theory. *Chinese Journal of Aeronautics* 24: 434–448.
54. Badriev, I.B. et al. 2015. Numerical solution of the issue about geometrically nonlinear behavior of sandwich plate with transversal soft filler. *Research Journal of Applied Sciences* 10 (8): 428–435. https://doi.org/10.36478/rjasci.2015.428.435.
55. Badriev, I.V., et al. 2015. Solving physically nonlinear equilibrium problems for sandwich plates with a transversally soft core. *Lobachevskii Journal of Mathematics* 36 (4): 474–481. https://doi.org/10.1134/S1995080215040216.
56. Belinha, J. et al. 2016. The analysis of laminated plates using distinct advanced discretization meshless techniques. *Composite Structures* (143): 165–179.
57. Belinha, J., and L.M. Dints. 2007. Nonlinear analysis of plates and laminates using the element free Galerkin method. *Composite Structures* 78 (3): 337–350.
58. Carrera, E. et al. 2022. Refined multilayered beam, plate and shell elements based on Jacobi polynomials. *Composite Structures* 304 (3). https://doi.org/10.1016/j.compstruct.2022.116275.
59. Carrera, E., Cinefra, M., and G. Li. 2017. Refined finite element solutions for anisotropic laminated plates. *Composite Structures* 183 (1): 63–76. https://doi.org/10.1016/j.compstruct.2017.01.014.
60. Carrera, E., A. Pagani, and S. Valvano. 2017. Shell elements with through-the-thickness variable kinematics for the analysis of laminated composite and sandwich structures. *Composites Part B: Engineering* 111: 294–314. https://doi.org/10.1016/j.compositesb.2016.12.001.
61. Carrera, E., Pagani, A., and St. Valvano. 2016. Shell elements with through-the-thickness variable kinematics for the analysis of laminated composite and sandwich structures. *Composites Part B: Engineering* 111. https://doi.org/10.1016/j.compositesb.2016.12.001.

62. Chigarev, A.V., and A.V. Borisov. 2014. Numerical estimates of stresses and deformations in a model of a multilayer human joint under static loading. *Aktual'nye Voprosy Mashinostroeniia* 3: 229–231 (in Russian).
63. Dzhagangirov, A.A. 2016. Load-bearing capacity of a three-layer fiber composite ring plate clamped at the edges. *Mechanics of Composite Materials* 52(2): 385–398 (in Russian).
64. Fallah, F., and A. Nosier2012. Nonlinear behavior of functionally graded circular plates with various boundary supports under asymmetric thermo-mechanical loading. *Composite Structures* 94 (9): 2834–2850.
65. Farhatnia, F., Saadat, R., and S. Oveissi. 2019. Functionally graded sandwich circular plate of non-uniform varying thickness with homogenous core resting on elastic foundation: investigation on bending via differential quadrature method. *American Journal of Mechanical Engineering* 7 (2): 68–78. https://doi.org/10.12691/ajme-7-2-3.
66. Frostig, Y., and G. Kardomateas. 2019. Geometrical nonlinear thermal response of sandwich panels with temperature dependent mechanical properties Extended high-order approach. *Journal of Sandwich Structures & Materials* 21 (5): 1700–1725.
67. Hohe, J., and L. Librescu. 2004. Advances in the structural modeling of elastic sandwich panels. *Mechanics of Advanced Materials and Structures* 11 (4/5): 395–424.
68. Jalali, S.K., and M. Heshmati. 2016. Buckling analysis of circular sandwich plates with tapered cores and functionally graded carbon nanotubes-reinforced composite face sheets. *Thin-Walled Structures* 100: 14–24. https://doi.org/10.1016/j.tws.2015.12.001.
69. Julien, D., and S. Karam. 2008. Limit analysis of multi-layered plates. Part II: The Homogenesized Love–Kirchhoff Model. *Journal of the Mechanics and Physics of Solids* 56 (2): 561–580.
70. Kaleel, I., Carrera, E., and M. Petrolo. 2019. Progressive delamination of laminated composites via 1D models. *Composite Structures* 235 (2). https://doi.org/10.1016/j.compstruct.2019.111799.
71. Karttunen, A., Reddy, J.N., and J. Romanoff. 2017. Micropolar modeling approach for periodic sandwich beams. *Composite Structures* 185. https://doi.org/10.1016/j.compstruct.2017.11.064.
72. Karttunen, A., Reddy, J.N., and J. Romanoff. 2018. Two-scale constitutive modeling of a lattice core sandwich beam. *Composites Part B: Engineering* 160. https://doi.org/10.1016/j.compositesb.2018.09.098.
73. Kasi, A. et al. 2013. Refined theory of nonlinear bending of three-layer plates made of functional-gradient material. *Prikladnaia Mekhanika i Tekhnicheskaia Fizika* 54(5): 187–198 (in Russian).
74. Keller, I.E., and D.S. Dudin. 2022. *Continuum mechanics. Conservation Laws*. Perm, Perm. National research Polytechnic univ., 142 p. (in Russian).
75. Khosravi, H., Khosravi, M., and M. Lezgy-Nazargah. 2018. Pseudo-spectral method for mechanical buckling analysis of circular plates with variable thickness made of bimorph FGMs. *Numerical Methods in Civil Engineering* 3(2): 57–69. https://doi.org/10.29252/nmce.3.2.57.
76. Kovalenko, A.D. 1970. *Fundamentals of Thermoelasticity*. Kyiv, Naukova Dumka, 307 p. (in Russian).
77. Lee, C.R., Sun, S.J., and T.Y. Kam. 2007. System parameters evaluation of flexibly supported laminated composite sandwich plates. *AIAA Journal* 45(9): 2312–2322.
78. Li, G., Cinefra, M., and E. Carrera. 2020. Coupled thermo-mechanical finite element models with node-dependent kinematics for multi-layered shell structures. *International Journal of Mechanical Sciences* 171(9). https://doi.org/10.1016/j.ijmecsci.2019.105379.
79. Mirzaei, S., Hejazi, M.,and R. Ansari. 2022. Isogeometric analysis for bending, buckling and free vibration of multi-directional functionally graded porous plates with variable thickness. *The European Physical Journal Plus* 137 (694). https://doi.org/10.1140/epjp/s13360-022-02902-5.
80. Nampally, P., Karttunen, A., and J.N. Reddy. 2018. Nonlinear finite element analysis of lattice core sandwich beams. *European Journal of Mechanics* 74 (1). https://doi.org/10.1016/j.euromechsol.2018.12.006.

81. Nampally, P., Karttunen, A., and J.N. Reddy. 2020. Nonlinear finite element analysis of lattice core sandwich plates. *International Journal of Non-Linear Mechanics* 121. https://doi.org/10.1016/j.ijnonlinmec.2020.103423.
82. Pagani, A. et al. 2021. Use of Lagrange polynomials to build refined theories for laminated beams, plates and shells. *Composite Structures* 276(3). https://doi.org/10.1016/j.compstruct.2021.114505.
83. Petrolo, M. et al. 2023. Evaluation of transverse shear stresses in layered beams/plates/shells via stress recovery accounting for various CUF-based theories. *Composite Structures* 307 (40): 116625. https://doi.org/10.1016/j.compstruct.2022.116625.
84. Piskunov, V.G., and A.O. Rasskazov. 2002. Evolution of the theory of laminated plates and shells. *International Applied Mechanics* 38 (2): 135–166.
85. Reinaldo, G.B. et al. 2017. Buckling and free vibration of shear-flexible sandwich beams using a couple-stress-based finite element. *Composite Structures* 165(2). https://doi.org/10.1016/j.compstruct.2017.01.033.
86. Romanoff, J., Reddy, J.N., and J. Jelovica. 2015. Using non-local Timoshenko beam theories for prediction of micro- and macro-structural responses. *Composite Structures* 156(3). https://doi.org/10.1016/j.compstruct.2015.07.010.
87. Roy, Ch.Sh., and J.N. Reddy. 2019. Geometrically exact micropolar Timoshenko beam and its application in modelling sandwich beams made of architected lattice core. *Composite Structures* 226. https://doi.org/10.1016/j.compstruct.2019.111228.
88. Ruocco, E., and J.N. Reddy. 2019. A closed-form solution for buckling analysis of orthotropic Reddy plates and prismatic plate structures. *Composites Part B: Engineering* 169(2). https://doi.org/10.1016/j.compositesb.2019.03.015.
89. Ruocco, E., and J.N. Reddy. 2021. A discrete differential geometry-based approach to buckling and vibration analyses of inhomogeneous Reddy plates. *Applied Mathematical Modelling* 100 (3). https://doi.org/10.1016/j.apm.2021.08.011.
90. Škec, L., and G. Jelenić. 2014. Analysis of a geometrically exact multi-layer beam with a rigid interlayer connection. *Acta Mechanica* 225 (2): 523–541.
91. Starovoitov, E.I., Yarovaya, A.V., and D.V. Leonenko. 2006. *Deformation of Three-Layer Structural Elements on an Elastic Base*. Moscow, FIZMATLIT, 380 p. (in Russian).
92. Starovoitov, E.I., Leonenko, D.V., and M. Suleyman. 2007. Deformation of a composite plate on an elastic foundation by local loads. *Mechanics of Composite Materials* 43 (1): 75–84.
93. Takele, K.T. 2018. Interfacial strain energy continuity assumption-based analysis of an orthotropic-skin sandwich plate using a refined layer-by-layer theory. *Mechanics of Composite Materials* 54 (3): 281–298.
94. Thai C.H. et al. 2012. Analysis of laminated composite plates using higher-order shear deformation plate theory and mode-based smoother discrete shear gap method. *Applied Mathematical Modeling* 36 (11): 5657–5677.
95. Than, C. et al. 2021. A multi-layer moving plate method for dynamic analysis of viscoelastically connected double-plate systems subjected to moving loads. *Advances in Structural Engineering* 24(1). https://doi.org/10.1177/1369433220982730.
96. Treshchev, A.A., Telichko, V.G., and P.Yu. Khodorovich. 2014. Modeling the stress-strain state of thick cylindrical shells made of materials with complex properties. *Fizika i Mekhanika Materialov* 21 (1): 38–50 (in Russian).
97. Xie, Z. 2018. An approximate solution to the plastic indentation of circular sandwich panels. *Mechanics of Composite Materials* 54 (2): 243–250.
98. Yang, L. et al. 2013. Comparison of bending properties for cellular core sandwich panels. *Materials Sciences and Applications* 4(8): 471–477.
99. Yankovskii, A.P. 2020. Modeling of non-isothermal viscoelastic-plastic behavior of flexible reinforced plates. *Vychislitel'naia Mekhanika Sploshnykh Sred* 13(3): 350–370 (in Russian).
100. Yankovskii, A.P. 2017. Refined deformation model for metal-composite plates of regular layered structure in bending under conditions of steady-state creep. *Mechanics of Composite Materials* 52 (6): 715–732.

101. Yarovaya, A.V. 2006. Thermoelastic bending of a sandwich plate on a deformable foundation. *International Applied Mechanics* 42 (2): 206–213.
102. Zadeh, H.V., and M. Tahani. 2017. Analytical bending analysis of a circular sandwich plate under distributed load. *International Journal of Recent Advances in Mechanical Engineering* 6 (1): 1–10.
103. Zenkour, A.M. 2015. Thermoelastic bending response of a laminated plate resting on elastic foundations. *Scientia Iranica. Transaction A. Civil Engineering* 22 (2): 287–298.
104. Zenkour, A.M., and H. El-Shahrany. 2020. Vibration suppression of magnetostrictive laminated beams resting on viscoelastic foundation. *Applied Mathematics and Mechanics* 41 (8): 1269–1286. https://doi.org/10.1007/s10483-020-2635-7.
105. Zenkour, A.M., and D. Mashat. 2022. Quasi-3D vibrational analyses of laminates and sandwich plates resting on elastic foundations. *The International Journal of Acoustics and Vibration* 27: 285–300. https://doi.org/10.20855/ijav.2022.27.31873.
106. Zenkour, A.M., D.S. Mashat, and R.A. Alghanmi. 2014. Hygrothermal analysis of antisymmetric cross-ply laminates using a refined plate theory. *International Journal of Mechanics and Materials in Design* 10: 213–226. https://doi.org/10.1007/s10999-014-9242-5.
107. Zhang, Z. et al. 2021. Analysis of temperature-dependent layered shells subjected to thermomechanical loading. *Mechanics of Advanced Materials and Structures*, 1–13.
108. Arefi, M., E.M. Bidgoli, and A.M Zenkour. 2018. Size-dependent free vibration and dynamic analyses of a sandwich microbeam based on higher-order sinusoidal shear deformation theory and strain gradient theory. *Smart Structures and Systems* 22 (1): 27–40. https://doi.org/10.12989/sss.2018.22.1.027.
109. Bakulin, V.N., Boitsova, D.A., and A.Ya. Nedbai. 2021. Parametric resonance of a three-layered cylindrical composite rib-stiffened shell. *Mechanics of Composite Materials* 57 (5): 623–634.
110. Bharati, R. et al. 2022. Flutter analysis of delaminated composite box-beam using higher-order kinematics. *Composite Structures* 301 (8). https://doi.org/10.1016/j.compstruct.2022.116145.
111. Bouazza, M., and A.M. Zenkour. 2018. Free vibration characteristics of multilayered composite plates in a hygrothermal environment via the refined hyperbolic theory. *The European Physical Journal Plus* 133 (6). https://doi.org/10.1140/epjp/i2018-12050-x.
112. Fedotenkov, G.V., Tarlakovsky, D.V., and Y.A. Vahterova. 2019. Identification of non-stationary load upon Timoshenko beam. *Lobachevskii Journal of Mathematics* 40 (4): 439–447.
113. García-Castillo, S.K., Buitrago, B.L., and E. Barbero. 2011. Behavior of sandwich structures and spaced plates subjected to high-velocity impacts. *Polymer Composites* 32 (2): 290–296.
114. Garg, A. et al. 2022. A parametric analysis of free vibration and bending behavior of sandwich beam containing an open-cell metal foam core. *Archives of Civil and Mechanical Engineering* 22(1). https://doi.org/10.1007/s43452-021-00368-3.
115. Garg, A. et al. 2023. On accurately capturing the through-thickness variation of transverse shear and normal stresses for composite beams using FSDT coupled with GPR. *Composite Structures* 305. https://doi.org/10.1016/j.compstruct.2022.116551.
116. Giovanny, L., and L. Stefano. 2007. Forced nonlinear oscillations of a semi-infinite beam resting on a unilateral elastic soil: analytical and numerical solutions. J*ournal of Computer and Nonlinear Dynamic* 2 (2): 155–166.
117. Hause, T. 2012. Elastic structural response of anisotropic sandwich plates with a first-order compressible core impacted by a Friedlander-type shock loading. *Composite Structures* 94 (5): 1634–1645.
118. Igumnov, L.A., et al. 2014. Soundproof properties of a one-dimensional three-layer plate. *Journal of Mathematical Sciences* 203: 104–113. https://doi.org/10.1007/s10958-014-2093-7.
119. Igumnov, L.A., et al. 2021. Interaction of harmonic waves of different types with the three-layer plate placed in the soil. *Advanced Structured Materials* 137: 111–124. https://doi.org/10.1007/978-3-030-53755-5_8.

120. Ivañez, I. et al. 2015. The oblique impact response of composite sandwich plates. *Composite Structures* 133: 1127–1136.
121. Kaplunov, J.D., Kossovich, L.Y., and E.V. Nolde. *Dynamics of thin Walled Elastic Bodies*. San Diego–London–Boston–New York–Sydney–Tokyo–Toronto, Academic Press, 226 p.
122. Kyrychok, I.F., Zhuk, Ya.O., and T.V. Karnaukhova. 2019. Resonance vibration and dissipative heating of a flexible viscoelastic beam with piezoactuators in the presence of shear strains. *Journal of Mathematical Sciences* 243: 73–84.
123. Lal, R., and N. Ahlawat. 2015. Buckling and vibration of functionally graded non-uniform circular plates resting on Winkler foundation. *Latin American Journal of Solids and Structures* 12: 2231–2258. https://doi.org/10.1590/1679-78251595.
124. Lal, R., and R. Rani. On radially symmetric vibrations of circular sandwich plates of non-uniform thickness. *International Journal of Mechanical sciences* 99:29–39. https://doi.org/10.1016/j.ijmecsci.2015.04.016.
125. Lal, R., and R. Rani. 2016. On the radially symmetric vibrations of circular sandwich plates with polar orthotropic facings and isotropic core of quadratically varying thickness by harmonic differential quadrature method. *Meccanica* 51: 611–634. https://doi.org/10.1007/s11012-015-0226-9.
126. Lal, R., and R. Rani. 2016. On the use of differential quadrature method in the study of free axisymmetric vibrations of circular sandwich plates of linearly varying thickness. *Journal of Vibration and Control* 22 (7): 1729–1748, https://doi.org/10.1177/1077546314544695.
127. Lokteva, N.A., and N.D. Phung. 2021. Unsteady dynamics of a sandwich plate under the influence of a cylindrical wave in an elastic medium. *Incas Bulletin* 13: 117–132. https://doi.org/10.13111/2066-8201.2021.13.S.12.
128. Mikhasev, G.I., et al. 2019. Assessment of dynamic characteristics of thin cylindrical sandwich panels with magnetorheological core. *Journal of Intelligent Material Systems and Structures* 30 (18–19): 2748–2769.
129. Mikhasev, G.I., and H. Altenbach. 2019. Free vibrations of elastic laminated beams, plates and cylindrical shells. *Advanced Structured Materials* 106: 157–198. https://doi.org/10.1007/978-3-030-12761-9_4.
130. Moleiro, F. et al. 2020. Evaluation of exact electro-elastic static and free vibration solutions of multilayered plates for benchmarking: Piezoelectric composite laminates and soft core sandwich plates. *Composites Part C: Open Access* 2. https://doi.org/10.1016/j.jcomc.2020.100038.
131. Natarajan, S., and M. Ganapathi. 2012. Bending and vibration of functionally graded material sandwich plates using an accurate theory. *Finite Elements in Analysis and Design* 57: 32–42.
132. Nemirovskii, Y.V., and S.V. Tikhonov. 2020. On the stability loss of a rapidly rotating polymetallic disc. *Mechanics of Solids* 55 (6): 767–775.
133. Paimushin, V.N., and R.K. Gazizullin. 2017. Static and monoharmonic acoustic impact on a laminated plate. *Mechanics of Composite Materials* 53 (3): 283–304.
134. Paimushin, V.N., V.A. Firsov, and V.M. Shishkin. 2017. Modeling the dynamic response of a carbon-fiber-reinforced plate at resonance vibrations considering the internal friction in the material and the external aerodynamic damping. *Mechanics of Composite Materials* 53 (4): 425–440.
135. Paimushin, V.N., V.A. Firsov, and V.M. Shishkin. 2020. Numerical modeling of resonant vibrations of an elongate plate with an integral damping coating. *Mechanics of Composite Materials* 56 (7): 149–168. https://doi.org/10.1007/s11029-020-09869-3.
136. Paimushin, V.N., et al. 2020. Vibration of the plate with integral layer damping: Experimental and theoretical studies. *Structural Integrity* 16: 254–260. https://doi.org/10.1007/978-3-030-47883-4_46.
137. Paimushin, V.N., et al. 2021. Sandwich shells with composite facings and a transversally flexible core: Refined equations and buckling modes of specimens under four-point bending tests. *Advanced Structured Materials* 141: 391–411. https://doi.org/10.1007/978-3-030-54928-2_29.

138. Paimushin, V.N., et al. 2020. Shear buckling mode and failure of sandwich specimen facing layer under four-point bending. *Structural Integrity* 16: 293–300. https://doi.org/10.1007/978-3-030-47883-4_52.
139. Rabboh, S., et al. 2013. The effect of functionally graded materials into the sandwich beam dynamic performance. *Materials Sciences and Applications* 4: 751–760.
140. Rani, R., and R. Lal. 2015. Radially symmetric vibrations of exponentially tapered clamped circular sandwich plate using harmonic differential quadrature method. *Mathematical Analysis and its Applications* 143: 633–643. https://doi.org/10.1007/978-81-322-2485-3_52.
141. Romanova, T.P. 2017. Modeling of dynamic bending of a rigid-plastic reinforced layered curved plate with a fixed round hole under blast loads. *Vestn. Permskogo Nats. Issled. Politekhn. Un-ta. Mekhanika* (3): 167–187 (in Russian).
142. Ryazantseva, M.Y., and E.I. Starovoitov. 2019. Static and dynamic models of bending for elastic sandwich plates. *Proceedings of the Second International Conference on Theoretical, Applied and Experimental Mechanics. Structural Integrity*. Springer Nature Switzerland AG 2019, pp. 294–297.
143. Ryazantseva, M.Yu., and F.K. Antonov. Harmonic running waves in sandwich plates. *International Journal of Engineering Science*, 59: 184–192.
144. Sobhy, M., and A.M. Zenkour. 2018. Magnetic field effect on thermomechanical buckling and vibration of viscoelastic sandwich nanobeams with CNT reinforced face sheets on a viscoelastic substrate. *Composites Part B: Engineering*. 154: 492–506. https://doi.org/10.1016/j.compositesb.2018.09.011.
145. Starovoitov, É.I., and D.V. Leonenko. 2010. Dynamics of circular three-layer plates with elastic base under axi-symmetric loading. *Strength of Materials* 42 (3): 322–330.
146. Starovoitov, É.I., Leonenko, D.V., and A.V. Yarovaya. 2002. Vibrations of round three-layer plates under the action of distributed local loads. *Strength of Materials* 34(5): 474–481.
147. Süsler, S., and H. Türkmeni. 2017. Nonlinear dynamic analysis of tapered sandwich plates with multi-layered faces subjected to air blast loading. *International Journal of Mechanics and Materials in Design* 13: 429–451. https://doi.org/10.1007/s10999-016-9346-1.
148. Tarlakovskii, D.V., and G.V. Fedotenkov. 2015. Nonstationary 3D motion of an elastic spherical shell. *Mechanics of Solids* 46(5): 779–787.
149. Tarlakovskii, D.V., and G.V. Fedotenkov. 2016. Non-stationary problems for elastic half-plane with moving point of changing boundary conditions. *PNRPU Mechanics Bulletin* (3): 188–206.
150. Tarlakovskii, D.V., and G.V. Fedotenkov. 2014. Two-Dimensional nonstationary contact of elastic cylindrical or spherical shells. J*ournal of Machinery Manufacture and Reliability* 43 (2): 145–152.
151. Tarlakovskiy, D.V., and G.V. Fedotenkov. 2009. Analytic investigation of features of stresses in plane nonstationary contact problems with moving boundaries. *Journal of Mathematical Sciences* 162(2): 246–253.
152. Wu, B., et al. 2019. Three-dimensional vibrations of multilayered hollow spheres submerged in a complex fluid. *Journal of Fluid Mechanics* 879: 682–715. https://doi.org/10.1017/jfm.2019.681.
153. Yoshihiro, N., Kenichi, F., and Zh. Xilu. 2002. Vibration analysis of shear-deformable circular and elliptical laminated composite plates. *JSME International Journal Series C* 45 (1): 113–120.
154. Yoshiki, O., and I. Toshihiro. The study of analytikal models for vibration of cross-ply laminated thick plates. *JSME International Journal Series C* 45 (1): 107–112.
155. Zhong, S. et al. 2020. Isogeometric vibration analysis of multi-directional functionally gradient circular, elliptical and sector plates with variable thickness. *Composite Structures* 250. https://doi.org/10.1016/j.compstruct.2020.112470.
156. Bauschinger, I. 1879. Über die Quercontraktion and Dilatation bei der Langenausdehnung und Zusammpndruckung pristiatiscner Körper. *Civilingenieur* 25: 81–124.
157. Bauschinger, I. 1881. Über die Veränderung der Elastizitatsgrenze und des Elastizitatsmoduls verschiedener Metall. *Civilingenieur* 27: 289–348.

158. Huber, M.T. 1904. *Die spezifische Formänderungsarbeit als Mass der Anstrengung eines Materials*. Crasopismo, tech., t. 15. Lemberg (Lwow)
159. Masing, G. 1926. Eigenspannungen und Verfestigung Beim Messing. *Proc. of the 2-nd International Congress of Application Mechanic*, 332–335.
160. Masing, G., and W. Mauksch. 1926. *Berechnung von Dehnungs- und Stauchungslinien auf Grund von inneren Spannungen*. Wissenschaftliche Veröffentlichungen aus dem Siemens-Konzern, b. V, pp. 142–155.
161. Mises, R. 1928. Mechanic der plastischen Formandenderung von Kristallen. *ZAMM* 8 (3): 161–184.
162. Mises, R. 1913. *Mechanik der festen Körper im plastisch- deformablen Zustand. Nachrichten von der Gesellschaft der Wissenschaften zu Göttingen*. Mathematisch-Physikalische Klasse, 582–592.
163. Ilyushin, A.A. 1948. *Plasticity. Part 1. Elastic-plastic Deformations*. Moscow, Gostekhizdat, 392 p. (in Russian).
164. Ilyushin, A.A., and P.M. Ogibalov. 1960. *Elastic-plastic deformations of hollow cylinders*. Moscow, State University, 224 p. (in Russian).
165. Moskvitin, V.V. 1981. *Cyclic loading of structural elements*. Moscow, Nauka, 344 p. (in Russian).
166. Moskvitin, V.V. 1965. *Plasticity under variable loads*. Moscow, Izd-vo MGU, 263 p. (in Russian).
167. Moskvitin, V.V. 1972. *Strength of viscoelastic materials*. Moscow, Nauka, 327 p. (in Russian).
168. Moskvitin, V.V., and A.V. Gorodetsky. 1983. To variable loading of elastoplastic bodies. *Mekhanika Deformiruemogo Tverdogo Tela*, 48–54 (in Russian).
169. Moskvitin, V.V., and É.I. Starovoitov. 1985. Deformation and variable loading of two-layer metal-polymer plates. *Mechanics of Composite Materials* 21(3): 267–273.
170. Moskvitin, V.V., and E.I. Starovojtov. To the investigation of stress-strain state of two-layered metal-polymer plates at cyclic loading. *Mechanics of Solids* (1): 116–121.
171. Bell, G.Ph. 1984. *Experimental foundations of deformable body mechanics Vol 1 and 2*. Moscow, Nauka, 600 p. (in Russian).
172. Makhutov, N.A. et al. 1986. *Mechanics of low-cycle fracture*. Moscow, Nauka, 264 p. (in Russian).
173. Vasin, R.A. 2011. Theory of elastoplastic processes and study of structure-mechanical properties of materials. *Mechanics of Solids* 46 (1): 15–20.
174. Goldman, A.Ya. 1982. *Volumetric deformation of plastics*. Leningrad, Mashinostroenie, 232 p. (in Russian).
175. Gusenkov, A.P. 1978. *Strength under Isothermal and Non-Isothermal Low-Cycle Loading*, 295. Moscow: Nauka.
176. Gusenkov, A.P., and P.I. Kotov. 1988. *Long-term and non-isothermal low-cycle strength of structural elements*. Moscow, Mashinostroenie, 264 p. (in Russian).
177. Rabotnov, Yu.N. 1979. *Mechanics of a deformable solid body*. Moscow, Nauka, 744 p. (in Russian).
178. Kulikov, I.S., Nesterenko, V.B., and B.E. Tverkovkin. 1990. *Strength of structural elements under irradiation*. Minsk, Navuka i technology, 144 p. (in Russian).
179. Moskvitin, G.V. 1994. On additional possibilities for using the equations of the generalized Masing principle. *Vestn. Mosk. Un-ta. Ser. 1. Matem., Mekh* (4): 47–53 (in Russian).
180. Platonov, P.A. 1971. *Effect of irradiation on the structure and properties of metals*. Moscow, Mashinostroenie, 40 p. (in Russian).
181. Rozumek, D., and M. Hepner. 2011. Influence of oxygenation time on crack growth in titanium alloy under cyclic bending. *Material Science* 47(1): 89–94.
182. Taleb, L., and G. Cailletaud. 2010. An updated version of the multimechanism model for cyclic plasticity. *International Journal Plastics* 26 (6): 859–874.
183. Yun Gun Jin, Shang Shen. 2011. A self-optimizing inverse analysis method for estimation of cyclic elasto-plasticity model parameters. *International Journal Plastics* 27 (4): 576–595.

184. Zang, S.L., et al. 2011. A model of one-surface cyclic plasticity and its application to springbuck prediction. *International Journal of Mechanical Sciences* 53 (6): 425–435.
185. Zhan-Li Liu et al. 2011. Bauschinger and size effect in thin-film plasticity due to defect-energy of geometrical necessary dislocations. *Acta Mechnical Since* 27 (2): 266–276.

Chapter 2
Some Basic Information From Solid Mechanics

The formulation of boundary value problems of elastic bodies is given. Deformation of physically nonlinear solids in thermoradiation fields is outlined in the framework of the theory of small elastoplastic deformations. Variational methods of solving problems and the method of elastic solutions are considered. The elasticity and physical nonlinearity parameters of some structural materials are given. The formula for calculating the temperature in a three-layer package under incident heat flux is obtained by averaging the thermophysical characteristics of the materials of the layers. The attenuation of neutron flux when passing through a three-layer element was analyzed.

2.1 Stress–strain State of an Elastic Solid Body

Under the action of given loads in the points of a solid elastic body a stress state occurs. To describe it, a symmetric stress tensor of the second rank with six independent components σ_{ij} is introduced. The first index indicates the normal to the coordinate area on which this stress acts, the second index indicates the axis to which it is parallel. Components with the same indices coincide with the stresses that are normal to the considered area. If the indices are different, σ_{ij} are tangential stresses. Tensile normal stresses are considered positive, compressive normal stresses are considered negative.

It is assumed that the stress components σ_{ij} are related to the volumetric external forces ρF_i (ρ is the material density) by three equations of equilibrium valid everywhere inside the deformed body:0

$$\sigma_{ij,j} + \rho F_i = 0. \tag{2.1}$$

S. Eduard et al., *Deformation of Three-layer Structural Elements in Thermal Radiation Fields*, https://doi.org/10.1007/978-981-97-7217-9_2

Hereinafter the comma in the lower index denotes the operation of differentiation by the coordinate following it. The summation is done by the repeating Latin index in the equation term, in this case $-j$ from 1 to 3.

There are coordinate sites at which the tangential stresses turn to zero and the normal stresses become extremes. They are denoted σ_1, σ_2, σ_3 and are called the *principal values* of the stress tensor. The corresponding coordinate axes are called the *principal axes* of the stress tensor. The principal axes are numbered so that the following conditions are fulfilled in the algebraic sense: $\sigma_1 \geq \sigma_2 \geq \sigma_3$.

To determine the magnitudes of the principal stresses, a cubic equation called the *secular equation* is used:

$$\sigma^3 - \sigma^2 J_1 + \sigma J_2 - J_3 = 0,$$

where

$$J_1 = s_{ii} s_{11} + s_{22} + s_{33} = s_1 + s_2 + s_3,$$

$$J_2 = (s_{ii} s_{jj} - s_{ij} s_{ij})/2 s_{11} s_{22} + s_{22} s_{33} + s_{33} s_{11} - == s_1 s_2 + s_2 s_3 + s_3 s_1,$$

$$J_3 = \left\| \sigma_{ij} \right\| \equiv \begin{vmatrix} \sigma_{11} & \sigma_{12} & \sigma_{13} \\ \sigma_{21} & \sigma_{22} & \sigma_{23} \\ \sigma_{31} & \sigma_{32} & \sigma_{33} \end{vmatrix} = \sigma_1 \sigma_2 \sigma_3 \tag{2.2}$$

It can be shown that all three roots of the secular equation are real roots. Their values are determined by the nature of the external load and do not depend on the initial orientation of the coordinate system. Therefore, the values of the coefficients J_1, J_2, J_3 in the secular equation defined by the formulas (2.2) must remain unchanged when rotating the axes. In this connection they are called *invariants of the stress tensor*.

On the areas parallel to the second principal axis and equally inclined to the first and third principal axes, the maximum tangential stresses act, which are determined by the following formula:

$$\tau_{\max} = \frac{1}{2}(\sigma_1 - \sigma_3)$$

The physical content of the stress tensor is determined by its three invariants, so the quantities that depend on it (e.g., plasticity criteria) must be functions of J_1, J_2, J_3 or other invariant quantities (e.g., principal stresses). This fact should be taken into account when constructing general theories of mechanical behaviour of materials.

Consider a stress state in which only three identical principal stresses σ, equal to the average stress at a given point of the body, act at three mutually perpendicular surfaces:

$$\sigma = (\sigma_{11} + \sigma_{22} + \sigma_{33})/3.$$

The tensor describing such a stress state is called a *spherical tensor* (the spherical part of the stress tensor).

Subtracting the spherical component from the stress tensor, we obtain the *deviator of the stress tensor*. Thus, the stress tensor at each point can be represented as a sum of the spherical (σ) and deviator parts (s_{ij}):

$$\sigma_{ij} = s_{ij} + \sigma\delta_{ij}(i, j = 1, 2, 3). \tag{2.3}$$

Here δ_{ij} is *metric tensor,* its components are called Kronecker symbols and in Cartesian coordinate system have the form

$$\delta_{ij} = \begin{cases} 1, i = j, \\ 0, i \neq j. \end{cases} \tag{2.4}$$

The decomposition of the stress tensor into spherical and deviator parts is of great fundamental importance in the study of the behaviour of elastic and non-elastic bodies under load. The spherical part distinguishes from the stress state a uniform all-round tension or compression in which only the volume of a given body element changes without changing its shape. The stress deviator characterises the shear state in which the shape of an element changes without changing its volume. As experiments show, materials respond differently to all-round compression and shear stresses.

The concept of stress tensor intensity, which was introduced by Ilyushin [1], is widely used in the theory of plasticity. It is formally defined as

$$\begin{aligned} \sigma_u &= \frac{\sqrt{2}}{2}\sqrt{(\sigma_{11} - \sigma_{22})^2 + (\sigma_{22} - \sigma_{33})^2 + (\sigma_{33} - \sigma_{11})^2 + 6(\sigma_{12}^2 + \sigma_{23}^2 + \sigma_{31}^2)} = \\ &= \frac{\sqrt{2}}{2}\sqrt{(\sigma_1 - \sigma_2)^2 + (\sigma_2 - \sigma_3)^2 + (\sigma_3 - \sigma_1)^2} \end{aligned} \tag{2.5}$$

The intensity of the stress tensor is an invariant quantity because it is expressed through the principal stresses. The numerical coefficient in formula (2.5) is chosen so that in the case of simple tension or compression ($\sigma_{11} = \sigma_1$, all other components are zero) the condition $\sigma_u = |\sigma_1|$ is fulfilled.

Under the action of external forces, all bodies change their shape, i.e. they are deformed. The points of the body change their position in space continuously, i.e. there are no discontinuities, cavities and cracks as a result of deformation of the body. The vector **u** (Fig. 2.1), which has its origin at the point A of the undeformed body and its end at the corresponding point A' of the deformed body, is called the *vector of total displacement* of the point. Its projections on the coordinate axes are denoted by $u_i \equiv u_i(x)$, where $x \equiv (x_1, x_2, x_3)$.

As a rule, in deformable solid mechanics, kinematically invariable systems are considered, which do not allow the body to move in space as a rigid whole. In this

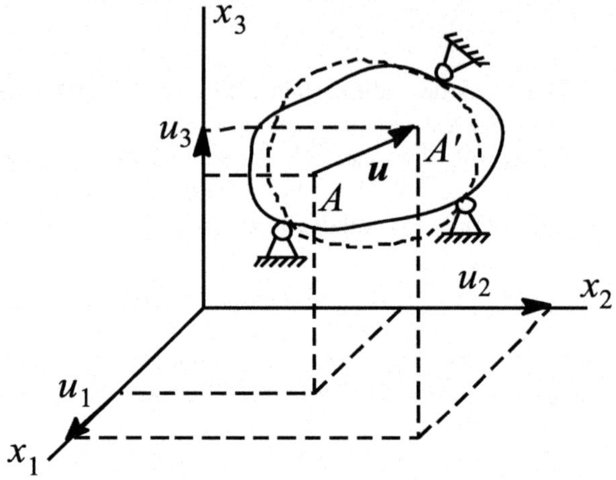

Fig. 2.1 Displacement vector

case, the introduced displacements for most of the considered systems are small compared to the geometrical dimensions of the body. In the general case, the strain components are related to small displacements by the *Cauchy relations*:

$$\varepsilon_{ij} = \frac{1}{2}(u_{i,j} + u_{j,i})(i, j = 1,\ 2,\ 3). \tag{2.6}$$

The strain components (2.6) form a tensor of the second rank, which, like the stress tensor, is symmetrical ($\varepsilon_{ij} = \varepsilon_{ji}$). It can be shown that it completely describes the deformation of an arbitrary fibre in the vicinity of the considered point.

There are three directions in which there are only linear strains along the axes and no shear. They correspond to the extreme values of relative strains, which are called *principal strains* and denoted ε_1, ε_2, ε_3. The *characteristic equation* can be used to determine the principal strains

$$\varepsilon^3 - \varepsilon^2 I_1 + \varepsilon I_2 - I_3 = 0. \tag{2.7}$$

The coefficients in Eq. (2.7) are *three invariants* independent of the orientation of the coordinate system:

$$I_1 = \varepsilon_{11} + \varepsilon_{22} + \varepsilon_{33} = \varepsilon_1 + \varepsilon_2 + \varepsilon_3,$$

$$I_2 = \varepsilon_{11}\varepsilon_{22} + \varepsilon_{11}\varepsilon_{22} + \varepsilon_{11}\varepsilon_{22} - = \varepsilon_1\varepsilon_2 + \varepsilon_2\varepsilon_3 + \varepsilon_3\varepsilon_1,$$

$$I_3 = \begin{vmatrix} \varepsilon_{11} & \varepsilon_{12} & \varepsilon_{13} \\ \varepsilon_{21} & \varepsilon_{22} & \varepsilon_{23} \\ \varepsilon_{31} & \varepsilon_{32} & \varepsilon_{33} \end{vmatrix} = \varepsilon_1\varepsilon_2\varepsilon_3. \tag{2.8}$$

The first invariant I_1 has a simple geometric meaning. It is the relative change in volume.

In the framework of the theory of elasticity the principal axes of stress and strain tensors for isotropic materials coincide.

The strain tensor can be represented as a sum of two tensors, like (2.3):

$$\varepsilon_{ij} = э_{ij} + \varepsilon\delta_{ij}(i, j = 1, 2, 3), \tag{2.9}$$

where $\varepsilon = \theta/3(\varepsilon_{11} + \varepsilon_{22} + \varepsilon_{33})/3$ is *spherical part* of the strain tensor; δ_{ij} are Kronecker symbols (2.4); $э_{ij}$ are components of the strain tensor *deviator*.

The spherical tensor in (2.9) describes the *volumetric strain.* Its first invariant coincides with the first invariant of the strain tensor (2.8). Deviator characterises the *deformation of the shape change*. Its first invariant is zero. *The intensity of the strain tensor* ε_u is introduced by analogy with (2.5):

$$\begin{aligned}\varepsilon_u &= \frac{\sqrt{2}}{3}\sqrt{(\varepsilon_{11} - \varepsilon_{22})^2 + (\varepsilon_{22} - \varepsilon_{33})^2 + (\varepsilon_{33} - \varepsilon_{11})^2 + 6(\varepsilon_{12}^2 + \varepsilon_{23}^2 + \varepsilon_{31}^2)} = \\ &= \frac{\sqrt{2}}{3}\sqrt{(\varepsilon_1 - \varepsilon_2)^2 + (\varepsilon_2 - \varepsilon_3)^2 + (\varepsilon_3 - \varepsilon_1)^2}\end{aligned} \tag{2.10}$$

In the case of simple tension or compression $\varepsilon_{11} = \varepsilon_1$, $\varepsilon_{22} = \varepsilon_{33} = -\nu\varepsilon_1$, $\varepsilon_{12} = \varepsilon_{23} = \varepsilon_{31} = 0$ (ν is Poisson's ratio).

Then it follows from (2.10): $\varepsilon_u = 2/3(1 + \nu)|\varepsilon_1|$, and for incompressible material ($\nu = 0{,}5$); $\varepsilon_u = |\varepsilon_1|$. The intensity of the strain tensor according to (2.10) is an invariant quantity, since it is defined through the invariant principal strains.

By the known three differentiable components of the displacement field $u_i(x)$, ($x \equiv x_1, x_2, x_3$) six independent components of the strain tensor are easily determined using the Cauchy relations (2.6). The inverse operation is difficult because not always six continuous components of $\varepsilon_{ij}(x)$ correspond to some continuous displacement field. If such a field exists, the strains are called *compatibile*, otherwise—*incompatible.*

Joint strains are related to each other by relations which are called *compatibility strain equations* (without summation by repeated Greek indexes):

$$\varepsilon_{\alpha\alpha,\beta\beta} + \varepsilon_{\beta\beta,\alpha\alpha} - 2\varepsilon_{\alpha\beta,\alpha\beta} = 0(\alpha, \beta = 1, 2, 3; \alpha\beta),$$

$$(\varepsilon_{\alpha\beta,\gamma} - \varepsilon_{\beta\gamma,\alpha} + \varepsilon_{\gamma\alpha,\beta})_{,\alpha} - \varepsilon_{\alpha\alpha,\beta\gamma} = 0(\alpha, \beta, \gamma = 1, 2, 3; \alpha\beta). \tag{2.11}$$

Thus, to find the corresponding displacement field by the six continuous components of the strain tensor, it is necessary to fulfill six differential compatibility Eq. (2.11) with respect to the six components of the strain tensor ε_{ij}. In the case of a single-connected region they are necessary and sufficient, but for a multi-connected region they are only necessary.

The relations of the *generalized Hooke's law* are valid for elastic isotropic linear bodies:

$$\sigma_{ij} = 2\mu\varepsilon_{ij} + \lambda\theta\delta_{ij}. \tag{2.12}$$

where λ and μ are *Lame parameters*; θ is the volume strain:

$$\theta = \varepsilon_{11} + \varepsilon_{22} + \varepsilon_{33} = \frac{1 - 2\nu}{E}(\sigma_{11} + \sigma_{22} + \sigma_{33}) \tag{2.13}$$

Lame parameters are related to shear modulus *G*, Poisson's ratio ν and Young's modulus *E* by the following relationships:

$$\lambda = \frac{\nu E}{(1 + \nu)(1 - 2\nu)}, \ldots \mu = G = \frac{E}{2(1 + \nu)}, \ldots \nu = \frac{\lambda}{2(\lambda + \mu)}$$

Hence, of the four elastic constants λ, μ, ν, *E* and the volume modulus K = E/3(1–2ν), only any two quantities are independent. The dimensions of λ, μ, *K*, *E* coincide with the dimension of stress. These parameters are positive. Poisson's ratio is a dimensionless quantity. The restriction of its possible values follows from the condition $K > 0$ and $\mu > 0$. Then: $-1 \leq \nu \leq 0{,}5$. The value $\nu = 0.5$ corresponds to incompressible material. Experience shows that for all known isotropic materials $\nu > 0$. Many attempts have been made to prove that the lower bound for ν is 0 rather than −1, but this has not been achieved [2].

In the future, Hooke's law will often be applied in deviator-spherical form:

$$s_{ij} = 2G\,\text{э}_{ij};\ \sigma = K\theta. \tag{2.14}$$

For incompressible materials instead of the second Eq. (2.14) the condition of zero volume strain is used (2.13).

If the body under study is in an inhomogeneous and unsteady temperature field $T(x_i, t)$, then, according to Neumann's hypothesis, the total linear strain ε_{ii} is summarized by the elastic strain from the force load ε'_{ii} and thermal expansion $\varepsilon''_{ii} = \alpha_0 T$ (α_0—*linear thermal expansion coefficient*):

$$\varepsilon_{ii} = \varepsilon'_{ii} + \varepsilon''_{ii} \tag{2.15}$$

The temperature terms for shear strains are zero. Therefore, Hooke's law (2.14) taking into account (2.15) takes the following form

$$s_{ij} = 2G\,\text{э}_{ij},\ \sigma = K(\theta - 3\alpha_0 T) \tag{2.16}$$

In thermoelasticity problems, in general, the heat distribution problem must be solved beforehand. The temperature $T(x_i, t)$ at each point of the body must satisfy the *heat conduction equation*

$$\frac{\partial T}{\partial t} = B\Delta T + \frac{Q}{c\gamma}, \tag{2.17}$$

where B is material diffusivity; Δ is Laplace operator,

$$\Delta = \frac{\partial^2}{\partial x_1^2} + \frac{\partial^2}{\partial x_2^2} + \frac{\partial^2}{\partial x_3^2} = (\ldots),_{ii};$$

Q is the quantity of heat produced by the source per unit volume and per unit time; c is specific heat capacity; γ is specific weight.

The initial conditions of temperature distribution $T(x_i, 0)$ should be added to Eq. (2.17). In addition, the conditions of heat exchange with the environment must be fulfilled at the boundary of the solid body.

2.2 Theory of Small Elastic–plastic Deformations

The ability of solids to deform under the action of external forces applied to it and to obtain permanent or temporary residual (plastic) deformations when the load is removed is called *plasticity.*

The constitutive relations of plasticity theory, obviously, should take into account not only the current values of stress and strain tensor components, but also the ways of their achievement. The theory of plasticity distinguishes two types of loading of bodies: *simple and complex.* Under simple loading, all components of the stress tensor at a given point of the body increase in proportion to one common parameter λ, i.e. $\sigma_{ij} = \lambda \sigma_{ij}{}^0$ (for example, time t). Otherwise, loading will be complex. Note that the concepts of simple and complex loading were introduced by Ilyushin [1].

Ilyushin proved a *theorem on sufficient conditions* under which a simple loading will take place. This requires proportionality of the external load to one certain general parameter and a power dependence of the intensity of stresses on the intensity of deformations:

$$\sigma_u = A\varepsilon_u^{\alpha},$$

where A and α are constant values.

In plasticity theories, it is assumed that at any moment of the deformation process, the total strain tensor is represented as the sum of the elastic and plastic strain tensors:

$$\varepsilon_{ij} = \varepsilon_{ij}^{e} + \varepsilon_{ij}^{p}. \quad (2.18)$$

Plastic strain in (2.18) is defined as a combination of components of the strain tensor, which are conserved in the considered point of the media, when all components of the stress tensor in this point turn to zero.

The deformation theory of plasticity is based on the hypotheses proposed by Huber, Mises, Henky and generalised to the case of material with Nadai hardening. It assumes that for elastic–plastic bodies one can establish unambiguous dependences between stresses and strains, similar to Hooke's law for elastic bodies. The development and justification of the theory of small elastic–plastic deformations is associated with the works of Ilyushin, so this theory is often called *Ilyushin's theory of plasticity*. The following hypotheses are accepted here:

Elasticity of volume deformation. The volume deformation of the body θ is considered elastic, it is directly proportional to the average normal stress σ and Hooke's law is valid for it:

$$\sigma = K\theta = 3K\varepsilon. \tag{2.19}$$

Otherwise, the hypothesis (2.19) can be formulated as follows: due to plastic deformation there is no change in the volume of the body.

Coaxiality of deviators. Components of the strain deviator $э_{ij}$ are proportional to components of the stress deviator s_{ij}:

$$s_{ij} = \frac{2}{3}\frac{\sigma_u}{\varepsilon_u}э_{ij}, \tag{2.20}$$

where σ_u, ε_u are stress and strain tensor intensities.

In components of stress and strain tensors expression (2.20) takes the following form:

$$\sigma_{ij} - \sigma\delta_{ij} = \frac{2}{3}\frac{\sigma_u}{\varepsilon_u}(\varepsilon_{ij} - \varepsilon\delta_{ij})\ldots(i, j = 1, 2, 3).$$

Hardening hypothesis. Regardless of the type of stress state, there is a universal relationship between stress intensity and strain intensity for each material:

$$\sigma_u = (\varepsilon_u). \tag{2.21}$$

For an elastic material, this relationship is expressed as a linear relationship

$$\sigma_u = 3G\varepsilon_u.$$

Experimental verification of these hypotheses gave reasonably good results for simple or slightly differing from simple loading. The deformation process should be active, without unloading.

Consider an elastoplastic body under the influence of mass forces ρF_i and surface loads R_i. To solve the problems of the theory of small elastoplastic deformations, i.e. to determine the unknown functions of displacements, strains and stresses (u_i, σ_{ij}, ε_{ij}), there are equilibrium equations, Cauchy relations, compatibility strain equations and boundary conditions:

$$\sigma_{ij,j} + \rho F_i = 0; \,..2\varepsilon_{ij} = u_{i,j} + u_{j,i};$$

$$\varepsilon_{ii,jj} + \varepsilon_{jj,ii} - 2\varepsilon_{ij,ij} = 0,\ (\varepsilon_{ij,k} - \varepsilon_{jk,i} + \varepsilon_{ki,j})_{,i} - \varepsilon_{ii,jk} = 0;$$

$$s_{ij} l_j = R_i \ on\ S_\sigma;\ u_i = u_{0i}\ on\ S_u (i, j = 1, 2, 3). \tag{2.22}$$

It is convenient to write the constitutive equations of state in the following form:

$$s_{ij} = 2G\varphi(\varepsilon_u)э_{ij}, \ldots .\sigma = 3K\varepsilon, \tag{2.23}$$

where the deviators are:

$$s_{ij} = \sigma_{ij} - \sigma\delta_{ij};\ э_{ij} = \varepsilon_{ij} - \varepsilon\delta_{ij};\ (i, j = 1, 2, 3).$$

Comparing (2.21) and (2.23), we find the expression of the plasticity function $\varphi(\varepsilon_u)$ through the previously introduced universal function $\Phi(\varepsilon_u)$:

$$\varphi(\varepsilon_u) \equiv \frac{\Phi(\varepsilon_u)}{3G\varepsilon_u}.$$

Equation (2.22) are valid only under loading. In the case of elastic unloading from Hooke's law (2.12) the following relations follow:

$$\sigma_{ij} - \sigma'_{ij} = 2\mu(\varepsilon_{ij} - \varepsilon'_{ij}) + \lambda(\theta - \theta'), \tag{2.24}$$

where σ_{ij}' и ε_{ij}' are stresses and strains before unloading.

The physical equations in the form (2.24) are preserved until new (secondary) plastic deformations appear during the unloading process.

The problem of plasticity theory is nonlinear, so the question arises about the *existence of a solution and its uniqueness.* It has been proved [3] that under the condition

$$3G \geq \frac{\sigma_u}{\varepsilon_u} \geq \frac{d\sigma_u}{d\varepsilon_u} > 0 \tag{2.25}$$

the system of Eqs. (2.22), (2.23) is of elliptic type, and the solution of this boundary value problem exists if the solution of the corresponding problem of linear elasticity theory exists.

The following theorem was proved by Ilyushin with respect to uniqueness [1]: at given volume forces ρF_i, surface forces R_i on a part of the boundary surface S_σ and displacements u_i on a part of the boundary surface S_u, the stressed and strained states of the body, i.e. u_i, σ_{ij}, ε_{ij}, are defined uniquely if the loading is simple.

The solution of plasticity theory problems is associated with the solution of a system of nonlinear partial differential equations, which is a rather complex mathematical problem, which is solved in analytical form, as a rule, in exceptional cases. Therefore, approximate methods are most often used, in particular the *method of elastic solutions proposed by Ilyushin*. Its essence consists in considering a sequence of linear problems of elasticity theory, the solutions of which converge to the required solution of the plasticity theory problem with increasing order number. To construct the algorithm of the method, let us represent the function $\varphi(\varepsilon_u)$ in the following form:

$$\varphi(\varepsilon_u) = 1-\omega(\varepsilon_u). \tag{2.26}$$

Substituting (2.26) into (2.22), we obtain

$$s_{ij} = 2G(1 - \omega(\varepsilon_u))Э_{ij}, \ldots \sigma = 3K\varepsilon, \tag{2.27}$$

where $0 \le \omega < 1$, herewith $\omega(\varepsilon_u) = 0$, if $\varepsilon_u \le$ (ε_y is yield strength).

Thus, at $\omega = 0$, Eq. (2.27) coincide with the relations of the linear theory of elasticity.

Let us write the differential equations of equilibrium and boundary conditions by decomposing the stress tensor into deviator and spherical parts:

$$s_{ij,j} + \sigma_{,i} + \rho F_i = 0,$$

$$s_{ij} l_j + \sigma l_i = R_i \, on \, S_\sigma, \, u_i = u_{0i} \, on \, S_u. \tag{2.28}$$

We solve the elastic-plasticity problem in displacements. For this purpose, we substitute the components (2.27) into the equilibrium equations and force boundary conditions (2.28) and take into account the Cauchy relations. As a result, we obtain a generalisation of the known Lame Eq. (2.22) and force boundary conditions:

$$(\lambda + \mu)\theta_{,i} + \mu \Delta u_i + \rho F_i - F_{\omega i} = 0;$$

$$\lambda \theta l_i + \mu\left(u_{i,j} + u_{j,i}\right) l_j = R_i + R_{\omega i} \, on \, S_\sigma,$$

where

$$F_{\omega i} = 2G(\omega Э_{ij})_{,j}; \, R_{\omega i} = 2G Э_{ij} l_j \omega.$$

As a zero approximation we take $\omega^{(0)} = 0$. Then the fictitious loads at the first step $F_{\omega i} = R_{\omega i} = 0$, and to determine the first approximation $u_i^{(1)}$ we have the usual problem of linear elasticity theory. According to the found displacements $u_i^{(1)}$ the following values are determined:

$$Э_{ij}^{(1)}, \varepsilon_u^{(1)}, \omega^{(1)} \equiv \omega(\varepsilon_u^{(1)}), F_{\omega i}^{(1)}, R_{\omega i}^{(1)}.$$

For any kth approximation the following relations take place:

$$(\lambda + \mu)\theta_{,i}^{(k)} + \mu \Delta u_i^{(k)} + \rho F_i - F_{\omega i}^{(k)} = 0,$$

$$\lambda\theta^{(k)} l_i + \mu\left(u_{i,j}^k + u_{j,i}^k\right) l_j = R_i + R_{\omega i}^{(k-1)} \ldots \; on \, S_\sigma. \tag{2.29}$$

The values $F_{\omega i}^{(k-1)}$, $R_{\omega i}^{(k-1)}$ are determined by the (k–1)th approximation.

The modified Lame equations and boundary conditions (2.29) are linear with respect to the unknown displacements $u_i^{(k)}$. They differ from the corresponding equations of the theory of elasticity in that they add fictitious forces $F_{\omega i}^{(k-1)}$, $R_{\omega i}^{(k-1)}$ to the external forces ρF_i, R_i.. This makes it possible to construct the solution of the elastic-plasticity problem in recurrence form using the known solution of the corresponding problem of elasticity theory. Examples of such iterative solutions for three-layer plates are given in the corresponding chapters below.

To ensure convergence of the elastic solution method, the plasticity and nonlinearity functions ω, related to the function $\varphi(\varepsilon_u)$ by the relation (2.26), must be small compared to unity. In this case, the following Ilyushin condition must be fulfilled [1]:

$$1 > \omega + \varepsilon_u \frac{d\omega}{d\varepsilon_u} \geq \omega \geq 0. \tag{2.30}$$

In further numerical calculations, duralumin (D16T) and fluoroplastic will be considered as layer materials. For them we take the plasticity function (physical nonlinearity) $\omega(\varepsilon_u)$ in the following form:

$$\omega(\varepsilon_u) = \begin{cases} 0, & \ldots \quad \varepsilon_u \leq \varepsilon_y; \\ A(1 - \varepsilon_y/\varepsilon_u)^\alpha, & \quad \varepsilon_u > \varepsilon_y. \end{cases} \tag{2.31}$$

Constants included in (2.31) for alloy D16T: $A = 0{,}96$, $\alpha = 2{,}34$, $\varepsilon_y = 0{,}735$.

Elastoplastic strains are considered small if they do not exceed 3–4 yield stresses in terms of strains ε_y. Therefore, let us check the fulfilment of condition (2.30) with a reserve by putting $\varepsilon_u = 5\varepsilon_y$. Then:

$$\omega = 0,96(1 - 1/5)^{2,34} = 0,569,$$

$$\varepsilon_u \frac{d\omega}{d\varepsilon_u} = A\alpha(1 - \varepsilon_y/\varepsilon_u)^{\alpha-1}\varepsilon_y/\varepsilon_u = 0,96 \cdot 2,34 \cdot 0,8^{1,34} \cdot 0,2 = 0,333.$$

Hence, $1 > 0,569 + 0,333 \geq 0,569 \geq 0$ and the convergence condition for duralumin is fulfilled.

For fluoroplastic: $A = 0{,}905$, $\alpha = 1{,}48$, $\varepsilon_s = 3{,}3\%$, ε_s is Physical nonlinearity limit (analogue of yield strength). Taking $\varepsilon_u = 5\varepsilon_s$, we obtain:

$$\omega = 0,905(1 - 1/5)^{1,48} = 0,65,$$

$$\varepsilon_u \frac{d\omega}{d\varepsilon_u} = A\alpha(1 - \varepsilon_s/\varepsilon_u)^{\alpha-1}\varepsilon_s/\varepsilon_u = 0,905 \cdot 1,48 \cdot 0,8^{0,48} \cdot 0,2 = 0,241.$$

Here: $1 > 0,65 + 0,241 \geq 0,65 \geq 0$, therefore the convergence condition for fluoroplastic is also fulfilled.

Thus, when D16T alloy and fluoroplastic-4 layers are used as materials, the method of elastic solutions should converge to the required solution within small strains, since in this case condition (2.30) is fulfilled.

2.3 Physically Nonlinear Materials in Thermoradiation Fields

In the case of thermal force on an elastic–plastic body, corrections must be made to the plasticity function, elastic modulus and volume strain. Physical equations of plasticity in the temperature field are assumed to be of the type (2.16), (2.23):

$$s_{ij} = 2G(T)f(\varepsilon_u, T)\,\text{э}_{ij},\ ..\sigma = 3K(T)(\varepsilon - \alpha_0 T). \tag{2.32}$$

where T is temperature, counted from some initial temperature T_0, α_0 is thermal expansion coefficient.

At radiation (neutron) irradiation of an elastic–plastic body, other corrections must be made to the plasticity function and volume strain. The physical equations of plasticity in the radiation field will be as follows:

$$s_{ij} = 2Gf(\varepsilon_u, I)\,\text{э}_{ij},\ \ldots\sigma = K(3\varepsilon - BI). \tag{2.33}$$

Integral neutron flux $I = \varphi t$ (φ—intensity of neutron flux, t—time of its exposure) within small strains, as a rule, leads to increase of radiation hardening of material and growth of yield point. Radiation increase of volume strain is accounted for by the value BI, where B is the material constant.

For the general formulation of the problem of loading from the natural state of an elastic–plastic body in thermoradiation fields it is necessary to join to Eqs. (2.32) or (2.33) the equilibrium equations, Cauchy relations (2.6) and boundary conditions (2.22). As a result, we obtain a nonlinear boundary value problem. For its solution it is supposed to use the method of elastic solutions of Ilyushin.

In the case of thermal influence the nonlinearity function has the form

$$f(\varepsilon_u, T) = 1 - \omega(\varepsilon_u, T), \tag{2.34}$$

where $0 \leq \omega < 1$, herewith $\omega_1 = 0$, if $\varepsilon_u \leq \varepsilon_y(T)$.

At neutron irradiation the nonlinearity function is taken as

$$f(\varepsilon_u, I) = 1 - \omega(\varepsilon_u, I), \tag{2.35}$$

where $0 \le \omega < 1$, herewith $\omega_1 = 0$, if $\varepsilon_u \le \varepsilon_y(I)$.

2.4 Thermomechanical Properties of Materials

At numerical approbation of solutions of concrete boundary value problems it is necessary to use mechanical characteristics of real materials or close to them. For this purpose, the necessary parameters for aluminum alloy D16T, and polytetrafluoroethylene (fluoroplastic-4, PTFE) were calculated from the known experimental data and are given below. Plasticity and physical nonlinearity functions are also proposed.

Aluminum alloy D16T. Experimental data are taken from the works of Gusenkov [4, 5] and Rabotnov [6]. The physical equations of state corresponding to an elastic–plastic material within the framework of the theory of small elastic–plastic deformations are taken in the following form:

$$s_{ij} = 2G(T)f_1(\varepsilon_u, T)э_{ij}, \ldots \sigma = 3K(T)(\varepsilon - \alpha_0 T). \tag{2.36}$$

Here s_{ij}, $э_{ij}$, σ, ε are deviator and spherical parts of stress and strain tensors; Ilyushin plasticity function $f_1(\varepsilon_u, T)$ is represented in the form (2.34); ε_u is strain intensity; α_0 is linear thermal expansion coefficient; T is inhomogeneous and unsteady temperature field.

To describe the temperature dependence of elastic moduli $G(T)$, $K(T)$ we use the linear formula proposed by Bell [7]:

$$\{G(T), K(T), E(T)\} = \{G(0), K(0), E(0)\}\, \varphi(T),$$

$$\varphi(T) = \begin{cases} 1, & 0 < T/T_m \le 0,06, \\ 1,03(1 - T/(2T_m)), & 0,06 < T/T_m \le 0,57. \end{cases} \tag{2.37}$$

Here T_m is material melting point; $G(0)$, $K(0)$, $E(0)$ are modulus values at so-called zero stress [7]. They can be determined from experiment: knowing G_0 at some initial temperature T_0 (for example, at room temperature 20°C), we obtain from (2.37) that $G(0) = G_0/\varphi(T_0)$. Poisson's ratio is assumed to be independent of temperature.

The analytical form of the plasticity function at a given constant temperature T_0 is taken as follows:

$$\omega_1(\varepsilon_u, T_0) = \begin{cases} 0, & \varepsilon_u \leq \varepsilon_{y0}, \\ A_1\left(1 - \frac{\varepsilon_{y0}}{\varepsilon_u}\right)^{\alpha_1}, & \varepsilon_u > \varepsilon_{y0}. \end{cases} \tag{2.38}$$

where $\varepsilon_{y0} = \varepsilon_y(T_0)$ is yield strength at temperature T_0.

As the temperature increases, the plastic properties of metals increase, so the graph of the function $\omega_1(\varepsilon_u, T)$ is assumed to be shifted along the abscissa axis relative to the graph $\omega_1(\varepsilon_u, T_0)$ by the value of the change in yield strength ε_{y0}–ε_y. Thus, generalizing the formula (2.38), we take the plasticity function under thermo-force loading from the natural state to be

$$\omega_1(\varepsilon_u, T) = \begin{cases} 0, & \varepsilon_u \leq \varepsilon_y, \\ A_1\left(1 - \frac{\varepsilon_{y0}}{\varepsilon_u + \varepsilon_{y0} - \varepsilon_y}\right)^{\alpha_1}, & \varepsilon_u > \varepsilon_y. \end{cases} \tag{2.39}$$

where the value $\varepsilon_y = \varepsilon_y(T)$ can be calculated by the formula introduced by N.A. Makhutov for steels [8]:

$$\sigma_y = \sigma_{y0} \exp\left\{\kappa\left(\frac{1}{T} - \frac{1}{T_0}\right)\right\}, \ldots \varepsilon_y(T) = \frac{\sigma_y(T)}{E(T)}.$$

The numerical values of the constants included in formulas (2.36)–(2.39) are given in Table 2.1. All thermomechanical characteristics of D16T alloy were obtained in the region of medium temperatures ($T \leq 543$ K).

When structural elements are irradiated with neutrons, mechanical properties of materials are changed. According to experimental data, the growth of neutron flux value $I = \varphi t$ (φ—neutron flux intensity, t—time) within the limits of small strains, as a rule, leads to an increase in radiation hardening of the material and, accordingly, to the growth of yield strength. For analytical description of radiation hardening of a material the following formula for yield strength is supposed to be used:

$$\sigma_y = \sigma_{y0}\left[1 + A\sqrt{1 - \exp(-\xi I)}\right], \tag{2.40}$$

Table 2.1 Elasticity and physical nonlinearity parameters of duralumin

Parameter	Parameter	Parameter	Value
$E(0)$, MPa	$0{,}829 \cdot 10^5$	ν	0,35
$G(0)$, MPa	$0{,}3075 \cdot 10^5$	A_1	0,96
$K(0)$, MPa	$0{,}9214 \cdot 10^5$	α_1	2,34
E_0, MPa	$0{,}72 \cdot 10^5$	ε_{y0}, %	0,735
G_0, MPa	$0{,}267 \cdot 10^5$	σ_{yo}, MPa	340
K_0, MPa	$0{,}8 \cdot 10^5$	κ, 1/K	301
α_0, 1/K	$24{,}3 \cdot 10^{-6}$	ρ_0, kg/m^3	2700
T_0, K	293	C_0, J/(kg · K)	880
T_m, K	933	k_0, J/(m · sec · K)	177

$$\varepsilon_y(I) = \varepsilon_{y0}\left[1 + A(1 - \exp(-\xi I))^{1/2}\right],$$

where σ_y, ε_y are yield strengths of irradiated and σ_{y0}, ε_{y0} are unirradiated materials; A, ξ—material constants obtained experimentally (for aluminum alloy 356: $A = 1{,}09$; $\xi = 9{,}73 \cdot 10^{-26}$ m^2/neutron).

Here it is also taken into account that saturation occurs as the irradiation dose increases, so the hardening of the material stops. In the absence of radiation ($I = 0$), the formula (2.40) gives no hardening ($\sigma_y = \sigma_{y0}$), which distinguishes it advantageously from the formula taken as a basis by Maykin and Minter[1)] for steel alloys. The influence of radiation on the elasticity parameters of materials is insignificant and is not taken into account further, the plasticity function is described by the relation (2.35).

Polytetrafluoroethylene. Experimental data on deformation of polytetrafluoroethylene are taken from monographs by A.Y. Goldman [9, 10]. The following equations of state are used for physically non-linear deformation of fluoroplastics:

$$s_x^{(3)} = 2G_3(T)f^{(3)}(\varepsilon_u^{(3)})\ \text{э}_x^{(3)}, \ldots s_{xz}^{(3)} = 2G_3(T)f^{(3)}(\varepsilon_u^{(3)})\ \text{э}_{xz}^{(3)},$$

$$\sigma^{(3)} = 3K_3(T)(\varepsilon^{(3)} - \alpha_{03}T), \tag{2.41}$$

where $f^{(3)}$ is universal function of physical nonlinearity

$$f^{(3)}(\varepsilon_u^{(3)}) = \begin{cases} 1, & \varepsilon_u^{(3)} \le \varepsilon_s^{(3)}(T), \\ 1 - \omega^{(3)}(\varepsilon_u^{(3)}), & \varepsilon_u^{(3)} > \varepsilon_s^{(3)}(T), \end{cases}$$

$\varepsilon_s^{(3)}$ is physical nonlinearity limit of fluoroplpsta-4.

Here the temperature dependence of elasticity parameters is taken in the following form:

$$G_3(T) = \frac{G_3(T_0)}{\varphi_3(T)}, ..K_3(T) = \frac{K_3(T_0)}{\varphi_3(T)},$$

$$\varphi_3(T) = 1 + B\left(\frac{T - T_0}{T_m}\right)^{\gamma}, \tag{2.42}$$

where B, γ are material constants obtained experimentally.

The analytical formula for the nonlinearity function will be as follows:

$$\omega(\varepsilon_u) = \begin{cases} 0, & \varepsilon_u \le \varepsilon_s, \\ A_1(1 - \varepsilon_s/\varepsilon_u)^{\alpha_1}, & \varepsilon_u > \varepsilon_s. \end{cases} \tag{2.43}$$

---

[1] *Makin M.J., Minter F.J.*, Acta Met., 8, 691 (1960).

The temperature dependences of the nonlinearity limit of fluoroplastic are described with a sufficiently high degree of accuracy by the relation

$$\sigma_s = \sigma_{s1}(1 - A_2((T - T_1)/T_m)^{\alpha_2}). \tag{2.44}$$

The averaged constants for fluoroplastic included in formulae (2.41)–(2.44) are given for temperature T_0 and summarised in Table 2.2.

The elastic properties of some other polymeric and composite materials are given in Table 2.3.

In the following, the material parameters given in Tables 2.1–2.3 will be used in the numerical realisation of the solutions of boundary value problems Table 2.4.

Table 2.2 Elasticity and physical nonlinearity parameters of fluoroplastic-4

Parameter	Value	Parameter	Value
A_1	0,905	ρ, kg/m^3	2150
α_1	1,48	T_0, K	313
ε_s, %	3,3	T_1, K	293
B	24,44	A_2, Pa$^{-\alpha 4}$	$2{,}525 \cdot 10^{-19}$
γ	1,27	α_2	3,226
G_0, MPa	90	p_0, MPa	800
K_0, MPa	4700	α_0,1/K	$1{,}78 \cdot 10^{-5}$
T_m, K	600	σ_{y1}, MPa	14,54
C, J/(kg · K)	2300	λ, Вт/(m · K)	0,25

Table 2.3 Elasticity properties of polymer and composite materials

Material	Young's module, E, MPa	Shear module, G, МПа	Poisson's coefficient ν
High-strength epoxy-bonded carbon fibres parallel to the rod axis	$1{,}265 \cdot 10^5$	$0{,}562 \cdot 10^4$	0,27
Aluminium-bonded boron fibres parallel to the rod axis	$2{,}32 \cdot 10^5$	$0{,}492 \cdot 10^4$	0,23
Ultra-high module epoxy-bonded carbon fibres	$3{,}16 \cdot 10^5$	$0{,}421 \cdot 10^4$	0,26
Polyurethane foam	12	4,5	0,33
Foam plastic	33	11,8	0,4

Table 2.4 Mechanical characteristics of materials

Material	G, MPa	K, MPa
Silicon nitride ceramics	$5{,}85 \cdot 10^5$	$12{,}7 \cdot 10^5$
Cordierite	$2{,}58 \cdot 10^5$	$5{,}58 \cdot 10^5$
Titanium	$0{,}41 \cdot 10^5$	$1{,}04 \cdot 10^5$

2.5 Lagrange's Variational Principle

Consider a body of volume V with surface $S = S_\sigma + S_u$, which is in equilibrium under the action of applied external surface forces R_i and volume forces F_i. Displacements u_i, stresses σ_{ij} and strains ε_{ij} occur in the body, which are functions of external loads. If we change the external forces by infinitesimal values dR_j, dF_j, then the displacement will get an infinitesimal real increment du_i. It will occur by changing the arguments, which are the loads. Let us call a *possible* or *virtual displacement* any infinitesimal imaginary displacement that a point can make at a given fixed moment of time according to the links imposed on it.

The possible displacement, in contrast to the real du_i, we will denote by δu_i, where the symbol δ is called *variation*, and the same rules are adopted for it as for the operator-differential d. Just remember that these rules do not apply to the arguments R_j, F_j of the function u_i. In other words, the variation of a function (in this case u_i) is a change in the form of the function itself at fixed coordinates of the point under consideration.

Let now displacements u_i have virtual increments δu_i, the consequence of which are *virtual strains* $\delta\varepsilon_{ij}$. We assume that variations of δu_i are sufficiently small and do not affect the equilibrium of external forces and internal stresses, they are compatible with the conditions of body fixation at the boundaries and continuity conditions inside the body. This means that δu_i are *kinematically admissible functions*, i.e. their variation at the boundary S_u where the stresses are given will be $\delta u_i = 0$. In other respects, the possible displacements can be arbitrary continuous functions.

A solid deformable body can be considered as a system of material points. Therefore, the *principle of possible displacements of Lagrange* can be applied to it: for the equilibrium of a system of material points with stationary non-releasing and ideal links, it is necessary and sufficient that the sum of elementary works of all active forces acting on the system at any possible displacements of the system is equal to zero.

The virtual work of external forces δA *and the virtual work of internal forces* δW on possible displacements and strains are defined by the following integrals:

$$\delta A = \int\limits_{S_\sigma} R_i \delta u_i dS + \int\limits_V \rho F_i \delta u_i dV, ..\delta W = \int\limits_V \sigma_{ij} \delta \varepsilon_{ij} dV.$$

It should be remembered that the actual forces are already fully applied to the body before the possible displacements occur and are not related to them. Thus, the virtual work of external forces acting on a body in equilibrium is equal to the work of internal stresses on the corresponding virtual strains:

$$\int_{S_\sigma} R_i \delta u_i dS + \int_V \rho F_i \delta u_i dV = \int_V \sigma_{ij} \delta \varepsilon_{ij} dV. \tag{2.45}$$

Equation (2.45) is valid for both elastic and inelastic bodies.

2.6 Temperature Field in a Three-Layer Element

When numerically examining analytical solutions of boundary value problems for three-layered structural elements in heat flows, it is often necessary to know the temperature distribution along their thicknesses. In this connection, we construct an approximate solution of the corresponding heat conduction problem for a three-layer package.

Consider a three-layered unbounded element of total thickness H, subjected to a heat flux of intensity q_t falling on it perpendicular to the external plane $z = c + h_1$ (Fig. 2.2).

The surface of the element $z = -c - h_2$ is assumed to be thermally insulated. Under the specified heat exchange conditions, the unsteady one-dimensional temperature field $T(z)$ according to (2.17) is described by the differential heat conduction equation

$$T_{,zz} = \frac{1}{a_k} \frac{\partial T}{\partial t} \tag{2.46}$$

at initial condition $\theta = 0$ $(t = 0)$ and boundary conditions on external planes of the element

$$\lambda_1 T_{,z} = -q \, at \, z = c + h_1,$$

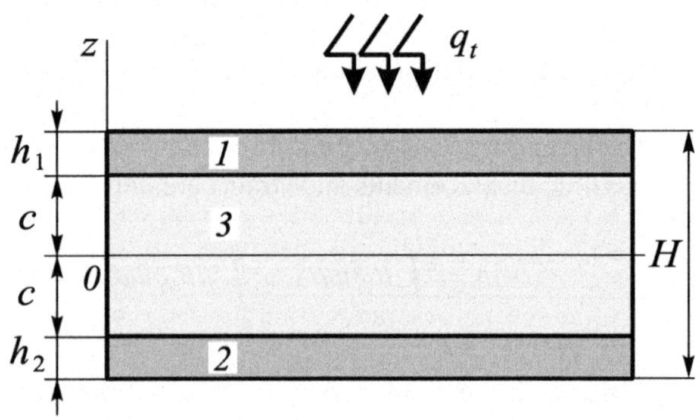

Fig. 2.2 Scheme of a three-layer element under the action of heat flow

$$T_{,z} = 0\, at\, z = -c - h_2, \tag{2.47}$$

where $a_k = \lambda_k / (C_k \rho_k)$ is thermal diffusivity of the kth layer; λ_k, C_k are thermal conductivity and heat capacity coefficients; ρ_k is material density.

Since the characteristics λ_k, C_k, ρ_k change discontinuously along the thickness of the three-layer package, in the exact formulation of the problem of determining the temperature field, Eq. (2.46) must be solved inside each homogeneous region (layer) independently, setting additional conditions of heat exchange and temperature equality at the layers' glues.

To simplify the solution of the above problem, we will carry out the procedure of averaging the thermophysical parameters over the thickness of a three-layer plate and, finally, reduce the problem to the determination of the temperature field in a homogeneous plate with modified characteristics.

Introduce the following parameter:

$$a = \lambda / C,$$

where

$$\lambda = \sum_{k=1}^{3} \lambda_k h_k / H, \,..C = \sum_{k=1}^{3} C_k h_k / H.$$

The problem of determining the temperature field in the plate in this case follows from (2.46), (2.47) after replacing a_k by a and λ_1 by λ.

Let us introduce dimensionless coordinate $s = z/H$ and «reduced» time $\tau = at/H$ (*Fourier number*). Then the heat conduction equation becomes as follows:

$$T_{,ss} = T_{,\tau}, \tag{2.48}$$

The initial conditions are assumed to be zero ($T = 0$ at $\tau = 0$). The following conditions must be fulfilled on the outer planes of the plate:

$$\frac{\lambda}{h} T_{,s} = -q \; s = (c + h_1)/H,$$

$$T_{,s} = 0 \; s = -(c + h_2)/H. \tag{2.49}$$

To solve Eq. (2.48) we apply the operational method based on the Laplace transform. Finding the image in this case reduces to solving the following equation:

$$T^*_{,ss} - pT^* = 0. \tag{2.50}$$

under the conditions following from (2.49):

$$\frac{\lambda}{H}T_{,s}^{*}=-\frac{q}{p}\ at\ s=(c+h_1)/H;$$

$$T_{,s}^{*}=0\ at\ s=-(c+h_2)/H \tag{2.51}$$

The general solution of Eq. (2.50) can be written out in hyperbolic functions:

$$T^{*}=C_1 ch\sqrt{p}s+C_2 sh\sqrt{p}s. \tag{2.52}$$

Having determined the integration constants C_1, C_2 from the conditions (2.51) and substituting their values into the solution (2.52), we obtain the following image for the temperature function in the three-layer plate:

$$T^{*}=A(p)/B(p), \tag{2.53}$$

where

$$A(p)=\frac{qH}{\lambda}ch\sqrt{p}\left(\frac{c+h_2}{H}+s\right),\ldots B(p)=p^2\frac{sh\sqrt{p}}{\sqrt{p}}$$

The roots of the equation $B(p)=0$ are the twofold root $p_1=0$ and the roots $p_{n+1}=-\pi^2n^2$ $(n=1,2,\ldots)$ of the equation

$$\frac{sh\sqrt{p}}{\sqrt{p}}=\frac{\sin i\sqrt{p}}{\sqrt{p}}=0.$$

Using theorems on determination of the original function by the image represented by relations of polynomials, taking into account multiplicity of roots, from (2.53) we find:

$$T=\frac{qH}{\lambda}\left\{\lim_{p\to 0}\left(\frac{d}{dp}\frac{\sqrt{p}ch\sqrt{p}(s+(c+h_2)/H)}{sh\sqrt{p}}e^{p\tau}\right)+\sum_{n=1}^{\infty}\frac{\mathrm{ch}\sqrt{p_{n+1}}\,(s+(c+h_2)/H)}{\left[\frac{d}{dp}(p\sqrt{p}\,\mathrm{sh}\sqrt{p})\right]_{p=p_{n+1}}}e^{p_{n+1}\tau}\right\} \tag{2.54}$$

Applying the well-known Lopital's rule to solve the uncertainties, we performthe following transformation of the first summand in curly brackets:

$$\lim_{p\to 0}\frac{d}{dp}\left[\frac{\sqrt{p}\,\mathrm{ch}\sqrt{p}\left(s+\frac{c+h_2}{H}\right)}{\mathrm{sh}\sqrt{p}}e^{p\tau}\right]=\tau+\lim_{p\to 0}\frac{\left(s+\frac{c+h_2}{H}\right)\mathrm{sh}\sqrt{p}\left(s+\frac{c+h_2}{H}\right)}{2\,\mathrm{sh}\sqrt{p}}+$$

$$+\lim_{p\to 0}\frac{ch\sqrt{p}\left(s+\frac{c+h_2}{H}\right)sh\sqrt{p}-\sqrt{p}ch\sqrt{p}}{2\sqrt{p}\,sh^2\sqrt{p}}=\tau+\frac{1}{2}\left(s+\frac{c+h_2}{H}\right)^2-\frac{1}{6}.$$

Substituting this expression into (2.54) and considering that

$$\left[\frac{d}{dp}(p\sqrt{p}sh\sqrt{p})\right]_{p=p_{n+1}} = \left(\frac{3}{2}\sqrt{p}\,sh\sqrt{p} + \frac{1}{2}p\,ch\sqrt{p}\right)_{p=p_{n+1}} = (-1)^{n+1}\frac{\pi^2 n^2}{2},$$

$$ch\sqrt{p_{n+1}}\left(s + \frac{c+h_2}{H}\right) = \cos\pi\, n\left(s + \frac{c+h_2}{H}\right),$$

we obtain the required formula for the temperature field in the considered three-layer element:

$$T = \frac{qH}{\lambda}\left\{\tau + \frac{1}{2}\left(s + \frac{c+h_2}{H}\right)^2 - \frac{1}{6} - \frac{2}{\pi^2}\sum_{n=1}^{\infty}\frac{(-1)^n}{n^2}\cos\left[\pi\, n\left(s + \frac{c+h_2}{H}\right)\right]e^{-n^2\pi^2\tau}\right\}, \tag{2.55}$$

where $\tau = at/H^2$; $s = z/H$

Note that at $h_1 = h_2 = 0$ the solution (2.55) coincides with the known Kovalenko solution for a homogeneous plate [11].

2.6.1 *Variation of Neutron Flux When Passing Through a Three-Layer Element*

For a thickness asymmetric three-layer element, the kinematic hypotheses of the broken line are adopted. The outer bearing layers are assumed to be thin, of thickness $h_1 \neq h_2$, and Kirchhoff's hypotheses of incompressibility, straightness and perpendicularity of deformed normals to the medial surface of the layers are valid for them. In a rigid sufficiently thick aggregate ($h_3 = 2c$) the work in the tangential direction is taken into account; here the hypothesis of straightness and incompressibility of the deformed normal, which is rotated with respect to the medial surface by some additional angle, holds true. The z-axis is directed upwards (Fig. 2.3).

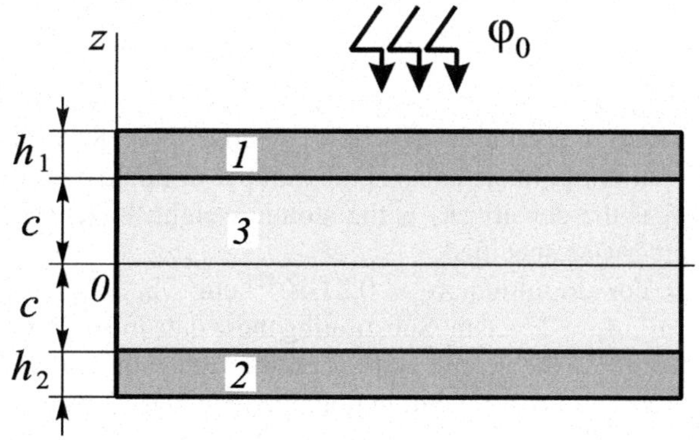

Fig. 2.3 Scheme of neutron flux irradiation of a three-layer element

It is assumed that a neutron flux of intensity $\varphi_0 = \text{const}$ in the direction opposite to the external normal is applied to the external surface $z = c + h_1$ of the considered element. Irradiation of structural elements with neutrons, ions, electrons changes mechanical properties of materials: hardness, yield strength, plasticity, creep. However, the influence of neutron irradiation on elasticity parameters—Young's modulus, Poisson's ratio—is insignificant [12, 13] and is not taken into account hereinafter.

As a result of neutron irradiation of solids, material swelling and volumetric strain occurs θ_I. According to experimental data at small strains in the linear approximation, it can be considered that the change of the material volume in the layers θ_{Ik} is directly proportional to the integral neutron flux $I_k(z)$ [14], which reaches the plane with coordinate z:

$$I_k(z, t) = \varphi_k(z)t, \tag{2.56}$$

where φ_k is the flux intensity (neutron/(m^2·s)) that has reached the surface with coordinate z during irradiation time t; $k = 1, 2, 3$ is layer number.

Consequently, the volume strain of the material in the kth layer due to neutron irradiation will be

$$\theta_{Ik} = B_k I_k(z, t), \tag{2.57}$$

where B_k is material constant obtained from experience.

We investigate the change in the intensity of neutron fluxe $\varphi_k(z)$ and integral fluxe $I_k(z)$ when passing through the thickness of the plate layers. Let on the external surface $z = c + h_1$, parallel to the z-axis fall neutrons with intensity φ_0. Then the intensity of neutron flux reaching the plane parallel to the boundary $z = \text{const}$ will decrease exponentially [14]. In our case, in the upper layer, the neutron flux intensity variation along the thickness will be described by the following equation:

$$\varphi_1(z) = \varphi_0 \exp(-\mu_1(h_1 + c - z)),$$

$$\mu_1 = \sigma_1 n_0 = \sigma_1 \frac{A_0 \rho_1}{A_1}, \tag{2.58}$$

where μ_1 is the macroscopic effective cross Sect. (1/cm); σ_1 is the effective cross section per nucleus; n_0 is the number of nuclei in 1 cm^3; A_0 is Avogadro's number; ρ_1 is the density; A_1 is the atomic weight; lower index is the layer number, unless otherwise specified.

For aluminium: $\sigma_1 = 0{,}21 \cdot 10^{-24}$ cm^2, $A_0 = 6{,}022 \cdot 10^{23}$ mol^{-1}, density $\rho_1 = 2{,}7$ g/cm^3, $A_1 = 27$ a.e.m. Substituting these data into (2.58) we obtain: $\mu_1 = 1, 26^{-1}$.Note that μ_1 is the inverse of the neutron free path, which for fast neutrons in aluminium is $\lambda_1 = 14.1$–15.9 cm, in polyethylene 5.5–13.9 cm.

Consequently, according to (2.56), by the moment t the integral neutron flux equal to:

$$I_1(z,t) = \varphi_0 t \exp(-\mu_1(c + h_1 - z)). \tag{2.59}$$

At $z = c + h_1$ the value $I_0 = \varphi_0 t$ gives the total neutron flux per unit surface area of the body. In reactors, φ_0 is of the order of 10^{17}–10^{18} neutrons/(m^{2}s), and I_0 reaches values of 10^{23}–10^{27} neutrons/m^2, and θ_I reaches values of the order of 0,1. Consequently, depending on the neutron energy and the irradiated material, the value of parameter B may be of the order of 10^{-28}–10^{-23} m^2/neutron.

At $z = c$, from formula (2.58) we obtain the neutron flux intensity φ_{13} at the boundary of transition from the first layer *1* to the filler *3*:

$$\varphi_{13} = \varphi_1(c) = \varphi_0 \exp(-\mu_1 h_1). \tag{2.60}$$

In the filler the neutron flux intensity $\varphi_3(z)$ on the plane $z =$ const and the integral flux $I_3(z,t)$ by the moment t will be:

$$\varphi_3(z) = \varphi_{13} \exp(-\mu_3(c - z),$$

$$I_3(z,t) = \varphi_{13} t \exp(-\mu_3(c - z)). \tag{2.61}$$

At the boundary $z = -c$, between the filler *3* and the lower bearing layer *2*, the neutron flux intensity value φ_{32} is obtained from the relation (2.61) at $z = 2c$:

$$\varphi_{32} = \varphi_3(-c) = \varphi_{13} \exp(-2\mu_3 c). \tag{2.62}$$

Consequently, in the lower layer the neutron flux intensity $\varphi_2(z)$ on the plane $z =$ const and the integral flux $I_2(z,t)$ by the moment t will be as follows:

$$\varphi_2(z) = \varphi_{32} \exp(\mu_2(c + z),$$

$$I_2(z,t) = \varphi_{32} t \exp(\mu_2(c + z)). \tag{2.63}$$

Thus, formulas (2.57)–(2.63) allow us to calculate the neutron flux intensity and integral neutron flux in a three-layer plate on planes with coordinate $z =$ const.

Numerical results hereinafter, unless otherwise stated, are obtained for an element whose layer materials are D16T-fluoroplastic-4-D16T. The material parameters were assumed to be as follows: $B_1 = B_2 = B_3 = 10^{\ 23}$ m^2/ neutron, which provides volumetric strain in layers of the order of 3–3,5%; for duralumin—$\mu_1 = 1{,}26$ m^{-1}; for fluoroplastic—$\mu_3 = 3{,}21$ m^{-1}; neutron flux time—$t_0 = 60$ min.

Figure 2.4 shows the variation of neutron flux magnitude along the height of the element at different thicknesses of the upper bearing layer: *1*–$h_1 = 0{,}01$ m; *2*–$h_1 = 0{,}02$ m; *3*–$h_1 = 0{,}04$ m. The thicknesses of the other layers are constant: $h_2 = 0{,}04$

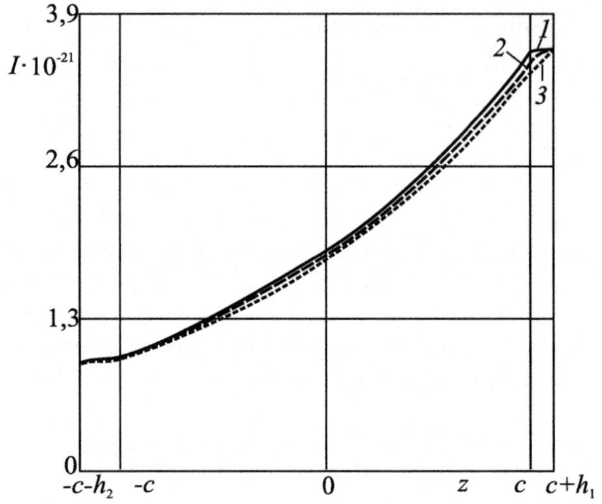

Fig. 2.4 Variation of neutron flux value by thickness of the three-layer element

m, $h_3 = 0{,}4$ m; flux rate is $\varphi_0 = 10^{18}$ neutron/(m^{2}sec). As follows from the above graphs, the increase in the thickness of the top bearing layer has a small effect on the change in the neutron flux value.

It follows from the obtained results that the value of the integral neutron flux decreases approximately by a factor of 3 when passing through the considered three-layer element.

References

1. Ilyushin, A.A. 1948. *Plasticity. Part 1. Elastic-plastic deformations.* Moscow, Gostekhizdat, 392 p. (in Russian).
2. Pobedrya, B.E. 1984. *Mechanics of composite materials.* Moscow, Izd-vo MGU, 336 p. (in Russian).
3. Moskvitin, V.V. 1981. *Cyclic loading of structural elements.* Moscow, Nauka, 344 p. (in Russian).
4. Gusenkov, A.P. 1978. *Strength under Isothermal and Non-Isothermal Low-Cycle Loading*, 295. Moscow: Nauka.
5. Gusenkov, A.P., and P.I. Kotov. 1988. *Long-term and non-isothermal low-cycle strength of structural elements.* Moscow, Mashinostroenie, 264 p. (in Russian).
6. Rabotnov, Yu.N. 1979. *Mechanics of a deformable solid body.* Moscow, Nauka, 744 p. (in Russian).
7. Bell, G.Ph. 1984. *Experimental foundations of deformable body mechanics Vol. 1 and 2.* Moscow, Nauka, 600 p. (in Russian).
8. Makhutov, N.A. et al. 1986. *Mechanics of low-cycle fracture.* Moscow, Nauka, 264 p. (in Russian).
9. Goldman, A.Ya. 1979. *Strength of engineering plastics.* Moscow, Mashinostroenie, 320 p. (in Russian).
10. Goldman, A.Ya. 1982. *Volumetric deformation of plastics.* Leningrad, Mashinostroenie, 232 p. (in Russian).

11. Kovalenko, A.D. 1970. *Fundamentals of thermoelasticity*. Kyiv, Naukova Dumka, 307 p. (in Russian).
12. Kulikov, I.S., V.B. Nesterenko, and B.E. Tverkovkin. 1990. *Strength of structural elements under irradiation*. Minsk, Navuka i technology, 144 p. (in Russian).
13. Platonov, P.A. 1971. *Effect of irradiation on the structure and properties of metals*. Moscow, Mashinostroenie, 40 p. (in Russian).
14. Ilyushin, A.A., and P.M. Ogibalov. 1960. *Elastic-plastic deformations of hollow cylinders*. Moscow, State University, 224 p. (in Russian).

Chapter 3
Elastic Three-Layer Rods with Non-compressible Filler

Deformation of thickness-asymmetrical elastic three-layer rods with a non-compressible filler, including an irregular boundary, is investigated. Tangential stresses in tangential direction are considered. Kinematic hypotheses are based on the hypothesis of the "broken" normal: Bernoulli hypotheses are accepted for external layers, the deformed normal is straight in a rigid filler. Types of loads: uniformly distributed, sinusoidal, parabolic, concentrated forces and moments.

3.1 Setting an Boundary Value Problem for an Elastic Rod

An asymmetric in thickness elastic three-layered rod with a rigid incompressible filler is considered (Fig. 3.1). The x, y, z coordinate system is associated with the median plane of the filler. The hypothesis of "broken" normal is used to describe the kinematics of the package: in thin bearing layers *1, 2* Bernoulli's hypotheses are valid, in incompressible in thickness comparatively thick filler *3* the normal remains rectilinear, it does not change length, but rotates by some additional angle $\psi(x)$. The strains are small.

Distributed loads $p(x)$, $q(x)$ apply to the outer layer of the rod. Through $w(x)$ and $u(x)$ we denote the deflection and longitudinal displacement of the median plane of the filler. Since the materials of all layers are incompressible in the transverse direction, the deflection at the remaining points of the rod coincides with $w(x)$. Rigid diaphragms are assumed at the ends, preventing relative shear of the layers, but not preventing deformation from their plane.

Let h_k be the thickness of the kth layer ($k = 1, 2, 3$—layer number), with $h_3 = 2c$. Using the introduced geometrical hypotheses, the longitudinal displacements in the layers $u^{(k)}$ are expressed through three target functions $u(x)$, $\psi(x)$ и $w(x)$:

$$u^{(1)} = u + c\psi - zw,_x \quad (c \leq z \leq c + h_1),$$

S. Eduard et al., *Deformation of Three-layer Structural Elements in Thermal Radiation Fields*, https://doi.org/10.1007/978-981-97-7217-9_3

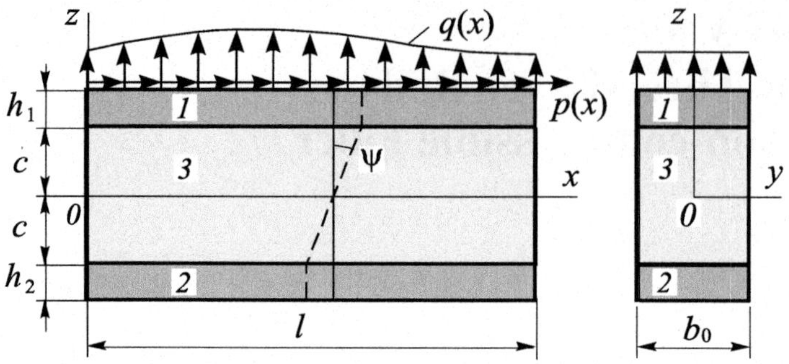

Fig. 3.1 Calculation scheme of a rod

$$
\begin{aligned}
u^{(3)} &= u + z\psi - zw_{,x} \quad (-c \le z \le c),\\
u^{(2)} &= u - c\psi - zw_{,x} \qquad (-c - h_2 \le z \le -c),
\end{aligned} \tag{3.1}
$$

where z is coordinate of the fiber under consideration; $(u + c\psi)$ is displacement value of the outer bearing layer *1* due to the deformation of the filler, $(u - c\psi)$ is displacement of bearing layer *2*; comma in the lower index denotes the differentiation operation on the coordinate following it.

From the Cauchy relations (1.6) and expressions (3.1) we obtain expressions for the components of the strain tensor in the layers:

$$
\begin{aligned}
\varepsilon_x^{(1)} &= u_{,x} + c\psi_{,x} - zw_{,xx}, \quad \varepsilon_{xz}^{(1)} = 0 \quad (c \le z \le c + h_1),\\
\varepsilon_x^{(3)} &= u_{,x} + z\psi_{,x} - zw_{,xx}, \quad \varepsilon_{xz}^{(3)} = \frac{1}{2}\psi \quad (-c \le z \le c),\\
\varepsilon_x^{(2)} &= u_{,x} - c\psi_{,x} - zw_{,xx}, \quad \varepsilon_{xz}^{(2)} = 0 \quad (-c - h_2 \le z \le -c).
\end{aligned} \tag{3.2}
$$

The spherical and deviator parts of strain tensors ($\text{э}_{ij} = \varepsilon_{ij} - \varepsilon\delta_{ij},\ i, j = x, y, z$) in the considered case have the following form:

$$
\varepsilon^{(k)} = \frac{1}{3}\varepsilon_x^{(k)}, \ \text{э}_x^{(k)} = \frac{2}{3}\varepsilon_x^{(k)}, \ \text{э}_{xz}^{(3)} = \varepsilon_{xz}^{(3)}. \tag{3.3}
$$

Internal forces and moments in the layers are introduced by the relations:

$$
N^{(k)} = b_0 \int_{h_k} \sigma_x^{(k)} \mathrm{d}z, \ M^{(k)} = b_0 \int_{h_k} \sigma_x^{(k)} z \, \mathrm{d}z, \ Q^{(3)} = b_0 \int_{h_3} \sigma_{xz}^{(3)} \mathrm{d}z, \tag{3.4}
$$

where $\sigma_x^{(k)}, \sigma_{xz}^{(3)}$ are stress tensor components, b_0 is rod cross-section width; integrals are taken over the thickness h_k of each layer.

The equations of equilibrium of the three-layer rod are obtained using Lagrange's principle of possible displacements (1.45):

$$\delta A = \delta W, \tag{3.5}$$

where δA, δW are variations of work of external forces and internal stresses.

In determining the work of external forces, it is assumed that arbitrary distributed loads are applied to the medial surface of the filler, and concentrated forces and moments are applied to the ends of the rod. The variation of the external surface load work will be:

$$\delta A_1 = \iint\limits_S (p\delta u + q\delta w)\mathrm{d}S = b_0 \int\limits_0^l (p\delta u + q\delta w)\mathrm{d}x.$$

Variation of the work of external forces and moments at the ends is:

$$\delta A_2 = (N_0 - N_l)\delta u + (Q_0 - Q_l)\delta w - (M_0 - M_l)\delta w_{,x},$$

where $N_0, Q_0, M_0, N_l, Q_l, M_l$ are given forces and moments at the rod ends ($x = 0$, $x = l$).

Variation of the work of the entire applied external load is:

$$\delta A = \delta A_1 + \delta A_2. \tag{3.6}$$

Variation of the work of elastic forces:

$$\delta W = \iint\limits_S \sum_{k=1}^{3} \int\limits_{h_k} (\sigma_x^{(k)}\delta\varepsilon_x^{(k)} + 2\sigma_{xz}^{(3)}\delta\varepsilon_{xz}^{(3)}\delta_{k3})\,\mathrm{d}z\,\mathrm{d}S =$$
$$= b_0 \int\limits_0^l \left[\sum_{k=1}^{3} \int\limits_{h_k} \sigma_x^{(k)}\delta\varepsilon_x^{(k)}\mathrm{d}z + 2\int\limits_{h_3} \sigma_{xz}^{(3)}\delta\varepsilon_{xz}^{(3)}\mathrm{d}z \right] \mathrm{d}x.$$

Since variations of displacements in layers are defined as

$$\delta u^{(1)} = \delta u + c\,\delta\psi - z\,\delta w_{,x} \quad (c \le z \le c + h_1),$$
$$\delta u^{(3)} = \delta u + z\,\delta\psi - z\,\delta w_{,x} \quad (-c \le z \le c),$$
$$\delta u^{(2)} = \delta u - c\,\delta\psi - z\,\delta w_{,x} \quad (-c - h_2 \le z \le -c),$$

then variations of strains will be written in the form:

$$\delta\varepsilon_x^{(1)} = \delta u_{,x} + c\,\delta\psi_{,x} - z\,\delta w_{,xx}, \quad \delta\varepsilon_{xz}^{(1)} = 0 \quad (c \le z \le c + h_1),$$

$$\delta\varepsilon_x^{(3)} = \delta u,_x + z\,\delta\psi,_x - z\,\delta w,_{xx}, \quad \delta\varepsilon_{xz}^{(3)} = \tfrac{1}{2}\delta\psi \quad (-c \le z \le c),$$
$$\delta\varepsilon_x^{(2)} = \delta u,_x - c\,\delta\psi,_x - z\,\delta w,_{xx}, \quad \delta\varepsilon_{xz}^{(2)} = 0 \quad (-c - h_2 \le z \le -c).$$

Let us consider the integrals over the thickness of the layers included in the expression for δW. For layer *1*, taking into account (3.4), we have:

$$\int_{h_1} \sigma_x^{(1)}\delta\varepsilon_x^{(1)}\mathrm{d}z = \int_c^{c+h_1} \sigma_x^{(1)}(\delta u,_x + c\delta\psi,_x - z\delta w,_{xx})\mathrm{d}z$$
$$= \frac{1}{b_0}\left(\delta u,_x N^{(1)} + c\delta\psi,_x N^{(1)} - \delta w,_{xx} M^{(1)}\right).$$

Similarly for layers *2* and *3*

$$\int_{h_2} \sigma_x^{(2)}\delta\varepsilon_x^{(2)}\mathrm{d}z = \frac{1}{b_0}\left(\delta u,_x N^{(2)} - c\delta\psi,_x N^{(2)} - \delta w,_{xx} M^{(2)}\right),$$
$$\int_{h_3} \sigma_x^{(3)}\delta\varepsilon_x^{(3)}\mathrm{d}z = \frac{1}{b_0}\left(\delta u,_x N^{(3)} + \delta\psi,_x M^{(3)} - \delta w,_{xx} M^{(3)}\right),$$
$$\int_{h_3} \sigma_{xz}^{(3)}\delta\varepsilon_{xz}^{(3)}\mathrm{d}z = \frac{1}{2b_0}\delta\psi\, Q^{(3)}.$$

Taking this into account, the variation of the elasticity forces work will be as follows:

$$\delta W = \int_0^l [\left(N^{(1)} + N^{(2)} + N^{(3)}\right)\delta u,_x - \left(M^{(1)} + M^{(2)} + M^{(3)}\right)\delta w,_{xx}$$
$$+ \left(M^{(3)} + cN^{(1)} - cN^{(2)}\right)\delta\psi,_x + Q^{(3)}\delta\psi]\mathrm{d}x.$$

Let us introduce total internal loads and moments

$$N = \sum_{k=1}^{3} N^{(k)}, \quad M = \sum_{k=1}^{3} M^{(k)}, \quad H = c(N^{(1)} - N^{(2)}) + M^{(3)}, \; Q = Q^{(3)}. \tag{3.7}$$

Substituting (3.7) into the relation for δW, we obtain:

$$\delta W = \int_0^l (N\delta u,_x - M\,\delta w,_{xx} + H\,\delta\psi,_x + Q\delta\psi)\,\mathrm{d}x. \tag{3.8}$$

Let us represent the terms of the integral expression (3.8) in the following form:

$$N\delta u_{,x} = (N\delta u)_{,x} - N_{,x}\delta u,\; H\delta\psi_{,x} = (H\delta\psi)_{,x} - H_{,x}\delta\psi,$$
$$M\delta w_{,xx} = (M\delta w_{,x})_{,x} - (M_{,x}\delta w)_{,x} + M_{,xx}\delta w.$$

Substituting these formulas into (3.8), we obtain the following expression of variation of work of internal elasticity forces:

$$\begin{aligned}\delta W = &-\int_0^l \left[N_{,x}\delta u + M_{,xx}\delta w + (H_{,x} - Q)\delta\psi\right]\mathrm{d}x \\ &- (N\delta u - M\delta w_{,x} + M_{,x}\delta w + H\delta\psi)|_{x=0} \\ &+ (N\delta u - M\delta w_{,x} + M_{,x}\delta w + H\delta\psi)|_{x=l}.\end{aligned} \tag{3.9}$$

If now the variations of external load work (3.6) and internal stresses (3.9) are substituted into Eq. (3.5), we get:

$$\begin{aligned}&\int_0^l \left[(bp + N_{,x})\delta u + (bq + M_{,xx})\delta w + (H_{,x} - Q)\delta\psi\right]\mathrm{d}x \\ &+ (N_l - N|_{x=l} - N_0 + N|_{x=0})\delta u + (Q_l - M_{,x}|_{x=l} - Q_0 + M_{,x}|_{x=0})\delta w \\ &- (M_l - M|_{x=l} - M_0 + M|_{x=0})\delta w_{,x} - (H|_{x=l} - H|_{x=0})\delta\psi.\end{aligned} \tag{3.10}$$

Equation (3.10) must be satisfied at any values of δu, $\delta\psi$ and δw. This can be achieved only if we equate to zero the coefficients at the mentioned variations. As a result, we obtain the system of differential equations of equilibrium of a three-layer rod in loading:

$$\begin{cases} N_{,x} + bp = 0\,, \\ H_{,x} - Q = 0\,, \\ M_{,xx} + bq = 0\,, \end{cases} \tag{3.11}$$

and force boundary conditions

$$\begin{aligned} &x = 0: N = N_0,\; M = M_0,\; M_{,x} = Q_0,\; H = 0, \\ &x = l: N = N_l,\; M = M_l,\; M_{,x} = Q_l,\; H = 0. \end{aligned} \tag{3.12}$$

If stresses are given at the ends of the rod, the forces and moments included in (3.12) can be represented as follows:

$$N_0 = \int_{-c-h_2}^{c+h_1} \sigma_{x(0)}\mathrm{d}z,\; M_0 = \int_{-c-h_2}^{c+h_1} \sigma_{x(0)}z\mathrm{d}z,\; Q_0 = \int_{-c-h_2}^{c+h_1} \sigma_{xz(0)}\mathrm{d}z,$$

$$N_l = \int\limits_{-c-h_2}^{c+h_1} \sigma_{x(l)}\, dz,\ M_l = \int\limits_{-c-h_2}^{c+h_1} \sigma_{x(l)}\, z\, dz,\ Q_l = \int\limits_{-c-h_2}^{c+h_1} \sigma_{xz(l)}\, dz,$$

where $\sigma_{x(0)}$, $\sigma_{xz(0)}$, $\sigma_{x(l)}$, $\sigma_{xz(l)}$ are stresses, given at the ends of the rod (at $x = 0$ and $x = l$, respectively).

3.2 Deformation by Continuous Surface Loading

Assume that the surface loads $p(x)$ and $q(x)$ are analytic functions along the entire length of the rod (see Fig. 3.1). To relate stresses and strains in layers we use the relations of Hooke's law (1.14) in deviator-spherical form:

$$s_{ij}^{(k)} = 2G_k\, э_{ij}^{(k)},\ \sigma^{(k)} = 3K_k \varepsilon^{(k)}, \tag{3.13}$$

where $s_{ij}^{(k)}$, $\sigma^{(k)}$ и $э_{ij}^{(k)}$, $\varepsilon^{(k)}$ are deviator and spherical parts of stress and strain tensors, G_k, K_k are shear and volume deformation modules, $k = 1, 2, 3$ is layer number, $i, j = x, y, z$.

Components of the stress tensor in layers taking into account the expressions (3.3) and (3.13) have the following form:

$$\begin{aligned} &\sigma_x^{(k)} = s_x^{(k)} + \sigma^{(k)} = 2G_k\, э_x^{(k)} + 3K_k \varepsilon^{(k)} = K_k^+ \varepsilon_x^{(k)}, \\ &\sigma_{xz}^{(3)} = s_{xz}^{(3)} = 2G_3\, э_{xz}^{(3)} = G_3 \psi,\ K_k^+ = K_k + \frac{4}{3} G_k. \end{aligned} \tag{3.14}$$

Applying relations (3.2), (3.4) and (3.14), we express internal loads and moments (3.8) through the target functions $u(x)$, $\psi(x)$, $w(x)$:

$$N^{(1)} = b_0 \int\limits_{c}^{c+h_1} \sigma_x^{(1)}\, dz = b_0 K_1^+ \left[h_1 (u,_x + c\psi,_x) - \left(c + \frac{1}{2} h_1 \right) h_1\, w,_{xx} \right],$$

$$N^{(2)} = b_0 \int\limits_{-c-h_2}^{-c} \sigma_x^{(2)}\, dz = b_0 K_2^+ \left[h_2 (u,_x - c\psi,_x) + \left(c + \frac{1}{2} h_2 \right) h_2\, w,_{xx} \right],$$

$$N^{(3)} = b_0 \int\limits_{-c}^{c} \sigma_x^{(3)}\, dz = 2 b_0 K_3^+\, c\, u,_x\,,\ M^{(1)} = b_0 \int\limits_{c}^{c+h_1} \sigma_x^{(1)}\, z\, dz$$

$$= b_0 K_1^+ \left[h_1 \left(c + \frac{1}{2} h_1 \right) (u,_x + c\psi,_x) - \left(c^2 + c h_1 + \frac{1}{3} h_1^2 \right) h_1\, w,_{xx} \right],$$

$$M^{(2)} = b_0 \int\limits_{-c-h_2}^{-c} \sigma_x^{(2)} z \, \mathrm{d}z$$

$$= b_0 K_2^+ \left[-h_2 \left(c + \frac{1}{2} h_2 \right) (u,_x - c\psi,_x) - \left(c^2 + c h_2 + \frac{4}{3} h_2^2 \right) h_2 \, w,_{xx} \right],$$

$$M^{(3)} = b_0 \int\limits_{-c}^{c} \sigma_x^{(3)} z \, \mathrm{d}z = \frac{2}{3} b_0 K_3^+ c^3 (\psi,_x - w,_{xx}),$$

$$Q^{(3)} = b_0 \int\limits_{-c}^{c} \sigma_{xz}^{(3)} \, \mathrm{d}z = 2 b_0 G^{(3)} c \psi . \tag{3.15}$$

Using expressions (3.15) and relations (3.8), let us express the generalized internal loads through the displacements. Substituting the result into Eqs. (3.11), we obtain a system of differential equations of equilibrium of a three-layer rod in displacements:

$$\begin{aligned} &a_1 u,_{xx} + a_6 \psi,_{xx} - a_7 w,_{xxx} = -p, \\ &a_6 u,_{xx} + a_2 \psi,_{xx} - a_3 w,_{xxx} - a_5 \psi = 0, \\ &a_7 u,_{xxx} + a_3 \psi,_{xxx} - a_4 w,_{xxxx} = -q. \end{aligned} \tag{3.16}$$

The coefficients here are as follows:

$$a_1 = K_1^+ h_1 + K_2^+ h_2 + 2K_3^+ c, \; a_2 = c^2 \left[K_1^+ h_1 + K_2^+ h_2 + \frac{2}{3} K_3^+ c \right],$$

$$a_3 = c \left[K_1^+ h_1 \left(c + \frac{1}{2} h_1 \right) + K_2^+ h_2 \left(c + \frac{1}{2} h_2 \right) + \frac{2}{3} K_3^+ c^2 \right],$$

$$a_4 = K_1^+ h_1 \left(c^2 + c h_1 + \frac{1}{3} h_1^2 \right) + K_2^+ h_2 \left(c + c h_2 + \frac{1}{3} h_2^2 \right) + \frac{2}{3} K_3^+ c^3, \; a_5 = 2 G_3 c,$$

$$a_6 = c \left[K_1^+ h_1 - K_2^+ h_2 \right], \; a_7 = K_1^+ h_1 \left(c + \frac{1}{2} h_1 \right) - K_2^+ h_2 \left(c + \frac{1}{2} h_2 \right). \tag{3.17}$$

Let us consider the procedure for solving the system of ordinary differential Eqs. (3.16). We perform a single integration of the third of these equations. From the second equation, we remove the longitudinal displacement $u(x)$ and deflection $w(\mathrm{x})$ using the other equations. As a result, we obtain an inhomogeneous linear differential equation for finding the shear function $\psi(x)$:

$$\psi,_{xx} - \beta^2 \psi = g(x). \tag{3.18}$$

Here

$$g(x) = \gamma_2 p + \gamma_1 \left(\int q \, \mathrm{d}x + C_1 \right), \tag{3.19}$$

where C_1 is integration constant,

$$\beta^2 = \frac{a_1 a_5 \alpha_2}{\alpha_2\alpha_3 - \alpha_1^2} > 0,\ \gamma_1 = \frac{a_1\alpha_1}{\alpha_2\alpha_3 - \alpha_1^2},\ \gamma_2 = \frac{a_6\alpha_2 - a_7\alpha_1}{\alpha_2\alpha_3 - \alpha_1^2},$$

$\alpha_2\alpha_3 - {\alpha_1}^2 \neq 0$, $\alpha_1 = a_1a_3 - a_6a_7$, $\alpha_2 = a_1a_4 - {a_7}^2$, $\alpha_3 = a_1a_2 - {a_6}^2$, coefficients a_1, …, a_7 are defined by the expressions (3.17).

The general solution $\psi_0(x)$ of the homogeneous second order differential equation corresponding to the inhomogeneous Eq. (3.18) can be written as [1]:

$$\psi_0 = A\ \mathrm{sh}(\beta x) + B\ \mathrm{ch}(\beta x),$$

where $\mathrm{sh}(\beta x)$, $\mathrm{ch}(\beta x)$ are hyperbolic sine and cosine.

We obtain the general solution of the inhomogeneous Eq. (3.16) by the method of variation of arbitrary constants:

$$\psi(x) = A(x)\ sh(\beta x) + B(x)\ ch(\beta x). \tag{3.20}$$

Parameters $A(x)$ and $B(x)$ are determined from the system of equations:

$$\begin{aligned} A_{,x}\, sh(\beta x) + B_{,x}\, ch(\beta x) &= 0, \\ A_{,x}\ \beta\, ch(\beta x) + B_{,x}\ \beta\, sh(\beta x) &= g(x). \end{aligned}$$

Then

$$A(x) = \frac{1}{\beta}\int g(x)\, ch(\beta x)\, \mathrm{d}x + C_2,\ B(x) = -\frac{1}{\beta}\int g(x)\, sh(\beta x)\, \mathrm{d}x + C_3. \tag{3.21}$$

Substituting the obtained expressions (3.21) into (3.20), we obtain the relative shear in the filler $\psi(x)$. Deflection $w(x)$ and longitudinal displacement $u(x)$ can be found from the remaining equations of the system (3.16). As a result, we have the following analytical solution to the problem of deformation of an elastic three-layer rod under the action of surface loads $p(x)$ and $q(x)$ distributed along its entire length:

$$\begin{aligned} \psi(x) =& C_2\mathrm{sh}(\beta x) + C_3\mathrm{ch}(\beta x) \\ &+ \frac{1}{\beta}\left[sh(\beta x) \int g(x) ch(\beta x)\mathrm{d}x - ch(\beta x) \int g(x) sh(\beta x)\mathrm{d}x \right], \\ w(x) =& \frac{1}{\alpha_2}\left[\alpha_1 \int \psi \mathrm{d}x - a_7 \mathrm{L}_3^{-1}(p) + a_1 \mathrm{L}_4^{-1}(q) + \frac{a_1 C_1}{6} x^3 \right] \\ &+ \frac{C_4}{2}x^2 + C_5 x + C_6, \\ u(x) =& \gamma_3\,\psi + \frac{1}{\alpha_2}\left[-a_4 \mathrm{L}_2^{-1}(p) + a_7 \mathrm{L}_3^{-1}(q) + \frac{a_7}{2} C_1 x^2 \right] + C_7 x + C_8. \end{aligned} \tag{3.22}$$

Here and hereinafter L_1^{-1}, L_2^{-1}, L_3^{-1}, L_4^{-1} are linear integral operators

$$L_1^{-1}(g) \equiv \int g\,dx, \quad L_2^{-1}(g) \equiv \iint g\,dx\,dx, \; L_3^{-1}(g) \equiv \iiint g\,dx\,dx\,dx,$$
$$L_4^{-1}(g) \equiv \int\iiint g\,dx\,dx\,dx\,dx, \; \gamma_3 = \frac{a_3 a_7 - a_4 a_6}{\alpha_2}. \tag{3.23}$$

Integration constants $C_1, \ldots, C_8$ are determined from boundary conditions. In this case it is assumed that there are rigid diaphragms at the ends, preventing the relative shear of the layers.

Let's consider two types of rod fixation.

In the case of *rigid embedding of the left end of the rod at the free right end*, the boundary conditions are as follows:

$$x = 0: \; w = w_{,x} = u = \psi = 0; \; x = l: \; \psi = N = M = M_{,x} = 0. \tag{3.24}$$

Substituting the solution (3.22) into the boundary conditions (3.24), we obtain a system of linear algebraic equations for determining the integration constants C_1, ..., C_8. After solving it we have:

$$C_1 = -L_1^{-1}(q)|_{x=l}, \; C_3 = \frac{1}{\beta}\int g(x)\,sh(\beta x)dx|_{x=0},$$
$$C_2 = \frac{1}{\beta}\Big[\frac{ch(\beta l)}{sh(\beta l)}\Big(\int g(x)\,sh(\beta x)dx|_{x=l} - \int g(x)\,sh(\beta x)dx|_{x=0}\Big)$$
$$- \int g(x)\,ch(\beta x)dx|_{x=l}\Big], \; C_4 = \frac{a_7}{\alpha_2}L_1^{-1}(p)|_{x=l}$$
$$- \frac{a_1}{\alpha_2}L_2^{-1}(q)|_{x=l} - \frac{a_1}{\alpha_2}C_1 l,$$
$$C_5 = \frac{a_7}{\alpha_2}L_2^{-1}(p)|_{x=0} - \frac{a_1}{\alpha_2}L_3^{-1}(q)|_{x=0},$$
$$C_6 = \frac{a_7}{\alpha_2}L_3^{-1}(p)\Big|_{x=0} - \frac{a_1}{\alpha_2}L_4^{-1}(q)\Big|_{x=0} - \frac{\alpha_1}{\alpha_2}\int \psi dx\Big|_{x=0},$$
$$C_7 = \frac{a_4}{\alpha_2}L_1^{-1}(p)|_{x=l} - \frac{a_7}{\alpha_2}L_2^{-1}(q)|_{x=l} - \frac{a_7}{\alpha_2}C_1 l,$$
$$C_8 = \frac{a_4}{\alpha_2}L_2^{-1}(p)|_{x=0} - \frac{a_7}{\alpha_2}L_3^{-1}(q)|_{x=0}. \tag{3.25}$$

For a *rod hinged at both ends*, the boundary conditions are as follows:

$$x = 0, \;\; l: u = \psi = w = M = 0 \tag{3.26}$$

The following integration constants satisfy these conditions:

$$
\begin{aligned}
C_1 =& \frac{1}{l}\left(\mathrm{L}_2^{-1}(q)|_{x=0} - \mathrm{L}_2^{-1}(q)|_{x=l}\right),\\
C_4 =& \frac{a_7}{a_4}(D_1|_{x=0} - D_1|_{x=l}) - \frac{1}{a_4}\mathrm{L}_2^{-1}(q)|_{x=0},\\
C_5 =& D_2|_{x=0} - D_2|_{x=l} - \frac{a_7 l}{2a_4}(D_1|_{x=0} - D_1|_{x=l}) - \frac{l}{2a_4}\mathrm{L}_2^{-1}(q)|_{x=0},\\
C_7 =& D_1|_{x=0} - D_1|_{x=l},\ D_1 = \frac{a_7}{l\alpha_2}\mathrm{L}_3^{-1}(q)\\
&- \frac{a_7 l}{2\alpha_2}\mathrm{L}_2^{-1}(q) - \frac{a_4}{l\alpha_2}\mathrm{L}_2^{-1}(p),\\
D_2 =& \frac{a_1}{l\alpha_2}\left(\int \psi \mathrm{d}x + \mathrm{L}_4^{-1}(q)\right) - \frac{a_1 l}{6\alpha_2}\mathrm{L}_2^{-1}(q) - \frac{a_7}{l\alpha_2}\mathrm{L}_3^{-1}(p).
\end{aligned}
\tag{3.27}
$$

The constants C_2, C_3, C_6, C_8 retain their previous form (3.25).

In the case of *rigid embedding of both ends* of the rod, the solution must satisfy the following boundary conditions:

$$
x = 0,\ l : w = u = \psi = w,_x = 0. \tag{3.28}
$$

The corresponding integration constants are:

$$
\begin{aligned}
C_4 &= \frac{1}{\alpha_2 l}\left[a_7 \mathrm{L}_2^{-1}(p')\big|_{x=l} - a_1 \mathrm{L}_3^{-1}(q')\big|_{x=l} - \frac{a_1 C_1 l^2}{2}\right] - \frac{C_5}{l},\\
C_7 &= \frac{1}{\alpha_2 l}\left[a_4 \mathrm{L}_2^{-1}(p)\big|_{x=l} - a_7 \mathrm{L}_3^{-1}(q)\big|_{x=l} - \frac{a_7 C_1 l^2}{2}\right] - \frac{C_8}{l}.
\end{aligned}
\tag{3.29}
$$

The constant $C_1^{(n)}$ is the same as in (3.27); the constants $C_2^{(n)}, C_3^{(n)}, C_5^{(n)}, C_6^{(n)}, C_8^{(n)}$ keep the form (3.25).

In the case of *constant distributed loads* $q = \text{const}$, $p = \text{const}$ at cantilever fixing of the rod we obtain:

$$
\begin{aligned}
\psi(x) &= C_2\mathrm{sh}(\beta x) + C_3\mathrm{ch}(\beta x) - \frac{1}{\beta^2}\left[\gamma_2 p + \gamma_1 q(x-l)\right],\\
u &= \gamma_3\psi + \frac{1}{\alpha_2}\left[-\frac{1}{2}a_4 p x^2 + \frac{1}{6}a_7 q x^3 - \frac{1}{2}a_7 q l x^2\right] + C_7 x,\\
w &= \frac{1}{\alpha_2}\left[\alpha_1 \int \psi\, \mathrm{d}x - \frac{1}{6}a_7 p x^3 + \frac{1}{6}a_1 q x^3\left(\frac{1}{4}x - l\right)\right] + \frac{1}{2}C_4 x^2 + C_6,
\end{aligned}
\tag{3.30}
$$

where

$$
\int \psi \mathrm{d}x = \frac{1}{\beta}(C_2\mathrm{ch}(\beta x) + C_3\mathrm{sh}(\beta x)) - \frac{x}{\beta^2}\left[\gamma_2 p + \gamma_1 q\left(\frac{1}{2}x - l\right)\right],
$$

$$C_2 = \frac{1}{\beta^2 sh(\beta l)}\left[\gamma_2 p(1 - ch(\beta l)) + \gamma_1 ql \cdot ch(\beta l)\right],$$

$$C_3 = \frac{1}{\beta^2}\left[\gamma_2 p - \gamma_1 ql\right],\ C_4 = \frac{1}{\alpha_2}\left[a_7 pl + \frac{1}{2}a_1 ql^2\right],$$

$$C_6 = -\frac{\alpha_1}{\beta\alpha_2}C_2,\ C_7 = \frac{1}{\alpha_2}\left[a_4 pl + \frac{1}{2}a_7 ql^2\right].$$

At rigid embedding of both ends of the rod and uniformly distributed loads from (3.22), (3.28), (3.29) we have:

$$\psi(x) = C_2 sh(\beta x) + C_3 ch(\beta x) - \frac{1}{\beta^2}\left[\gamma_2 p + \gamma_1 q\left(x - \frac{l}{2}\right)\right],$$

$$u(x) = \gamma_3\,\psi + \frac{1}{\alpha_2}\left[-\frac{1}{2}a_4 px^2 + \frac{1}{6}a_7 qx^3 - \frac{a_7}{4}qlx^2\right] + C_7 x,$$

$$w = \frac{1}{\alpha_2}\left[\alpha_1 \int \psi\,\mathrm{d}x - \frac{1}{6}a_7 px^3 + \frac{1}{12}a_1 qx^3\left(\frac{x}{2} - l\right)\right] + \frac{1}{2}C_4 x^2 + C_6,$$

where $\int \psi \mathrm{d}x = \frac{1}{\beta}(C_2 \mathrm{ch}(\beta x) + C_3 \mathrm{sh}(\beta x)) - \frac{x}{2\beta^2}(2\gamma_2 p + \gamma_1 q(x - l))$,

$$C_2 = \frac{1}{2\beta^2 sh(\beta l)}\left[2\gamma_2 p(1 - ch(\beta l)) + \gamma_1 ql(1 + ch(\beta l))\right],$$

$$C_3 = \frac{1}{2\beta^2}\left[2\gamma_2 p - \gamma_1 ql\right],\ C_4 = \frac{1}{2\alpha_2}\left(a_7 pl + \frac{1}{6}a_1 ql^2\right),$$

$$C_6 = -\frac{\alpha_1}{\beta\alpha_2}C_2,\ C_7 = \frac{1}{2\alpha_2}\left[a_4 pl + \frac{1}{6}a_7 ql^2\right].$$

Thus, expressions (3.22)–(3.27) give an analytical solution to the problem of deformation of an elastic three-layer rod for various types of boundary conditions. It should be noted that at $p = 0$, $q = \mathrm{const}$ the solution (33.30coincides with the known solution.

Numerical results. Here and further in this section, during numerical calculations it was assumed that the bearing layers of the rod are made of D16T alloy (duralumin), the filler is fluoroplastic (polytetrafluoroethylene). Mechanical characteristics of these materials are given in Tables 1.1–1.3. When implementing the solution (3.30), the following values of relative layer thicknesses were assumed $c = 0.09$; $h_1 + h_2 = 0.06$; distributed load intensity $q = -1{,}5$ МПа, $p = 0$.

Figures 3.2, 3.3 show the shear in the filler and deflection of the three-layer rod as a function of its asymmetry: *1*—$h_1 = h_2 = 0.03$; *2*—$h_1 = 0.04$, $h_2 = 0.02$; *3*—$h_1 = 0.02$, $h_2 = 0.04$; *4*—$h_1 = 0.05$, $h_2 = 0.01$; *5*—$h_1 = 0.01$, $h_2 = 0.5$.

According to the displacements $\psi(x)$, $w(x)$ calculated for a rod of constant mass with asymmetric material distribution of bearing layers, we can conclude that the most rigid in bending is a two-layer rod in which the mass of bearing layers is concentrated in one layer. A symmetrical in thickness three-layer rod (sandwich) has

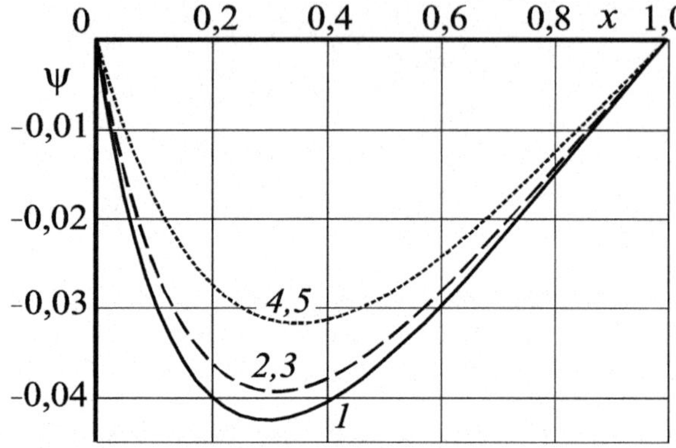

Fig. 3.2 Variation of shear in the filler as a function of packet asymmetry

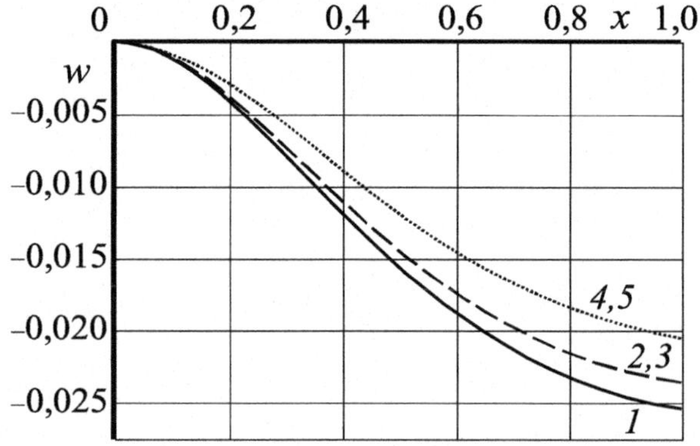

Fig. 3.3 Deflection variation as a function of packet asymmetry

the highest deflection, i.e. the lowest bending stiffness. For rods of inverse symmetry (curves 2, 3 and 4, 5), the value of deflection and relative shear are the same.

Figure 3.4 illustrates the normal stress distribution σ_x in the embedment ($x = 0$). The numbering of the curves remains the same. As the asymmetry of the rod increases, the stresses change to a greater extent in the thinning layer, where they increase in modulus. In the other layer, the stresses are not much different from the case of a symmetrical rod. By varying the thickness of the layers, we can shift the stresses to the positive or negative region as required for practice. In the places where the layers are glued together, the stresses undergo a rupture due to the difference in elastic characteristics of the materials.

Figures 3.5, 3.6, 3.7, 3.8, 3.9, 3.10, 3.11 and 3.12 show the results of displacements and stresses in a symmetrical three-layer rod under different boundary conditions and different materials of bearing layers and filler material.

Figure 3.5 shows the variation of deflection w and shear ψ along the axis of an elastic three-layer rod (D16T-fluoroplastic-D16T) under different boundary conditions: *1*—left end embedded, right end free, *2*—hinge-hinge, *3*—embedded-embedded. As the stiffness of the rod fixation increases, the magnitude of the

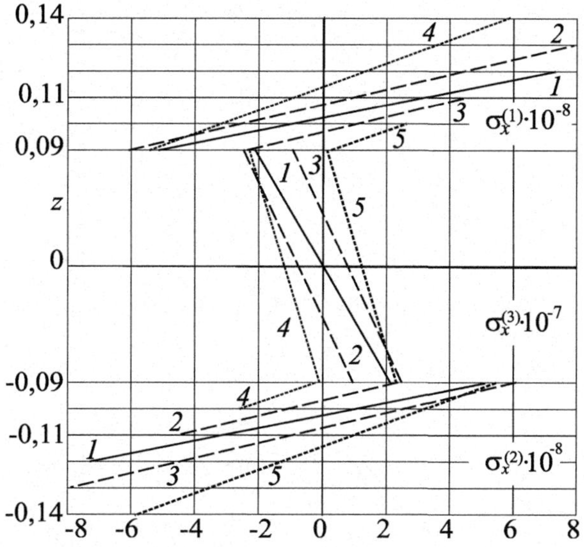

Fig. 3.4 Distribution of normal stresses σ_x in the embedment ($x = 0$) by thickness

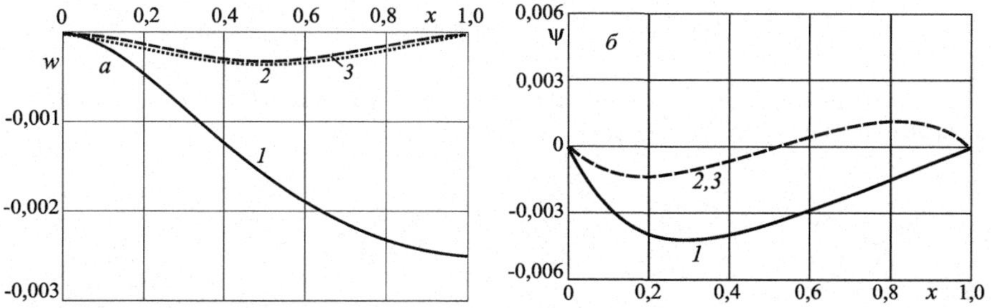

Fig. 3.5 Variation of deflection w (a) and shear ψ (b) along the rod axis under different boundary conditions

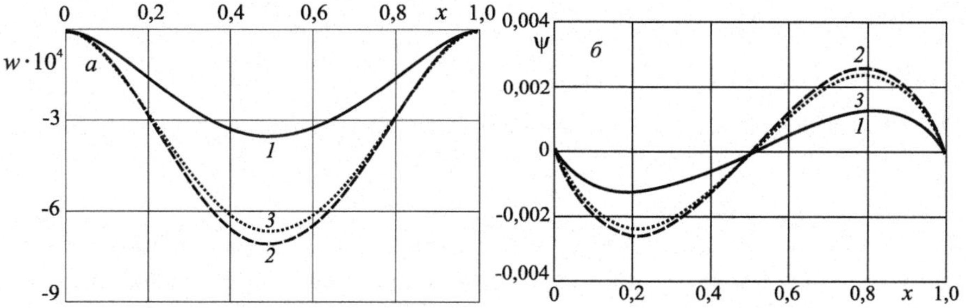

Fig. 3.6 Variation of deflection w (a) and shear ψ (b) along the rod axis for different filler materials

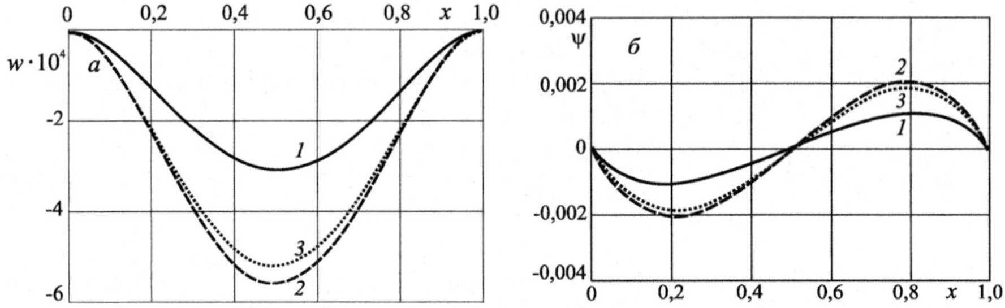

Fig. 3.7 Variation of deflection w (a) and shear ψ (b) along the rod axis (bearing layers—boron fibres on aluminium binder)

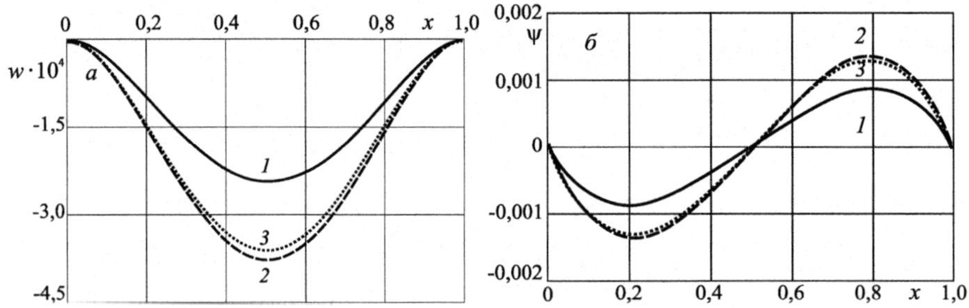

Fig. 3.8 Variation of deflection w (a) and shear ψ (b) along the rod axis (bearing layers—ultra-high modulus carbon fibres on epoxy binder)

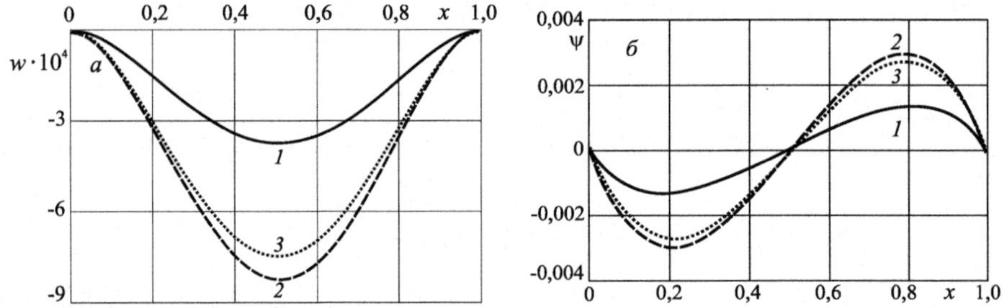

Fig. 3.9 Variation of deflection w (a) and shear ψ (b) along the rod axis (bearing layers—high-strength carbon fibers on epoxy binder)

maximum deflection decreases by a factor of 6–8 and the maximum shear by a factor of 3. In this case, the differences in the hinge-hinge and embedded-embedded cases are negligible.

Figure 3.6 illustrates the dependence of deflection (a) and shear (b) in an elastic rod (boundary conditions is "embedded-embedded", bearing layers is D-16T alloy) on the filler material. In this case, there is a natural increase in displacements as the elastic parameters of the filler material decrease.

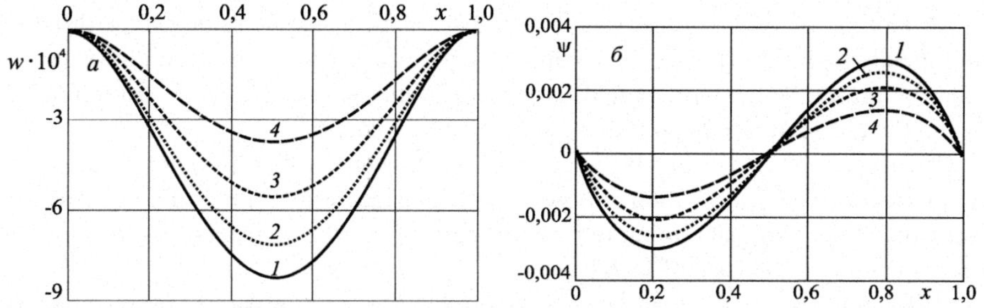

Fig. 3.10 Variation of deflection w (a) and shear ψ (b) along the rod axis depending on the material of bearing layers

Fig. 3.11 Variation of normal stresses along the thickness at different filler materials

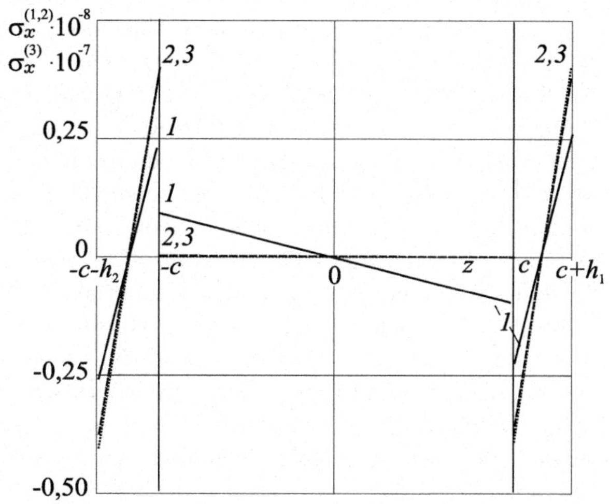

Fig. 3.12 Variation of normal stresses along the thickness at different materials of bearing layers

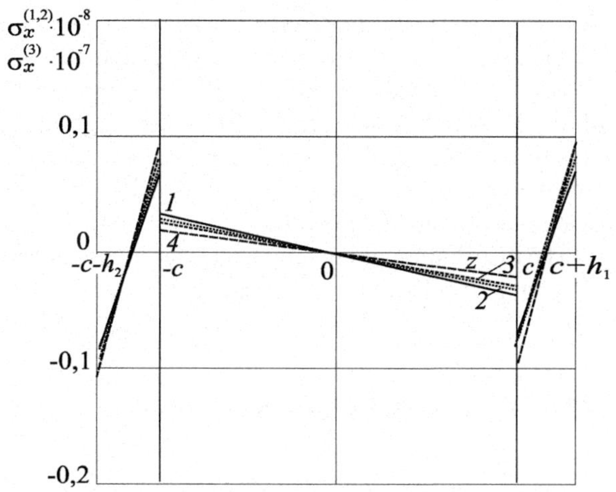

A similar picture is observed in the case of bearing layers made of different composite materials (Figs. 3.7, 3.8 and 3.9) in the case of boundary conditions at the ends of the "embedded-embedded" type at different filler materials: *1*—fluoroplastic-4 (PTFE), *2*—polyurethane foam (PU), *3*—foam plastic.

Figure 3.7 shows the displacement graphs when the load-bearing layers are made of composite material, which is boron fibres on aluminium binder. In this case, the bearing layers demonstrate more rigid properties and the displacements are smaller in value than in the case of D16T alloy.

If the material of the load-bearing layers is ultra-high modulus carbon fibres on epoxy binder (Fig. 3.8), the displacements are further reduced.

In the case of bearing layers made of composite—high-strength carbon fibres on epoxy binder (Fig. 3.9)—the displacements are the largest of the considered ones.

Figure 3.10 shows the variation of deflection (*a*) and shear (*b*) along the axis of an elastic three-layer rod depending on the material of the bearing layers (boundary conditions of the "embedded-embedded" type, filler is polyurethane foam): *1*—high-strength carbon fibers, *2*—D-16T alloy, *3*—boron fibers, *4*—ultra-high-modulus carbon fibers. Although the Young's modulus of D-16T alloy is the lowest, however, due to its higher shear modulus, its bending stiffness is higher than that of the composite "high-strength carbon fibers on epoxy binder".

Figure 3.11 illustrates the *variation of normal stresses* along the thickness at the left end of the rod at different materials of filler (boundary conditions is "embedding-embedding", bearing layers is alloy D-16T): *1*—fluoroplastic-4 (PTFE), *2*—polyurethane foam (PU), *3*—foam. The graphs show that rigid filler (fluoroplastic) takes up almost half of the normal stresses of the bearing layers compared to lightweight fillers (polyurethane foam, foam plastic). In this case, the stresses in the embedment reach a maximum, which agrees well with the stress distribution in a homogeneous rod.

Figure 3.12 shows the dependence of *normal stresses* in the cross-section of a three-layer rod on the material of the bearing layers at rigid filler. The graphs illustrate the change of stresses along the thickness at the left end of the rod (boundary conditions is "embedded-embedded", filler is fluoroplastic-4): *1*—high-strength carbon fibers, *2*—D16T, 3—boron fibers, *4*—ultra-high-modulus carbon fibers. As the stiffness of the load-bearing layers increases, the stresses in them increase and those in the filler decrease. Consequently, the load-bearing layers take on more load in this case, thus off-loading the filler.

The distribution of tangential stresses along the axis of the filler depending on its material is similar to the shear in Fig. 3.10.

3.3 Deformation by Local Loads

Action of a local uniformly distributed (rectangular) load inside the segment $[b, a]$ on the outer surface of a three-layer rod (Fig. 3.13). In this case, the load components are represented as

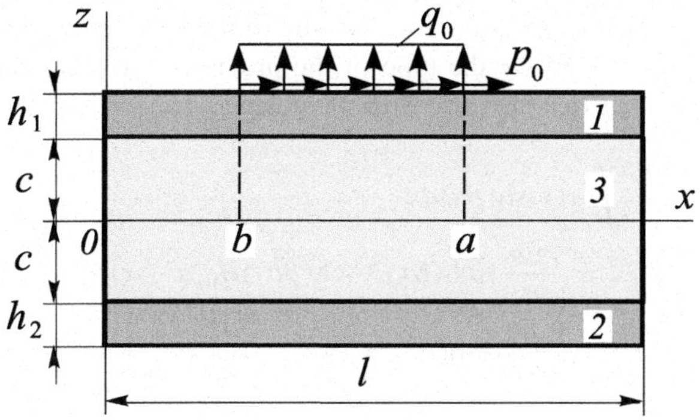

Fig. 3.13 Calculation diagram of a rod under local loading

$$q(x) = q_0(H_0(a-x)-H_0(b-x)),$$
$$p(x) = p_0(H_0(a-x)-H_0(b-x)).$$

Here q_0, p_0 are constant quantities, $H_0(z)$ is one of the family of Heaviside functions (n is integer):

$$H_n(z) = \begin{cases} \dfrac{z^n}{n!}, & z \geq 0, \\ 0, & z < 0, \end{cases} \tag{3.31}$$

When constructing the solution of the corresponding boundary value problem, we use the solution (3.22) obtained for the case of surface loading of analytical form. The main problem here is the integral operators (3.22) from the external loads and the functions into which these loads are included. Therefore, we will preliminarily analyze some key fragments of the solution.

The function $g(x)$ (3.19) in the considered case will be as follows:

$$g(x) = \gamma_2 p_0(H_0(a-x) - H_0(b-x)) + \gamma_1(q_0(H_1(b-x) - H_1(a-x)) + C_1).$$

The indefinite integral from the product of this function by the hyperbolic cosine is reduced to the form

$$\begin{aligned} &\int g(x)\mathrm{ch}(\beta x)dx \\ &= \frac{\gamma_2 p_0}{\beta}((\mathrm{sh}(\beta x) - \mathrm{sh}(\beta a))H_0(a-x) - (\mathrm{sh}(\beta x) - \mathrm{sh}(\beta b))H_0(b-x)) \\ &- \frac{\gamma_1 q_0}{\beta^2}((\mathrm{ch}(\beta x) + (a\beta - \beta x)\mathrm{sh}(\beta x) - \mathrm{ch}(\beta a))H_0(a-x) \\ &-(\mathrm{ch}(\beta x) + (\beta b - \beta x)\mathrm{sh}(\beta x) - \mathrm{ch}(\beta b))H_0(b-x)) + \gamma_1 C_1 \frac{sh(\beta x)}{\beta}. \end{aligned}$$

The expression for the integral from the product of functions $g(x)\mathrm{sh}(\beta x)$ is constructed by the type of the previous one, you only need to replace hyperbolic cosines by sines and sines by cosines:

$$\int g(x)\mathrm{sh}(\beta x)dx$$
$$= \frac{\gamma_2 p_0}{\beta}((\mathrm{ch}(\beta x) - \mathrm{ch}(\beta a))H_0(a-x) - (\mathrm{ch}(\beta x) - \mathrm{ch}(\beta b))H_0(b-x))$$
$$- \frac{\gamma_1 q_0}{\beta^2}((\mathrm{sh}(\beta x) + (a\beta - \beta x)\mathrm{ch}(\beta x) - \mathrm{sh}(\beta a))H_0(a-x)$$
$$-(\mathrm{sh}(\beta x) + (\beta b - \beta x)\mathrm{ch}(\beta x) - \mathrm{sh}(\beta b))H_0(b-x)) + \gamma_1 C_1 \frac{ch(\beta x)}{\beta}.$$

As a result, substituting the integrals into (3.22), we obtain the solution to the problem of a three-layered rod deformation under the action of a linear load locally distributed on the segment $[b, a]$:

$$\psi(x) = C_2\mathrm{sh}(\beta x) + C_3\mathrm{ch}(\beta x) + \frac{1}{\beta^2}\big[\gamma_2 p_0((\mathrm{ch}(\beta x - a\beta) - 1)H_0(a-x)$$
$$- (\mathrm{ch}(\beta x - b\beta) - 1)H_0(b-x))$$
$$- \frac{\gamma_1 q_0}{\beta}((\beta x - \beta a - \mathrm{sh}(\beta x - \beta a))H_0(a-x)$$
$$-(\beta x - \beta b - \mathrm{sh}(\beta x - \beta b))H_0(b-x)) - \gamma_1 C_1\big]$$
$$u(x) = \gamma_3\psi + \frac{1}{\alpha_2}\Big[-a_4 p_0(H_2(a-x) - H_2(b-x))$$
$$- a_7 q_0(H_3(a-x) - H_3(b-x)) + \frac{a_7 C_1}{2}x^2\Big] + C_7 x + C_8,$$
$$w(x) = \frac{1}{\alpha_2}\Big[\alpha_1 \int \psi dx + a_7 p_0(H_3(a-x) - H_3(b-x))$$
$$+ a_1 q_0(H_4(a-x) - H_4(b-x)) + \frac{1}{6}a_1 C_1 x^3\Big] + \frac{C_4}{2}x^2 + C_5 x + C_6, \qquad (3.32)$$

where $H_n(z)$ $(n = 0, 2, 3, 4)$ is Heaviside functions (3.31).

In the case of rigid embedding of the left end of the rod, the boundary conditions (3.24) will be satisfied by the following integration constants obtained using the expressions (3.25):

$$C_1 = C_4 = C_7 = 0,$$
$$C_2 = \frac{1}{\beta}\frac{ch(\beta l)}{sh(\beta l)}\left(\frac{\gamma_1 q_0}{\beta^2}(a\beta - b\beta + sh(\beta b) - sh(\beta a)) - \frac{\gamma_2 p_0}{\beta}(ch(\beta b) - ch(\beta a))\right),$$
$$C_3 = \frac{\gamma_2 p_0}{\beta^2}(ch(\beta b) - ch(\beta a)) - \frac{\gamma_1 q_0}{\beta^3}(a\beta - b\beta + sh(\beta b) - sh(\beta a)),$$

$$C_5 = \frac{1}{\alpha_2}\left(a_7 p_0 \frac{a^2 - b^2}{2} + a_1 q_0 \frac{a^3 - b^3}{6}\right),$$

$$C_6 = \frac{1}{\alpha_2}\left(a_7 p_0 \frac{b^3 - a^3}{6} - a_1 q_0 \frac{a^4 - b^4}{24} - \alpha_1 \int \psi \mathrm{d}x\Big|_{x=0}\right),$$

$$C_8 = \frac{1}{\alpha_2}\left(a_4 p_0 \frac{a^2 - b^2}{2} - a_7 q_0 \frac{b^3 - a^3}{6}\right),$$

where the integral of the relative shear is

$$\int \psi dx|_{x=0} = \frac{C_2}{\beta} + \frac{1}{\beta^2}\left[\frac{\gamma_2 p_0}{\beta}(\mathrm{sh}(\beta b) - \mathrm{sh}(\beta a)\right.$$
$$\left. + \frac{\gamma_1 q_0}{\beta^2}\mathrm{ch}(\beta a) - \mathrm{ch}(\beta b)\right) - \gamma_1 q_0 \frac{a^2 - b^2}{2}\Bigg].$$

Numerical results. At numerical realization of the solution (3.32), the following values of relative layer thicknesses were assumed $h_1 = 0.02$, $h_2 = 0.04$, $c = 0.09$. The load of intensity $q_0 = -1.5$ MPa is distributed on the segment [0, a]. Layer materials: D16T alloy, fluoroplastic-4.

Figures 3.14, 3.15 show the variation of shear in the filler and deflection along the axis of a unit-length three-layer rod as a function of the length of the section where the distributed load is applied: *1*—$a = 0.25$, *2*—$a = 0.50$, *3*—$a = 0.75$, *4*—$a = 1$. By analyzing the curves, it can be concluded that the maximum values of deflection and relative shear increase nonlinearly with increasing linear load area.

Action of concentrated force. Suppose that the previously considered distributed surface loads $q(x)$, $p(x)$ act in some small neighbourhood of a point with coordinate $x = a$ (Fig. 3.16).

Let us denote the radius of this neighbourhood by ξ. Then the distributed loads can be written using the Heaviside functions (3.31) as follows:

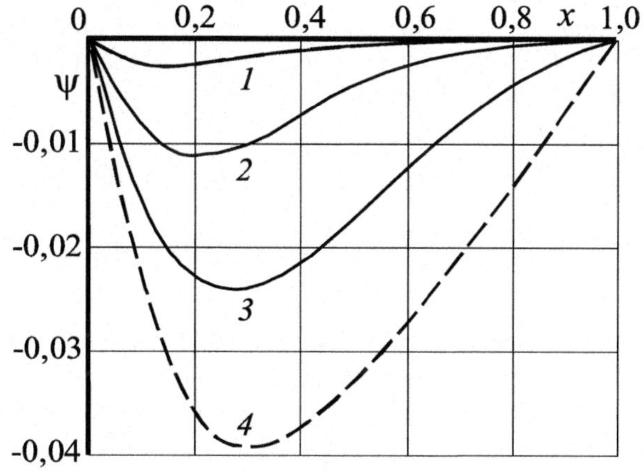

Fig. 3.14 Variation of shear in filler as a function of the load action length

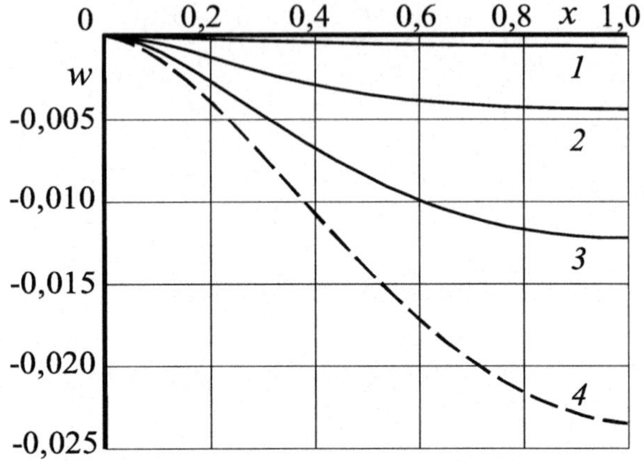

Fig. 3.15 Deflection variation as a function of the load action length

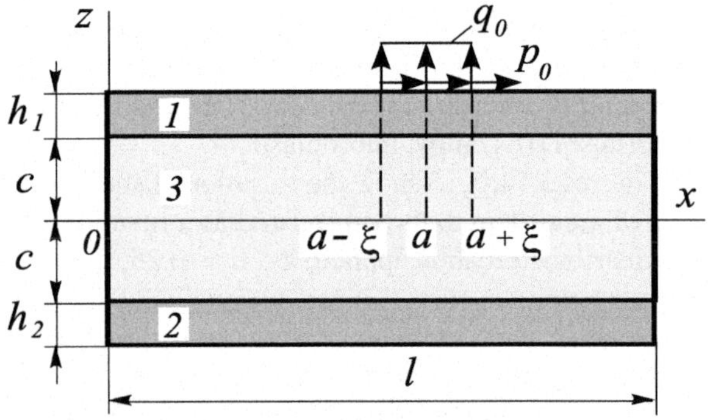

Fig. 3.16 Calculation diagram of a rod under concentrated load

$$q(x) = q_0(H_0(a + \xi - x) - H_0(a - \xi - x)),$$
$$p(x) = p_0(H_0(a + \xi - x) - H_0(a - \xi - x)).$$

Introducing in these expressions values of the type of concentrated forces $Q = 2q_0\xi$, $P = 2p_0\xi$, let us represent the external forces in a form convenient for the following transformations:

$$q(x) = \frac{Q}{2\xi}(H_0(a + \xi - x) - H_0(a - \xi - x)),$$
$$p(x) = \frac{P}{2\xi}(H_0(a + \xi - x) - H_0(a - \xi - x)). \tag{3.33}$$

Let's introduce relations (3.33) into the solution for the locally distributed load (3.32) and set the length of the section ξ to zero, leaving the values of Q and P constant. Then we obtain the corresponding partial solution for the shear $\psi(x)$:

$$\lim_{\xi\to 0}\left[\frac{1}{\beta^2}\left(\frac{\gamma_2 P}{2\xi}((ch(\beta x-(a+\xi)\beta)-1)H_0(a+\xi-x)\right.\right.$$
$$-(ch(\beta x-(a-\xi)\beta)-1)\times H_0(a-\xi-x))$$
$$-\frac{\gamma_1 Q}{2\xi\beta}((\beta x-(a+\xi)\beta-sh(\beta x-(a+\xi)\beta))H_0(a+\xi-x)$$
$$\left.\left.-(\beta x-(a-\xi)\beta-sh(\beta x-(a-\xi)\beta))H_0(a-\xi-x))-\gamma_1 C_1\right)\right].$$

Let us calculate this limit, leaving the magnitudes of the concentrated forces Q and P constant. After that, we perform a similar operation with expressions for other displacements in the solution (3.32). As a result, we obtain a solution to the problem of deformation of an elastic three-layer rod by a concentrated force (with components Q and P) applied in the section with coordinate $x=a$:

$$\psi(x)=C_2 sh(\beta x)+C_3 ch(\beta x)+\frac{1}{\beta^2}\left[\gamma_2 P\beta sh((\beta a-\beta x)H_0(a-x)\right.$$
$$\left.+\gamma_1 Q(1-\mathrm{ch}(\beta x-\beta a))H_0(a-x)-\gamma_1 C_1\right],$$
$$u(x)=\gamma_3\psi+\frac{1}{\alpha_2}\left[-a_4 PH_1(a-x)-a_7 QH_2(a-x)+\frac{a_7 C_1}{2}x^2\right]+C_7 x+C_8,$$
$$w(x)=\frac{1}{\alpha_2}\left(\alpha_1\int\psi\mathrm{d}x-a_7 PH_2(a-x)-a_1 QH_3(a-x)\right.$$
$$\left.+\frac{a_1 C_1}{6}x^3\right)+\frac{C_4}{2}x^2+C_5 x+C_6, \qquad (3.34)$$

where $H_n(z)$ $(n=0, 1, 2, 3)$ is Heaviside functions (3.31).

At rigid embedding of the left section of the rod, the integration constants are:

$$C_1=C_4=C_7=0,$$
$$C_2=\lim_{\xi\to 0}\left[\frac{1}{\beta}\left(\frac{ch(\beta l)}{sh(\beta l)}\left(-\frac{\gamma_2 P}{2\beta\xi}(ch(\beta(a-\xi))-ch(\beta(a+\xi)))+\frac{\gamma_1 Q}{2\beta^2\xi}((a+\xi)\beta\right.\right.\right.$$
$$\left.\left.\left.-(a-\xi)\beta+sh(\beta(a-\xi))-sh(\beta(a+\xi)))+\frac{\gamma_1 C_1}{\beta}(1-ch(\beta l))\right)+\gamma_1 C_1\frac{sh(\beta l)}{\beta}\right)\right]$$
$$=\frac{1}{\beta}\left[\frac{ch(\beta l)}{sh(\beta l)}\left(\gamma_2 P sh(\beta a)-\frac{\gamma_1 Q}{\beta}(ch(\beta a)-1)+\frac{\gamma_1 C_1}{\beta}(1-ch(\beta l))\right)+\gamma_1 C_1\frac{sh(\beta l)}{\beta}\right],$$
$$C_3=\lim_{c\to 0}\left[\frac{\gamma_2 P}{2\beta^2 c}(ch(\beta(a-c))-ch(\beta(a+c)))-\frac{\gamma_1 Q}{2\beta^3 c}((a+c)\beta-(a-c)\beta\right.$$
$$\left.+sh(\beta(a-\xi))-sh(\beta(a+\xi)))+\gamma_1 C_1\frac{1}{\beta^2}\right]$$
$$=-\frac{\gamma_2 P}{\beta}sh(\beta a)+\frac{\gamma_1 Q}{\beta^2}(ch(\beta a)-1)+\gamma_1 C_1\frac{1}{\beta^2}$$
$$C_5=\lim_{\xi\to 0}\left[\frac{1}{\alpha_2}\left(a_7 P\frac{(a+\xi)^2-(a-\xi)^2}{4\xi}\right.\right.$$

$$+a_1Q\frac{(a+\xi)^3-(a-\xi)^3}{12\xi}\Bigg)\Bigg]=\frac{1}{\alpha_2}\left(a_7Pa+a_1Q\frac{a^2}{2}\right),$$

$$C_6=\lim_{\xi\to 0}\left[\frac{1}{\alpha_2}\left(a_7P\frac{(a-\xi)^3-(a+\xi)^3}{12\xi}-a_1Q\frac{(a+\xi)^4-(a-\xi)^4}{48\xi}\right.\right.$$

$$\left.\left.-\alpha_1\int\psi dx\Big|_{x=0}\right)\right]=\frac{1}{\alpha_2}\left(-a_7P\frac{a^2}{2}-a_1Q\frac{a^3}{6}-\alpha_1\int\psi dx\Big|_{x=0}\right),$$

$$C_8=\lim_{\xi\to 0}\left[\frac{1}{\alpha_2}\left(a_4P\frac{(a+\xi)^2-(a-\xi)^2}{4\xi}-a_7Q\frac{(a-\xi)^3-(a+\xi)^3}{12\xi}\right)\right]$$

$$=\frac{1}{\alpha_2}\left(a_4Pa+a_7Q\frac{a^2}{2}\right),$$

where the integral of the relative shear is

$$\int\psi\mathrm{d}x\Big|_{x=0}=\frac{C_2}{\beta}+\frac{1}{\beta^2}\left[\gamma_2P(1-ch(\beta a))+\gamma_1Q\left(\frac{sh(\beta a)}{\beta}-a\right)\right].$$

Numerical results. The geometrical parameters and materials of the layers are the same as before, and the values of the applied concentrated forces is as follows $P = 0$, $Q = -1{,}5\cdot 10^6$ N.

Figures 3.17, 3.18 show the variation of relative shear in the filler and deflection (3.34) along the length of an elastic three-layer rod as a function of the coordinate $x = a$ of the concentrated force application: *1*—$a = 0.25$, *2*—$a = 0.50$, *3*—$a = 0.75$, *4*—$a = 1.00$. The fourth curves correspond to the case of force application at the cantilever end of the rod. From these curves it is possible to estimate the maximum value of the investigated displacements.

It should be noted that compared to the case of locally distributed load, the curves on the graphs differ in shape.

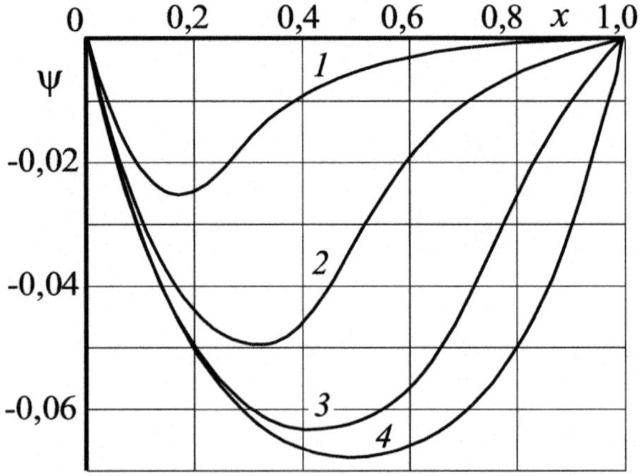

Fig. 3.17 Variation of relative shear in the filler along the rod from the coordinate of concentrated force application

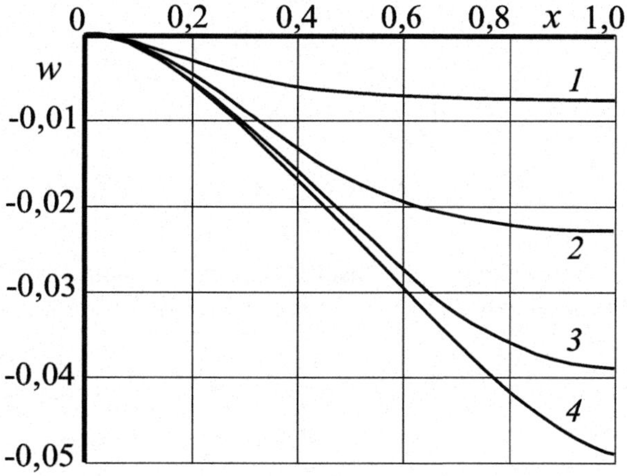

Fig. 3.18 Variation of deflection along the rod from the coordinate of concentrated force application

Action of concentrated moment (Fig. 3.19). Since the integral operators involved in the mathematical calculations are linear, a load of the concentrated moment type can be represented as the sum of two concentrated forces, equal in magnitude, oppositely directed and located at a small distance from each other. The force is positive when directed upwards, negative when directed downwards.

Thus, the solution (3.34) for two normal forces Q located at the points with coordinates $a-\xi$ and $a+\xi$, under the condition $P=0$, will be:

$$\begin{aligned}\psi(x) &= C_2 sh(\beta x) + C_3 ch(\beta x)\\ &+ \frac{1}{\beta^2}\big[\gamma_1 Q((1-\mathrm{ch}(\beta x - \beta(a+\xi)))H_0(a+\xi-x)\\ &-(1-ch(\beta x-\beta(a-\xi)))H_0(a-\xi-x)) - \gamma_1 C_1\big]\end{aligned}$$

$$w(x) = \frac{1}{\alpha_2}\left[\alpha_1 \int \psi \mathrm{d}x - a_1 Q(H_3(a+\xi-x)\right.$$

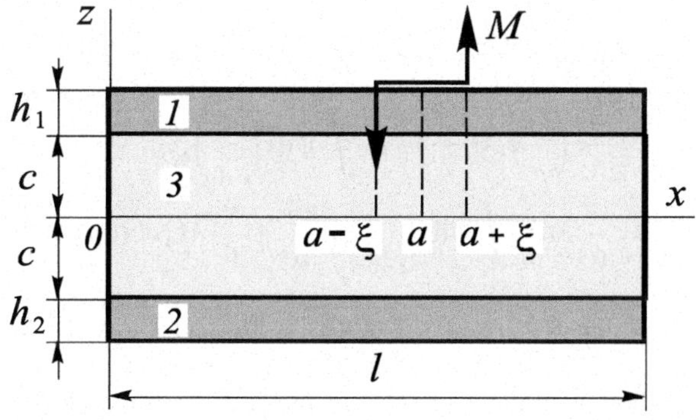

Fig. 3.19 Calculation diagram of the rod under the action of the moment

$$-H_3(a-\xi-x))+\frac{a_1C_1}{6}x^3\Big]+\frac{1}{2}C_4x^2+C_5x+C_6,$$

$$u(x)=\gamma_3\psi+\frac{1}{\alpha_2}\big[-a_7Q(H_2(a+\xi-x)$$

$$-H_2(a-\xi-x))+\frac{a_7C_1}{2}x^2\Big]+C_7x+C_8.$$

Introducing in the last relations the moment $M=2\xi Q$ and taking ξ to zero ($\xi\to 0$) in all expressions, we obtain the required solution:

$$\begin{aligned}
\psi(x)&=C_2sh(\beta x)+C_3ch(\beta x)\\
&\quad+\frac{1}{\beta^2}\big[\gamma_1M\beta sh(\beta a-\beta x)H_0(a-x)-\gamma_1C_1\big],\\
u(x)&=\gamma_3\psi+\frac{1}{\alpha_2}\left(\frac{a_7C_1}{2}x^2-a_7MH_1(a-x)\right)+C_7x+C_8,\\
w(x)&=\frac{1}{\alpha_2}\left[\alpha_1\int\psi\mathrm{d}x+a_1MH_2(a-x)+\frac{a_1C_1}{6}x^3\right]\\
&\quad+\frac{1}{2}C_4x^2+C_5x+C_6.
\end{aligned}\tag{3.35}$$

The integration constants for a rigid embedding of the left section of the rod and a free right end are equal to:

$$\begin{aligned}
C_1&=C_4=C_7=0,\\
C_2&=\lim_{\xi\to0}\left[\frac{1}{\beta}\left(\frac{ch(\beta l)}{sh(\beta l)}\left(\frac{\gamma_1M}{2\beta\xi}(ch(\beta(a+\xi))-ch(\beta(a-\xi)))+\frac{\gamma_1C_1}{\beta}(1-ch(\beta l))\right)\right.\right.\\
C_3&=\lim_{\xi\to0}\left[\frac{\gamma_1M}{2\beta^2\xi}(ch(\beta(a+\xi))-ch(\beta(a-\xi)))+\gamma_1C_1\frac{1}{\beta^2}\right]\\
&=\frac{\gamma_1M}{\beta}sh(\beta a)+\gamma_1C_1\frac{1}{\beta^2},\\
C_5&=\lim_{\xi\to0}\left(\frac{a_1M}{\alpha_2}\frac{(a+\xi)^2-(a-\xi)^2}{4\xi}\right)=\frac{a_1Ma}{\alpha_2},\\
C_6&=\frac{1}{\alpha_2}\left(-a_1M\frac{a^2}{2}-\alpha_1\int\psi\mathrm{d}x\Big|_{x=0}\right),\\
C_8&=\frac{a_7}{\alpha_2}Ma,\ \int\psi\mathrm{d}x\Big|_{x=0}=\frac{C_2}{\beta}+\frac{\gamma_1}{\beta^2}M\,(ch(\beta a)-1).
\end{aligned}$$

Numerical results. Figures 3.20, 3.21 show the shear in the filler and deflection of the three-layer rod calculated by formulas (3.35). The curve numbers correspond to the coordinates ($x=a$) of the section in which the concentrated moment $M=-$

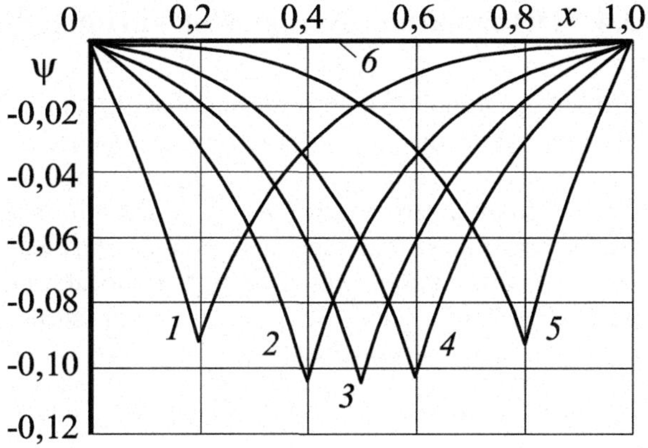

Fig. 3.20 Variation of relative shear in the filler along the rod from the coordinate of concentrated moment application

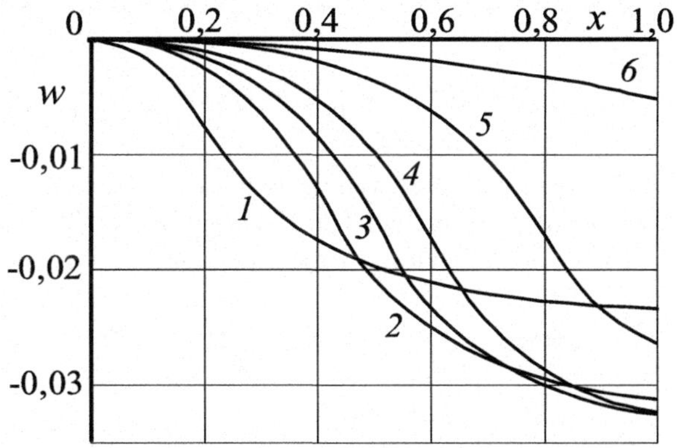

Fig. 3.21 Variation of deflection along the rod from the coordinate of concentrated moment application

7.5 10^5 Нм is applied: *1*—$a = 0.2$; *2*—$a = 0.4$; *3*—$a = 0.5$; *4*—$a = 0.6$; *5*—$a = 0.8$; *6*—$a = 1.0$.

It should be noted that there are breaks in the curves ψ with vertices at the points of moment application. Moreover, if the moment is applied at the cantilever end of the rod, the appearance of shear in the filler is prevented by a rigid diaphragm, and the deflection is minimised in this case. The maximum values of the investigated parameters are reached when the moment is applied in the middle of the rod.

The obtained solutions in displacements (3.32), (3.34), (3.35) allow us to describe the stress–strain state of an elastic three-layered rod with a rigid filler under the action of local uniformly distributed loads, concentrated forces and moments. For any combination of these forces, the corresponding solutions can be obtained by a combination of the above results.

3.4 Deformation by Loads of Various Shapes

In the above solutions of the problems of deformation of an elastic three-layer rod with incompressible filler, the surface load was assumed to be uniformly distributed inside the impact area. Its shape was *rectangular*. The rod bending induced by surface loads of other shapes, particularly *sinusoidal and parabolic*, is also of interest. Consider the impact of these loads and make comparisons with previous results.

Action of sinusoidal load. Suppose that a local static surface load distributed along the half-wave of a sinusoidal wave acts on the surface of the considered three-layer rod in the region $x \le b$ (Fig. 3.22):

$$q(x) = q_0 H_0(b-x)\sin\left(\frac{\pi x}{b}\right),\ p(x) = 0. \tag{3.36}$$

The function $g(x)$ (3.19) under the load (3.36) takes the following form:

$$g(x) = \gamma_1\left[-\frac{q_0 b}{\pi}\left(\cos\left(\frac{\pi x}{b}\right)+1\right)H_0(b-x)+C_1\right].$$

To construct an analytical solution to the problem, it is necessary to calculate several integrals included in the solution (3.22). The indefinite integral from the product of this function by the hyperbolic cosine will be:

$$\int g(x)\mathrm{ch}(\beta x)\mathrm{d}x = -\frac{\gamma_1 q_0 b H_0(b-x)}{\pi\left(\pi^2+\beta^2 b^2\right)}$$
$$\left(b\pi\sin\left(\frac{\pi x}{b}\right)ch(\beta x)+\beta b^2\cos\left(\frac{\pi x}{b}\right)sh(\beta x)\right.$$
$$\left.+\beta b^2 sh(\beta b)\right)-\frac{\gamma_1 q_0 b H_0(b-x)}{\pi}(sh(\beta x)-sh(\beta b))+\gamma_1 C_1\frac{\mathrm{sh}(\beta x)}{\beta}.$$

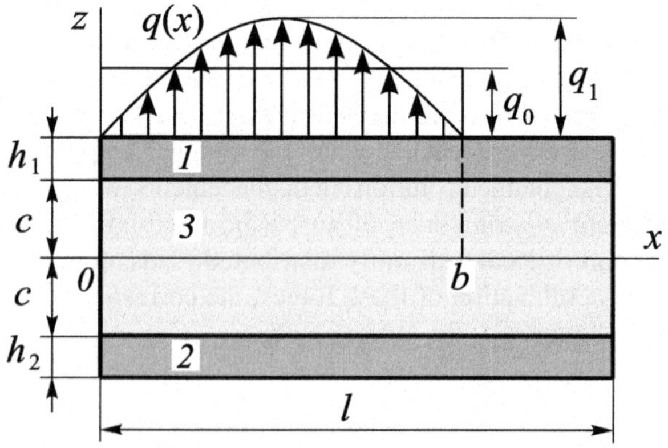

Fig. 3.22 Calculation diagram of a rod under the action of sinusoidal load

The expression for the integral from the product of functions $g(x)$ and $\mathrm{sh}(\beta x)$ is constructed similarly to the previous one. In this case it is only necessary to swap the hyperbolic cosines and sines:

$$\int g(x)\mathrm{sh}(\beta x)\mathrm{d}x$$
$$= -\frac{\gamma_1 q_0 b H_0(b-x)}{\pi\left(\pi^2+\beta^2 b^2\right)}\left(b\pi \sin\left(\frac{\pi x}{b}\right)sh(\beta x) + \beta b^2 \cos\left(\frac{\pi x}{b}\right)ch(\beta x)\right.$$
$$\left.+\beta b^2 ch(\beta b)\right) - \frac{\gamma_1 q_0 b H_0(b-x)}{\pi}(ch(\beta x) - \mathrm{ch}(\beta b)) + \gamma_1 C_1 \frac{\mathrm{ch}(\beta x)}{\beta}.$$

As a result, the solution of the problem of deformation of the considered three-layer rod under the action of sinusoidal load is obtained in the form of:

$$\psi(x) = C_2\mathrm{sh}(\beta x) + C_3\mathrm{ch}(\beta x) - \frac{\gamma_1 C_1}{\beta^2} +$$
$$+\frac{\gamma_1 q_0 b H_0(b-x)}{\beta^2\pi}\left(\frac{\beta^2 b^2\cos\left(\frac{\pi x}{b}\right) + \beta^2 b^2 ch(\beta x - \beta b)}{\pi^2+\beta^2 b^2} - ch(\beta x - \beta b) + 1\right),$$
$$u(x) = \gamma_3\psi + \frac{1}{\alpha_2}\left(a_7 \mathrm{L}_3^{-1}(q) + \frac{a_7 C_1}{2}x^2\right) + C_7 x + C_8,$$
$$w(x) = \frac{1}{\alpha_2}\left[\alpha_1\int \psi \mathrm{d}x + a_1 \mathrm{L}_4^{-1}(q) + \frac{a_1 C_1}{6}x^3\right] + \frac{C_4}{2}x^2 + C_5 x + C_6, \tag{3.37}$$

where

$$\mathrm{L}_3^{-1}(q) = \frac{q_0 b^3}{\pi^3}\left[\cos\left(\frac{\pi x}{b}\right)+1\right]H_0(b-x) - \frac{q_0 b}{\pi}H_2(b-x),$$
$$\mathrm{L}_4^{-1}(q) = \frac{q_0 b^4}{\pi^4}\sin\left(\frac{\pi x}{b}\right)H_0(b-x) - \frac{q_0 b^3}{\pi^3}H_1(b-x) + \frac{q_0 b}{\pi}H_3(b-x).$$

In the case of rigid embedding of the left end of the rod, the following integration constants will satisfy the boundary conditions:

$$C_1 = 0,\ C_2 = \frac{\gamma_1 q_0 b ch(\beta l)}{\pi\beta^2(\pi^2+\beta^2 b^2)sh(\beta l)}\left(2\beta^2 b^2 + \pi^2 - \pi^2 ch(\beta x - \beta b)\right),$$
$$C_3 = \frac{\gamma_1 q_0 b}{\pi\beta^2(\pi^2+\beta^2 b^2)}\left(\pi^2 ch(\beta x - \beta b) - 2\beta^2 b^2 - \pi^2\right)$$
$$C_4 = 0,\ C_5 = -\frac{a_1}{\alpha_2}L_3^{-1}(q)|_{x=0} = \frac{a_1 q_0 b^3(\pi^2-4)}{2\alpha_2\pi^3},$$
$$C_6 = \frac{a_1 q_0 b^4(6-\pi^2)}{6\alpha_2\pi^3} - \frac{\alpha_1}{\alpha_2}\int \psi \mathrm{d}x\bigg|_{x=0},$$
$$C_7 = 0,\ C_8 = \frac{a_7 q_0 b^3(\pi^2-4)}{2\alpha_2\pi^3},$$

herewith

$$\int \psi dx\Big|_{x=0} = \frac{C_2}{\beta} - \frac{\gamma_1 q_0 b^3 sh(\beta b)}{\pi\beta(\pi^2 + \beta^2 b^2)} + \frac{\gamma_1 q_0 b\, sh(\beta b)}{\pi\beta^3} - \frac{\gamma_1 q_0 b^2}{\pi\beta^2}.$$

Numerical results. Hereinafter in this paragraph, the following values for relative layer thicknesses $h_1 = 0.02$, $h_2 = 0.04$, $c = 0.09$ were assumed; the intensity of the distributed load is $q_0 = 1.5 \cdot 10^6$ Pa; the materials of the layers are the same as in the previous examples.

The value of the amplitude of the sinusoidal load q_1, resultant force of which is equivalent to the rectangular load q_0 acting on the same surface area of the rod (see Fig. 3.22) is equal to

$$q_1 = q_0 l \Big/ \int_0^l \sin\left(\frac{\pi x}{l}\right) dx = \frac{1}{2}\pi q_0. \tag{3.38}$$

Figures 3.23, 3.24 show the variation of relative shear in the filler and deflection along the axis of the rod calculated using formulae (3.37). The first three curves were calculated when the load was applied to the entire external surface of the rod ($b = l$): *1*—sinusoidal q_0, *2*—rectangular q_0, *3*—sinusoidal q_1. For the rest it was taken ($b = 0.5$): *4*—sinusoidal q_0, *5*—rectangular q_0, *5*—sinusoidal q_1.

For the same load amplitude (curves *1, 2* and *4, 5*), the displacements from a uniformly distributed load are much larger in magnitude. If the equal forces are identical according to (3.38), the maximum displacements are slightly larger from sinusoidal loading.

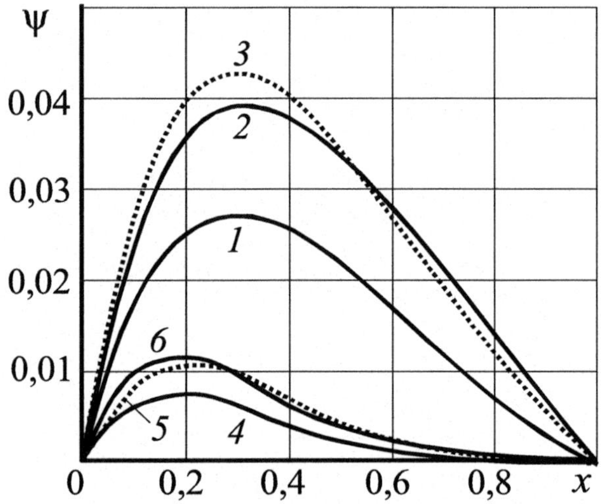

Fig. 3.23 Variation of relative shear in the filler along the rod

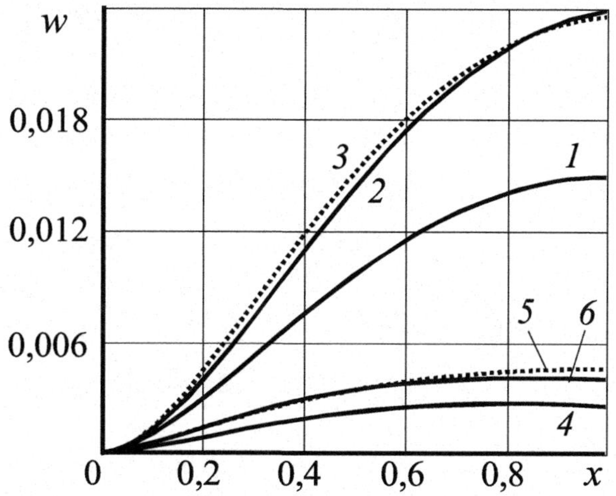

Fig. 3.24 Variation of deflection along the rod

Action of a local "convex" parabolic load. Consider the case corresponding to the direction of the parabola with the *convexity upwards* (Fig. 3.25). In the region $x \leq b$ such a load can be written as

$$q(x) = \frac{4q_0 x}{b}\left(1 - \frac{x}{b}\right)H_0(b - x), p(x) = 0. \tag{3.39}$$

The function $g(x)$ (3.19) in the case of convex parabolic load (3.39) will be

$$g(x) = \gamma_1\left[-\frac{2q_0}{3b^2}(3bx^2 - 2x^3 - b^3)H_0(b - x) + C_1\right].$$

The indefinite integral from the product of this function by the hyperbolic cosine is reduced to the form

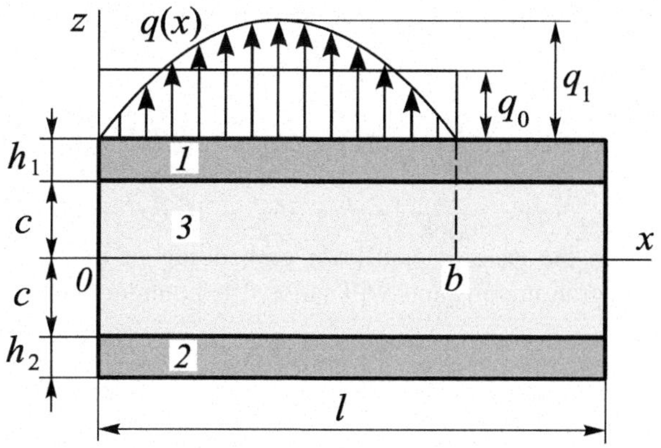

Fig. 3.25 Calculation diagram of a rod under the action of "convex" parabolic load

$$\int g(x)\mathrm{ch}(\beta x)dx = -\frac{2\gamma_1 q_0 H_0(b-x)}{3\beta^4 b^2}$$
$$(6\beta^2 bx\mathrm{ch}(\beta x) - \beta^3 bx^2\mathrm{sh}(\beta x) - 6\beta b\mathrm{sh}(\beta x)$$
$$+ 2\beta^3 x^3\mathrm{sh}(\beta x) - 6\beta^2 x^2\mathrm{ch}(\beta x)$$
$$+ 12\beta x\mathrm{sh}(\beta x) - 12\mathrm{ch}(\beta x) + \beta^3 b^3\mathrm{sh}(\beta x)$$
$$-6\beta b\mathrm{sh}(\beta b) + 12\mathrm{ch}(\beta b)) + \gamma_1 C_1\frac{\mathrm{sh}(\beta x)}{\beta}.$$

Similarly, we obtain:

$$\int g(x)\mathrm{sh}(\beta x)dx = -\frac{2\gamma_1 q_0 H_0(b-x)}{3\beta^4 b^2}$$
$$(6\beta^2 bx\mathrm{sh}(\beta x) - \beta^3 bx^2\mathrm{ch}(\beta x) - 6\beta b\mathrm{ch}(\beta x) + 2\beta^3 x^3\mathrm{ch}(\beta x)$$
$$- 6\beta^2 x^2\mathrm{sh}(\beta x) + 12\beta x\mathrm{ch}(\beta x) - 12\mathrm{sh}(\beta x) + \beta^3 b^3\mathrm{ch}(\beta x)$$
$$-6\beta b\mathrm{ch}(\beta b) + 12\mathrm{sh}(\beta b)) + \gamma_1 C_1\frac{\mathrm{ch}(\beta x)}{\beta}$$

As a result, the solution of the problem of deformation of a three-layer rod with incompressible filler under the action of a convex parabolic load is obtained in the form of

$$\psi(x) = C_2 sh(\beta x) + C_3 ch(\beta x) - \frac{\gamma_1 C_1}{\beta^2}$$
$$+ \frac{2\gamma_1 q_0 H_0(b-x)}{3\beta^5 b^2}\left(\beta^3 b^3 - 3\beta^3 bx^2 - 6\beta b - 6\beta b ch(\beta x - \beta b)\right.$$
$$\left.+2\beta^3 x^3 + 12\beta x - 12 sh(\beta x - \beta b)\right),$$
$$u(x) = \gamma_3\psi + \frac{1}{\alpha_2}\left(a_7 L_3^{-1}(q) + \frac{a_7 C_1}{2}x^2\right) + C_7 x + C_8,$$
$$w(x) = \frac{1}{\alpha_2}\left[\alpha_1\int\psi dx + a_1 L_4^{-1}(q) + \frac{a_1 C_1}{6}x^3\right] + \frac{C_4}{2}x^2 + C_5 x + C_6, \tag{3.40}$$

where

$$\mathrm{L}_3^{-1}(q) = \frac{q_0}{30b^2}\left[-2x^5 + 5bx^4 - 10b^3x^2 + 10b^4x - 3b^5\right]H_0(b-x),$$
$$\mathrm{L}_4^{-1}(q) = \frac{q_0}{900b^2}\left[-x^6 + 3bx^5 - 10b^3x^3 + 15b^4x^2 - 9b^5x + 2b^6\right]H_0(b-x).$$

In the case of cantilever embedding of the left end of the rod, the following integration constants will satisfy the boundary conditions (3.24):

$$C_1 = 0,\ C_2 = \frac{2\gamma_1 q_0 ch(\beta l)}{3\beta^5 b^2 sh(\beta l)}\left(\beta^3 x^3 - 6\beta b - 6\beta b ch(\beta b) + 12 sh(\beta b)\right),$$
$$C_3 = \frac{2\gamma_1 q_0}{3\beta^5 b^2}\left(-\beta^3 x^3 + 6\beta b + 6\beta b ch(\beta b) - 12 sh(\beta b)\right),$$

$$C_4 = -\frac{a_1}{\alpha_2}\mathrm{L}_2^{-1}(q)|_{x=l} - \frac{a_1}{\alpha_2}C_1 l = 0,\ C_5 = \frac{a_1 q_0 b^3}{10\alpha_2},$$

$$C_6 = -\frac{a_1 q_0 b^4}{45\alpha_2} - \frac{\alpha_1}{\alpha_2}\int \psi \mathrm{d}x\Big|_{x=0},\ C_7 = -\frac{a_7}{\alpha_2}\mathrm{L}_2^{-1}(q)|_{x=l} - \frac{a_7}{\alpha_2}C_1 l = 0,$$

$$C_8 = -\frac{a_7}{\alpha_2}\mathrm{L}_3^{-1}(q)|_{x=0} = \frac{a_7 q_0 b^3}{10\alpha_2},$$

where the integral of the relative shear

$$\int \psi \mathrm{d}x\Big|_{x=0} = \frac{C_2}{\beta} + \frac{\gamma_1 q_0}{3\beta^6 b^2}\left(-\beta^4 b^4 + 12\beta b sh(\beta b) + 24 - 24ch(\beta b)\right).$$

Numerical results. Figures 3.26, 3.27 show the variation of relative shear in the filler and deflection along the axis of the three-layer rod calculated using formulae (3.40). The first three curves correspond to the action of the load on the entire external surface of the rod ($b = l$): *1*—parabolic q_0, *2*—rectangular q_0, *3*—equivalent parabolic $q_1 = 1{,}5\ q_0$. For the other curves, the load acts on half of the rod ($b = 0.5\ l$): *4*—parabolic q_0, *5*—rectangular q_0, *6*—equivalent parabolic q_1. For the same load amplitude (curves *1*, *2* and *4*, *5*), the displacements from a uniformly distributed load are much larger in value. If, according to (3.38), the resultant forces are the same, the maximum displacements are slightly larger from parabolic load.

It should be noted, that in this case, the deflections and shears in magnitude and shape do not differ practically from the corresponding displacements under sinusoidal loading, which is explained by the closeness of the shapes of these loads.

Action of a "concave" parabolic load. Suppose that the *local parabolic load is directed convexly downwards* and acts on the rod surface, going to zero at the centre and reaching a maximum at the ends of the section x $\leq b$ (Fig. 3.28). Then the load

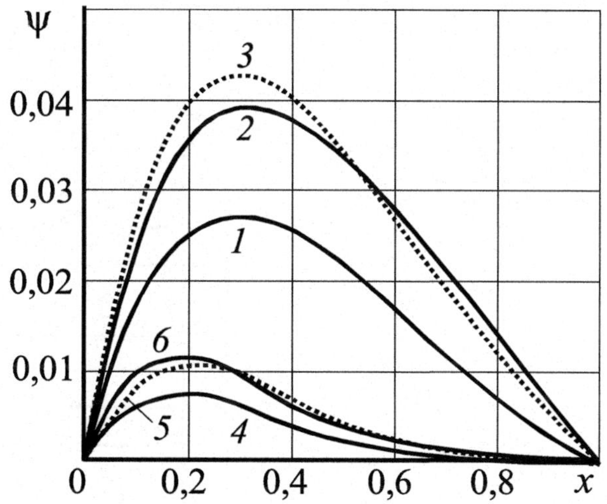

Fig. 3.26 Variation of relative shear in the filler along the rod

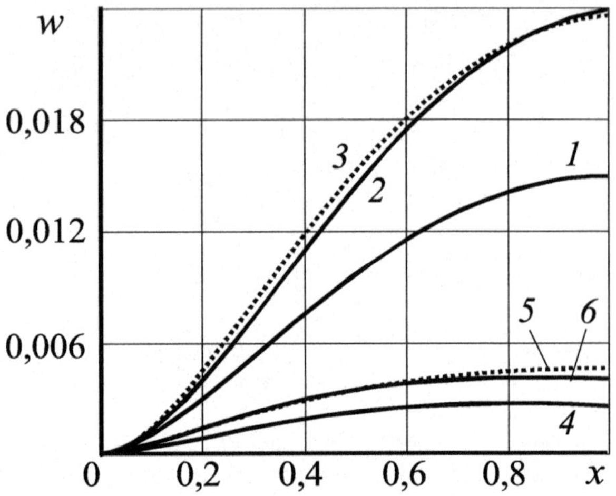

Fig. 3.27 Variation of deflection along the rod

can be written in the form

$$q(x) = 4q_0\left(\frac{x}{b} - \frac{1}{2}\right)^2 H_0(b - x),\ p(x) = 0. \tag{3.41}$$

The function $g(x)$ corresponding to the "concave" parabolic load (3.41) will be

$$g(x) = \gamma_1\left[\frac{q_0}{3b^2}(4x^3 + 3xb^2 - 6bx^2 - b^3)H_0(b - x) + C_1\right].$$

The indefinite integral from the product of this function by the hyperbolic cosine is reduced to the form

$$\int g(x)\mathrm{ch}(\beta x)\mathrm{d}x = -\frac{\gamma_1 q_0 H_0(b - x)}{3\beta^4 b^2}$$

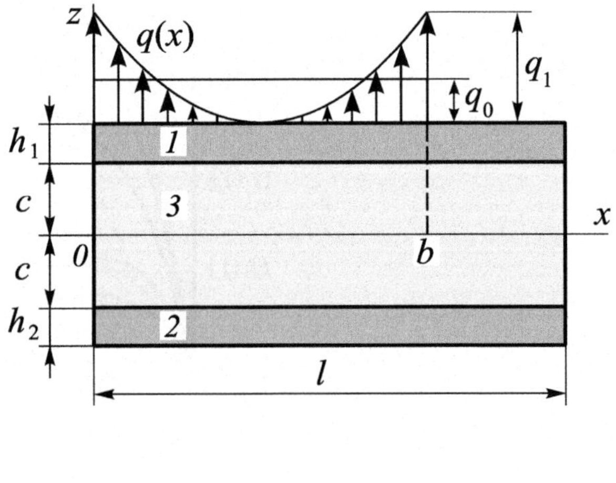

Fig. 3.28 Calculation diagram of a rod under the action of "concave" parabolic load

$$\times \big(-12\beta^2 bxch(\beta x) + 6\beta^3 bx^2 sh(\beta x) + 12\beta bsh(\beta x)$$
$$- 4\beta^3 x^3 sh(\beta x) + 12\beta^2 x^2 ch(\beta x) - 24\beta xsh(\beta x)$$
$$+ 24ch(\beta x) + \beta^3 b^3 sh(\beta x) + 12\beta bsh(\beta b)$$
$$-24ch(\beta b) - 3\beta^3 b^2 xsh(\beta x) + 3\beta^2 b^2 ch(\beta x) - 3\beta^2 b^2 ch(\beta b)\big) + \gamma_1 C_1 \frac{\mathrm{sh}(\beta x)}{\beta}.$$

By analogy we get:

$$\int g(x)\mathrm{sh}(\beta x)\mathrm{d}x = -\frac{\gamma_1 q_0 H_0(b-x)}{3\beta^4 b^2}$$
$$\times \big(-12\beta^2 bxsh(\beta x) + 6\beta^3 bx^2 ch(\beta x) + 12\beta bch(\beta x) - 4\beta^3 x^3 ch(\beta x)$$
$$+ 12\beta^2 x^2 sh(\beta x) - 24\beta xch(\beta x) + 24sh(\beta x) + \beta^3 b^3 ch(\beta x) + 12\beta bch(\beta b)$$
$$-24sh(\beta b) - 3\beta^3 b^2 xch(\beta x) + 3\beta^2 b^2 sh(\beta x) - 3\beta^2 b^2 sh(\beta b)\big) + \gamma_1 C_1 \frac{\mathrm{ch}(\beta x)}{\beta}.$$

As a result, the solution of the problem of deformation of a three-layer rod under the action of a local concave parabolic load is obtained in the form of

$$\begin{aligned}
\psi(x) =& C_2 \mathrm{sh}(\beta x) + C_3 \mathrm{ch}(\beta x) - \frac{\gamma_1 C_1}{\beta^2} - \frac{\gamma_1 q_0 H_0(b-x)}{3\beta^5 b^2} \\
& \big(-\beta^3 b^3 - 6\beta^3 bx^2 - 12\beta b - 12\beta bch(\beta x - \beta b) + 3\beta^3 b^2 x \\
& +4\beta^3 x^3 + 24\beta x - 24sh(\beta x - \beta b) - 3\beta^2 x^2 sh(\beta x - \beta b)\big), \\
u(x) =& \gamma_3 \psi + \frac{1}{\alpha_2}\left(a_7 \mathrm{L}_3^{-1}(q) + \frac{a_7 C_1}{2} x^2\right) + C_7 x + C_8, \\
w(x) =& \frac{1}{\alpha_2}\left[\alpha_1 \int \psi \mathrm{d}x + a_1 \mathrm{L}_4^{-1}(q) + \frac{a_1 C_1}{6} x^3\right] + \frac{C_4}{2} x^2 + C_5 x + C_6,
\end{aligned} \qquad (3.42)$$

where

$$\mathrm{L}_3^{-1}(q) = \frac{q_0}{120b^2}\left[4x^5 - 5bx^4 - 10b^3 x^2 + 20b^4 x - 9b^5\right] H_0(b-x),$$
$$\mathrm{L}_4^{-1}(q) = \frac{q_0}{360b^2}\left[2x^6 - 3bx^5 - 10b^3 x^3 + 30b^4 x^2 - 27b^5 x + 8b^6\right] H_0(b-x).$$

In the case of cantilever embedding of the left end of the rod, the following integration constants will satisfy the boundary conditions (3.24):

$$C_1 = C_4 = C_7 = 0,$$
$$C_2 = \frac{\gamma_1 q_0 ch(\beta l)}{3\beta^5 b^2 sh(\beta l)}\big(\beta^3 x^3 + 12\beta b + 12\beta bch(\beta b) - 24sh(\beta b) - 3\beta^2 b^2 sh(\beta b)\big),$$
$$C_3 = \frac{\gamma_1 q_0}{3\beta^5 b^2}\big(-\beta^3 x^3 - 12\beta b - 12\beta bch(\beta b) + 24sh(\beta b) + 3\beta^2 b^2 sh(\beta b)\big),$$
$$C_5 = -\frac{a_1}{\alpha_2}\mathrm{L}_3^{-1}(q)|_{x=0} = \frac{3a_1 q_0 b^3}{40\alpha_2},$$

$$C_6 = -\frac{a_1 q_0 b^4}{45\alpha_2} - \frac{\alpha_1}{\alpha_2}\int \psi dx\Big|_{x=0}, \quad C_8 = -\frac{a_7}{\alpha_2}\mathrm{L}_3^{-1}(q)|_{x=0} = \frac{3a_7 q_0 b^3}{40\alpha_2},$$

where the integral of the relative shear

$$\int \psi dx\Big|_{x=0} = \frac{C_2}{\beta} + \frac{\gamma_1 q_0}{6\beta^6 b^2}\big(6\beta^2 b^2 ch(\beta b) - \beta^4 b^4 - 24\beta b sh(\beta b)$$
$$-48 + 48ch(\beta b) - 6\beta b^2\big).$$

Numerical results. Figures 3.29, 3.30 show the variation of relative shear in the filler and deflection along the axis of the three-layer rod, calculated using formulas (3.42). The first three curves correspond to the action of the load on the entire external surface of the rod ($b = l$): *1*—parabolic q_0, *2*—rectangular q_0, *3*—equivalent parabolic $q_1 = 3q_0$. For the other curves, the load acts on half of the rod ($b = 0.5\ l$): *4*—parabolic q_0, *5*—rectangular q_0, *6*—equivalent parabolic q_1.

For the same amplitude of loads (curves *1, 2* and *4, 5*) the displacements from a uniformly distributed load are much larger in magnitude. If the resultant forces, according to (3.38), are the same, the maximum deflections are practically the identical.

Consequently, under local static loading in the elastic deformation region, the displacements in the three-layer rod are determined mainly by the static equivalent of the distributed load rather than by its shape.

Thus, the proposed solutions and research methodology can be successfully applied to study the stress–strain state of elastic three-layer rods with rigid incompressible filler under various types and forms of external loading.

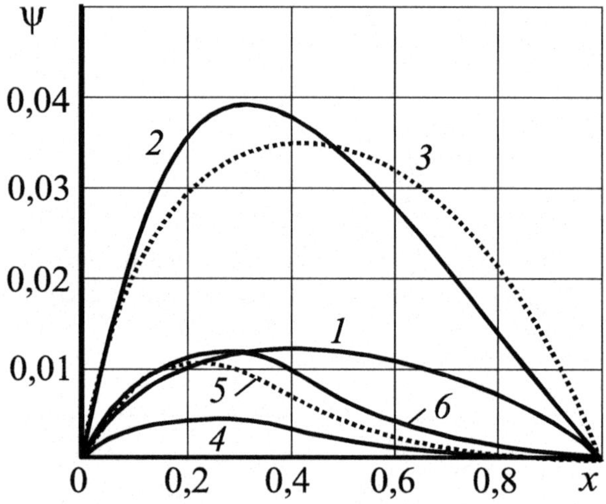

Fig. 3.29 Variation of relative shear in the filler along the rod

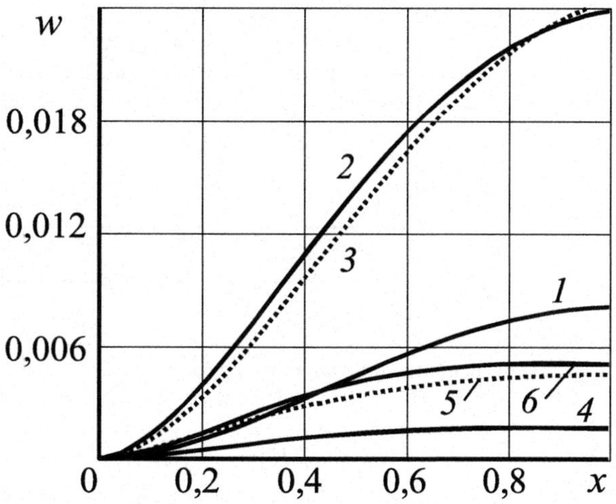

Fig. 3.30 Variation of deflection along the rod

3.5 Step Rod[1]

Within the framework of the hypotheses introduced earlier, we consider the deformation of an asymmetric in thickness stepped elastic three-layered rod (Fig. 3.31). Distributed loads $p(x)$, $q(x)$ act on the outer layer of the rod. Through $w(x)$ and $u(x)$ we denote the deflection and longitudinal displacement of the median plane of the filler material. Since the materials of all layers are incompressible in the transverse direction, the deflection at the remaining points of the rod coincides with $w(x)$. Rigid diaphragms are assumed at the ends, preventing relative shear of the layers, but not preventing deformation from their plane. Through h_k denotes the thickness of the kth layer ($k = 1, 2, 3$), and

$$h_3 = 2c,\ h_2 = h_{20} + h_{21}H_0(x - x_0). \tag{3.43}$$

The introduced geometrical hypotheses allow us to express longitudinal displacements in layers $u^{(k)}$ through the three required functions $u(x)$, $\psi(x)$ and $w(x)$ using the relations (3.1), (3.43). The strains are determined by relations (3.2), (3.3), internal forces and moments by relations (3.4). The equations of equilibrium of the three-layer rod in forces (3.11) and the force boundary conditions (3.12) are valid in this case as well. Hooke's law relations (3.13) and (3.14) are used to relate stresses and strains in layers. The internal forces and moments are expressed through the required functions u $u(x)$, $\psi(x)$, $w(x)$ by the formulas (3.15). The system of differential equations of equilibrium of an elastic three-layer step rod in displacements (3.16) takes the following form:

$$(a_1 + a_{10}H(x - x_0))u,_{xx} + (a_6 + a_{60}H(x - x_0))\psi,_{xx}$$

[1] This section was written in collaboration with E.E. Starovoitova.

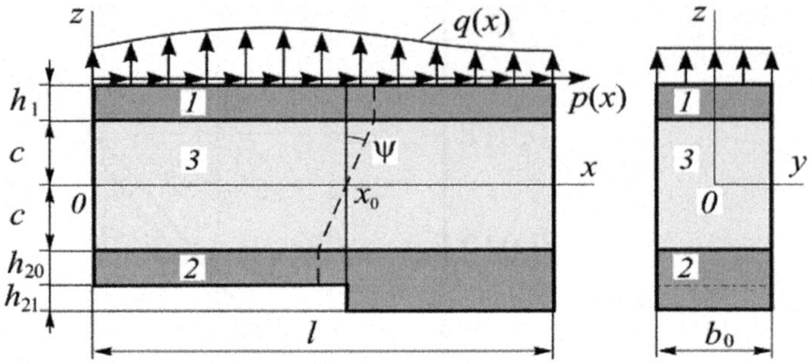

Fig. 3.31 Calculation diagram of a step rod

$$
\begin{aligned}
&- (a_7 + a_{70}H(x - x_0))w,_{xxx} = -p,\\
&(a_6 + a_{60}H(x - x_0))u,_{xx} + (a_2 + a_{20}H(x - x_0))\psi,_{xx} -,\\
&- (a_3 + a_{30}H(x - x_0))w,_{xxx} - a_5\psi = 0,\\
&(a_7 + a_{70}H(x - x_0))u,_{xxx} + (a_3 + a_{30}H(x - x_0))\psi,_{xxx}\\
&- (a_4 + a_{40}H(x - x_0))w,_{xxxx} = -q,
\end{aligned}
\tag{3.44}
$$

where the coefficients a_k ($k = 1, 2, \ldots, 7$) are described by the formulas (3.17),

$$
\begin{aligned}
&a_{10} = K_2^+ h_{21},\ a_{20} = c^2 K_2^+ h_{21},\ a_{30} = cK_2^+ h_{21}\left(c + h_{20} + \frac{1}{2}h_{21}\right),\\
&a_{40} = K_2^+ h_{21}\left(c(c + 2h_{20} + h_{21}) + h_{20}h_{21} + h_{20}^2 + \frac{1}{3}h_{21}^2\right),\\
&a_{60} = -cK_2^+ h_{21}\ a_{70} = -K_2^+ h_{21}\left(c + h_{20} + \frac{1}{2}h_{21}\right).
\end{aligned}
$$

The system of ordinary differential Eqs. (3.44) contains discontinuous coefficients (due to the Heaviside function), so we will solve it separately in the regions of continuity of these coefficients.

In the first of them ($x < x_0$), the Heaviside function converges to zero and $h_2 = h_{20}$. Displacements in this area are marked with the index "1" at the bottom. The system (3.44) is transformed to the form (3.16), so its solution will not differ from (3.22):

$$
\begin{aligned}
\psi_1(x) &= C_{21}sh(\beta_1 x) + C_{31}ch(\beta_1 x)\\
&+ \frac{1}{\beta_1}\left[sh(\beta_1 x)\int g_1(x)ch(\beta_1 x)\mathrm{d}x - ch(\beta_1 x)\int g_1(x)sh(\beta_1 x)\mathrm{d}x\right],\\
w_1(x) &= \frac{1}{\alpha_{21}}\left[\alpha_{11}\int \psi_1\mathrm{d}x - a_7L_3^{-1}(p) + a_1L_4^{-1}(q) + \frac{a_1C_{11}}{6}x^3\right]\\
&+ \frac{C_{41}}{2}x^2 + C_{51}x + C_{61},
\end{aligned}
$$

$$u_1(x) = \gamma_{31}\,\psi_1 + \frac{1}{\alpha_{21}}\left[-a_4L_2^{-1}(p) + a_7L_3^{-1}(q) + \frac{a_7}{2}C_{11}x^2\right] + C_{71}x + C_{81}. \tag{3.45}$$

Here C_{11}, ..., C_{81} are integration constants; linear integral operators (3.23) are denoted through L_1^{-1}, L_2^{-1}, L_3^{-1}, L_4^{-1};

$$\begin{aligned}
&g_1(x) = \gamma_{21}p + \gamma_{11}\left(\int\limits_0^x q\ \mathrm{d}x + C_{11}\right),\\
&\beta_1^2 = \frac{a_1a_5\alpha_{21}}{\alpha_{21}\alpha_{31} - \alpha_{11}^2} > 0,\ \gamma_{11} = \frac{a_1\alpha_{11}}{\alpha_{21}\alpha_{31} - \alpha_{11}^2},\\
&\gamma_{21} = \frac{a_6\alpha_{21} - a_7\alpha_{11}}{\alpha_{21}\alpha_{31} - \alpha_{11}^2},\ \gamma_{31} = \frac{a_3a_7 - a_4a_6}{\alpha_{21}},\\
&\alpha_{21}\alpha_{31} - \alpha_{11}^2 \neq 0,\ \alpha_{11} = a_1a_3 - a_6a_7,\\
&\alpha_{21} = a_1a_4 - a_7^2,\ \alpha_{31} = a_1a_2 - a_6^2.
\end{aligned}$$

In the second region ($x \geq x_0$, after the step), the Heaviside function is equal to one and $h_2 = h_{20} + h_{21}$. Displacements in this area are marked with the index "2" at the bottom. The system (3.44) is transformed to the form:

$$\begin{aligned}
&a_{11}u_{2,xx} + a_{61}\psi_{2,xx} - a_{71}w_{2,xxx} = -p,\\
&a_{61}u_{2,xx} + a_{21}\psi_{2,xx} - a_{31}w_{2,xxx} - a_5\psi_2 = 0,\\
&a_{71}u_{2,xxx} + a_{31}\psi_{2,xxx} - a_{41}w_{2,xxxx} = -q,
\end{aligned} \tag{3.46}$$

where

$$\begin{aligned}
&a_{11} = a_1 + a_{10},\ a_{21} = a_2 + a_{20},\ a_{31} = a_3 + a_{30},\\
&a_{41} = a_4 + a_{40},\ a_{61} = a_6 + a_{60},\ a_{71} = a_7 + a_{70}.
\end{aligned}$$

The system (3.46) does not differ from (3.18), so its solution is:

$$\begin{aligned}
\psi_2(x) &= C_{22}sh(\beta_2x) + C_{32}ch(\beta_2x)\\
&\quad + \frac{1}{\beta_2}\left[sh(\beta_2x)\int g_2(x)ch(\beta_2x)\mathrm{d}x - ch(\beta_2x)\int g_2(x)sh(\beta_2x)\mathrm{d}x\right]\\
w_2(x) &= \frac{1}{\alpha_{22}}\left[\alpha_{12}\int \psi_2\mathrm{d}x - a_{71}L_3^{-1}(p) + a_{11}L_4^{-1}(q) + \frac{a_{11}C_{12}}{6}x^3\right]\\
&\quad + \frac{C_{42}}{2}x^2 + C_{52}x + C_{62},\\
u_2(x) &= \gamma_{32}\,\psi_2 + \frac{1}{\alpha_{22}}\left[-a_{41}L_2^{-1}(p) + a_{71}L_3^{-1}(q) + \frac{a_{71}}{2}C_{12}x^2\right] + C_{72}x + C_{82}.
\end{aligned} \tag{3.47}$$

where C_{12}, …, C_{82} are integration constants, L_1^{-1}, L_2^{-1}, L_3^{-1}, L_4^{-1} are linear integral operators (3.23),

$$g_2(x) = \gamma_{22}p + \gamma_{12}\left(\int_0^x q\,\mathrm{d}x + C_{12}\right),\ \beta_2^2 = \frac{a_{11}a_{51}\alpha_{21}}{\alpha_{22}\alpha_{32} - \alpha_{12}^2} > 0,$$

$$\gamma_{12} = \frac{a_{11}\alpha_{12}}{\alpha_{22}\alpha_{32} - \alpha_{12}^2},\ \gamma_{22} = \frac{a_{61}\alpha_{22} - a_{71}\alpha_{12}}{\alpha_{22}\alpha_{32} - \alpha_{12}^2},$$

$$\gamma_{32} = \frac{a_{31}a_{71} - a_{41}a_{61}}{\alpha_{22}},\ \alpha_{22}\alpha_{32} - \alpha_{12}^2 \neq 0,$$

$$\alpha_{12} = a_{11}a_{31} - a_{61}a_{71},\ \alpha_{22} = a_{11}a_{41} - a_{71}^2,\ \alpha_{32} = a_{11}a_{21} - a_{61}^2.$$

The solution of the starting system (3.44) can now be written in the form:

$$\begin{aligned}
\psi(x) &= \psi_1(x) + (\psi_2(x) - \psi_1(x))H(x - x_0),\\
u(x) &= u_1(x) + (u_2(x) - u_1(x))H(x - x_0),\\
w(x) &= w_1(x) + (w_2(x) - w_1(x))H(x - x_0).
\end{aligned} \quad (3.48)$$

The integration constants C_{11}, …, C_{81}, C_{12}, …, C_{82} are determined from the 8 boundary conditions at the rod ends and additional 8 conditions of equality of displacements and their derivatives at the point of boundary irregularity $x = x_0$:

$$\begin{gathered}
\langle\psi_2(x) - \psi_1(x)\rangle|_{x=x_0} = 0,\ \langle\psi_{2,x}(x) - \psi_{1,x}(x)\rangle|_{x=x_0} = 0,\\
\langle\psi_{2,xx}(x) - \psi_{1,xx}(x)\rangle|_{x=x_0} = 0,\\
\langle w_2(x) - w_1(x)\rangle|_{x=x_0} = 0,\ \langle w_{2,x}(x) - w_{1,x}(x)\rangle|_{x=x_0} = 0,\\
\langle w_{2,xx}(x) - w_{1,xx}(x)\rangle|_{x=x_0} = 0,\\
\langle u_2(x) - u_1(x)\rangle|_{x=x_0} = 0,\ \langle u_{2,x}(x) - u_{1,x}(x)\rangle|_{x=x_0} = 0.
\end{gathered} \quad (3.49)$$

In this case, it is assumed that there are rigid diaphragms at the ends of the rod to prevent the relative shear of the layers. Let us consider three types of fixation of the rod ends.

In the case of *rigid embedding of the left end of the rod with a free right end*, the boundary conditions are as follows:

$$\begin{aligned}
x = 0 &: w_1 = w_{1,x} = u_1 = \psi_1 = 0;\\
x = l &: \psi_2 = N_2 = M_2 = M_{2,x} = 0.
\end{aligned} \quad (3.50)$$

Substituting the solution (3.48) into the boundary conditions (3.49) and (3.50), we obtain a system of 16 linear algebraic equations to determine the 16 integration constants, C_{11}, …, C_{81}, C_{12}, …, C_{82}.

For a *rod hinged at both ends*, the following boundary conditions should be used instead of (3.50):

$$x = 0,\ l:\ u_{(1,2)} = \psi_{(1,2)} = w_{(1,2)} = M_{(1,2)} = 0, \tag{3.51}$$

where the lower index indicates the application area of the required solution.

If the *rod is clamped at both ends*, the following conditions must be fulfilled:

$$x = 0,\ l:\ u_{(1,2)} = \psi_{(1,2)} = w_{(1,2)} = w_{,x(1,2)} = 0. \tag{3.52}$$

Let's write out an algebraic system of eight equations, "stitching" together solutions at the point $x = x_0$ in the general case. For this purpose, we require eight conditions to be fulfilled (3.49). The result is:

$$\begin{aligned}
&C_{21}sh(\beta_1 x_0) + C_{31}ch(\beta_1 x_0) - C_{22}sh(\beta_2 x_0) - C_{32}ch(\beta_2 x_0)\\
&= \frac{1}{\beta_2}\left[sh(\beta_2 x)\int g_2(x)ch(\beta_2 x)\mathrm{d}x - ch(\beta_2 x)\int g_2(x)sh(\beta_2 x)\mathrm{d}x\right]\Bigg|_{x=x_0}\\
&- \frac{1}{\beta_1}\left[sh(\beta_1 x)\int g_1(x)ch(\beta_1 x)\mathrm{d}x - ch(\beta_1 x)\int g_1(x)sh(\beta_1 x)\mathrm{d}x\right]\Bigg|_{x=x_0}\\
&C_{21}\beta_1 ch(\beta_1 x_0) + C_{31}\beta_1 sh(\beta_1 x_0) - C_{22}\beta_2 sh(\beta_2 x_0) - C_{32}\beta_2 ch(\beta_2 x_0)\\
&= \frac{1}{\beta_2}\left[sh(\beta_2 x)\int g_2(x)ch(\beta_2 x)\mathrm{d}x - ch(\beta_2 x)\int g_2(x)sh(\beta_2 x)\mathrm{d}x\right]_{,x}\Bigg|_{x=x_0}\\
&- \frac{1}{\beta_1}\left[sh(\beta_1 x)\int g_1(x)ch(\beta_1 x)\mathrm{d}x - ch(\beta_1 x)\int g_1(x)sh(\beta_1 x)\mathrm{d}x\right]_{,x}\Bigg|_{x=x_0}\\
&C_{21}\beta_1^2 sh(\beta_1 x_0) + C_{31}\beta_1^2 ch(\beta_1 x_0) - C_{22}n\beta_2^2 sh(\beta_2 x_0) - C_{32}\beta_2^2 ch(\beta_2 x_0)\\
&= \frac{1}{\beta_2}\left[sh(\beta_2 x)\int g_2(x)ch(\beta_2 x)\mathrm{d}x - ch(\beta_2 x)\int g_2(x)sh(\beta_2 x)\mathrm{d}x\right]_{,xx}\Bigg|_{x=x_0}\\
&- \frac{1}{\beta_1}\left[sh(\beta_1 x)\int g_1(x)ch(\beta_1 x)\mathrm{d}x - ch(\beta_1 x)\int g_1(x)sh(\beta_1 x)\mathrm{d}x\right]_{,xx}\Bigg|_{x=x_0}\\
&\frac{1}{\alpha_{21}}\left[\alpha_{11}\int \psi_1\mathrm{d}x - a_7 L_3^{-1}(p) + a_1 L_4^{-1}(q) + \frac{a_1 C_{11}}{6}x^3\right]\Bigg|_{x=x_0} + \frac{C_{41}}{2}x_0^2 + C_{51}x_0 + C_{61}\\
&- \frac{1}{\alpha_{22}}\left[\alpha_{12}\int \psi_2\mathrm{d}x - a_{71} L_3^{-1}(p) + a_{11} L_4^{-1}(q) + \frac{a_{11} C_{12}}{6}x^3\right]\Bigg|_{x=x_0}\\
&- \frac{C_{42}}{2}x_0^2 - C_{52}x_0 - C_{62} = 0,\\
&\frac{1}{\alpha_{21}}\left[\alpha_{11}\psi_1 - a_7 L_2^{-1}(p) + a_1 L_3^{-1}(q) + \frac{a_1 C_{11}}{2}x^2\right]\Bigg|_{x=x_0} + C_{41}x_0 + C_{51}\\
&- \frac{1}{\alpha_{22}}\left[\alpha_{12}\psi_2 - a_{71} L_2^{-1}(p) + a_{11} L_3^{-1}(q) + \frac{a_{11} C_{12}}{2}x^2\right]\Bigg|_{x=x_0} - C_{42}x_0 - C_{52} = 0,\\
&\frac{1}{\alpha_{21}}\left[\alpha_{11}\psi_{1,x} - a_7 L_1^{-1}(p) + a_1 L_2^{-1}(q) + a_1 C_{11}x\right]\Big|_{x=x_0} + C_{41}\\
&- \frac{1}{\alpha_{22}}\left[\alpha_{12}\psi_{2,x} - a_{71} L_1^{-1}(p) + a_{11} L_2^{-1}(q) + a_{11} C_{12}x\right]\Big|_{x=x_0} - C_{42} = 0,\\
&\gamma_{31}\psi_1|_{x=x_0} + \frac{1}{\alpha_{21}}\left[-a_4 L_2^{-1}(p) + a_7 L_3^{-1}(q) + \frac{a_7}{2}C_{11}x^2\right]\Bigg|_{x=x_0} + C_{71}x_0 + C_{81}\\
&- \gamma_{32}\psi_2|_{x=x_0} - \frac{1}{\alpha_{22}}\left[-a_{41} L_2^{-1}(p) + a_{71} L_3^{-1}(q) + \frac{a_{71}}{2}C_{12}x^2\right]\Bigg|_{x=x_0} - C_{72}x_0 - C_{82} = 0,
\end{aligned}$$

$$\gamma_{31}\,\psi_{1},_x|_{x=x_0} + \frac{1}{\alpha_{21}}\left[-a_4L_1^{-1}(p) + a_7L_2^{-1}(q) + a_7C_{11}x\right]\Big|_{x=x_0} + C_{71}$$
$$-\gamma_{32}\,\psi_{2},_x|_{x=x_0} - \frac{1}{\alpha_{22}}\left[-a_{41}L_1^{-1}(p) + a_{71}L_2^{-1}(q) + a_{71}C_{12}x\right]\Big|_{x=x_0} - C_{72} = 0. \tag{3.53}$$

Thus, the solution of the boundary value problem is closed by joining to the displacements (3.48) and Eqs. (3.53) eight boundary conditions at the rod ends (of the type (3.50)–(3.52)).

In the case of *constant linear loads* $q = \text{const}, p = \text{const}$ when both ends of the rod are embedded, the required displacements (3.45) in region I are written out explicitly as follows:

$$\begin{aligned}
\psi_1(x) =& C_{21}\text{sh}(\beta_1 x) + C_{31}\text{ch}(\beta_1 x) - \frac{1}{\beta_1^2}\left[\gamma_{21}p + \gamma_{11}(qx + C_{11})\right],\\
w_1(x) =& \frac{1}{\alpha_{21}}\left[\frac{\alpha_{11}}{\beta_1}\left[C_{21}\text{ch}(\beta_1 x) + C_{31}\text{sh}(\beta_1 x) - \frac{x}{\beta_1}\left(\gamma_{21}p + \gamma_{11}\left(\frac{1}{2}qx + C_{11}\right)\right)\right]\right.\\
& \left. -a_7p\frac{x^3}{6} + a_1q\frac{x^4}{24} + \frac{a_1C_{11}}{6}x^3\right] + \frac{C_{41}}{2}x^2 + C_{51}x + C_{61},\\
u_1(x) =& \gamma_{31}\,\psi_1 + \frac{1}{\alpha_{21}}\left[-a_4p\frac{x^2}{2} + a_7q\frac{x^3}{6} + \frac{a_7}{2}C_{11}x^2\right] + C_{71}x + C_{81}.
\end{aligned} \tag{3.54}$$

The corresponding solution (3.47) in region II takes the form:

$$\begin{aligned}
\psi_2 =& C_{22}\text{sh}(\beta_2 x) + C_{32}\text{ch}(\beta_2 x) - \frac{1}{\beta_2^2}\left[\gamma_{22}p + \gamma_{12}(qx + C_{12})\right],\\
w_2(x) =& \frac{1}{\alpha_{22}}\left[\alpha_{12}\int \psi_2 dx - a_{71}p\frac{x^3}{6} + a_{11}q\frac{x^4}{24} + \frac{a_{11}C_{12}}{6}x^3\right]\\
& + \frac{C_{42}}{2}x^2 + C_{52}x + C_{62}\\
=& \frac{1}{\alpha_{22}}\left[\frac{\alpha_{12}}{\beta_2}\left[C_{22}\text{ch}(\beta_2 x) + C_{32}\text{sh}(\beta_2 x) - \frac{x}{\beta_2}\left(\gamma_{22}p + \gamma_{12}\left(\frac{1}{2}qx + C_{12}\right)\right)\right]\right.\\
& \left. -a_{71}p\frac{x^3}{6} + a_{11}q\frac{x^4}{24} + \frac{a_{11}C_{12}}{6}x^3\right] + \frac{C_{42}}{2}x^2 + C_{52}x + C_{62},\\
u_2(x) =& \gamma_{32}\,\psi_2 + \frac{1}{\alpha_{22}}\left[-a_{41}p\frac{x^2}{2} + a_{71}q\frac{x^3}{6} + \frac{a_{71}}{2}C_{12}x^2\right] + C_{72}x + C_{82},
\end{aligned} \tag{3.55}$$

where $C_{12}, \ldots, C_{82}$ are integration constants.

By combining (3.54), (3.55) we obtain the required solution:

$$\begin{aligned}
\psi(x) &= \psi_1(x) + (\psi_2(x) - \psi_1(x))H(x - x_0),\\
u(x) &= u_1(x) + (u_2(x) - u_1(x))H(x - x_0),\\
w(x) &= w_1(x) + (w_2(x) - w_1(x))H(x - x_0).
\end{aligned}$$

When both ends of the rod are embedded, the 16 algebraic equations following from the boundary conditions at the ends of the rod (3.52) and the conjugation requirements (3.53) must be satisfied:

1. $(x = 0, \psi_1 = 0) \Rightarrow \gamma_{11}C_{11} - \beta_1^2 C_{31} = -\gamma_{21}p;$
2. $(x = 0 : u_1 = 0) \Rightarrow C_{81} = 0;$
3. $(x = 0 : w_1 = 0) \Rightarrow \frac{\alpha_{11}}{\alpha_{21}\beta_1}C_{21} + C_{61} = 0;$
4. $(x = 0 : w_{1,x} = 0) \Rightarrow C_{51} = 0;$

   $(x = l : \psi_2 = 0) \Rightarrow$
5. $$-\frac{\gamma_{12}}{\beta_2^2}C_{12} + C_{22}\mathrm{sh}(\beta_2 l) + C_{32}\mathrm{ch}(\beta_2 l) = \frac{1}{\beta_2^2}[\gamma_{22}p + \gamma_{12}ql];$$
6. $(x = l : u_2 = 0) \Rightarrow \frac{a_{71}l^2}{2\alpha_{22}}C_{12} + C_{72}l + C_{82} = \frac{1}{6\alpha_{22}}[3a_{41}pl^2 - a_{71}ql^3];$

   $(x = l : w_2 = 0) \Rightarrow \left(\frac{a_{11}l^3}{6\alpha_{22}} - \frac{\alpha_{12}\gamma_{12}l}{\alpha_{22}\beta_2^2}\right)C_{12}$
7. $$+\frac{\alpha_{12}}{\alpha_{22}\beta_2}\mathrm{ch}(\beta_2 l)C_{22} + \frac{\alpha_{12}}{\alpha_{22}\beta_2}\mathrm{sh}(\beta_2 l)C_{32}$$
   $$+\frac{l^2}{2}C_{42} + lC_{52} + C_{62}$$
   $$= \frac{1}{\alpha_{22}}\left(\frac{\alpha_{12}l}{\beta_2^2}\left(\gamma_{22}p + \frac{\gamma_{12}}{2}ql\right) + a_{71}p\frac{l^3}{6} - a_{11}q\frac{l^4}{24}\right);$$

   $(x = l : w_{2,x} = 0) \Rightarrow$
8. $$\frac{a_{11}l^2}{2\alpha_{22}}C_{12} + C_{42}l + C_{52} = \frac{1}{6\alpha_{22}}(3a_{71}pl^2 - a_{11}ql^3);$$

   $\langle\psi_2(x) - \psi_1(x)\rangle|_{x=x_0} = 0 \Rightarrow$
9. $$\frac{\gamma_{11}}{\beta_1^2}C_{11} - \mathrm{sh}(\beta_1 x_0)C_{21} - C_{31}\mathrm{ch}(\beta_1 x_0)$$
   $$-\frac{\gamma_{12}}{\beta_2^2}C_{12} + \mathrm{sh}(\beta_2 x_0)C_{22} + \mathrm{ch}(\beta_2 x_0)C_{32}$$
   $$= \frac{1}{\beta_2^2}[\gamma_{22}p + \gamma_{12}qx_0] - \frac{1}{\beta_1^2}[\gamma_{21}p + \gamma_{11}qx_0];$$

   $\langle\psi_{2,x}(x) - \psi_{1,x}(x)\rangle|_{x=x_0} = 0$
10. $$-\beta_1\mathrm{ch}(\beta_1 x_0)C_{21} - \beta_1\mathrm{sh}(\beta_1 x_0)C_{31} + \beta_2\mathrm{ch}(\beta_2 x_0)C_{22}$$
    $$+\beta_2\mathrm{sh}(\beta_2 x_0)C_{32} = \left(\frac{\gamma_{12}}{\beta_2^2} - \frac{\gamma_{11}}{\beta_1^2}\right)q;$$

    $\langle\psi_{2,xx}(x) - \psi_{1,xx}(x)\rangle|_{x=x_0} = 0$
11. $$-\beta_1^2\mathrm{sh}(\beta_1 x_0)C_{21} - \beta_1^2\mathrm{ch}(\beta_1 x_0)C_{31}$$
    $$+\beta_2^2\mathrm{sh}(\beta_2 x_0)C_{22} + \beta_2^2\mathrm{ch}(\beta_2 x_0)C_{32} = 0;$$

$$\langle w_2(x) - w_1(x)\rangle|_{x=x_0} = 0 \Rightarrow$$

12.
$$\begin{aligned}
&\left(\frac{\alpha_{11}\gamma_{11}x_0}{\alpha_{21}\beta_1^2} - \frac{a_1x_0^3}{6\alpha_{21}}\right)C_{11} - \frac{\alpha_{11}}{\alpha_{21}\beta_1}\mathrm{ch}(\beta_1x_0)C_{21}\\
&- \frac{\alpha_{11}}{\alpha_{21}\beta_1}\mathrm{sh}(\beta_1x_0)C_{31} - \frac{x_0^2}{2}C_{41}\\
&- x_0C_{51} - C_{61} + \left(\frac{a_{11}x_0^3}{6\alpha_{22}} - \frac{\alpha_{12}\gamma_{12}x_0}{\alpha_{22}\beta_2^2}\right)C_{12}\\
&+ \frac{\alpha_{12}}{\alpha_{22}\beta_2}\mathrm{ch}(\beta_2x_0)C_{22} + \frac{\alpha_{12}}{\alpha_{22}\beta_2}\mathrm{sh}(\beta_2x_0)\\
&\times C_{32} + \frac{x_0^2}{2}C_{42} + x_0C_{52} + C_{62}\\
&= \frac{1}{\alpha_{21}}\left(-\frac{\alpha_{11}x_0}{\beta_1^2}\left(\gamma_{21}p + \frac{\gamma_{11}qx_0}{2}\right) - \frac{a_7px_0^3}{6} + \frac{a_1qx_0^4}{24}\right)\\
&+ \frac{1}{\alpha_{22}}\left(\frac{\alpha_{12}x_0}{\beta_2^2}\left(\gamma_{22}p + \frac{\gamma_{12}}{2}qx_0\right) + \frac{a_{71}px_0^3}{6} - \frac{a_{11}qx_0^4}{24}\right);
\end{aligned}$$

$$\langle w_{2,x}(x) - w_{1,x}(x)\rangle|_{x=x_0} = 0 \Rightarrow$$

13.
$$\begin{aligned}
&\left(\frac{\gamma_{11}}{\beta_1^2}\frac{\alpha_{11}}{\alpha_{21}} - \frac{a_1x_0^2}{2\alpha_{21}}\right)C_{11} - \frac{\alpha_{11}}{\alpha_{21}}\mathrm{sh}(\beta_1x_0)C_{21}\\
&- \frac{\alpha_{11}}{\alpha_{21}}\mathrm{ch}(\beta_1x_0)C_{31} - x_0C_{41} - C_{51}\\
&+ \left(-\frac{\gamma_{12}}{\beta_2^2}\frac{\alpha_{12}}{\alpha_{22}} + \frac{a_{11}x_0^2}{2\alpha_{22}}\right)C_{12} + \frac{\alpha_{12}}{\alpha_{22}}\mathrm{sh}(\beta_2x_0)C_{22}\\
&+ \frac{\alpha_{12}}{\alpha_{22}}\mathrm{ch}(\beta_2x_0)C_{32} + x_0C_{42} + C_{52}\\
&= \frac{1}{\alpha_{21}}\left[-a_7p\frac{x_0^2}{2} + a_1q\frac{x_0^3}{6}\right] - \frac{1}{\alpha_{22}}\left[-a_{71}p\frac{x_0^2}{2} + a_{11}q\frac{x_0^3}{6}\right]\\
&- \frac{1}{\beta_1^2}\frac{\alpha_{11}}{\alpha_{21}}[\gamma_{21}p + \gamma_{11}qx_0] + \frac{1}{\beta_2^2}\frac{\alpha_{12}}{\alpha_{22}}[\gamma_{22}p + \gamma_{12}qx_0];
\end{aligned}$$

$$\langle w_{2,xx}(x) - w_{1,xx}(x)\rangle\big|_{x=x_0} = 0 \Rightarrow$$

14.
$$\begin{aligned}
&- \frac{a_1x_0}{\alpha_{21}}C_{11} - \frac{\alpha_{11}}{\alpha_{21}}\beta_1\mathrm{ch}(\beta_1x_0)C_{21} - \frac{\alpha_{11}}{\alpha_{21}}\beta_1\mathrm{sh}(\beta_1x_0)C_{31} - C_{41}\\
&+ \frac{a_{11}x_0}{\alpha_{22}}C_{12} + \frac{\alpha_{12}}{\alpha_{22}}\beta_2\mathrm{ch}(\beta_2x_0)C_{22} + \frac{\alpha_{12}}{\alpha_{22}}\beta_2\mathrm{sh}(\beta_2x_0)C_{32} + C_{42}\\
&= \frac{1}{\alpha_{21}}\left[-a_7px_0 + a_1q\frac{x_0^2}{2}\right] - \frac{1}{\alpha_{22}}\left[-a_{71}px_0 + a_{11}q\frac{x_0^2}{2}\right]\\
&+ \left[\frac{\alpha_{12}}{\alpha_{22}}\frac{\gamma_{12}}{\beta_2^2} - \frac{\alpha_{11}}{\alpha_{21}}\frac{\gamma_{11}}{\beta_1^2}\right]q;
\end{aligned}$$

$$\langle u_2(x) - u_1(x)\rangle|_{x=x_0} = 0 \Rightarrow$$

15.
$$\left(\frac{\gamma_{11}\gamma_{31}}{\beta_1^2} - \frac{a_7 x_0^2}{2\alpha_{21}}\right)C_{11} - \gamma_{31}\mathrm{sh}(\beta_1 x_0)C_{21}$$
$$- \gamma_{31}\mathrm{ch}(\beta_1 x_0)C_{31} - x_0 C_{71} - C_{81}$$
$$- \left(\frac{\gamma_{12}\gamma_{32}}{\beta_2^2} - \frac{a_{71}x_0^2}{2\alpha_{22}}\right)C_{12} + \gamma_{32}\mathrm{sh}(\beta_2 x_0)C_{22}$$
$$+ \gamma_{32}\mathrm{ch}(\beta_2 x_0)C_{32} + x_0 C_{72} + C_{82}$$
$$= \gamma_{32}\left(\frac{1}{\beta_2^2}[\gamma_{22}p + \gamma_{12}qx_0]\right) - \gamma_{31}\left(\frac{1}{\beta_1^2}[\gamma_{21}p + \gamma_{11}qx_0]\right)$$
$$+ \frac{1}{\alpha_{21}}\left[-a_4 p\frac{x_0^2}{2} + a_7 q\frac{x_0^3}{6}\right] - \frac{1}{\alpha_{22}}\left[-a_{41}p\frac{x_0^2}{2} + a_{71}q\frac{x_0^3}{6}\right];$$

$$\langle u_{2,x}(x) - u_{1,x}(x)\rangle|_{x=x_0} = 0 \Rightarrow$$

16.
$$- \frac{a_7 \quad x_0}{\alpha_{21}}C_{11} - \beta_1\gamma_{31}\mathrm{ch}(\beta_1 x_0)C_{21} - \beta_1\gamma_{31}\mathrm{sh}(\beta_1 x_0)C_{31} - C_{71}$$
$$+ \frac{a_{71} \quad x_0}{\alpha_{22}}C_{12} + \beta_2\gamma_{32}\mathrm{ch}(\beta_2 x_0)C_{22} + \beta_2\gamma_{32}\mathrm{sh}(\beta_2 x_0)C_{32} + C_{72}$$
$$= \left[\frac{\gamma_{32}\gamma_{12}}{\beta_2^2} - \frac{\gamma_{31}\gamma_{11}}{\beta_1^2}\right]q - \frac{1}{\alpha_{22}}\left[-a_{41}px_0 + a_{71}q\frac{x_0^2}{2}\right] \tag{3.56}$$
$$+ \frac{1}{\alpha_{21}}\left[-a_4 px_0 + a_7 q\frac{x_0^2}{2}\right].$$

In the case of a rod hinged at both ends, only the 4th and 8th equations in the system (3.56) will change, and they will take the following form:

$$(x = 0 : M_1 = 0) \Rightarrow$$

4.
$$\beta_1\left(a_7\gamma_{31} + a_3 - \frac{a_4\alpha_{11}}{\alpha_{21}}\right)C_{21} - a_4 C_{41} + a_7 C_{71} = \frac{\gamma_{11}}{\beta_1^2}q\left(a_7\gamma_{31} + a_3 - \frac{a_4\alpha_{11}}{\alpha_{21}}\right);$$

$$(x = l : M_2 = 0) \Rightarrow$$

8.
$$\frac{\left(a_{71}^2 - a_{41}a_{11}\right)l}{\alpha_{22}}C_{12} + \beta_2\left(a_{71}\gamma_{32} + a_{31} - \frac{a_{41}\alpha_{12}}{\alpha_{22}}\right)\mathrm{ch}(\beta_2 l)C_{22} +$$
$$+ \beta_2\left(a_{71}\gamma_{32} + a_{31} - \frac{a_{41}\alpha_{12}}{\alpha_{22}}\right)\mathrm{sh}(\beta_2 l)C_{32} - a_{41}C_{42} + a_{71}C_{72} = \tag{3.57}$$
$$= \frac{\gamma_{12}}{\beta_2^2}q\left(a_{71}\gamma_{32} + a_{31} - \frac{a_{41}\alpha_{12}}{\alpha_{22}}\right) - \frac{ql^2}{2\alpha_{22}}\left(a_{71}^2 - a_{11}a_{41}\right).$$

Numerical results are obtained for a three-layer rod of unit length ($l = 1$ m), the bearing layers of which are made of duralumin (D16T), the filler is fluoroplastic (PTFE). Geometric characteristics and displacements are referred to the length of the rod, whereby $c = 0.09$, $h_1 = 0.03$, $h_{20} = 0.02$, $h_{21} = 0.01$; step start coordinate $x_0 = 0.5$ unless otherwise specified. Load intensities $p = 0$, $q = -1$ MPa.

Figure 3.32 shows the change in the magnitude and shape of the deflection w (*a*), relative shear ψ (*b*) along the x-axis depending on the value of the elastic characteristics of the material of the load-bearing layers of the rod under consideration. When their elastic modules are increased by a factor of 5, the displacements (curve *1*) decrease by a factor of about 3.6–3.8 (*2*); if these characteristics are increased by a factor of 10, the displacements decrease by a factor of about 5 (*3*). This indicates a significant influence of the filler on the stiffness of the system. At the ends of the rod the boundary conditions of rigid embedment are assumed.

Figure 3.33 illustrates the variation of deflection w (*a*) and relative shear ψ (*b*) along the x-axis as a function of step height, while the total material quantity in layer 2 remains constant: *1*—$h_{20} = 0$, $h_{21} = 0.03$; *2*—$h_{20} = 0.01$, $h_{21} = 0.02$; *3*—$h_{20} = 0.015$, $h_{21} = 0.015$; *4*—$h_{20} = 0.02$, $h_{21} = 0.01$; *5*—$h_{20} = 0.03$, $h_{21} = 0$. Rigid embedding conditions are assumed at the ends of the rod. Consequently, the effect of the step height on the deflection and relative shear is negligible when the amount of material of the load-bearing layers is unchanged.

Figure 3.34 illustrates the variation of deflection w (*a*) and relative shear ψ (*b*) along the x-axis depending on the step length with a standard three-layer package: *1*—$x_0 = 1$; *2*—$x_0 = 0.75$, *3*—$x_0 = 0.5$; *4*—$x_0 = 0.25$; *5*—$x_0 = 0$. Rigid embedding conditions are assumed at the ends of the rod. Curve (*1*) corresponds to the complete

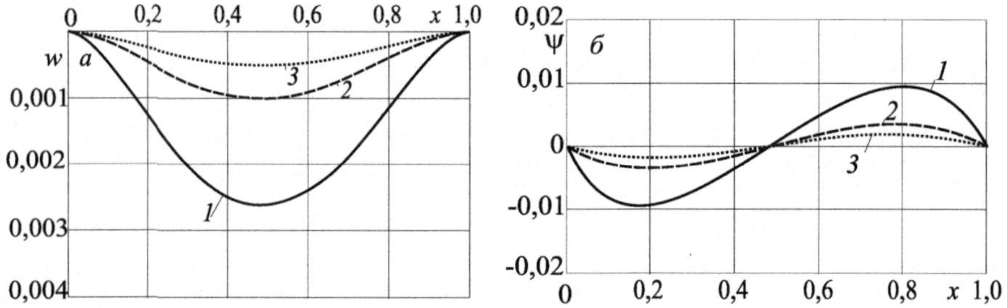

Fig. 3.32 Changing of deflection w (*a*) and relative shear ψ (*b*) along the axis depending on elastic characteristics of the material of bearing layers

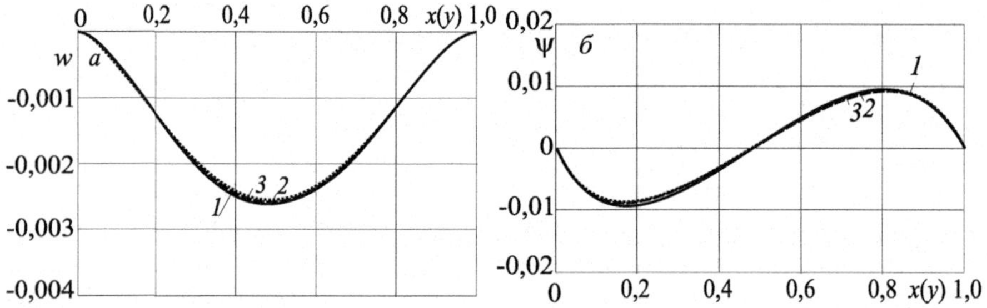

Fig. 3.33 Variation of deflection w (*a*) and relative shear ψ (*b*) along the axis as a function of step height

absence of a step, resulting in minimum stiffness and consequently maximum deflection. An increase in step length leads to an increase in rod stiffness and a decrease in deflection. The relative shear decreases only slightly in this case. If the step is spread along the entire length of a thickness-symmetric rod, the stiffness is maximized, the deflection is minimized, and the longitudinal displacement goes to zero.

Similar graphs for hinged support of the rod ends are shown in Fig. 3.35: *1*—$x_0 = 1$; *2*—$x_0 = 0.75$; *3*—$x_0 = 0.5$; *4*—$x_0 = 0.25$; *5*—$x_0 = 0$. Increasing the step length results in increased rod stiffness and decreased deflection. Relative shear changes insignificantly. The maximum deflection is larger by approximately 5% than when the rod ends are embedded.

If the rod is cantilevered at the left end with the right end free, the corresponding displacement graphs as a function of step length for a standard three-layer package are illustrated by Fig. 3.36 (load is $q = -0.5$ MПa): *1*—$x_0 = 1$; *2*—$x_0 = 0.75$, *3*—$x_0 = 0.5$; *4*—$x_0 = 0.25$; *5*—$x_0 = 0$. Increasing the step length results in a reduction in deflection and relative shear of approximately 10%. It should be noted that if the load is increased by a factor of 3, the obtained deflection *5* will exactly coincide with the deflection calculated earlier for a three-layer rod with a regular boundary under the same boundary conditions.

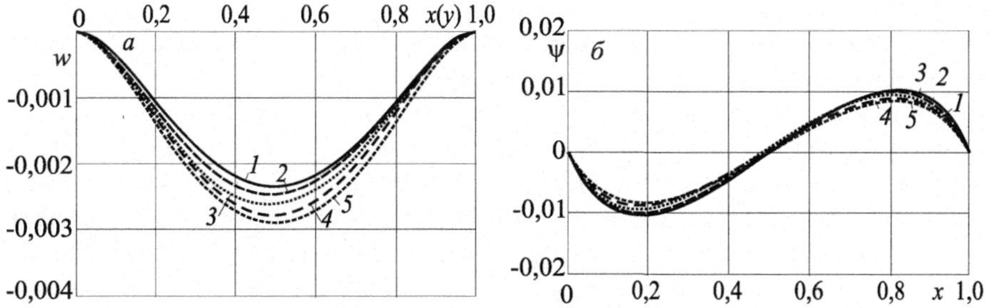

Fig. 3.34 Variation of deflection w (a) and relative shear ψ (b) along the axis depending on the length of the step (rigid embedding)

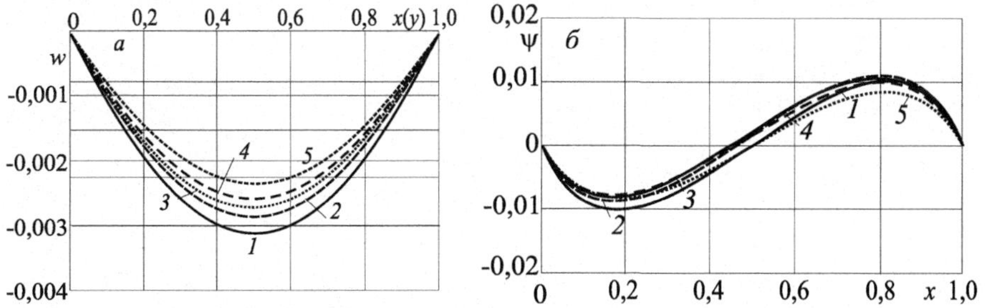

Fig. 3.35 Variation of deflection w (a) and relative shear ψ (b) along the axis depending on the length of the step (hinged support)

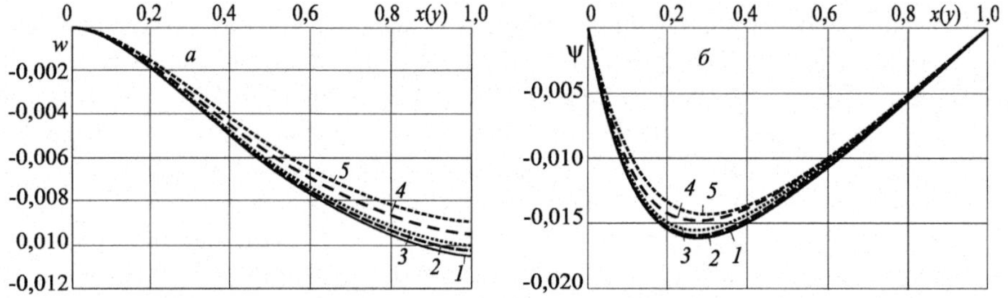

Fig. 3.36 Variation of deflection w (a) and relative shear ψ (b) along the axis depending on the length of the step (cantilever end termination—free face)

Figure 3.37 shows the variation of shear stresses $\sigma_{xz}^{(3)}$ in the filler at different positions of the "irregularity point": *1*—$x_0 = 0.25$; *2*—$x_0 = 0.5$; *3*—$x_0 = 0.75$. As the step length decreases, the overall stiffness of the rod decreases and the tangential stresses in the filler increase, in accordance with the relative shear.

It should be noted that the failure of laminated structural elements is often observed in soft layers (fillers), so it is important to know the stress fields in them. At the ends of the rod, hereinafter (Figs. 3.37, 3.38, 3.39 and 3.40), the conditions of rigid embedment are assumed.

Figure 3.38 shows the variation of normal stresses $\sigma_{xx}^{(k)}$ on the surfaces of the rod layers at the coordinate of the irregularity point $x_0 = 0.25$: *1*—$z = h_2 - c$; *2*—$z = -c$ (layer *1*); *3*—$z = -c$ (filler); *4*—$z = c$ (filler); *4*—$z = c$(layer *2*); *4*—$z = c + h_2$. The layers of the rod deform almost uniformly: i.e., in regions where the upper surface of the layer is tensile, the lower surface is compressed, and vice versa. Stresses change signs in cross-sections x = 0.18; 0.82. At the right end of the rod the stresses on the outer surfaces of the bearing layers are equal in modulus. On the left end, the bottom bearing layer is thinner and the stresses in it are 23% higher.

Figure 3.39 shows the variation of normal stresses $\sigma_{xx}^{(k)}$ on the surfaces of the rod layers at the coordinate of the irregularity point $x_0 = 0.5$. Here, the stresses increase slightly at the left end on the bottom surface of the rod compared to the previous case due to the reduced stiffness of the bottom layer.

Fig. 3.37 Variation of shear stresses $\sigma_{xz}^{(3)}$ in filler at different positions of the irregularity point

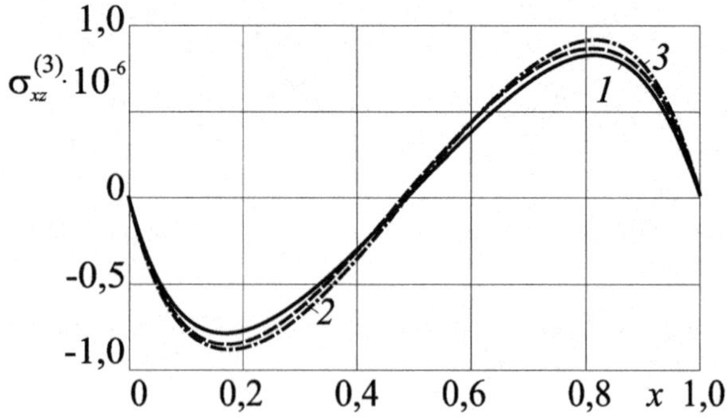

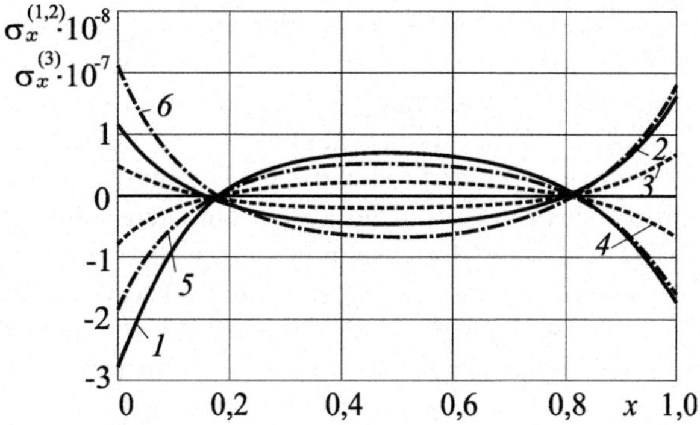

Fig. 3.38 The variation of normal stresses $\sigma_{xx}^{(k)}$ on the surfaces of the rod layers at the coordinate of the irregularity point $x_0 = 0.25$

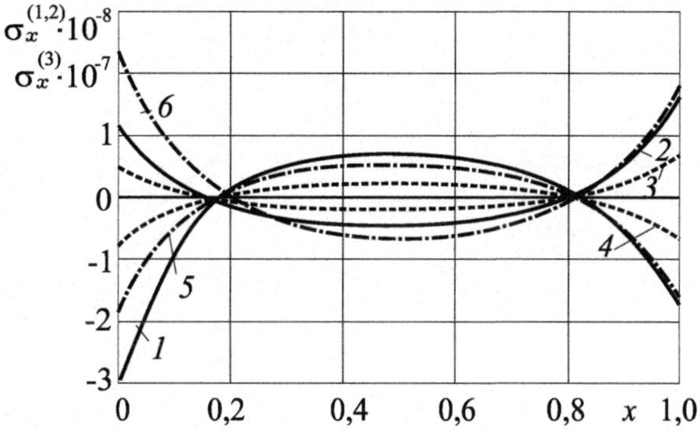

Fig. 3.39 The variation of normal stresses $\sigma_{xx}^{(k)}$ on the surfaces of the rod layers at the coordinate of the irregularity point $x_0 = 0.5$

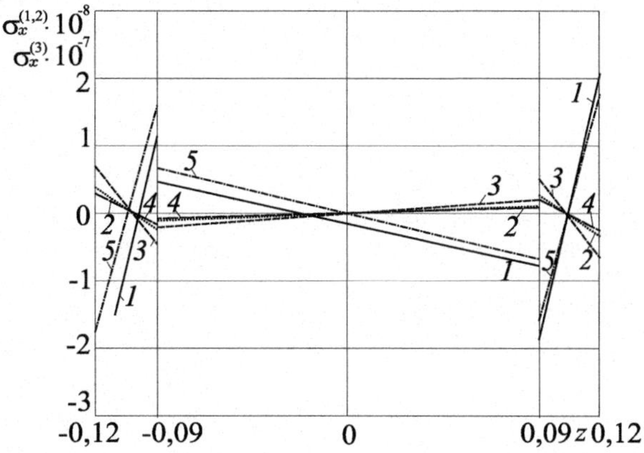

Fig. 3.40 The variation of normal stresses $\sigma_{xx}^{(k)}$ in cross sections with coordinate x_s (rigid embedding)

Figure 3.40 illustrates the variation of normal stresses $\sigma_{xx}^{(k)}$ in cross sections with coordinate x_s of a rod rigidly embedded by both ends at the coordinate of the irregularity point $x_0 = 0.25$: *1*—$x_s = 0$; *2*—$x_s = 0.25$, *3*—$x_s = 0.5$; *4*—$x_s = 0.75$; *5*—$x_s = 1$. The stresses inside the layers, in accordance with the accepted deformation hypotheses, change linearly, changing sign approximately along the layer midline (the displacement occurs due to the presence of a step). Along the length of the rod, the stresses vary similarly to the previous plots.

Figure 3.41 shows similar plots of normal stress $\sigma_{xx}^{(k)}$ variation in the cross sections of a hinge-supported rod at the same point of irregularity $x_0 = 0.25$. The maximum difference of stresses in the layers, compared to the case of rod end embedded, reaches 10%, while they decrease in the upper bearing layer and increase in the lower one. In the filler, a similar stress increase is observed at the lower boundary surface.

If the rod is cantilevered at the left end with the right end free, the corresponding stresses in the same cross-section are illustrated in Fig. 3.42: *1*—$x_s = 0$; *2*—$x_s = 0.25$, *3*—$x_s = 0.5$; *4*—$x_s = 0.75$; *5*—$x_s = 1$.

Tensile of the upper surface is observed in sections up to the step. Further, in all sections the upper plane is compressed, which is explained by the presence of a step and a rigid diaphragm on the right end of the rod.

Fig. 3.41 The variation of normal stresses $\sigma_{xx}^{(k)}$ in cross sections with coordinate x_s (hinged support)

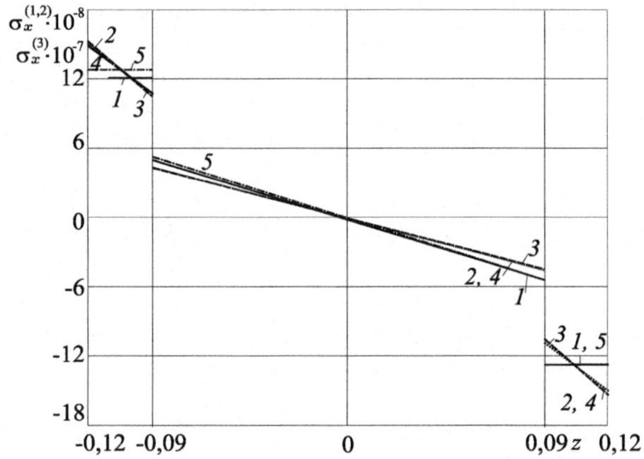

Fig. 3.42 The variation of normal stresses $\sigma_{xx}^{(k)}$ in cross sections with coordinate x_s (cantilever end termination—free face)

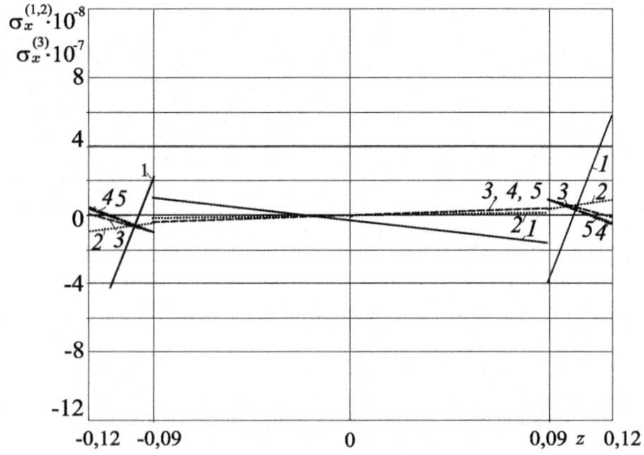

3.6 Stiffening Rod with Rigid Lining

Suppose that the previously considered three-layer rod has a lining (reinforcement) of width $\Delta l = x_1 - x_0$ (Fig. 3.43). Then the thicknesses of layers can be written in the form:

$$h_1, \ \ h_2 = h_{20} + h_{21}[H_0(x_1 - x) - H_0(x_0 - x)], \ h_3 = 2c. \tag{3.58}$$

Repeating the mathematical operations of the previous Sect. 3.5, we obtain a system of differential equilibrium equations of the rod under consideration with respect to the required displacements u, ψ, w, of the type (3.44):

$$\begin{aligned}
&(a_1 + a_{10}H(x_0, x_1))u,_{xx} + (a_6 + a_{60}H(x_0, x_1))\psi,_{xx} \\
&\quad - (a_7 + a_{70}H(x_0, x_1))w,_{xxx} = -p, \\
&(a_7 + a_{70}H(x_0, x_1))u,_{xxx} + (a_3 + a_{30}H(x_0, x_1))\psi,_{xxx} \\
&\quad - (a_4 + a_{40}H(x_0, x_1))w,_{xxxx} = -q, \\
&(a_6 + a_{60}H(x_0, x_1))u,_{xx} + (a_2 + a_{20}H(x_0, x_1))\psi,_{xx} \\
&\quad - (a_3 + a_{30}H(x_0, x_1))w,_{xxx} - a_5\psi = 0.
\end{aligned} \tag{3.59}$$

All parameters used in these equations are defined in (3.44). The combination of Heaviside functions is defined as follows:

$$H(x_0, x_1) \equiv H_0(x_1 - x) - H_0(x_0 - x).$$

The system of ordinary differential Eq. (3.59) contains coefficients having two breakpoints, by virtue of (3.58). As a consequence, its solution must be carried out separately in all continuity areas of these coefficients.

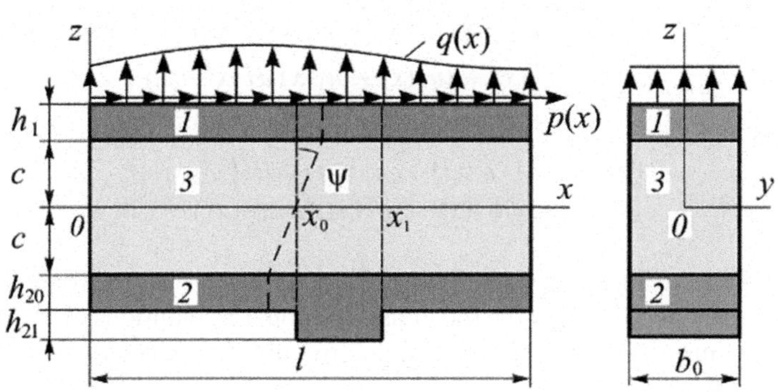

Fig. 3.43 Calculation diagram of a rod with lining

According to the solution procedure, we consider the system of Eq. (3.59) in three domains: I—before the thickness change point ($x < x_0$); II—at the protruding step ($x_0 \leq x < x_1$); III—after the second point ($x \geq x_1$).

In the first and third domains, the function $H(x_0, x_1) = 0$ and $h_2 = h_{20}$. Displacements in these areas are marked with the lower indices "*1*" and "*3*" respectively. The system (3.59) is transformed to a form of the type (3.18). As a result, in the first domain we have an analytical solution of the problem under analysis identical to (3.45). In the third domain, it is as follows:

$$
\begin{aligned}
&\psi_3(x) = C_{23} sh(\beta_1 x) + C_{33} ch(\beta_1 x) \\
&+ \frac{1}{\beta_1}\left[sh(\beta_1 x) \int g_1(x) ch(\beta_1 x) dx - ch(\beta_1 x) \int g_1(x) sh(\beta_1 x) dx \right] \\
&w_3(x) = \frac{1}{\alpha_{21}}\left[\alpha_{11} \int \psi_3 dx - a_7 L_3^{-1}(p) + a_1 L_4^{-1}(q) + \frac{a_1 C_{13}}{6} x^3 \right] \\
&+ \frac{C_{43}}{2} x^2 + C_{53} x + C_{63}, \\
&u_3(x) = \gamma_{31}\, \psi_3 + \frac{1}{\alpha_{21}}\left[-a_4 L_2^{-1}(p) + \right. \\
&\left. a_7 L_3^{-1}(q) + \frac{a_7}{2} C_{13} x^2 \right] + C_{73} x + C_{83},
\end{aligned} \tag{3.60}
$$

where C_{13}, ..., C_{83} are integration constants, other parameters are introduced in Sect. 3.1.

In the second domain, the function $H(x_0, x_1) = 1$ and $h_2 = h_{20} + h_{21}$. Displacements in this domain are marked with the lower index "2". Their analytical form coincides with the solution of (3.47).

The required solution of the initial system (3.59) now can be written in the form:

$$
\begin{aligned}
\psi(x) =& \psi_1(x) + (\psi_2(x) - \psi_1(x)) H(x - x_0) \\
&+ (\psi_3(x) - \psi_2(x)) H(x - x_1), \\
u(x) =& u_1(x) + (u_2(x) - u_1(x)) H(x - x_0) \\
&+ (\psi_3(x) - \psi_2(x)) H(x - x_1), \\
w(x) =& w_1(x) + (w_2(x) - w_1(x)) H(x - x_0) \\
&+ (\psi_3(x) - \psi_2(x)) H(x - x_1).
\end{aligned} \tag{3.61}
$$

The integration constants $C_{11}, \ldots, C_{81}, C_{12}, \ldots, C_{82}, C_{13}, \ldots, C_{83}$ are determined from 8 boundary conditions at the rod ends, additional 8 conditions (3.49) of equality of displacements and their derivatives at the irregularity point $x = x_0$, and 8 similar conditions at the point $x = x_1$:

$$
\begin{aligned}
&\langle \psi_2(x) - \psi_3(x) \rangle|_{x=x_1} = 0, \ \langle \psi_{2,x}(x) - \psi_{3,x}(x) \rangle|_{x=x_1} = 0, \\
&\langle \psi_{2,xx}(x) - \psi_{3,xx}(x) \rangle|_{x=x_1} = 0,
\end{aligned}
$$

$$
\begin{aligned}
&\langle w_2(x) - w_3(x)\rangle|_{x=x_1} = 0,\ \langle w_2,_x(x) - w_3,_x(x)\rangle|_{x=x_1} = 0,\\
&\langle w_2,_{xx}(x) - w_3,_{xx}(x)\rangle|_{x=x_1} = 0,\\
&\langle u_2(x) - u_3(x)\rangle|_{x=x_1} = 0,\ \langle u_2,_x(x) - u_3,_x(x)\rangle|_{x=x_1} = 0.
\end{aligned}
\tag{3.62}
$$

Let's consider three types of fixation of the rod ends. In the case of *rigid embedding of the left end of the rod with a free right end*, the boundary conditions are as follows:

$$
\begin{aligned}
&x = 0 : w_1 = w_1,_x = u_1 = \psi_1 = 0;\\
&x = l : \psi_3 = N_3 = M_3 = M_3,_x = 0.
\end{aligned}
\tag{3.63}
$$

By substituting the solution of (3.48) into the boundary conditions (3.49), (3.62) and (3.63), we obtain a system of 24 linear algebraic equations to determine the 24 integration constants $C_{11}, \ldots, C_{81}, C_{12}, \ldots, C_{82}, C_{13}, \ldots, C_{83}$.

For a *rod hinged at both ends*, instead of (3.63) one should use the following boundary conditions:

$$
x = 0,\ l:\ u_{(1,3)} = \psi_{(1,3)} = w_{(1,3)} = M_{(1,3)} = 0. \tag{3.64}
$$

When *both ends of the rod are embedded*, the boundary conditions are as follows:

$$
x = 0,\ l:\ u_{(1,3)} = \psi_{(1,3)} = w_{(1,3)} = w,_{x(1,3)} = 0. \tag{3.65}
$$

The algebraic system of eight equations that stitches together the solutions at the point x = x 0 for the first two domains of regularity generally is the same as (3.53). Let's write out the system of equations of conjugation of solutions at the point $x = x_1$. For this purpose, we require eight conditions (3.62) to be fulfilled. As a result, we have:

$$
\begin{aligned}
&C_{23}sh(\beta_1 x_0) + C_{33}ch(\beta_1 x_0) - C_{22}sh(\beta_2 x_0) - C_{32}ch(\beta_2 x_0) =\\
&= \frac{1}{\beta_2}\left[sh(\beta_2 x)\int g_2(x)ch(\beta_2 x)\mathrm{d}x - ch(\beta_2 x)\int g_2(x)sh(\beta_2 x)\mathrm{d}x\right]\Bigg|_{x=x_1} -\\
&- \frac{1}{\beta_1}\left[sh(\beta_1 x)\int g_1(x)ch(\beta_1 x)\mathrm{d}x - ch(\beta_1 x)\int g_1(x)sh(\beta_1 x)\mathrm{d}x\right]\Bigg|_{x=x_1},\\
&C_{23}\beta_1 ch(\beta_1 x_0) + C_{33}\beta_1 sh(\beta_1 x_0) - C_{22}\beta_2 sh(\beta_2 x_0) - C_{32}\beta_2 ch(\beta_2 x_0) =\\
&= \frac{1}{\beta_2}\left[sh(\beta_2 x)\int g_2(x)ch(\beta_2 x)\mathrm{d}x - ch(\beta_2 x)\int g_2(x)sh(\beta_2 x)\mathrm{d}x\right],_x\Bigg|_{x=x_1} -\\
&- \frac{1}{\beta_1}\left[sh(\beta_1 x)\int g_1(x)ch(\beta_1 x)\mathrm{d}x - ch(\beta_1 x)\int g_1(x)sh(\beta_1 x)\mathrm{d}x\right],_x\Bigg|_{x=x_1}\\
&C_{23}\beta_1^2 sh(\beta_1 x_0) + C_{33}\beta_1^2 ch(\beta_1 x_0) - C_{22}\beta_2^2 sh(\beta_2 x_0) - C_{32}\beta_2^2 ch(\beta_2 x_0) =\\
&= \frac{1}{\beta_2}\left[sh(\beta_2 x)\int g_2(x)ch(\beta_2 x)\mathrm{d}x - ch(\beta_2 x)\int g_2(x)sh(\beta_2 x)\mathrm{d}x\right],_{xx}\Bigg|_{x=x_1} -\\
&- \frac{1}{\beta_1}\left[sh(\beta_1 x)\int g_1(x)ch(\beta_1 x)\mathrm{d}x - ch(\beta_1 x)\int g_1(x)sh(\beta_1 x)\mathrm{d}x\right],_{xx}\Bigg|_{x=x_1},
\end{aligned}
$$

$$\frac{1}{\alpha_{21}}\left[\alpha_{11}\int\psi_3\mathrm{d}x - a_7L_3^{-1}(p) + a_1L_4^{-1}(q) + \frac{a_1C_{13}}{6}x^3\right]\Bigg|_{x=x_1} + \frac{C_{43}}{2}x_1^2 + C_{53}x_1 + C_{63} -$$
$$-\frac{1}{\alpha_{22}}\left[\alpha_{12}\int\psi_2\mathrm{d}x - a_{71}L_3^{-1}(p) + a_{11}L_4^{-1}(q) + \frac{a_{11}C_{12}}{6}x^3\right]\Bigg|_{x=x_1} -$$
$$-\frac{C_{42}}{2}x_1^2 - C_{52}x_1 - C_{62} = 0,$$
$$\frac{1}{\alpha_{21}}\left[\alpha_{11}\psi_3 - a_7L_2^{-1}(p) + a_1L_3^{-1}(q) + \frac{a_1C_{13}}{2}x^2\right]\Bigg|_{x=x_1} + C_{43}x_1 + C_{53} -$$
$$-\frac{1}{\alpha_{22}}\left[\alpha_{12}\psi_2 - a_{71}L_2^{-1}(p) + a_{11}L_3^{-1}(q) + \frac{a_{11}C_{12}}{2}x^2\right]\Bigg|_{x=x_1} - C_{42}x_1 - C_{52} = 0$$
$$\frac{1}{\alpha_{21}}\left[\alpha_{11}\psi_{3,x} - a_7L_1^{-1}(p) + a_1L_2^{-1}(q) + a_1C_{13}x\right]\Big|_{x=x_1} + C_{43} -$$
$$-\frac{1}{\alpha_{22}}\left[\alpha_{12}\psi_{2,x} - a_{71}L_1^{-1}(p) + a_{11}L_2^{-1}(q) + a_{11}C_{12}x\right]\Big|_{x=x_1} - C_{42} = 0,$$
$$\gamma_{31}\psi_3|_{x=x_1} + \frac{1}{\alpha_{21}}\left[-a_4L_2^{-1}(p) + a_7L_3^{-1}(q) + \frac{a_7}{2}C_{13}x^2\right]\Big|_{x=x_1} + C_{73}x_1 + C_{83} -$$
$$-\gamma_{32}\psi_2|_{x=x_1} - \frac{1}{\alpha_{22}}\left[-a_{41}L_2^{-1}(p) + a_{71}L_3^{-1}(q) + \frac{a_{71}}{2}C_{12}x^2\right]\Big|_{x=x_1} - C_{72}x_1 - C_{82} = 0,$$
$$\gamma_{31}\psi_{3,x}|_{x=x_1} + \frac{1}{\alpha_{21}}\left[-a_4L_1^{-1}(p) + a_7L_2^{-1}(q) + a_7C_{13}x\right]\Big|_{x=x_1} + C_{73} -$$
$$-\gamma_{32}\psi_{2,x}|_{x=x_1} - \frac{1}{\alpha_{22}}\left[-a_{41}L_1^{-1}(p) + a_{71}L_2^{-1}(q) + a_{71}C_{12}x\right]\Big|_{x=x_1} - C_{72} = 0. \quad (3.66)$$

Thus, the solution of the boundary value problem for a rod with two irregularity sections is closed by joining to the displacements (3.61), taking into account (3.65), (3.47) and (3.60), eight boundary conditions (3.63)–(3.65) at the ends and 16 additional relations (3.62), (3.66), at the points of boundary change.

In the case of *constant linear loads* $q =$ const, $p =$ const when *both ends of the rod are embedded*, the solution in domain I is described by relations (3.54) and in domain II by relations (3.55). In the third domain, the solution is as follows:

$$\psi_3(x) = C_{23}\mathrm{sh}(\beta_1 x) + C_{33}\mathrm{ch}(\beta_1 x) - \frac{1}{\beta_1^2}\left[\gamma_{21}p + \gamma_{11}(qx + C_{13})\right],$$
$$w_3(x) = \frac{1}{\alpha_{21}}\left[\frac{\alpha_{11}}{\beta_1}\left[C_{23}\mathrm{ch}(\beta_1 x) + C_{33}\mathrm{sh}(\beta_1 x) - \frac{x}{\beta_1}\left(\gamma_{21}p + \gamma_{11}\left(\frac{1}{2}qx + C_{13}\right)\right)\right]\right.$$
$$\left. -a_7p\frac{x^3}{6} + a_1q\frac{x^4}{24} + \frac{a_1C_{13}}{6}x^3\right] + \frac{C_{43}}{2}x^2 + C_{53}x + C_{63},$$
$$u_3(x) = \gamma_{31}\psi_3 + \frac{1}{\alpha_{21}}\left[-a_4p\frac{x^2}{2} + a_7q\frac{x^3}{6} + \frac{a_7}{2}C_{13}x^2\right] + C_{73}x + C_{83}. \quad (3.67)$$

where $C_{13}, \ldots, C_{83}$ are integration constants.

Summing (3.54), (3.55), (3.67) in the form (3.61), we obtain the required solution of the boundary value problem for a rod with two points of boundary irregularity.

The following 24 algebraic equations must be fulfilled when both ends of the rod are embedded in the considered case:

1. $(x = 0, \psi_1 = 0) \Rightarrow \gamma_{11} C_{11} - \beta_1^2 C_{31} = -\gamma_{21} p;$
2. $(x = 0 : u_1 = 0) \Rightarrow C_{81} = 0;$
3. $(x = 0 : w_1 = 0) \Rightarrow \frac{\alpha_{11}}{\alpha_{21}\beta_1} C_{21} + C_{61} = 0;$
4. $(x = 0 : w_{1,x} = 0 \Rightarrow C_{51} =;$

$(x = l : \psi_3 = 0) \Rightarrow$

5. $$-\frac{\gamma_{11}}{\beta_1^2} C_{13} + C_{23} \mathrm{sh}(\beta_1 l) + C_{33} \mathrm{ch}(\beta_1 l) = \frac{1}{\beta_1^2} [\gamma_{21} p + \gamma_{11} q l];$$

$(x = l : u_3 = 0) \Rightarrow$

6. $$\frac{a_7 l^2}{2\alpha_{21}} C_{13} + C_{73} l + C_{83} = \frac{1}{6\alpha_{21}} [3 a_4 p l^2 - a_7 q l^3];$$

$(x = l : w_3 = 0) \Rightarrow$

7. $$\left(\frac{a_1 l^3}{6\alpha_{21}} - \frac{\alpha_{11} \gamma_{11} l}{\alpha_{21} \beta_1^2} \right) C_{13} + \frac{\alpha_{11}}{\alpha_{21} \beta_1} \mathrm{ch}(\beta_1 l) C_{23} + \frac{\alpha_{11}}{\alpha_{21} \beta_1} \mathrm{sh}(\beta_1 l) C_{33} +;$$
$$+ \frac{l^2}{2} C_{43} + l C_{53} + C_{63} = \frac{1}{\alpha_{21}} \left(\frac{\alpha_{11} l}{\beta_1^2} \left(\gamma_{21} p + \frac{\gamma_{11}}{2} q l \right) + a_7 p \frac{l^3}{6} - a_1 q \frac{l^4}{24} \right);$$

$(x = l : w_{3,x} = 0) \Rightarrow$

8. $$\frac{a_1 l^2}{2\alpha_{21}} C_{13} + C_{43} l + C_{53} = \frac{1}{6\alpha_{21}} (3 a_7 p l^2 - a_1 q l^3)$$

$\langle \psi_2(x) - \psi_1(x) \rangle |_{x = x_0} = 0 \Rightarrow$

9. $$\frac{\gamma_{11}}{\beta_1^2} C_{11} - \mathrm{sh}(\beta_1 x_0) C_{21} - C_{31} \mathrm{ch}(\beta_1 x_0)$$
$$-\frac{\gamma_{12}}{\beta_2^2} C_{12} + \mathrm{sh}(\beta_2 x_0) C_{22} + \mathrm{ch}(\beta_2 x_0) C_{32}$$
$$= \frac{1}{\beta_2^2} [\gamma_{22} p + \gamma_{12} q x_0] - \frac{1}{\beta_1^2} [\gamma_{21} p + \gamma_{11} q x_0];$$

$\langle \psi_{2,x}(x) - \psi_{1,x}(x) \rangle |_{x = x_0} = 0 \Rightarrow$

10. $$-\beta_1 \mathrm{ch}(\beta_1 x_0) C_{21} - \beta_1 \mathrm{sh}(\beta_1 x_0) C_{31}$$
$$+ \beta_2 \mathrm{ch}(\beta_2 x_0) C_{22} + \beta_2 \mathrm{sh}(\beta_2 x_0) C_{32} = \left(\frac{\gamma_{12}}{\beta_2^2} - \frac{\gamma_{11}}{\beta_1^2} \right) q;$$

$\langle \psi_{2,xx}(x) - \psi_{1,xx}(x) \rangle |_{x = x_0} = 0$

11. $$-\beta_1^2 \mathrm{sh}(\beta_1 x_0) C_{21} - \beta_1^2 \mathrm{ch}(\beta_1 x_0) C_{31}$$
$$+ \beta_2^2 \mathrm{sh}(\beta_2 x_0) C_{22} + \beta_2^2 \mathrm{ch}(\beta_2 x_0) C_{32} = 0;$$

$$\langle w_2(x) - w_1(x)\rangle|_{x=x_0} = 0 \Rightarrow$$

12.
$$\frac{\alpha_{11}}{\alpha_{21}}\left[\left(\frac{\gamma_{11}x_0}{\beta_1^2} - \frac{a_1x_0^3}{6\alpha_{21}}\right)C_{11} - \frac{1}{\beta_1}\text{ch}(\beta_1x_0)C_{21} - \frac{1}{\beta_1}\text{sh}(\beta_1x_0)C_{31}\right]$$
$$-\frac{x_0^2}{2}C_{41} - x_0C_{51}$$
$$-C_{61} + \left(\frac{a_{11}x_0^3}{6\alpha_{22}} - \frac{\alpha_{12}\gamma_{12}x_0}{\alpha_{22}\beta_2^2}\right)C_{12} + \frac{\alpha_{12}}{\alpha_{22}\beta_2}\text{ch}(\beta_2x_0)C_{22}$$
$$+\frac{\alpha_{12}}{\alpha_{22}\beta_2}\text{sh}(\beta_2x_0)C_{32} + \frac{x_0^2}{2}C_{42} + x_0C_{52} + C_{62}$$
$$= \frac{1}{\alpha_{21}}\left(-\frac{\alpha_{11}x_0}{\beta_1^2}\left(\gamma_{21}p + \frac{\gamma_{11}qx_0}{2}\right) - \frac{a_7px_0^3}{6} + \frac{a_1qx_0^4}{24}\right)$$
$$+\frac{1}{\alpha_{22}}\left(\frac{\alpha_{12}x_0}{\beta_2^2}\left(\gamma_{22}p + \frac{\gamma_{12}}{2}qx_0\right) + \frac{a_{71}px_0^3}{6} - \frac{a_{11}qx_0^4}{24}\right);$$

$$\langle w_{2,x}(x) - w_{1,x}(x)\rangle|_{x=x_0} = 0 \Rightarrow$$

13.
$$\left(\frac{\gamma_{11}}{\beta_1^2}\frac{\alpha_{11}}{\alpha_{21}} - \frac{a_1x_0^2}{2\alpha_{21}}\right)C_{11} - \frac{\alpha_{11}}{\alpha_{21}}\text{sh}(\beta_1x_0)$$
$$C_{21} - \frac{\alpha_{11}}{\alpha_{21}}\text{ch}(\beta_1x_0)C_{31} - x_0C_{41} - C_{51}$$
$$+\left(-\frac{\gamma_{12}}{\beta_2^2}\frac{\alpha_{12}}{\alpha_{22}} + \frac{a_{11}x_0^2}{2\alpha_{22}}\right)C_{12} + \frac{\alpha_{12}}{\alpha_{22}}\text{sh}(\beta_2x_0)C_{22}$$
$$+\frac{\alpha_{12}}{\alpha_{22}}\text{ch}(\beta_2x_0)C_{32} + x_0C_{42} + C_{52}$$
$$= \frac{1}{\alpha_{21}}\left[-a_7p\frac{x_0^2}{2} + a_1q\frac{x_0^3}{6}\right] - \frac{1}{\alpha_{22}}\left[-a_{71}p\frac{x_0^2}{2} + a_{11}q\frac{x_0^3}{6}\right]$$
$$-\frac{1}{\beta_1^2}\frac{\alpha_{11}}{\alpha_{21}}[\gamma_{21}p + \gamma_{11}qx_0] + \frac{1}{\beta_2^2}\frac{\alpha_{12}}{\alpha_{22}}[\gamma_{22}p + \gamma_{12}qx_0];$$

$$\langle w_{2,xx}(x) - w_{1,xx}(x)\rangle\big|_{x=x_0} = 0 \Rightarrow$$

14.
$$-\frac{a_1x_0}{\alpha_{21}}C_{11} - \frac{\alpha_{11}}{\alpha_{21}}\beta_1\text{ch}(\beta_1x_0)C_{21} - \frac{\alpha_{11}}{\alpha_{21}}\beta_1\text{sh}(\beta_1x_0)C_{31} - C_{41}$$
$$+\frac{a_{11}x_0}{\alpha_{22}}C_{12} + \frac{\alpha_{12}}{\alpha_{22}}\beta_2\text{ch}(\beta_2x_0)C_{22} + \frac{\alpha_{12}}{\alpha_{22}}\beta_2\text{sh}(\beta_2x_0)C_{32} + C_{42}$$
$$= \frac{1}{\alpha_{21}}\left[-a_7px_0 + a_1q\frac{x_0^2}{2}\right] - \frac{1}{\alpha_{22}}\left[-a_{71}px_0 + a_{11}q\frac{x_0^2}{2}\right]$$
$$+\left[\frac{\alpha_{12}}{\alpha_{22}}\frac{\gamma_{12}}{\beta_2^2} - \frac{\alpha_{11}}{\alpha_{21}}\frac{\gamma_{11}}{\beta_1^2}\right]q;$$

15.
$$\langle u_2(x) - u_1(x)\rangle|_{x=x_0} = 0 \Rightarrow$$
$$\left(\frac{\gamma_{11}\gamma_{31}}{\beta_1^2} - \frac{a_7 x_0^2}{2\alpha_{21}}\right)C_{11} - \gamma_{31}\mathrm{sh}(\beta_1 x_0)C_{21}$$
$$- \gamma_{31}\mathrm{ch}(\beta_1 x_0)C_{31} - x_0 C_{71} - C_{81} - \left(\frac{\gamma_{12}\gamma_{32}}{\beta_2^2} - \frac{a_{71} x_0^2}{2\alpha_{22}}\right)C_{12}$$
$$+ \gamma_{32}\mathrm{sh}(\beta_2 x_0)C_{22} + \gamma_{32}\mathrm{ch}(\beta_2 x_0)C_{32} + x_0 C_{72} + C_{82}$$
$$= \gamma_{32}\left(\frac{1}{\beta_2^2}[\gamma_{22}p + \gamma_{12}qx_0]\right) - \gamma_{31}\left(\frac{1}{\beta_1^2}[\gamma_{21}p + \gamma_{11}qx_0]\right)$$
$$+ \frac{1}{\alpha_{21}}\left[-a_4 p\frac{x_0^2}{2} + a_7 q\frac{x_0^3}{6}\right] - \frac{1}{\alpha_{22}}\left[-a_{41}p\frac{x_0^2}{2} + a_{71}q\frac{x_0^3}{6}\right]$$

16.
$$\langle u_{2,x}(x) - u_{1,x}(x)\rangle|_{x=x_0} = 0 \Rightarrow$$
$$- \frac{a_7 \quad x_0}{\alpha_{21}}C_{11} - \beta_1\gamma_{31}\mathrm{ch}(\beta_1 x_0)C_{21}$$
$$- \beta_1\gamma_{31}\mathrm{sh}(\beta_1 x_0)C_{31} - C_{71} + \frac{a_{71} \quad x_0}{\alpha_{22}}C_{12}$$
$$+ \beta_2\gamma_{32}\mathrm{ch}(\beta_2 x_0)C_{22} + \beta_2\gamma_{32}\mathrm{sh}(\beta_2 x_0)C_{32} + C_{72}$$
$$= \left[\frac{\gamma_{32}\gamma_{12}}{\beta_2^2} - \frac{\gamma_{31}\gamma_{11}}{\beta_1^2}\right]q - \frac{1}{\alpha_{22}}\left[-a_{41}px_0 + a_{71}q\frac{x_0^2}{2}\right]$$
$$+ \frac{1}{\alpha_{21}}\left[-a_4 px_0 + a_7 q\frac{x_0^2}{2}\right];$$

17.
$$\langle \psi_2(x) - \psi_3(x)\rangle|_{x=x_1} = 0 \Rightarrow$$
$$- \frac{\gamma_{12}}{\beta_2^2}C_{12} + \mathrm{sh}(\beta_2 x_1)C_{22} + \mathrm{ch}(\beta_2 x_1)C_{32}$$
$$+ \frac{\gamma_{11}}{\beta_1^2}C_{13} - \mathrm{sh}(\beta_1 x_1)C_{23} - C_{33}\mathrm{ch}(\beta_1 x_1)$$
$$= \frac{1}{\beta_2^2}[\gamma_{22}p + \gamma_{12}qx_1] - \frac{1}{\beta_1^2}[\gamma_{21}p + \gamma_{11}qx_1];$$

18.
$$\langle \psi_{2,x}(x) - \psi_{3,x}(x)\rangle|_{x=x_1} = 0 \Rightarrow$$
$$\beta_2\mathrm{ch}(\beta_2 x_1)C_{22} + \beta_2\mathrm{sh}(\beta_2 x_1)C_{32}$$
$$- \beta_1\mathrm{ch}(\beta_1 x_1)C_{23} - \beta_1\mathrm{sh}(\beta_1 x_1)C_{33} = \left(\frac{\gamma_{12}}{\beta_2^2} - \frac{\gamma_{11}}{\beta_1^2}\right)q;$$

19.
$$\langle \psi_{2,xx}(x) - \psi_{3,xx}(x)\rangle|_{x=x_1} = 0 \Rightarrow$$
$$\beta_2^2\mathrm{sh}(\beta_2 x_1)C_{22} + \beta_2^2\mathrm{ch}(\beta_2 x_1)C_{32}$$
$$- \beta_1^2\mathrm{sh}(\beta_1 x_1)C_{23} - \beta_1^2\mathrm{ch}(\beta_1 x_1)C_{33} = 0;$$

$$\langle w_2(x) - w_3(x)\rangle|_{x=x_1} = 0 \Rightarrow$$

20.
$$\begin{aligned}
&-\left(\frac{\alpha_{12}\gamma_{12}x_1}{\alpha_{22}\beta_2^2} - \frac{a_{11}x_1^3}{6\alpha_{22}}\right)C_{12} + \frac{\alpha_{12}}{\alpha_{22}\beta_2}\mathrm{ch}(\beta_2 x_1)C_{22}\\
&+\frac{\alpha_{12}}{\alpha_{22}\beta_2}\mathrm{sh}(\beta_2 x_1)C_{32} + \frac{x_1^2}{2}C_{42}\\
&+x_1C_{52} + C_{62} + \left(\frac{\alpha_{11}\gamma_{11}x_1}{\alpha_{21}\beta_1^2} - \frac{a_1x_1^3}{6\alpha_{21}}\right)C_{13}\\
&-\frac{\alpha_{11}}{\alpha_{21}\beta_1}\mathrm{ch}(\beta_1 x_1)C_{23} - \frac{\alpha_{11}}{\alpha_{21}\beta_1}\mathrm{sh}(\beta_1 x_1)C_{33}\\
&-\frac{x_1^2}{2}C_{43} - x_1C_{53} - C_{63} = \frac{1}{\alpha_{21}}\Bigg(-\frac{\alpha_{11}x_1}{\beta_1^2}\\
&\left(\gamma_{21}p + \frac{\gamma_{11}qx_1}{2}\right) - \frac{a_7px_1^3}{6} + \frac{a_1qx_1^4}{24}\Bigg)\\
&+\frac{1}{\alpha_{22}}\left(\frac{\alpha_{12}x_1}{\beta_2^2}\left(\gamma_{22}p + \frac{\gamma_{12}}{2}qx_1\right) + \frac{a_{71}px_1^3}{6} - \frac{a_{11}qx_1^4}{24}\right);
\end{aligned}$$

$$\langle w_{2,x}(x) - w_{3,x}(x)\rangle|_{x=x_1} = 0 \Rightarrow$$

21.
$$\begin{aligned}
&-\left(\frac{\gamma_{12}}{\beta_2^2}\frac{\alpha_{12}}{\alpha_{22}} - \frac{a_{11}x_1^2}{2\alpha_{22}}\right)C_{12} + \frac{\alpha_{12}}{\alpha_{22}}\mathrm{sh}(\beta_2 x_1)C_{22}\\
&+\frac{\alpha_{12}}{\alpha_{22}}\mathrm{ch}(\beta_2 x_1)C_{32} + x_1C_{42} + C_{52}\\
&+\left(\frac{\gamma_{11}}{\beta_1^2}\frac{\alpha_{11}}{\alpha_{21}} - \frac{a_1x_1^2}{2\alpha_{21}}\right)C_{13} - \frac{\alpha_{11}}{\alpha_{21}}\mathrm{sh}(\beta_1 x_1)C_{23}\\
&-\frac{\alpha_{11}}{\alpha_{21}}\mathrm{ch}(\beta_1 x_1)C_{33} - x_1C_{43} - C_{53}\\
&=\frac{1}{\alpha_{21}}\left[-a_7p\frac{x_1^2}{2} + a_1q\frac{x_1^3}{6}\right] - \frac{1}{\alpha_{22}}\left[-a_{71}p\frac{x_1^2}{2} + a_{11}q\frac{x_1^3}{6}\right]\\
&-\frac{1}{\beta_1^2}\frac{\alpha_{11}}{\alpha_{21}}[\gamma_{21}p + \gamma_{11}qx_1] + \frac{1}{\beta_2^2}\frac{\alpha_{12}}{\alpha_{22}}[\gamma_{22}p + \gamma_{12}qx_1]
\end{aligned}$$

$$\left\langle w_{2,xx}(x) - w_{3,xx}(x)\right\rangle\Big|_{x=x_1} = 0 \Rightarrow$$

22.
$$\begin{aligned}
&\frac{a_{11}x_1}{\alpha_{22}}C_{12} + \frac{\alpha_{12}}{\alpha_{22}}\beta_2\mathrm{ch}(\beta_2x_1)C_{22} \\
&+ \frac{\alpha_{12}}{\alpha_{22}}\beta_2\mathrm{sh}(\beta_2x_1)C_{32} + C_{42} \\
&- \frac{a_1x_1}{\alpha_{21}}C_{13} - \frac{\alpha_{11}}{\alpha_{21}}\beta_1\mathrm{ch}(\beta_1x_1)C_{23} \\
&- \frac{\alpha_{11}}{\alpha_{21}}\beta_1\mathrm{sh}(\beta_1x_1)C_{33} - C_{43} \\
&= \frac{1}{\alpha_{21}}\left[-a_7px_1 + a_1q\frac{x_1^2}{2}\right] - \frac{1}{\alpha_{22}}\left[-a_{71}px_1 + a_{11}q\frac{x_1^2}{2}\right] \\
&+ \left[\frac{\alpha_{12}}{\alpha_{22}}\frac{\gamma_{12}}{\beta_2^2} - \frac{\alpha_{11}}{\alpha_{21}}\frac{\gamma_{11}}{\beta_1^2}\right]q;
\end{aligned}$$

$$\langle u_2(x) - u_3(x)\rangle|_{x=x_1} = 0 \Rightarrow$$

23.
$$\begin{aligned}
&-\left(\frac{\gamma_{12}\gamma_{32}}{\beta_2^2} - \frac{a_{71}x_1^2}{2\alpha_{22}}\right)C_{12} + \gamma_{32}\mathrm{sh}(\beta_2x_1)C_{22} \\
&+ \gamma_{32}\mathrm{ch}(\beta_2x_1)C_{32} + x_1C_{72} + C_{82} \\
&+ \left(\frac{\gamma_{11}\gamma_{31}}{\beta_1^2} - \frac{a_7x_1^2}{2\alpha_{21}}\right)C_{13} - \gamma_{31}\mathrm{sh}(\beta_1x_1)C_{23} \\
&- \gamma_{31}\mathrm{ch}(\beta_1x_1)C_{33} - x_1C_{73} - C_{83} \\
&= \gamma_{32}\left(\frac{1}{\beta_2^2}[\gamma_{22}p + \gamma_{12}qx_1]\right) - \gamma_{31}\left(\frac{1}{\beta_1^2}[\gamma_{21}p + \gamma_{11}qx_1]\right) \\
&+ \frac{1}{\alpha_{21}}\left[-a_4p\frac{x_1^2}{2} + a_7q\frac{x_1^3}{6}\right] - \frac{1}{\alpha_{22}}\left[-a_{41}p\frac{x_1^2}{2} + a_{71}q\frac{x_1^3}{6}\right];
\end{aligned}$$

$$\langle u_{2,x}(x) - u_{3,x}(x)\rangle|_{x=x_1} = 0 \Rightarrow$$

24.
$$\begin{aligned}
&\frac{a_{71}\quad x_1}{\alpha_{22}}C_{12} + \beta_2\gamma_{32}\mathrm{ch}(\beta_2x_1)C_{22} \\
&+ \beta_2\gamma_{32}\mathrm{sh}(\beta_2x_1)C_{32} + C_{72} \\
&- \frac{a_7\quad x_1}{\alpha_{21}}C_{13} - \beta_1\gamma_{31}\mathrm{ch}(\beta_1x_1)C_{23} \\
&- \beta_1\gamma_{31}\mathrm{sh}(\beta_1x_1)C_{33} - C_{73} \\
&= \left[\frac{\gamma_{32}\gamma_{12}}{\beta_2^2} - \frac{\gamma_{31}\gamma_{11}}{\beta_1^2}\right]q - \frac{1}{\alpha_{22}}\left[-a_{41}px_1 + a_{71}q\frac{x_1^2}{2}\right] \\
&+ \frac{1}{\alpha_{21}}\left[-a_4px_1 + a_7q\frac{x_1^2}{2}\right].
\end{aligned} \tag{3.68}$$

In the case of a rod hinged at both ends, only the fourth and eighth equations in Eqs. (3.68) will change, which will take the form:

$(x = 0 : M_1 = 0) \Rightarrow$

4.
$$\beta_1\left(a_7\gamma_{31} + a_3 - \frac{a_4\alpha_{11}}{\alpha_{21}}\right)C_{21} - a_4C_{41} + a_7C_{71} = \frac{\gamma_{11}}{\beta_1^2}q\left(a_7\gamma_{31} + a_3 - \frac{a_4\alpha_{11}}{\alpha_{21}}\right);$$

8.

$$(x = l : M_3 = 0 \Rightarrow$$

$$\frac{(a_7^2 - a_4a_1)l}{\alpha_{21}}C_{13} + \beta_1\left(a_7\gamma_{31} + a_3 - \frac{a_4\alpha_{11}}{\alpha_{21}}\right)\mathrm{ch}(\beta_1 l)C_{23} +$$

$$+ \beta_1\left(a_7\gamma_{31} + a_3 - \frac{a_4\alpha_{11}}{\alpha_{21}}\right)\mathrm{sh}(\beta_1 l)C_{33} - a_4C_{43} + a_7C_{73} =$$

$$= \frac{\gamma_{11}}{\beta_1^2}q\left(a_7\gamma_{31} + a_3 - \frac{a_4\alpha_{11}}{\alpha_{21}}\right) - \frac{ql^2}{2\alpha_{21}}\left(a_7^2 - a_1a_4\right). \quad (3.69)$$

Numerical results were obtained under the same conditions as before: the bearing layers of the rod are made of duralumin alloy (D16T), the filler (*3*) is fluoroplastic (PTFE); geometric characteristics of the layers: $c = 0.09$, $h_1 = 0.03$, $h_{20} = 0.02$, $h_{21} = 0.01$; loads intensity: $p = 0$, $q = -1$ MPa.

Figure 3.44 shows the variation of deflection w (*a*), relative displacement ψ (*b*) along the x-axis at different positions of the lining of length $\Delta l = 0.1$: *1*—$x_0 = 0$; *2*—$x_0 = 0.2$; *3*—$x_0 = 0.4$; *4*—$x_0 = 0.7$; *5*—$x_0 = 0.9$. Boundary conditions are rigid embedding of the rod ends. Maximum deflections and relative shears change insignificantly—within 3–4%.

Figure 3.45 illustrates the variation of the same parameters along the x-axis as a function of the width of the lining $\Delta l = x_1 - x_0$, located symmetrically about the centre of the rod: *1*—$\Delta l = 0.05$; *2*—$\Delta l = 0.1$; *3*—$\Delta l = 0.25$; *4*—$\Delta l = 0.5$; *5*—$\Delta l = 0.075$. Rigid embedding on both ends of the rod. Increasing the width of the lining has a significant effect on deflection, the relative shear changes to a much lesser extent.

Similar investigations have been carried out for the stresses in the cross section of the rod ($x = 0.5$), (see Fig. 3.46): *1*—$\Delta l = 0.5$; *2*—$\Delta l = 0.1$; *3*—$\Delta l = 0.5$; *4*—$\Delta l = 0.075$. In this case, increasing the lining length leads to stiffening of layer *2* and a reduction of stresses in it.

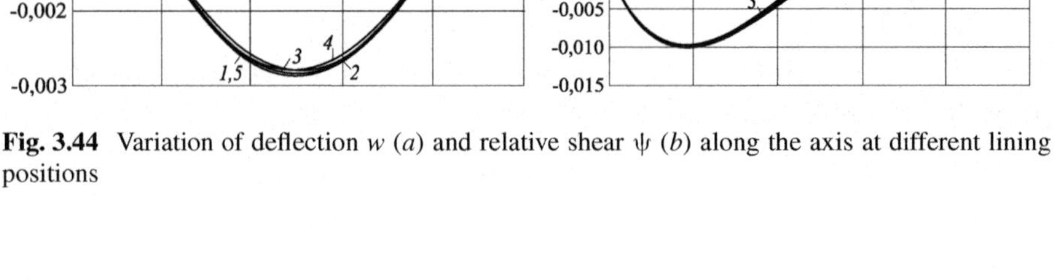

Fig. 3.44 Variation of deflection w (*a*) and relative shear ψ (*b*) along the axis at different lining positions

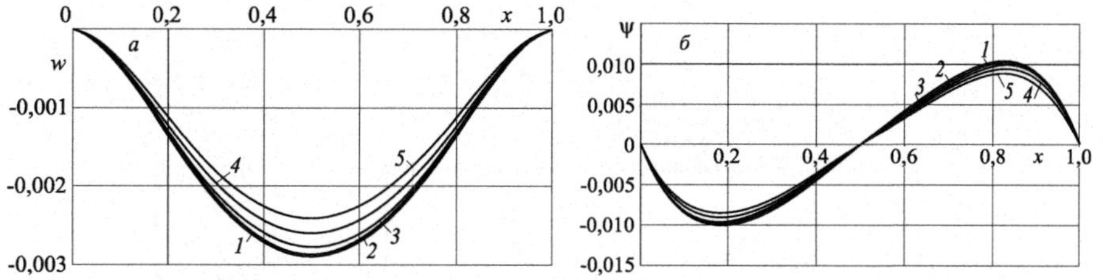

Fig. 3.45 Variation of deflection w (a) and relative shear ψ (b) along the axis for different lining widths

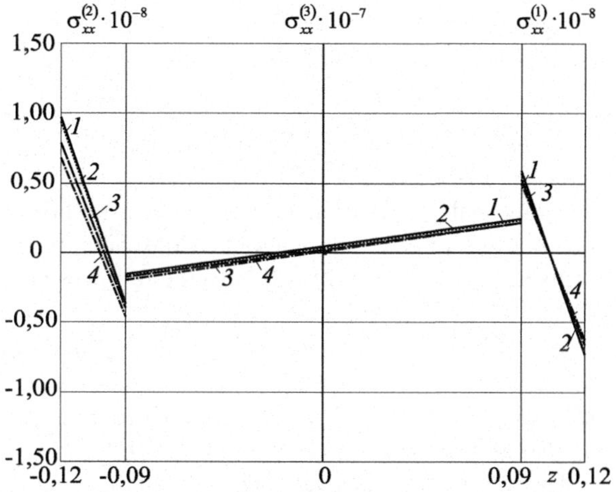

Fig. 3.46 Variation of stresses in the cross-section of the rod ($x = 0.5$) at different lining widths

It should be noted that the longitudinal displacement u is much more dependent on the presence of the lining. But its magnitude is 2 orders of magnitude smaller than the other displacements and does not significantly affect the stress–strain state of the rod.

3.7 Composite Rod

Within the framework of the previously accepted hypotheses, we consider the bending of an asymmetric in thickness elastic composite three-layer rod with a rigid filler (Fig. 3.47).

Rigid diaphragms are assumed at the ends, preventing relative shear of the layers, but not preventing deformation from their plane. All displacements and linear dimensions are referred to the length of the rod l. Through h_k denotes the thickness of the kth layer ($k = 1, 2, 3$—layer number). In the considered case we can write:

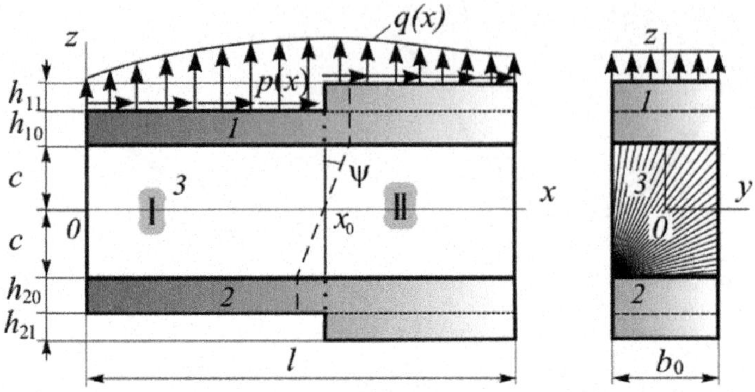

Fig. 3.47 Calculation diagram of a composite rod

$$h_1 = h_{10} + h_{11}H_0(x - x_0),\ h_3 = 2c,\ h_2 = h_{20} + h_{21}H_0(x - x_0). \tag{3.70}$$

Additionally, it is assumed that the bearing layers of the constituent parts of the rod are made of different materials. Consequently, the volume modulus of elasticity for these layers ($k = 1, 2$) will have the following form:

$$K_k^+ = K_{k0}^+ + (K_{k1}^+ - K_{k0}^+)H_0(x - x_0),$$

where $K_{ki}^+ = K_{ki} + \frac{4}{3}G_{ki}, K_{ki},\ G_{ki}$ are elastic constants of the kth layer material in the first and second parts of the rod ($i = 0, 1$).

The longitudinal displacements $u^{(k)}$ in the layers are expressed through the three functions $u(x)$, $\psi(x)$ and $w(x)$ using the relations (3.2), (3.3), and the internal forces and moments are expressed using relations (3.4). The equations of equilibrium of the three-layer rod in stress (3.11) and the force boundary conditions (3.12) will be valid here as well. The relations of Hooke's law (3.13) and (3.14) are used to relate stresses and strains in the layers. Internal stresses and moments are expressed through the required functions $u(x)$, $\psi(x)$, $w(x)$ by formulas (3.15). The system of differential equations of equilibrium of elastic composite three-layer rod in displacements (3.16) is as follows:

$$\begin{aligned}
&(a_1 + a_{10}H(x - x_0))u,_{xx} + (a_6 + a_{60}H(x - x_0))\psi,_{xx}\\
&\quad - (a_7 + a_{70}H(x - x_0))w,_{xxx} = -p,\\
&(a_7 + a_{70}H(x - x_0))u,_{xxx} + (a_3 + a_{30}H(x - x_0))\psi,_{xxx}\\
&\quad - (a_4 + a_{40}H(x - x_0))w,_{xxxx} = -q,\\
&(a_6 + a_{60}H(x - x_0))u,_{xx} + (a_2 + a_{20}H(x - x_0))\psi,_{xx}\\
&\quad - (a_3 + a_{30}H(x - x_0))w,_{xxx} - a_5\psi = 0.
\end{aligned} \tag{3.71}$$

Here the coefficients a_k are as follows:

$$a_1 = K_{10}^+ h_{10} + K_{20}^+ h_{20} + 2K_3^+ c,\ a_{10} = K_{11}^+ h_{11} + K_{21}^+ h_{21},$$

$$a_2 = c^2 \left[K_{10}^+ h_{10} + K_{20}^+ h_{20} + \frac{2}{3} K_3^+ c \right],\ a_{20} = c^2 \left[K_{11}^+ h_{11} + K_{21}^+ h_{21} \right],$$

$$a_3 = c \left[K_{10}^+ h_{10} \left(c + \frac{1}{2} h_{10} \right) + K_{20}^+ h_{20} \left(c + \frac{1}{2} h_{20} \right) + \frac{2}{3} K_3^+ c^2 \right],$$

$$a_{30} = c \left[K_{11}^+ h_{11} \left(c + h_{10} + \frac{1}{2} h_{11} \right) + K_{21}^+ h_{21} \left(c + h_{20} + \frac{1}{2} h_{21} \right) \right],$$

$$a_4 = K_{10}^+ h_{10} \left(c^2 + c h_{10} + \frac{1}{3} h_{10}^2 \right) + K_{20}^+ h_{20} \left(c^2 + c h_{20} + \frac{1}{3} h_{20}^2 \right) + \frac{2}{3} K_3^+ c^3,$$

$$a_{40} = K_{11}^+ h_{11} \left(c(c + 2h_{10} + h_{11}) + h_{10} h_{11} + h_{10}^2 + \frac{1}{3} h_{11}^2 \right) + K_{21}^+ h_{21} \left(c(c + 2h_{20} + h_{21}) + h_{20} h_{21} + h_{20}^2 + \frac{1}{3} h_{21}^2 \right),$$

$$a_5 = 2G_3 c,\ a_6 = c \left[K_{10}^+ h_{10} - K_{20}^+ h_{20} \right],\ a_{60} = -c \left[K_{11}^+ h_{11} + K_{21}^+ h_{21} \right],$$

$$a_7 = K_{10}^+ h_{10} \left(c + \frac{1}{2} h_{10} \right) - K_{20}^+ h_{20} \left(c + \frac{1}{2} h_{20} \right),$$

$$a_{70} = -K_{11}^+ h_{11} \left(c + h_{10} + \frac{1}{2} h_{11} \right) - K_{21}^+ h_{21} \left(c + h_{20} + \frac{1}{2} h_{21} \right). \tag{3.72}$$

We solve the system of differential Eq. (3.71) separately in all regions of continuity of its coefficients: I—up to the conjugation point of the components ($x < x_0$); II—after it ($x \geq x_0$). In the first domain the Heaviside function turns to zero and for the bearing layers: $h_1 = h_{10}$, $h_2 = h_{20}$, $K_k^+ = K_{k0}^+$. As before, mark the required displacements in this area with the index "1" at the bottom. The system (3.71) is transformed to the form (3.16):

$$\begin{aligned} a_1 u_{1,xx} + a_6 \psi_{1,xx} - a_7 w_{1,xxx} &= -p, \\ a_6 u_{1,xx} + a_2 \psi_{1,xx} - a_3 w_{1,xxx} - a_5 \psi_1 &= 0, \\ a_7 u_{1,xxx} + a_3 \psi_{1,xxx} - a_4 w_{1,xxxx} &= -q. \end{aligned} \tag{3.73}$$

Applying the previously developed procedure for solving such systems of differential equations, we obtain the following analytical solution of the problem in the domain I:

$$\psi_1(x) = C_{21} sh(\beta_1 x) + C_{31} ch(\beta_1 x) + \frac{1}{\beta_1} \left[sh(\beta_1 x) \int g_1(x) ch(\beta_1 x) dx \right.$$

$$
\begin{aligned}
&-ch(\beta_1 x)\int g_1(x) sh(\beta_1 x)\mathrm{d}x\Bigg],\\
w_1(x) &= \frac{1}{\alpha_{21}}\Bigg[\alpha_{11}\int \psi_1 \mathrm{d}x - a_7 L_3^{-1}(p)\\
&+a_1 L_4^{-1}(q) + \frac{a_1 C_{11}}{6}x^3\Bigg] + \frac{C_{41}}{2}x^2 + C_{51}x + C_{61},\\
u_1(x) &= \gamma_{31}\,\psi_1 + \frac{1}{\alpha_{21}}\Big[-a_4 L_2^{-1}(p)\\
&+a_7 L_3^{-1}(q) + \frac{a_7}{2}C_{11}x^2\Big] + C_{71}x + C_{81}.
\end{aligned}
\tag{3.74}
$$

Here $C_{11}, \ldots, C_{81}$ are integration constants, $L_1^{-1},\ L_2^{-1},\ L_3^{-1},\ L_4^{-1}$ are linear integral operators (3.24),

$$
\begin{aligned}
&g_1(x) = \gamma_{21}p + \gamma_{11}\left(\int q\mathrm{d}x + C_{11}\right),\\
&\beta_1^2 = \frac{a_1 a_5 \alpha_{21}}{\alpha_{21}\alpha_{31} - \alpha_{11}^2} > 0,\ \gamma_{11} = \frac{a_1\alpha_{11}}{\alpha_{21}\alpha_{31} - \alpha_{11}^2}\\
&\gamma_{21} = \frac{a_6\alpha_{21} - a_7\alpha_{11}}{\alpha_{21}\alpha_{31} - \alpha_{11}^2},\ \alpha_{21}\alpha_{31} - \alpha_{11}^2 \neq 0,\\
&\alpha_{11} = a_1 a_3 - a_6 a_7,\ \alpha_{21} = a_1 a_4 - a_7^2,\ \alpha_{31} = a_1 a_2 - a_6^2.
\end{aligned}
$$

Now consider the system of equations of equilibrium of the rod (3.71) in the second domain, where the Heaviside function is equal to unity ($x \geq x_0$, after the point of conjugation of the parts). In this domain for bearing layers we have: $K_k^+ = K_{k1}^+$, $h_1 = h_{10} + h_{11}$, $h_2 = h_{20} + h_{21}$. Displacements in this area are marked with the index "2" at the bottom. The system (3.71) is transformed to the form:

$$
\begin{aligned}
&a_{11}u_{2,xx} + a_{61}\psi_{2,xx} - a_{71}w_{2,xxx} = -p,\\
&a_{61}u_{2,xx} + a_{21}\psi_{2,xx} - a_{31}w_{2,xxx} - a_5\psi_2 = 0,\\
&a_{71}u_{2,xxx} + a_{31}\psi_{2,xxx} - a_{41}w_{2,xxxx} = -q,
\end{aligned}
\tag{3.75}
$$

where

$$
\begin{aligned}
&a_{11} = a_1 + a_{10},\ a_{21} = a_2 + a_{20},\ a_{31} = a_3 + a_{30},\\
&a_{41} = a_4 + a_{40},\ a_{61} = a_6 + a_{60},\ a_{71} = a_7 + a_{70}.
\end{aligned}
$$

The system (3.75) does not differ from (3.73) in its form, so its solution can be written out at once by the type (3.74):

$$
\psi_2(x) = C_{22}sh(\beta_2 x) + C_{32}ch(\beta_2 x)
$$

$$+\frac{1}{\beta_2}\left[sh(\beta_2 x)\int g_2(x)ch(\beta_2 x)\mathrm{d}x - ch(\beta_2 x)\int g_2(x)sh(\beta_2 x)\mathrm{d}x\right],$$

$$u_2(x) = \gamma_{32}\,\psi_2 + \frac{1}{\alpha_{22}}\Big[-a_{41}L_2^{-1}(p) + a_{71}L_3^{-1}(q)$$

$$+\frac{a_{71}}{2}C_{12}x^2\Big] + C_{72}x + C_{82},$$

$$w_2(x) = \frac{1}{\alpha_{22}}\left[\alpha_{12}\int \psi_2\mathrm{d}x - a_{71}L_3^{-1}(p) + a_{11}L_4^{-1}(q) + \frac{a_{11}C_{12}}{6}x^3\right]$$

$$+\frac{C_{42}}{2}x^2 + C_{52}x + C_{62}, \tag{3.76}$$

where $C_{12}, \ldots, C_{82}$ are integration constants, $L_1^{-1}, \ L_2^{-1}, \ L_3^{-1}, \ L_4^{-1}$ are linear integral operators (3.23),

$$g_2(x) = \gamma_{22}p + \gamma_{12}\left(\int q\mathrm{d}x + C_{12}\right),$$

$$\beta_2^2 = \frac{a_{11}a_{51}\alpha_{21}}{\alpha_{22}\alpha_{32} - \alpha_{12}^2} > 0,\ \gamma_{12} = \frac{a_{11}\alpha_{12}}{\alpha_{22}\alpha_{32} - \alpha_{12}^2},$$

$$\gamma_{22} = \frac{a_{61}\alpha_{22} - a_{71}\alpha_{12}}{\alpha_{22}\alpha_{32} - \alpha_{12}^2},\ \gamma_{32} = \frac{a_{31}a_{71} - a_{41}a_{61}}{\alpha_{22}},$$

$$\alpha_{22}\alpha_{32} - \alpha_{12}^2 \neq 0,\ \alpha_{12} = a_{11}a_{31} - a_{61}a_{71},$$

$$\alpha_{22} = a_{11}a_{41} - a_{71}^2,\ \alpha_{32} = a_{11}a_{21} - a_{61}^2.$$

The required solution of the initial system (3.71) takes the form:

$$\psi(x) = \psi_1(x) + (\psi_2(x) - \psi_1(x))H(x - x_0),$$

$$u(x) = u_1(x) + (u_2(x) - u_1(x))H(x - x_0),$$

$$w(x) = w_1(x) + (w_2(x) - w_1(x))H(x - x_0). \tag{3.77}$$

The integration constants $C_{11}, \ldots, C_{81}, C_{12}, \ldots, C_{82}$ are determined from the 8 boundary conditions (3.63)–(3.65) at the rod ends and additional 8 conditions (3.69) of equality of displacements and their derivatives at the point of boundary irregularity $x = x_0$.

The algebraic system of equations for determining the integration constants obtained by "stitching" the two solutions at the conjugate point $x = x_0$ coincides in form with (3.66), the coefficients of which are calculated by the formulas (3.72).

In the case of *constant linear loads* $q =$ const, $p =$ const when both ends of the composite rod are embedded, we write the solution (3.74) in region I in the following form:

$$\psi_1(x) = C_{21}\mathrm{sh}(\beta_1 x) + C_{31}\mathrm{ch}(\beta_1 x)$$

$$
-\frac{1}{\beta_1^2}\left[\gamma_{21}p+\gamma_{11}(qx+C_{11})\right],
$$

$$
\begin{aligned}
u_1(x) &= \gamma_{31}\,\psi_1+\frac{1}{\alpha_{21}}\left[-a_4p\frac{x^2}{2}+a_7q\frac{x^3}{6}+\frac{a_7}{2}C_{11}x^2\right]+C_{71}x+C_{81},\\
w_1(x) &= \frac{1}{\alpha_{21}}\left[\frac{\alpha_{11}}{\beta_1}\left[C_{21}\mathrm{ch}(\beta_1x)+C_{31}\mathrm{sh}(\beta_1x)-\frac{x}{\beta_1}\left(\gamma_{21}p+\gamma_{11}\left(\frac{1}{2}qx+C_{11}\right)\right)\right]\right.\\
&\left.-a_7p\frac{x^3}{6}+a_1q\frac{x^4}{24}+\frac{a_1C_{11}}{6}x^3\right]+\frac{C_{41}}{2}x^2+C_{51}x+C_{61}.
\end{aligned}
\tag{3.78}
$$

The corresponding solution (3.76) in region II takes the form:

$$
\begin{aligned}
\psi_2 &= C_{22}\mathrm{sh}(\beta_2x)+C_{32}\mathrm{ch}(\beta_2x)-\frac{1}{\beta_2^2}\left[\gamma_{22}p+\gamma_{12}(qx+C_{12})\right],\\
u_2(x) &= \gamma_{32}\,\psi_2+\frac{1}{\alpha_{22}}\left[-a_{41}p\frac{x^2}{2}+a_{71}q\frac{x^3}{6}+\frac{a_{71}}{2}C_{12}x^2\right]+C_{72}x+C_{82},\\
w_2(x) &= \frac{1}{\alpha_{22}}\left[\frac{\alpha_{12}}{\beta_2}\left[C_{22}\mathrm{ch}(\beta_2x)+C_{32}\mathrm{sh}(\beta_2x)-\frac{x}{\beta_2}\left(\gamma_{22}p+\gamma_{12}\left(\frac{1}{2}qx+C_{12}\right)\right)\right]\right.\\
&\left.-a_{71}p\frac{x^3}{6}+a_{11}q\frac{x^4}{24}+\frac{a_{11}C_{12}}{6}x^3\right]+\frac{C_{42}}{2}x^2+C_{52}x+C_{62},
\end{aligned}
\tag{3.79}
$$

where C_{12}, ..., C_{82} are integration constants.

Combining (3.78), (3.79) we obtain the required solution of the boundary value problem:

$$
\begin{aligned}
\psi(x) &= \psi_1(x)+(\psi_2(x)-\psi_1(x))H(x-x_0),\\
u(x) &= u_1(x)+(u_2(x)-u_1(x))H(x-x_0),\\
w(x) &= w_1(x)+(w_2(x)-w_1(x))H(x-x_0).
\end{aligned}
\tag{3.80}
$$

Thus, the solution (3.80), taking into account (3.78), (3.79) fully describes the deformation of a composite three-layer rod with an irregular boundary in the case of loading by uniformly distributed loads.

Numerical results for the bending of a composite three-layer rod with a uniformly distributed load are obtained under the previous conditions when the rod ends are rigidly embedded.

Figure 3.48 shows the variation of deflection (*a*) and shear in the filler (*b*) along the axis of the rod at different coordinates x_0: *1*—$x_0 = 1$; 2—$x_0 = 0.75$, *3*—$x_0 = 0.5$; *4*—$x_0 = 0.25$; *5*—$x_0 = 0$. Curve 5 corresponds to a thickness symmetric solid rod with a larger cross section. The appearance of a smaller cross-sectional component leads to a decrease in the stiffness of the rod, an increase in its deflection and relative shear.

Figure 3.49 shows the variation of normal stresses $\sigma_{xx}^{(k)}$ on the surfaces of the composite rod layers at the coordinate of the irregularity point $x_0 = 0.5$: *1*—$z = h_2 - c$;

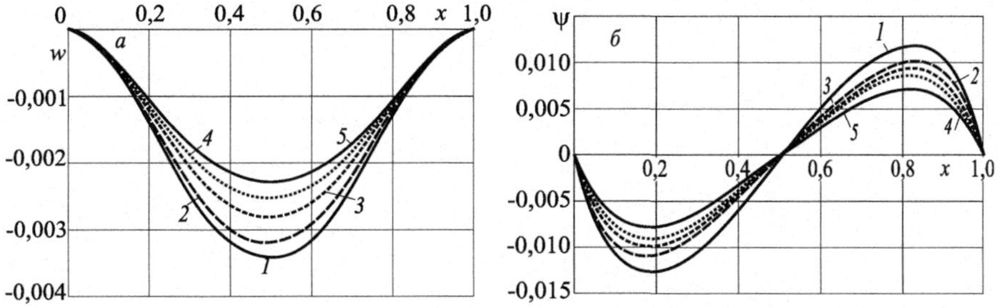

Fig. 3.48 Variation of deflection w (a) and relative shear ψ (b) along the axis at different coordinates x_0

Fig. 3.49 Variation of normal stresses $\sigma_{xx}^{(k)}$ on the surfaces of composite rod layers at the coordinate of the irregularity point $x_0 = 0.5$

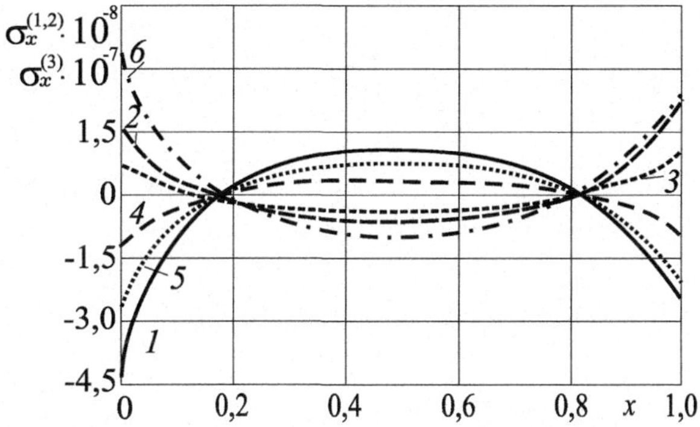

2—$z = -c$ (layer *1*); *3*—$z = -c$ (filler); *4*—$z = c$ (filler); *4*—$z = c$ (layer *2*); *4*—$z = c + h_2$.

The layers of the rod deform almost uniformly. The stresses change signs in sections x = 0.18; 0.82. At the right end of the rod, the stress distribution on the outer surfaces of the bearing layers is symmetric.

The formulations and solutions of boundary value problems presented in this section allow us to study the stress–strain state of elastic three-layered rods with rigid incompressible filler with smooth and irregular boundaries under continuous and local loads.

Reference

1. Korn, G., and T. Korn T. 1978. *Mathematical handbook for scientists and engineers*, 832. Moscow: Nauka. (in Russian).

Chapter 4
Isothermal Deformation of Elastoplastic Three-Layered Rods

Single and alternating isothermal deformation of asymmetric in thickness elastoplastic three-layer rods with rigid incompressible fillers is investigated. Loads are uniformly distributed and localised. The kinematic hypotheses are based on the "broken" normal hypothesis. The work of tangential stresses in the filler is taken into account. Numerical analysis of the obtained analytical solutions was carried out.

4.1 Formulation and Solution of the Boundary Value Problem

A single isothermal loading of an asymmetric in thickness three-layered rod by a continuous distributed load is considered. The outer bearing layers of the rod are made of elastic–plastic material, and the incompressible in thickness inner layer (filler) is nonlinearly elastic. To describe the kinematics of the package, as before, the hypotheses of the broken normal are adopted: in the bearing layers the Bernoulli hypotheses are valid, in the filler the normal remains rectilinear, does not change its length, but rotates by some angle ψ.

Let us use the coordinate system introduced earlier in Sect. 2.1 (see Fig. 2.1). The outer layer of the rod is subjected to distributed force loads $p(x)$, $q(x)$. All displacements and linear dimensions are referred to the length of the rod.

The physical equations of state of Ilyushin's theory of small elastoplastic deformations are used in the layers of the rod (1.27):

$$s_{ij}^{(k)} = 2G_k f^{(k)}\, э_{ij}^{(k)} = 2G_k(1 - \omega^{(k)})\, э_{ij}^{(k)},$$

$$\sigma^{(k)} = 3K_k \varepsilon^{(k)} (k = 1, 2, 3;\ i, j = x, y, z). \quad (4.1)$$

S. Eduard et al., *Deformation of Three-layer Structural Elements in Thermal Radiation Fields*, https://doi.org/10.1007/978-981-97-7217-9_4

Here $s_{ij}^{(k)}$, $\sigma^{(k)}$ и $э_{ij}^{(k)}$, $\varepsilon^{(k)}$ are deviator and spherical parts of stress and strain tensors, G_k, K_k are shear and volume deformation modules, $\varepsilon_u^{(k)}$ is strain intensity, $f^{(k)}$ is Ilyushin's plasticity function ($k = 1, 2$)

$$f^{(k)}(\varepsilon_u^{(k)}) = \begin{cases} 1, & \varepsilon_u^{(k)} \le \varepsilon_y^{(k)}, \\ 1 - \omega^{(k)}(\varepsilon_u^{(k)}), & \varepsilon_u^{(k)} > \varepsilon_y^{(k)}, \end{cases}$$

$f^{(3)}$ is universal function of physical nonlinearity of the filler

$$f^{(3)}(\varepsilon_u^{(3)}) = \begin{cases} 1, & \varepsilon_u^{(3)} \le \varepsilon_{u0}^{(3)}, \\ 1 - \omega^{(3)}(\varepsilon_u^{(3)}), & \varepsilon_u^{(3)} > \varepsilon_{u0}^{(3)}, \end{cases}$$

where $\varepsilon_y^{(k)}$ is strain yield strength, $\varepsilon_{u0}^{(3)}$ is the limiting strain intensity up to which creep is physically linear.

Let us consider the methodology for solving the formulated boundary value problem. Recall that the force equations of equilibrium (2.11) and boundary conditions (2.12) were obtained independently of the physical equations of state. Therefore, we can use them in the case under consideration as well. If in the internal force factors (2.8) included in Eq. (2.11), we express stresses through strains using relations (2.1) and then strains through three linearly independent functions $u(x)$, $\psi(x)$, $w(x)$ using formulas (2.2) and (2.3), the result is a system of nonlinear differential equations. The exact solution of the formulated boundary value problem cannot be constructed in the considered case. Therefore, we will use the method of "elastic" solutions of Ilyushin (see Sect. 1.7), which we will extend to the studied layered systems.

The components of the stress tensor in the layers, using (2.1), let us represent through the deviator and the spherical part of the strain tensor in the following form:

$$\sigma_x^{(k)} = s_x^{(k)} + \sigma^{(k)} = 2G_k\, э_x^{(k)} + 3K_k\varepsilon^{(k)} - 2G_k\, э_x^{(k)}\, \omega^{(k)}(\varepsilon_u^{(k)}),$$

$$\sigma_{xz}^{(3)} = s_{xz}^{(3)} = 2G_3\, э_{xz}^{(3)} - 2G_3\, э_{xz}^{(3)}\, \omega^{(3)}(\varepsilon_u^{(3)}).$$

Let us separate elastic (with index "e") and inelastic (with index "ω") terms in the stress tensor:

$$\sigma_x^{(k)} = \sigma_x^{(k)e} - \sigma_x^{(k)\omega}, \sigma_{xz}^{(3)} = \sigma_{xz}^{(3)e} - \sigma_{xz}^{(3)\omega}, \sigma_x^{(k)}$$

where

$$\sigma_x^{(k)e} = 2G_k\, э_x^{(k)} + 3K_k\varepsilon^{(k)}, \sigma_x^{(k)\omega} = 2G_k\, э_x^{(k)}\, \omega^{(k)}(\varepsilon_u^{(k)}),$$

$$\sigma_{xz}^{(3)e} = 2G_3\, э_{xz}^{(3)}, \sigma_{xz}^{(k)\omega} = 2G_3\, э_{xz}^{(3)}\, \omega^{(3)}(\varepsilon_u^{(3)}). \quad (4.2)$$

Carrying out a similar operation with the internal forces (2.4), we obtain:

$$N^{(k)} = N^{(k)e} - N^{(k)\omega}, M^{(k)} = M^{(k)e} - M^{(k)\omega},$$

$$Q^{(3)} = Q^{(3)e} - Q^{(3)\omega}. \tag{4.3}$$

The elastic (with index "e") and inelastic (with index "ω") components in (4.3) are calculated by the formulas:

$$N^{(k)e} = b_0 \int_{h_k} \sigma_x^{(k)e} dz, N^{(k)\omega} = b_0 \int_{h_k} \sigma_x^{(k)\omega} dz,$$

$$M^{(k)e} = b_0 \int_{h_k} \sigma_x^{(k)e} z dz, M^{(k)\omega} = b_0 \int_{h_k} \sigma_x^{(k)\omega} z dz,$$

$$Q^{(3)e} = b_0 \int_{-c}^{c} \sigma_{xz}^{(3)e} dz, Q^{(3)\omega} = b_0 \int_{-c}^{c} \sigma_{xz}^{(3)\omega} dz. \tag{4.4}$$

Substituting the internal forces (4.3) into the equations of equilibrium (2.11), we reduce the system to the form:

$$\begin{cases} N,_x^e = -b_0 p + N,_x^\omega, \\ H,_x^e - Q^e = H,_x^\omega - Q^\omega, \\ M,_{xx}^e = -b_0 q + M,_{xx}^\omega, \end{cases} \tag{4.5}$$

where

$$N^e = \sum_{k=1}^{3} N^{(k)e}, M^e = \sum_{k=1}^{3} M^{(k)e}, Q^e = Q^{(3)e},$$

$$H^e = c(N^{(1)e} - N^{(2)e}) + M^{(3)e}, N^\omega = \sum_{k=1}^{3} N^{(k)\omega}, M^\omega = \sum_{k=1}^{3} M^{(k)\omega},$$

$$Q^\omega = Q^{(3)\omega}, H^\omega = c(N^{(1)\omega} - N^{(2)\omega}) + M^{(3)\omega}.$$

The final formulas for calculating the inelastic components in (4.5) are obtained using relations (4.2) and (4.4):

$$N^\omega = \tfrac{4}{3} b_0 \sum_{k=1}^{3} G_k \int_{h_k} \omega_k \varepsilon_x^{(k)} dz, M^\omega = \tfrac{4}{3} b_0 \sum_{k=1}^{3} G_k \int_{h_k} \omega_k \varepsilon_x^{(k)} z dz,$$

$$Q^{\omega} = 2b_0G_3 \int_{-c}^{c} \omega_3 \psi dz. \tag{4.6}$$

The system of equations for determining the required functions is obtained by substituting into the equilibrium Eq. (4.5) the relations (2.15) expressing the elastic components of the internal forces by u, ψ and w:

$$\begin{cases} a_1 u,_{xx} + a_6 \psi,_{xx} - a_7 w,_{xxx} = -p + p_{\omega}, \\ a_6 u,_{xx} + a_2 \psi,_{xx} - a_3 w,_{xxx} - a_5 \psi = h_{\omega}, \\ a_7 u,_{xxx} + a_3 \psi,_{xxx} - a_4 w,_{xxxx} = -q + q_{\omega}. \end{cases} \tag{4.7}$$

Here the coefficients $a_1, \ldots, a_7$ are determined by the formulas (2.17),

$$p_{\omega} = \frac{1}{b_0} N,_x^{\omega}, \; h_{\omega} = \frac{1}{b_0}(H,_x^{\omega} - Q^{\omega}), \; q_{\omega} = \frac{1}{b_0} M,_{xx}^{\omega}. \tag{4.8}$$

Suppose that (4.7), (4.8) contain a small parameter (e.g. all $\omega_k < 1$). Then the method of iterations is possible, at which for any nth approximation we obtain such system of equations:

$$\begin{cases} a_1 u,_{xx}^{(n)} + a_6 \psi,_{xx}^{(n)} - a_7 w,_{xxx}^{(n)} = -p + p_{\omega}^{(n-1)}, \\ a_6 u,_{xx}^{(n)} + a_2 \psi,_{xx}^{(n)} - a_3 w,_{xxx}^{(n)} - a_5 \psi^{(n)} = h_{\omega}^{(n-1)}, \\ a_7 u,_{xxx}^{(n)} + a_3 \psi,_{xxx}^{(n)} - a_4 w,_{xxxx}^{(n)} = -q + q_{\omega}^{(n-1)}. \end{cases} \tag{4.9}$$

The values $p_{\omega}^{(n-1)}$, $h_{\omega}^{(n-1)}$, $q_{\omega}^{(n-1)}$, corresponding to inelastic components (4.6), at the first step ($n = 1$) are taken equal to zero, further they are calculated according to the results of the previous approximation and are called *additional "external" loads*. They represent corrections for plasticity and physical nonlinearity of the layer materials:

$$p_{\omega}^{(n-1)} = \frac{1}{b_0} N,_x^{\omega(n-1)}, \; q_{\omega}^{(n-1)} = \frac{1}{b_0} M,_{xx}^{\omega(n-1)}, \; h_{\omega}^{(n-1)} = \frac{1}{b_0}\left(H,_x^{\omega(n-1)} - Q^{\omega(n-1)}\right),$$

$$N^{\omega(n-1)} = \sum_{k=1}^{3} N^{(k)(n-1)} = \frac{4}{3} b_0 \sum_{k=1}^{3} \int_{h_k} G_k \omega^{(k)} (\varepsilon_u^{(k)(n-1)}) \varepsilon_x^{(k)(n-1)} dz,$$

$$M^{\omega(n-1)} = \sum_{k=1}^{3} M^{(k)(n-1)} = \frac{4}{3} b_0 \sum_{k=1}^{3} \int_{h_k} G_k \omega^{(k)} (\varepsilon_u^{(k)(n-1)}) \varepsilon_x^{(k)(n-1)} z dz,$$

$$H^{\omega(n-1)} = c(N^{(1)\omega(n-1)} - N^{(2)\omega(n-1)}) + M^{(3)\omega(n-1)},$$

$$Q^{\omega(n-1)} = 2b_0 \int_{-c}^{c} G_3\omega^{(3)}(\varepsilon_u^{(3)(n-1)})\psi^{(n-1)}dz. \tag{4.10}$$

Application of the above method of successive approximations allows at each step of approximation the considered problem to be reduced to a linear problem of the theory of elasticity with additional "external" loads.

Let us consider the loading of a three-layer elastoplastic rod without pre-deformation. The procedure of obtaining the solution of the system of Eq. (4.9) does not differ from the one applied for the corresponding problem of elasticity theory (2.18)–(2.22). As a result, we obtain the required solution in recurrence form:

$$\begin{aligned}
\psi^{(n)}(x) &= C_2^{(n)} sh(\beta x) + C_3^{(n)} ch(\beta x) \\
&+ \frac{1}{\beta}\left[sh(\beta x)\int g^{(n)} ch(\beta x)\mathrm{d}x - ch(\beta x)\int g^{(n)} sh(\beta x)dx\right],
\end{aligned}$$

$$\begin{aligned}
u^{(n)}(x) =& \gamma_3\psi^{(n)} + \frac{1}{\alpha_2}\Big[-a_4\mathrm{L}_2^{-1}(p - p_\omega^{(n-1)}) + a_7\mathrm{L}_3^{-1}(q - q_\omega^{(n-1)}) \\
&+ \frac{a_7}{2}C_1^{(n)}x^2\Big] + C_7^{(n)}x + C_8^{(n)}
\end{aligned}$$

$$\begin{aligned}
w^{(n)}(x) =& \frac{1}{\alpha_2}\Big[\alpha_1\int\psi^{(n)}\mathrm{d}x - a_7\mathrm{L}_3^{-1}(p - p_\omega^{(n-1)}) \\
&+ a_1\mathrm{L}_4^{-1}(q - q_\omega^{(n-1)}) + \frac{1}{6}a_1C_1^{(n)}x^3\Big] \\
&+ \frac{1}{2}C_4^{(n)}x^2 + C_5^{(n)}x + C_6^{(n)},
\end{aligned} \tag{4.11}$$

$$\begin{aligned}
g_\omega^{(n)}(x) =& \frac{\alpha_2}{\alpha_1}\gamma_1 h_\omega^{(n-1)} + \gamma_2(p - p_\omega^{(n-1)}) \\
&+ \gamma_1\left(\int(q - q_\omega^{(n-1)})\mathrm{d}x + C_1^{(n)}\right),
\end{aligned}$$

where the coefficients $\alpha_1, \alpha_2, \alpha_3, \beta^2, \gamma_1, \gamma_2$ are calculated by the formulas (2.17), (2.19); $C_1{}^{(n)}$ is integration constant; linear integral operators $\mathrm{L}_2^{-1}, L_3^{-1}, L_4^{-1}$ and coefficient γ_3 are determined by the formulas (2.23).

Integration constants $C_1^{(n)}$, ..., $C_8^{(n)}$ are determined from the conditions of rod clamping. Satisfying by the solution (4.11) at each step *the boundary conditions for the cantilever* (2.24), we obtain:

$$C_1^{(n)} = -\mathrm{L}_1^{-1}(q - q_\omega^{(n-1)})|_{x=l},$$

$$C_2^{(n)} = \frac{1}{\beta}\left[\frac{ch(\beta l)}{sh(\beta l)}\left(\int g_\omega^{(n)} sh(\beta x)\mathrm{d}x|_{x=l} - \int g_\omega^{(n)} sh(\beta x)\mathrm{d}x|_{x=0}\right)\right.$$

$$- \int g_{\omega}^{(n)} ch(\beta x) \mathrm{d}x|_{x=l} \Big],$$

$$C_3^{(n)} = \frac{1}{\beta} \int g_{\omega}^{(n)}(x) sh(\beta x) \mathrm{d}x|_{x=0},$$

$$\begin{aligned} C_4^{(n)} =& \frac{a_7}{\alpha_2} \mathrm{L}_1^{-1}(p - p_{\omega}^{(n-1)})|_{x=l} \\ &- \frac{a_1}{\alpha_2} \mathrm{L}_2^{-1}(q - q_{\omega}^{(n-1)})|_{x=l} - \frac{a_1}{\alpha_2} C_1^{(n)} l, \end{aligned}$$

$$C_5^{(n)} = \frac{a_7}{\alpha_2} \mathrm{L}_2^{-1}(p - p_{\omega}^{(n-1)})|_{x=0} - \frac{a_1}{\alpha_2} \mathrm{L}_3^{-1}(q - q_{\omega}^{(n-1)})|_{x=0},$$

$$\begin{aligned} C_6^{(n)} =& \frac{a_7}{\alpha_2} \mathrm{L}_3^{-1}(p - p_{\omega}^{(n-1)})\Big|_{x=0} - \frac{a_1}{\alpha_2} \mathrm{L}_4^{-1}(q - q_{\omega}^{(n-1)})\Big|_{x=0} \\ &- \frac{\alpha_1}{\alpha_2} \int \psi^{(n)} \mathrm{d}x \Big|_{x=0}, \end{aligned}$$

$$C_7^{(n)} = \frac{a_4}{\alpha_2} \mathrm{L}_1^{-1}(p - p_{\omega}^{(n-1)})|_{x=l} - \frac{a_7}{\alpha_2} \mathrm{L}_2^{-1}(q - p_{\omega}^{(n-1)})|_{x=l} - \frac{a_7}{\alpha_2} C_1^{(n)} l,$$

$$C_8^{(n)} = \frac{a_4}{\alpha_2} L_2^{-1}(p - p_{\omega}^{(n-1)})|_{x=0} - \frac{a_7}{\alpha_2} L_3^{-1}(q - q_{\omega}^{(n-1)})|_{x=0}. \tag{4.12}$$

For a *hinged rod*, the boundary conditions are of the type (2.26). In this case the integration constants are:

$$C_1^{(n)} = \frac{1}{l} (\mathrm{L}_2^{-1}(q - q_{\omega}^{(n-1)})|_{x=0} - \mathrm{L}_2^{-1}(q - q_{\omega}^{(n-1)})|_{x=l}),$$

$$C_4^{(n)} = \frac{a_7}{a_4} \left(D_1^{(n)}|_{x=0} - D_1^{(n)}|_{x=l} \right) - \frac{1}{a_4} \mathrm{L}_2^{-1}(q - q_{\omega}^{(n-1)})|_{x=0},$$

$$\begin{aligned} C_5^{(n)} =& D_2^{(n)}|_{x=0} - D_2^{(n)}|_{x=l} - \frac{a_7 l}{2a_4} \left(D_1^{(n)}|_{x=0} - D_1^{(n)}|_{x=l} \right) \\ &- \frac{l}{2a_4} \mathrm{L}_2^{-1}(q - q_{\omega}^{(n-1)})|_{x=0}, \end{aligned}$$

$$C_7^{(n)} = D_1^{(n)}|_{x=0} - D_1^{(n)}|_{x=l}, \tag{4.13}$$

where

$$\begin{aligned} D_1^{(n)} =& \frac{a_7}{l\alpha_2} \mathrm{L}_3^{-1}(q - q_{\omega}^{(n-1)}) \\ &- \frac{a_7 l}{2\alpha_2} \mathrm{L}_2^{-1}(q - q_{\omega}^{(n-1)}) - \frac{a_4}{l\alpha_2} \mathrm{L}_2^{-1}(p - p_{\omega}^{(n-1)}), \end{aligned}$$

$$D_2^{(n)} = \frac{a_1}{l\alpha_2}\left(\int \psi^{(n)}\mathrm{d}x + L_4^{-1}(q - q_\omega^{(n-1)})\right)$$
$$- \frac{a_1 l}{6\alpha_2} L_2^{-1}(q - q_\omega^{(n-1)}) - \frac{a_7}{l\alpha_2} L_3^{-1}(p - p_\omega^{(n-1)}).$$

The constants $C_2^{(n)}$, $C_3^{(n)}$, $C_6^{(n)}$, $C_8^{(n)}$ retain the form (4.12).

In the case of *rigid embedding of both ends* of the rod, the solution must satisfy the boundary conditions (2.28), then

$$C_4^{(n)} = \frac{1}{\alpha_2 l}\Big[a_7 L_2^{-1}(p' - p_\omega'^{(n-1)})\big|_{x=l}$$
$$-a_1 L_3^{-1}(q' - q_\omega'^{(n-1)})\big|_{x=l} - \frac{a_1 C_1^{(n)} l^2}{2}\Big] - \frac{C_5^{(n)}}{l},$$

$$C_7^{(n)} = \frac{1}{\alpha_2 l}\Big[a_4 L_2^{-1}(p' - p_\omega'^{(n-1)})\big|_{x=l} - a_7 L_3^{-1}(q' - q_\omega'^{(n-1)})\big|_{x=l}$$
$$- \frac{a_7 C_1^{(n)} l^2}{2}\Big] - \frac{C_8^{(n)}}{l}. \tag{4.14}$$

The constant $C_1^{(n)}$ is the same as in (4.13); $C_2^{(n)}, C_3^{(n)}, C_5^{(n)}, C_6^{(n)}, C_8^{(n)}$ retain the form (4.12).

Thus, formulas (4.11)–(4.14) give in iterations an analytical solution of the problem of deformation of a three-layer elastic–plastic rod under isothermal quasi-static loads for various types of boundary conditions.

Numerical results here and later in this section, unless otherwise stated, are obtained for a three-layer rod of unit length whose layers are made of D16T-fluoroplastic-D16T materials. The elastic and elastic–plastic characteristics of these and other materials under consideration under direct and repeated loading are given in Tables 1.2–1.3. Geometric parameters of layers: $h_1 = h_2 = 0.03$, $c = 0.09$. *Boundary conditions*: left end termination with free right end of the rod (cantilever termination); hinge support of both ends (hinge-hinge); termination of both ends (termination-termination). The plasticity function was taken as (1.38).

Numerical analysis confirmed the practical convergence of the elastic solution method for the considered three-layer physically nonlinear rod under different boundary conditions: termination-termination ($q = -10$ MPa); left cantilever ($q = -3.5$ MPa).

In case *both ends are embedded*, the 4th approximation (Fig. 4.1), which differs from the previous one by less than 0.1%, is taken as the required numerical solution. The curve number is the same as the iteration number. The intensity of strains in the bearing layers reached 1.12%, in the filler −5.22%. The final difference between the maximum nonlinear displacements and the elastic displacements is 9–12%.

The corresponding distribution of plasticity and physical nonlinearity regions of the layer materials (gray fill) along the longitudinal section of the three-layer rod is

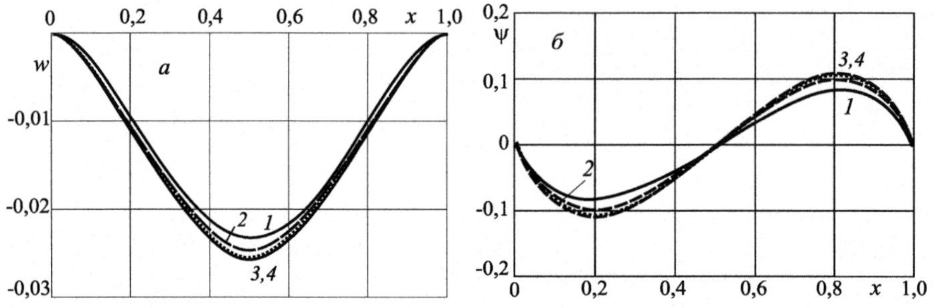

Fig. 4.1 Variation of deflection w (a) and relative shear ψ (b) along the axis at different number of iterations

shown in Fig. 4.2: *a*—load $q = -10$ MPa; *b*—$q = -20$ MPa. In the first case, the regions of nonlinearity in the filler make up 80% of the volume, and in the bearing layers—14%. The strain intensities in this case reached values of 5.3% and 1.2%, respectively. When the load is increased by a factor of 2, the areas of nonlinearity in the filler increase by 10% of the volume, in the bearing layers—by 6%. The maximum strain intensities in the bearing layers are 2.3%, in the filler—10.6%, which is of the order of 3 yield strengths of D16-T alloy and physical nonlinearity of fluoroplastic-4.

In the case of elastic filler, the displacement extremums are reduced by 2–3%. This indicates a relatively small contribution of the physical nonlinearity of the filler to the deformation of the rod.

Figure 4.3 shows the calculated displacements deflection (*a*), shear (*b*) in the filler in a three-layer rod with elastoplastic bearing layers: 1—elastic rod, 2—elastic–plastic bearing layers, elastic filler, 3—elastic–plastic bearing layers, nonlinear filler. Taking into account the physical nonlinearity of materials increases the calculated deflection under loading from the natural state by about 9%, shear—by 12%.

A similar pattern of the process convergence is observed for the cantilever rod (Fig. 4.4). In this case, it took 6 iterations at load $q = -3.5$ MPa to obtain the final result. The intensity of strains in the bearing layers reached 1.2%, in the filler—6.9%.

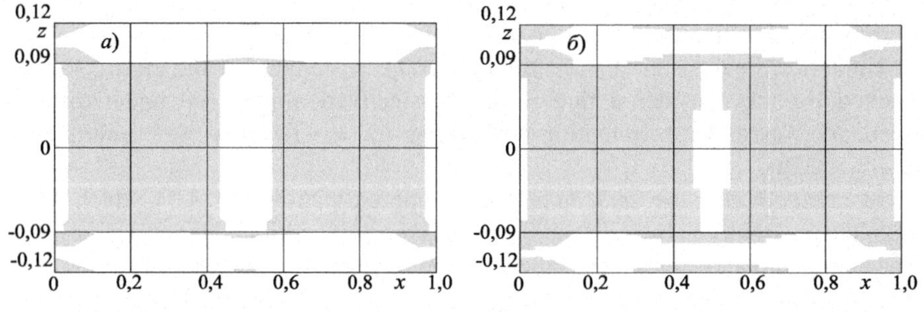

Fig. 4.2 Distribution of plasticity and physical nonlinearity regions at load $q = -10$ MPa (*a*) and $q = -20$ MPa (*b*)

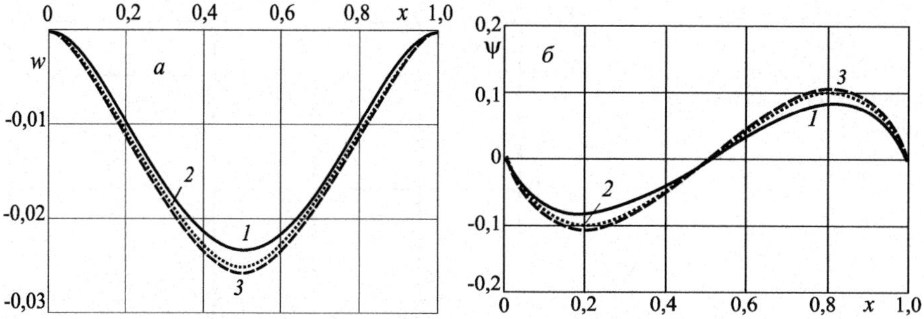

Fig. 4.3 Variation of deflection w (a) and relative shear ψ (b) along the axis for different layer properties (hinge-to-hinge)

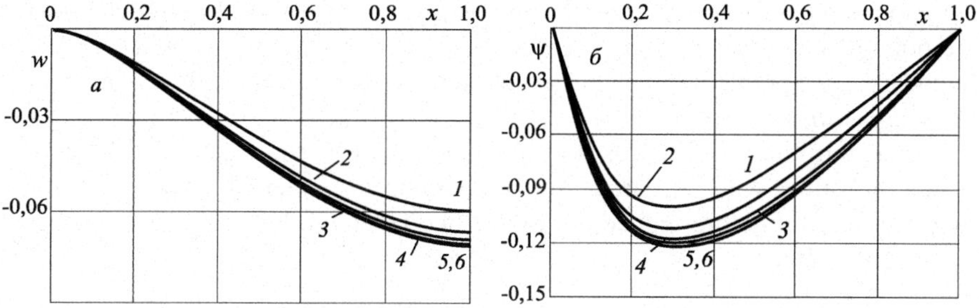

Fig. 4.4 Variation of deflection w (a) and relative shear ψ (b) along the axis for different layer properties (embedding-free edge)

Taking into account the physical nonlinearity of materials increases the calculated displacements by about 16–20%.

The corresponding distribution of plasticity and physical nonlinearity zones of the layer materials (gray fill) along the longitudinal cross-section of the three-layer rod at the cantilevered fixed left end is shown in Fig. 4.5. In this case, the regions of nonlinearity in the filler and the plasticity zones in the bearing layers are constricted to the embedment and occupy a smaller volume than before.

4.2 Repeated Alternating Loading

Let, starting from the moment $t = t_1$, there is an instantaneous unloading and re-loading with loads of the opposite sign p'', q'', changing according to the same law as at loading from the natural state. These loads will create in the kth layer of the rod the fields of displacements u'', ψ'', w'', strains $\varepsilon_x''^{(k)}$ $\varepsilon''^{(k)}$, $э_x''^{(k)}$, and stresses $\sigma_x''^{(k)}$, $\sigma''^{(k)}$, $\sigma_{xz}''^{(3)}$. In this case we assume that the corresponding loading trajectories belong to the class of simple trajectories. The difficulty of finding the parameters of the

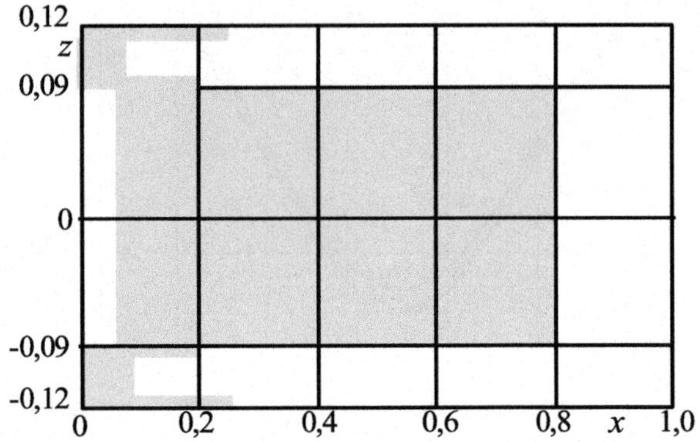

Fig. 4.5 Distribution of areas of plasticity and physical nonlinearity of layer materials (embedding-free edge)

corresponding stress–strain state lies in the dependence of the required solution on the point of unloading ($\varepsilon'_{u1}, \sigma'_{u1}$), i.e. at each point it is necessary to set a different boundary value problem and obtain its solution. *Consider one possibility for avoiding these difficulties.*

Let us introduce differences for all values of stress–strain state and loading, in which the values with one stroke are stresses, strains, external loads and displacements in the rod before unloading, two stroke label similar parameters during the second half cycle:

$$\sigma_x^{(k)*} = \sigma_x^{\prime(k)} - \sigma_x^{\prime\prime(k)} s_x^{(k)*} = s_x^{\prime(k)} - s_x^{\prime\prime(k)}, \sigma^{(k)*} = \sigma^{\prime(k)*} - \sigma^{\prime\prime(k)},$$

$$\sigma_{xz}^{(3)*} = \sigma_{xz}^{\prime(3)} - \sigma_{xz}^{\prime\prime(3)}, \varepsilon_x^{(k)*} = \varepsilon_x^{\prime(k)} - \varepsilon_x^{\prime\prime(k)},$$

$$\text{э}_x^{(k)*} = \text{э}_x^{(k)} - \text{э}_x^{\prime\prime(k)}, \varepsilon^{(k)*} = \varepsilon' - \varepsilon^{\prime\prime(k)}, q^* = q' - q'', p^* = p' - p'',$$

$$u^* = u' - u'', \psi^* = \psi' - \psi'', w^* = w' - w''. \tag{4.15}$$

Due to the linearity of the relation between strains and displacements in the rod (2.2), similar relations will be valid for the values with asterisks as well:

$$\varepsilon_x^{(1)*} = u,_x^* + c\psi,_x^* - zw,_{xx}^*, \varepsilon_{xz}^{(1)*} = 0 (c \le z \le c + h_1),$$

$$\varepsilon_x^{(3)*} = u,_x^* + z\psi,_x^* - zw,_{xx}^*, \varepsilon_{xz}^{(3)*} = \frac{1}{2}\psi^*(-c \le z \le c),$$

$$\varepsilon_x^{(2)*} = u,_x^* - c\psi,_x^* - zw,_{xx}^*, \varepsilon_{xz}^{(2)*} = 0(-c - h_2 \le z \le -c). \tag{4.16}$$

For stresses and strains (4.15), (4.16), marked with asterisks, we assume physical equations of state of the type (4.1):

$$s_x^{(k)*} = 2G_k f^{(k)*}\, э_x^{(k)*},\ \sigma^{(k)*} = 3K_k \varepsilon^{(k)*} (k = 1, 2, 3),$$

$$s_{xz}^{(3)*} = 2G_3 f^{(3)*}\, э_{xz}^{(3)*} \tag{4.17}$$

According to Moskvitin's hypothesis [1], we assume the corresponding universal plasticity functions $f^{(k)*}$ in bearing layers to be expressed through the plasticity functions under loading from the natural state:

$$f^{(k)*}(\varepsilon_u^{(k)}) = \begin{cases} 1,\ \varepsilon_u^{(k)*} \le \varepsilon_y^{(k)*}, \\ 1 - \omega^{(k)*},\ \varepsilon_u^{(k)*} > \varepsilon_y^{(k)*}, \end{cases} \quad \omega^{(k)*} = \omega\prime^{(k)}(\varepsilon_u^{(k)*}, \varepsilon_y^{(k)*}, d_m^*). \tag{4.18}$$

Here d_m^*—experimental parameters of the material included in the approximation formula of the nonlinearity function (4.18). For example, for alloy D16T by analogy with (1.38) we take:

$$\omega^{(k)*} = \begin{cases} 0,\ \varepsilon_u^{(k)*} \le \varepsilon_y^{(k)*}, \\ A_k^*\left(1 - \frac{\varepsilon_y^{(k)*}}{\varepsilon_u^{(k)*}}\right)^{\alpha_k^*},\ \varepsilon_u^{(k)*} > \varepsilon_y^{(k)*}, \end{cases} \tag{4.19}$$

where $A_k^* = 0.924$, $\alpha_k^* = 2.27$, $\varepsilon_y^{(k)*} = \alpha_2^{(k)} \varepsilon_y^{(k)}(T_1)$, $\alpha_2^{(k)} = 2.02$; the case $\alpha_2^{(k)} > 2$ corresponds to cyclically strengthening material, $\alpha_2^{(k)} < 2$—to unstrengthening material, $\alpha_2^{(k)} = 2$—for cyclically ideal material (the lower index "2" denotes the second half-cycle).

Stress components with asterisks, using the physical relations (4.17), are represented as the difference of elastic and nonlinear components

$$\sigma_x^{(k)*} = \sigma_{xe}^{(k)*} - \sigma_{x\omega}^{(k)*},\ \sigma_{xz}^{(3)*} = \sigma_{xze}^{(3)*} - \sigma_{xz\omega}^{(3)*}, \tag{4.20}$$

where

$$\sigma_{xe}^{(k)*} = 2G_k\, э_x^{(k)*} + 3K_k \varepsilon^{(k)*},\ \sigma_{x\omega}^{(k)*} = 2G_k\, э_x^{(k)*}\, \omega^{(k)*}$$

$$\sigma_{xze}^{(3)*} = 2G_3\, э_{xz}^{(3)*},\ \sigma_{xz\omega}^{(k)*} = 2G_3\, э_{xz}^{(3)*}\, \omega^{(3)*}.$$

Carrying out a similar operation with the values of the type of internal forces (2.4), we obtain:

$$N^{(k)*} = N_e^{(k)*} - N_\omega^{(k)*},\ M^{(k)*} = M_e^{(k)*} - M_\omega^{(k)*},\ Q^* = Q_e^* - Q_\omega^*, \tag{4.21}$$

The elastic (index "e") and inelastic (index "ω") components in expressions (4.21) are calculated through the corresponding components of the stress tensor (4.20) by the formulas:

$$N_e^{(k)*} = b_0 \int_{h_k} \sigma_{xe}^{(k)*} dz,\ N_\omega^{(k)*} = b_0 \int_{h_k} \sigma_{x\omega}^{(k)*} dz,\ M_e^{(k)*} = b_0 \int_{h_k} \sigma_{xe}^{(k)*} z dz,$$

$$M_\omega^{(k)*} = b_0 \int_{h_k} \sigma_{x\omega}^{(k)*} z dz,\ Q_e^* = b_0 \int_{-c}^{c} \sigma_{xze}^{(3)*} dz,\ Q_\omega^* = b_0 \int_{-c}^{c} \sigma_{xz\omega}^{(3)*} dz. \tag{4.22}$$

The iteration equations of equilibrium (4.9) for the values with asterisks are written in the form

$$a_1 u,_{xx}^{(n)*} + a_6 \psi,_{xx}^{(n)*} - a_7 w,_{xxx}^{(n)*} = -p^* + p_\omega^{(n-1)*},$$

$$a_6 u,_{xx}^{(n)*} + a_2 \psi,_{xx}^{(n)*} - a_3 w,_{xxx}^{(n)*} - a_5 \psi^{(n)*} = h_\omega^{(n-1)*},$$

$$a_7 u,_{xxx}^{(n)*} + a_3 \psi,_{xxx}^{(n)*} - a_4 w,_{xxxx}^{(n)*} = -q^* + q_\omega^{(n-1)*}. \tag{4.23}$$

The values $p_\omega^{(n-1)*}$, $h_\omega^{(n-1)*}$, $q_\omega^{(n-1)*}$ corresponding to inelastic components in (4.22), at the first step ($n = 1$) are assumed to be equal to zero, and further are calculated according to the results of the previous approximation by formulas like (4.10) and are called additional "external" loads. They are used as corrections for plasticity and physical nonlinearity of the layer materials under repeated loading:

$$p_\omega^{(n-1)*} = \frac{1}{b_0} N_\omega^{(n-1)*},_x,\ h_\omega^{(n-1)*} = \frac{1}{b_0}\left(H_\omega^{(n-1)*},_x - Q_\omega^{(n-1)*}\right),$$

$$q_\omega^{(n-1)*} = \frac{1}{b_0} M_\omega^{(n-1)*},_{xx},$$

$$N_\omega^{(n-1)*} = \sum_{k=1}^{3} N_\omega^{(k,\,n-1)*} = \frac{4}{3} b_0 \sum_{k=1}^{3} \int_{h_k} G_k \omega^{(k)*}(\varepsilon_u^{(k,\,n-1)*}) \varepsilon_x^{(k,\,n-1)*} dz,$$

$$M_\omega^{(n-1)*} = \sum_{k=1}^{3} M_\omega^{(k,\,n-1)*} = \frac{4}{3} b_0 \sum_{k=1}^{3} \int_{h_k} G_k \omega^{(k)*}(\varepsilon_u^{(k,\,n-1)*}) \varepsilon_x^{(k,\,n-1)*} z dz,$$

$$H_\omega^{(n-1)*} = c(N_\omega^{(1,\,n-1)*} - N_\omega^{(2,\,n-1)*}) + M_\omega^{(3,\,n-1)*},$$

$$Q_\omega^{(n-1)*} = 2b_0 \int_{-c}^{c} G_3 \omega^{(3)*}(\varepsilon_u^{(3,\,n-1)*}) \psi^{(n-1)*} dz. \tag{4.24}$$

Boundary conditions (2.24) at *cantilevered embedding of the left end of the rod and free right end* for the values with asterisks take the following form:

$$x = 0 : w^{(n)*} = w,_x^{(n)*} = u^{(n)*} = \psi^{(n)*} = 0,$$

$$x = l : N^{(n)*} = M,_x^{(n)*} = M^{(n)*} = \psi^{(n)*} = 0. \tag{4.25}$$

For a rod *hinged at both ends*, the boundary conditions (2.26) are as follows

$$x = 0, l : w^{(n)*} = M^{(n)*} = u^{(n)*} = \psi^{(n)*} = 0 \tag{4.26}$$

In the case of *rigid embedding of both ends of the rod*, the boundary conditions of the type (2.28) must be fulfilled:

$$x = 0, l : w^{(n)*} = u^{(n)*} = \psi^{(n)*} = w,_x^{(n)*} = 0 \tag{4.27}$$

Thus, the system of differential equations for the quantities with asterisks (4.23), (4.24) (with accuracy of notation) coincides with the corresponding system (4.9), (4.10) when the rod is loaded from the natural state. Therefore, its analytical solution will have the following recurrence form:

$$\begin{aligned}
\psi^{(n)*}(x) &= C_2^{(n)*} sh(\beta x) + C_3^{(n)*} ch(\beta x) \\
&+ \frac{1}{\beta}\left[sh(\beta x) \int g_\omega^{(n)*} ch(\beta x) \mathrm{d}x - ch(\beta x) \int g_\omega^{(n)*} sh(\beta x) \mathrm{d}x \right],
\end{aligned}$$

$$\begin{aligned}
u^{(n)*}(x) =& \gamma_3 \psi^{(n)*} + \frac{1}{\alpha_2}\left[-a_4 \mathrm{L}_2^{-1}\left(p^* - p_\omega^{(n-1)*}\right) \right. \\
& \left. + a_7 \mathrm{L}_3^{-1}\left(q - q_\omega^{(n-1)*}\right) \right] \\
& + \frac{a_7}{2\alpha_2} C_1^{(n)*} x^2 + C_7^{(n)*} x + C_8^{(n)*},
\end{aligned}$$

$$\begin{aligned}
w^{(n)*}(x) =& \frac{1}{\alpha_2}\left[\alpha_1 \int \psi^{(n)*} \mathrm{d}x - a_7 \mathrm{L}_3^{-1}\left(p^* - p_\omega^{(n-1)*}\right) \right. \\
& \left. + a_1 \mathrm{L}_4^{-1}\left(q^* - q_\omega^{(n-1)*}\right) + \frac{1}{6} a_1 C_1^{(n)*} x^3 \right] \\
& + \frac{1}{2} C_4^{(n)*} x^2 + C_5^{(n)*} x + C_6^{(n)*}.
\end{aligned} \tag{4.28}$$

Here the coefficients $\alpha_1, \alpha_2, \alpha_3, \beta^2, \gamma_1, \gamma_2, \gamma_3$ and linear integral operators $\mathrm{L}_2^{-1}, \mathrm{L}_3^{-1}, \mathrm{L}_4^{-1}$ are defined by the formulas (2.19), (2.23); integration constants $C_1^{(n)*}, \ldots, C_8^{(n)*}$ at each step of the approximation follow from the conditions of rod fixation; integrand function is:

$$
\begin{aligned}
g_{\omega}^{(n)*}(x) =& \frac{\alpha_2}{\alpha_1}\gamma_1 h_{\omega}^{(n-1)*} + \gamma_2(p^* - p_{\omega}^{(n-1)*}) \\
&+ \gamma_1\left(\int (q^* - q_{\omega}^{(n-1)*})\mathrm{d}x + C_1^{(n)*}\right).
\end{aligned}
$$

Satisfying by the solution (4.28) the boundary conditions for the cantilever (4.25), we obtain:

$$
C_1^{(n)*} = -\mathrm{L}_1^{-1}(q^* - q_{\omega}^{(n-1)*})|_{x=l}\,,
$$

$$
\begin{aligned}
C_2^{(n)*} =& \frac{1}{\beta}\left[\frac{ch(\beta l)}{sh(\beta l)}\left(\int g_{\omega}^{(n)*} sh(\beta x)\mathrm{d}x|_{x=l} - \int g_{\omega}^{(n)*} sh(\beta x)\mathrm{d}x|_{x=0}\right)\right. \\
&\left. - \int g_{\omega}^{(n)*} ch(\beta x)\mathrm{d}x|_{x=l}\right],
\end{aligned}
$$

$$
C_3^{(n)*} = \frac{1}{\beta}\int g_{\omega}^{(n)*}(x) sh(\beta x)\mathrm{d}x|_{x=0}\,,
$$

$$
\begin{aligned}
C_4^{(n)*} =& \frac{a_7}{\alpha_2}\mathrm{L}_1^{-1}(p^* - p_{\omega}^{(n-1)*})|_{x=l} \\
&- \frac{a_1}{\alpha_2}\mathrm{L}_2^{-1}(q^* - q_{\omega}^{(n-1)*})|_{x=l} - \frac{a_1}{\alpha_2}C_1^{(n)*}l,
\end{aligned}
$$

$$
C_5^{(n)*} = \frac{a_7}{\alpha_2}\mathrm{L}_2^{-1}(p^* - p_{\omega}^{(n-1)*})|_{x=0} - \frac{a_1}{\alpha_2}\mathrm{L}_3^{-1}(q^* - q_{\omega}^{(n-1)*})|_{x=0}\,,
$$

$$
\begin{aligned}
C_6^{(n)*} =& \frac{a_7}{\alpha_2}\mathrm{L}_3^{-1}(p^* - p_{\omega}^{(n-1)*})\Big|_{x=0} \\
&- \frac{a_1}{\alpha_2}\mathrm{L}_4^{-1}(q^* - q_{\omega}^{(n-1)*})\Big|_{x=0} - \frac{\alpha_1}{\alpha_2}\int \psi^{(n)*}\mathrm{d}x\Big|_{x=0}\,,
\end{aligned}
$$

$$
\begin{aligned}
C_7^{(n)*} =& \frac{a_4}{\alpha_2}\mathrm{L}_1^{-1}(p^* - p_{\omega}^{(n-1)*})|_{x=l} \\
&- \frac{a_7}{\alpha_2}\mathrm{L}_2^{-1}(q^* - p_{\omega}^{(n-1)*})|_{x=l} - \frac{a_7}{\alpha_2}C_1^{(n)*}l,
\end{aligned}
$$

$$
C_8^{(n)*} = \frac{a_4}{\alpha_2}\mathrm{L}_2^{-1}(p^* - p_{\omega}^{(n-1)*})|_{x=0} - \frac{a_7}{\alpha_2}\mathrm{L}_3^{-1}(q^* - q_{\omega}^{(n-1)*})|_{x=0}\,. \tag{4.29}
$$

For a hinged rod from the boundary conditions (4.26) we have:

$$
C_1^{(n)*} = \frac{1}{l}\left(\mathrm{L}_2^{-1}(q^* - q_{\omega}^{(n-1)*})|_{x=0} - \mathrm{L}_2^{-1}(q^* - q_{\omega}^{(n-1)*})|_{x=l}\right),
$$

$$
C_4^{(n)*} = \frac{a_7}{a_4}\left(D_1^{(n)*}|_{x=0} - D_1^{(n)*}|_{x=l}\right) - \frac{1}{a_4}\left(\mathrm{L}_2^{-1}(q^* - q_{\omega}^{(n-1)*})|_{x=0}\right),
$$

$$
\begin{aligned}
C_5^{(n)*} =& D_2^{(n)*}|_{x=0} - D_2^{(n)*}|_{x=l} - \frac{a_7 l}{2a_4}\left(D_1^{(n)*}|_{x=0} - D_1^{(n)*}|_{x=l}\right) \\
&- \frac{l}{2a_4}\left(\mathrm{L}_2^{-1}(q^* - q_\omega^{(n-1)*})|_{x=0}\right),\ C_7^{(n)*} = D_1^{(n)*}|_{x=0} - D_1^{(n)*}|_{x=l}, \qquad (4.30)
\end{aligned}
$$

where

$$
\begin{aligned}
D_1^{(n)*} &= \frac{a_7}{l\alpha_2}\mathrm{L}_3^{-1}(q^* - q_\omega^{(n-1)*}) \\
&- \frac{a_7 l}{2\alpha_2}\mathrm{L}_2^{-1}(q^* - q_\omega^{(n-1)*}) - \frac{a_4}{l\alpha_2}\mathrm{L}_2^{-1}(p^* - p_\omega^{(n-1)*}), \\
D_2^{(n)*} &= \frac{a_1}{l\alpha_2}\left(\int \psi^{(n)*}\mathrm{d}x + \mathrm{L}_4^{-1}(q^* - q_\omega^{(n-1)*})\right) \\
&- \frac{a_1 l}{6\alpha_2}\mathrm{L}_2^{-1}(q^* - q_\omega^{(n-1)*}) - \frac{a_7}{l\alpha_2}\mathrm{L}_3^{-1}(p^* - p_\omega^{(n-1)*}).
\end{aligned}
$$

The constants $C_2^{(n)*}$, $C_3^{(n)*}$, $C_6^{(n)*}$, $C_8^{(n)*}$ retain the form (4.29).

In the case of *rigid embedding of both ends* from (4.27) we obtain:

$$
\begin{aligned}
C_4^{(n)*} =& \frac{1}{\alpha_2 l}\Big[a_7\mathrm{L}_2^{-1}(p^* - p_\omega^{(n-1)*})\big|_{x=l} - a_1\mathrm{L}_3^{-1}(q^* - q_\omega^{(n-1)*})\big|_{x=l} \\
&- \frac{a_1 C_1^{(n)*} l^2}{2}\Big] - \frac{C_5^{(n)*}}{l}, \\
C_7^{(n)*} =& \frac{1}{\alpha_2 l}\Big[a_4\mathrm{L}_2^{-1}(p^* - p_\omega^{(n-1)*})\big|_{x=l} - a_7\mathrm{L}_3^{-1}(q^* - q_\omega^{(n-1)*})\big|_{x=l} \\
&- \frac{a_7 C_1^{(n)*} l^2}{2}\Big] - \frac{C_8^{(n)*}}{l}, \\
D_3^{(n)} =& a_1\int \psi^{(n)*}\mathrm{d}x + a_1\mathrm{L}_4^{-1}(q^* - q_\omega^{(n-1)*}) \\
&- a_7\mathrm{L}_3^{-1}(p^* - p_\omega^{(n-1)*}) - \frac{l a_1}{2}\mathrm{L}_3^{-1}(q^* - q_\omega^{(n-1)*}) \\
&+ \frac{l a_7}{2}\mathrm{L}_2^{-1}(p^* - p_\omega^{(n-1)*}). \qquad (4.31)
\end{aligned}
$$

The constant $C_1^{(n)}$ is the same as for the hinge support (4.30); $C_2^{(n)}$, $C_3^{(n)}$, $C_5^{(n)} C_6^{(n)}$, $C_8^{(n)}$ retain the form (4.29).

Thus, Eqs. (4.28)–(4.31) describe in iterations the analytical solution of the boundary value problem in displacements for values with asterisks under repeated alternating loading for a cantilevered, hinged and rigidly embedded at both ends of a three-layer rod.

The corresponding parameters of solving the problem during the second half-cycle are obtained from relations (4.15) taking into account solutions (4.11)–(4.14) and (4.28)–(4.30):

$$\sigma_x^{\prime\prime(k)}(x,z) = \sigma_x^{\prime(k)}(x,z) - \sigma_x^{(k)*}(x,z), \sigma_{xz}^{\prime\prime(3)}(x,z)$$
$$= \sigma_{xz}^{\prime(3)}(x,z) - \sigma_{xz}^{(3)*}(x,z),$$

$$\varepsilon_x^{\prime\prime(k)}(x,z) = \varepsilon_x^{\prime(k)}(x,z) - \varepsilon_x^{(k)*}(x,z), u''(x) = u'(x) - u^*(x),$$

$$\psi''(x) = \psi'(x) - \psi^*(x), w''(x) = w'(x) - w^*(x), \tag{4.32}$$

where values with one stroke are stresses, strains and displacements in the rod before unloading.

Thus, the obtained result (4.32) and the procedure (4.15)–(4.30) can be regarded as an extension of the theory of Moskvitin on repeated alternating isothermal loading of elastoplastic bodies in case of three-layer rods.

4.3 Cyclic Deformation

In a completely analogous way we obtain the solution for any mth cyclic deformation by a continuous load of the considered three-layer elastic–plastic rod. Let, after (m–1) alternating half-cycles of loading, starting from the moment t_{m-1}, an instantaneous unloading and a new loading by forces of the opposite sign $p^{(m)}$, $q^{(m)}$ (the index "m" is the number of loading) is carried out. These loads will create in the kth layer of the rod a displacement field $u^{(m)}$, $\psi^{(m)}$, $w^{(m)}$, strains $\varepsilon_x^{(m,k)}$, $э_x^{(m,k)}$, $\varepsilon^{(m,k)}$ and stresses $\sigma_x^{(m,k)}$, $s_x^{(m,k)}$, $\sigma^{(m,k)}$, $\sigma_{xz}^{(m,3)}$. We will assume that the corresponding loading trajectories still belong to the class of simple trajectories (according to Ilyushin).

As before, we introduce differences for all values of stress–strain state and load, where values with index (m–1) are stresses, strains, displacements and loads in the rod before unloading, and values with index m denote similar parameters during the mth cycle:

$$\sigma_x^{(m,k)*} = (-1)^n(\sigma_x^{(m-1,k)} - \sigma_x^{(m,k)}), \varepsilon_x^{(m,k)*} = (-1)^n(\varepsilon_x^{(m-1,k)} - \varepsilon_x^{(m,k)}),$$

$$u^{(m)*} = (-1)^n(u^{(m-1)} - u^{(m)}), \psi^{(m)*} = (-1)^n(\psi^{(m-1)} - \psi^{(m)}),$$

$$w^{(m)*} = (-1)^n(w^{(m-1)} - w^{m}),$$

$$q^{(m)*} = (-1)^n(q^{(m-1)} - q^{(m)}), p^{(m)*} = (-1)^n(p^{(m-1)} - p^{(m)}). \tag{4.33}$$

The relations (4.16), which relate strains and displacements to the asterisks in (4.33), will be valid here as well.

For the stresses and strains marked with an asterisk, we assume physical equations of state of the type (4.17):

$$s_x^{(m,k)*} = 2G_k\, э_x^{(m,k)*} f^{(m,k)*}(\varepsilon_u^{(m,k)*}),\ \sigma^{(m,k)*} = 3K_k \varepsilon^{(m,k)*} (k = 1, 2, 3).$$

$$s_{xz}^{(m,3)*} = 2G_3\, э_{xz}^{(m,3)*} f^{(m,k)*}(\varepsilon_u^{(m,3)*}). \tag{4.34}$$

The corresponding universal nonlinearity functions in the bearing layers at the mth half-cycle *are assumed to be expressed through the plasticity functions under loading from the natural state*:

$$f^{(m,k)*}(\varepsilon_u^{(m,k)*}) = \begin{cases} 1, & \varepsilon_u^{(m,k)*} \le \varepsilon_y^{(m,k)*}, \\ 1 - \omega^{(m,k)*}(\varepsilon_u^{(m,k)*}), & \varepsilon_u^{(m,k)*} > \varepsilon_y^{(m,k)*}, \end{cases}$$

$$\omega^{(m,k)*} = \omega'^{(k)}(\varepsilon_u^{(m,k)*}, \varepsilon_y^{(m,k)*}, a_i^*), \tag{4.35}$$

where $\varepsilon_y^{(m,k)*}$ is value corresponding to the yield strength at the mth half-cycle, a_i^* are experimental material parameters included in the approximation formula of the type (4.19).

The physical nonlinearity of the filler under mth alternating loading is still described by relations (4.1).

In the components of stresses and strains with asterisks, as before, let us identify the nonlinear components using the physical relations (4.34):

$$\sigma_x^{(m,k)*} = \sigma_{xe}^{(m,k)*} - \sigma_{x\omega}^{(m,k)*},\ \sigma_{xz}^{(m,3)*} = \sigma_{xze}^{(m,3)*} - \sigma_{xz\omega}^{(m,k)*}, \tag{4.36}$$

where

$$\sigma_{xe}^{(m,k)*} = 2G_k\, э_x^{(m,k)*} + 3K_k \varepsilon^{(m,k)*},\ \sigma_{x\omega}^{(m,k)*} = 2G_k\, э_x^{(m,k)*}\, \omega^{(m,k)*},$$

$$\sigma_{xze}^{(m,3)*} = 2G_3\, э_{xz}^{(m,3)*},\ \sigma_{xz\omega}^{(m,3)*} = 2G_3\, э_{xz}^{(m,k)*}\, \omega^{(m,3)*}.$$

Here the nonlinearity function is calculated by formula (1.38).

Carrying out a similar operation with the values of the type of internal forces (2.4), we obtain:

$$N^{(m,k)*} = N_e^{(m,k)*} - N_\omega^{(m,k)*},\ M^{(m,k)*} = M_e^{(m,k)*} - M_\omega^{(m,k)*},$$

$$Q^{(m,3)*} = Q_e^{(m,3)*} - Q_\omega^{(m,3)*}. \tag{4.37}$$

The elastic (index "e") and inelastic (index "ω") components in (4.37) are calculated through the values of (4.36) by the formulas:

$$N_e^{(m,k)*} = b_0 \int \sigma_{xe}^{(m,k)*} dz,\ N_\omega^{(m,k)*} = b_0 \int_{h_k} \sigma_{x\omega}^{(m,k)*} dz,$$

$$M_e^{(m,k)*} = b_0 \int_{h_k} \sigma_{xe}^{(m,k)*} z dz,\ M_\omega^{(m,k)*} = b_0 \int_{h_k} \sigma_{x\omega}^{(m,k)*} z dz,$$

$$Q_e^{(m,3)*} = b_0 \int_{-c}^{c} \sigma_{xze}^{(m,3)*} dz,\ Q_\omega^{(m,k)*} = b_0 \int_{-c}^{c} \sigma_{xz\omega}^{(m,3)*} dz. \tag{4.38}$$

The corresponding iterative equilibrium Eq. (4.23) at the nth step for the values with asterisks at the mth half-cycle will be:

$$a_1 u,_{xx}^{(m,n)*} + a_6 \psi,_{xx}^{(m,n)*} - a_7 w,_{xxx}^{(m,n)*} = -p^{(m)*} + p_\omega^{(m,n-1)*},$$

$$a_6 u,_{xx}^{(m,n)*} + a_2 \psi,_{xx}^{(m,n)*} - a_3 w,_{xxx}^{(m,n)*} - a_5 \psi^{(m,n)*} = h_\omega^{(m,n-1)*},$$

$$a_7 u,_{xxx}^{(m,n)*} + a_3 \psi,_{xxx}^{(m,n)*} - a_4 w,_{xxxx}^{(m,n)*} = -q^{(m)*} + q_\omega^{(m,n-1)*}. \tag{4.39}$$

The values $p_\omega^{(m,n-1)*}$, $h_\omega^{(m,n-1)*}$, $q_\omega^{(m,n-1)*}$, corresponding to inelastic components in (4.38), at the first step ($n = 1$) are assumed to be equal to zero, and further they are calculated according to the results of the previous approximation and are called additional "external" loads:

$$p_\omega^{(m,n-1)*} = \frac{1}{b_0} N_\omega^{(m,n-1)*},_x;\ q_\omega^{(m,n-1)*} = \frac{1}{b_0} M_\omega^{(m,n-1)*},_{xx},$$

$$h_\omega^{(m,n-1)*} = \frac{1}{b_0}(H_\omega^{(m,n-1)*},_x - Q_\omega^{(m,n-1)*}),$$

$$N_\omega^{(m,n-1)*} = \sum_{k=1}^{3} N_\omega^{(m,n-1,k)*} = \frac{4}{3} b_0 \sum_{k=1}^{3} \int_{h_k} G_k \omega^{(k)*}(\varepsilon_u^{(m,n-1,k)*}) \varepsilon_x^{(m,n-1,k)*} dz,$$

$$M_\omega^{(m,n-1)*} = \sum_{k=1}^{3} M_\omega^{(m,n-1,k)*} = \frac{4}{3} b_0 \sum_{k=1}^{3} \int_{h_k} G_k \omega^{(k)*}(\varepsilon_u^{(m,n-1,k)*}) \varepsilon_x^{(m,n-1,k)*} z dz,$$

$$H_\omega^{(m,n-1)*} = c(N_\omega^{(m,n-1,1)*} - N_\omega^{(m,n-1,2)*}) + M_\omega^{(m,n-1,3)*},$$

$$Q_{\omega}^{(m,n-1)*} = 2b_0 \int_{-c}^{c} G_3 \omega^{(3)*}(\varepsilon_u^{(m,n-1,3)*}) \psi^{(m,n-1)*} dz. \tag{4.40}$$

The boundary conditions (4.25) at the *cantilever termination of the left end and free right end of the rod* for the values with asterisks have the form:

$$x = 0 : w^{(m,n)*} = w,_x^{(m,n)*} = u^{(m,n)*} = \psi^{(m,n)*} = 0,$$

$$x = l : N^{(m,n)*} = M,_x^{(m,n)*} = M^{(m,n)*} = \psi^{(m,n)*} = 0. \tag{4.41}$$

For a *rod hinged at both ends*, the boundary conditions (4.26) take the form:

$$x = 0, l : w^{(m,n)*} = M^{(m,n)*} = u^{(m,n)*} = \psi^{(m,n)*} = 0 \tag{4.42}$$

In case of *rigid embedding of both ends of the rod*, the boundary conditions of the type (4.27) must be fulfilled:

$$x = 0, l : w^{(m,n)*} = u^{(m,n)*} = \psi^{(m,n)*} = w,_x^{(m,n)*} = 0. \tag{4.43}$$

Thus, the boundary value problem for quantities with asterisks (4.35), (4.39)–(4.43) coincides with the boundary value problem (4.18), (4.23)–(4.26) with exactness of notation. Therefore its recurrence analytical solution will be:

$$\begin{aligned}
\psi^{(m,n)*}(x) &= C_2^{(m,n)*} sh(\beta x) + C_3^{(m,n)*} ch(\beta x) \\
&+ \frac{1}{\beta}\left[sh(\beta x) \int g_{\omega}^{(m,\,n)*} ch(\beta x) dx - ch(\beta x) \int g_{\omega}^{(m,\,n)*} sh(\beta x) dx \right],
\end{aligned}$$

$$\begin{aligned}
u^{(m,n)*}(x) &= \frac{1}{\alpha_2}\left[-a_4 L_2^{-1}\left(p^{(m)*} - p_{\omega}^{(m,n)*}\right) + a_7 L_3^{-1}\left(q^{(m)*} - q_{\omega}^{(m,n)*}\right)\right] \\
&+ \frac{a_7}{2\alpha_2} C_1^{(m,n)*} x^2 + C_7^{(m,n)*} x + C_8^{(m,n)*} + \gamma_3 \psi^{(m,n)*},
\end{aligned}$$

$$\begin{aligned}
w^{(m,n)*}(x) =& \frac{1}{\alpha_2}\left[\alpha_1 \int \psi^{(m,n)*} dx - a_7 L_3^{-1}\left(p^{(m)*} - p_{\omega}^{(m,n)*}\right) \right. \\
& \left. + a_1 L_4^{-1}\left(q^{(m)*} - q_{\omega}^{(m,n)*}\right) + \frac{1}{6} a_1 C_1^{(m,n)*} x^3 \right] \\
& + \frac{1}{2} C_4^{(m,n)*} x^2 + C_5^{(m,n)*} x + C_6^{(m,n)*}.
\end{aligned} \tag{4.44}$$

In (4.44) we introduce the function

$$g_{\omega}^{(m,\,n)*}(x) = \frac{\alpha_2}{\alpha_1}\gamma_1 h_{\omega}^{(m,n-1)*} + \gamma_2(p^{(m)*} - p_{\omega}^{(m,n-1)*}) + \gamma_1\left(\int (q^{(m)*} - q_{\omega}^{(m,n-1)*})\mathrm{d}x + C_1^{(m,n)*}\right); \quad (4.45)$$

coefficients $\alpha_1, \alpha_2, \alpha_3, \beta^2, \gamma_1, \gamma_2, \gamma_3$ and linear integral operators $\mathrm{L}_2^{-1}, L_3^{-1}, L_4^{-1}$ are defined by formulas (2.19), (2.23); integration constants $C_1^{(m,n)*}, \ldots, C_8^{(m,n)*}$ follow from boundary conditions.

Satisfying by the solution (4.44) the boundary conditions for the *cantilever* (4.41), we obtain:

$$C_1^{(m,n)*} = -\mathrm{L}_1^{-1}(q^{(m)*} - q_{\omega}^{(m,n-1)*})|_{x=l},$$

$$C_2^{(m,n)*} = \frac{1}{\beta}\left[\frac{ch(\beta l)}{sh(\beta l)}\left(\int g_{\omega}^{(m,n)*}sh(\beta x)\mathrm{d}x|_{x=l} - \int g_{\omega}^{(m,n)*}sh(\beta x)\mathrm{d}x|_{x=0}\right) - \int g_{\omega}^{(m,n)*}ch(\beta x)\mathrm{d}x|_{x=l}\right],$$

$$C_3^{(m,n)*} = \frac{1}{\beta}\int g^{(m,n)*}(x)sh(\beta x)\mathrm{d}x|_{x=0},$$

$$C_4^{(m,n)*} = \frac{a_7}{\alpha_2}\mathrm{L}_1^{-1}(p^{(m)*} - p_{\omega}^{(m,n-1)*})|_{x=l} - \frac{a_1}{\alpha_2}\mathrm{L}_2^{-1}(q^{(m)*} - q_{\omega}^{(m,n-1)*})|_{x=l} - \frac{a_1}{\alpha_2}C_1^{(m,n)*}l,$$

$$C_5^{(m,n)*} = \frac{a_7}{\alpha_2}\mathrm{L}_2^{-1}(p^{(m)*} - p_{\omega}^{(n-1)*})|_{x=0} - \frac{a_1}{\alpha_2}\mathrm{L}_3^{-1}(q^{(m)*} - q_{\omega}^{(m,n-1)*})|_{x=0},$$

$$C_6^{(m,n)*} = \frac{a_7}{\alpha_2}\mathrm{L}_3^{-1}(p^{(m)*} - p_{\omega}^{(m,n-1)*})\Big|_{x=0} - \frac{a_1}{\alpha_2}\mathrm{L}_4^{-1}(q^{(m)*} - q_{\omega}^{(m,n-1)*})\Big|_{x=0} - \frac{\alpha_1}{\alpha_2}\int \psi^{(m,n)*}\mathrm{d}x\Big|_{x=0},$$

$$C_7^{(m,n)*} = \frac{a_4}{\alpha_2}\mathrm{L}_1^{-1}(p^{(m)*} - p_{\omega}^{(m,n-1)*})|_{x=l} - \frac{a_7}{\alpha_2}\mathrm{L}_2^{-1}(q^{(m)*} - p_{\omega}^{(m,n-1)*})|_{x=l} - \frac{a_7 l}{\alpha_2}C_1^{(m,n)*},$$

$$\begin{aligned} C_8^{(m,n)*} =& \frac{a_4}{\alpha_2} \mathrm{L}_2^{-1} (p^{(m)*} - p_\omega^{(m,n-1)*})|_{x=0} \\ &- \frac{a_7}{\alpha_2} \mathrm{L}_3^{-1} (q^{(m)*} - q_\omega^{(m,n-1)*})|_{x=0} \,. \end{aligned} \tag{4.46}$$

For a *hinged rod* from the conditions (4.42) we have:

$$C_1^{(m,n)*} = \frac{1}{l} \big(\mathrm{L}_2^{-1} (q^{(m)*} - q_\omega^{(m,n-1)*})|_{x=0} - \mathrm{L}_2^{-1} (q^{(m)*} - q_\omega^{(m,n-1)*})|_{x=l} \big),$$

$$C_4^{(m,n)*} = \frac{a_7}{a_4} \Big(D_1^{(m,n)*}|_{x=0} - D_1^{(m,n)*}|_{x=l} \Big) - \frac{1}{a_4} \big(\mathrm{L}_2^{-1} (q^{(m)*} - q_\omega^{(m,n-1)*})_{|x=0} \big),$$

$$\begin{aligned} C_5^{(m,n)*} =& D_2^{(m,n)*}|_{x=0} - D_2^{(m,n)*}|_{x=l} \\ &- \frac{a_7 l}{2a_4} \Big(D_1^{(m,n)*}|_{x=0} - D_1^{(m,n)*}|_{x=l} \Big) \\ &- \frac{l}{2a_4} \big(\mathrm{L}_2^{-1} (q^{(m)*} - q_\omega^{(m,n-1)*})|_{x=0} \big), \\ C_7^{(m,n)*} =& D_1^{(m,n)*}|_{x=0} - D_1^{(m,n)*}|_{x=l} \,, \end{aligned}$$

$$\begin{aligned} D_1^{(m,n)*} =& \frac{a_7}{l\alpha_2} \mathrm{L}_3^{-1} (q^{(m)*} - q_\omega^{(m,n-1)*}) \\ &- \frac{a_7 l}{2\alpha_2} \mathrm{L}_2^{-1} (q^{(m)*} - q_\omega^{(m,n-1)*}) \\ &- \frac{a_4}{l\alpha_2} \mathrm{L}_2^{-1} (p^{(m)*} - p_\omega^{(m,n-1)*}), \end{aligned}$$

$$\begin{aligned} D_2^{(m,n)*} =& \frac{a_1}{l\alpha_2} \Big(\int \psi^{(m,n)*} \mathrm{d}x + \mathrm{L}_4^{-1} (q^{(m)*} - q_\omega^{(m,n-1)*}) \Big) \\ &- \frac{a_1 l}{6\alpha_2} \mathrm{L}_2^{-1} (q^{(m)*} - q_\omega^{(m,n-1)*}) \\ &- \frac{a_7}{l\alpha_2} \mathrm{L}_3^{-1} (p^{(m)*} - p_\omega^{(m,n-1)*}). \end{aligned} \tag{4.47}$$

The constants $C_2^{(m,\,n)*}$, $C_3^{(m,\,n)*}$, $C_6^{(m,\,n)*}$, $C_8^{(m,\,n)*}$ retain the form (4.46).
In the case of *rigid embedding of both ends*, from (4.43) we obtain:

$$C_4^{(m,n)*} = \frac{1}{\alpha_2 l} \Big[a_7 \mathrm{L}_2^{-1} \big(p^{(m)*} - p_\omega^{(m,\,n-1)*} \big) \big|_{x=l}$$

$$\mathrm{L}_3^{-1} \big(q^{(m)*} - q_\omega^{(m,n-1)*} \big) \big|_{x=l} - a_1 - \frac{a_1 C_1^{(m,n)*} l^2}{2} \Big] - \frac{C_5^{(m,n)*}}{l},$$

$$C_7^{(m,n)*} = \frac{1}{\alpha_2 l}\Big[a_4 \mathrm{L}_2^{-1}\big(p^{(m)*} - p_\omega^{(m,\,n-1)*}\big)\big|_{x=l}$$
$$-a_7 L_3^{-1}\big(q^{(m)*} - q_\omega^{(m,n-1)*}\big)\big|_{x=l} - \frac{a_7 C_1^{(m,n)*} l^2}{2}\Big] - \frac{C_8^{(m,n)*}}{l},$$

$$\begin{aligned}D_3^{(m,n)*} =& a_1 \int \psi^{(m,n)*} \mathrm{d}x + a_1 \mathrm{L}_4^{-1}(q^{(m)*} - q_\omega^{(m,\,n-1)*})\\ &- a_7 \mathrm{L}_3^{-1}(p^{(m)*} - p_\omega^{(m,n-1)*})\\ &- \frac{l a_1}{2}\mathrm{L}_3^{-1}(q^{(m)*} - q_\omega^{(m,n-1)*})\\ &+ \frac{l a_7}{2}\mathrm{L}_2^{-1}(p^{(m)*} - p_\omega^{(m,n-1)*}).\end{aligned} \tag{4.48}$$

The constant $C_1^{(m,n)*}$ is the same as for the hinge support (4.47); $C_2^{(m,n)*}$, $C_3^{(m,n)*}$, $C_5^{(m,n)*}$, $C_6^{(m,n)*}$, $C_8^{(m,n)*}$ retain the form (4.46).

The required solution of the problem at the mth half-cycle is obtained from relations (4.33) taking into account solutions (4.11)–(4.14) and (4.44)–(4.48):

$$\sigma_x^{(m,k)}(x,z) = \sigma\prime_x^k(x,z) - \sum_{l=2}^{m}(-1)^k \sigma_x^{(l,k)*}(x,z),$$

$$\sigma_{xz}^{(m,3)}(x,z) = \sigma\prime_x^{(3)}(x,z) - \sum_{l=2}^{m}(-1)^k \sigma_x^{(l,3)*}(x,z),$$

$$\varepsilon_x^{(m,k)}(x,z) = \varepsilon\prime_{ij}^k(x,z) - \sum_{l=2}^{m}(-1)^k \varepsilon_x^{(l,k)*}(x,z),$$

$$u^{(m)}(x) = u'(x) - \sum_{l=2}^{m}(-1)^l u^{(l)*}(x),$$

$$\psi^{(m)}(x) = \psi'(x) - \sum_{l=2}^{m}(-1)^l \psi^{(l)*}(x),$$

$$w^{(m)}(x) = w'(x) - \sum_{l=2}^{m}(-1)^l w^{(l)*}(x), \tag{4.49}$$

where values with one stroke are stresses, strains and displacements in the rod before first unloading.

Numerical results on the deformation of a three-layer rod under variable loads are obtained under the assumptions introduced in Sect. 4.1.

Figure 4.6 shows the variation of deflection w and shear ψ along the axis of a three-layer rod (embedding-embedding) under alternating loading: *1*—elastic rod, *2*—elastoplastic rod, 3^X—elastoplastic rod at the 10th half-cycle. Curves with one stroke correspond to loading from the natural state, with two strokes—repeated bending by alternating loading ($q = 10$ MPa). The displacement extremums at repeated loading decrease by 1–2%, at the 10th half-cycle—by another 4%, which is explained by cyclic hardening of the material of bearing layers (see Table 1.1).

In practice, loads up to 10^5 cycles are considered to be low-cycle loads. Therefore, small changes in displacements from cycle to cycle accumulate and may subsequently lead to failure of the structural element.

The corresponding distribution of normal stresses σ_{xx} across the cross-section ($x = 0$) is illustrated in Fig. 4.7: *1*—elastic rod, 2—elastoplastic rod at the first half-cycle, 3—elastoplastic under reverse loading. Accounting for nonlinearity leads to an increase in the maximum stresses in the outer layers by 4%, in the filler—by 12%.

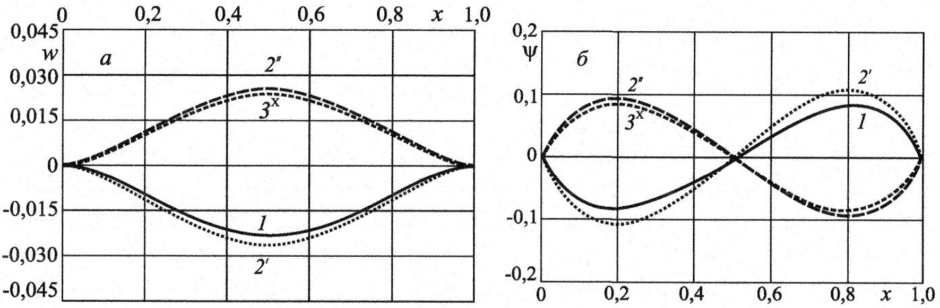

Fig. 4.6 Variation of deflection w (a) and shear ψ (b) along the axis (embedding-embedding) under alternating loading

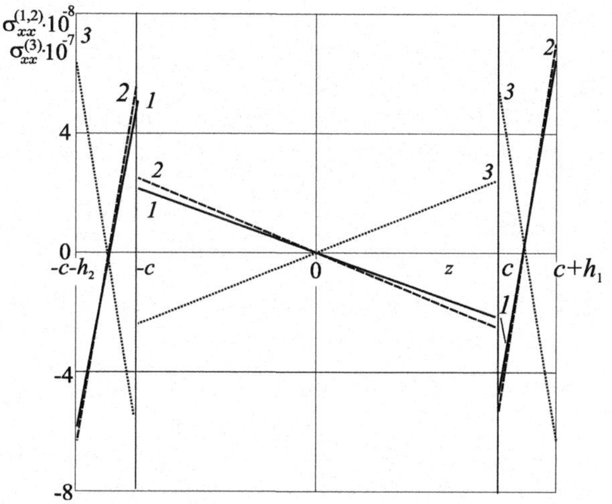

Fig. 4.7 Distribution of normal stresses σ_{xx} across the cross-section ($x = 0$) (embedding-embedding)

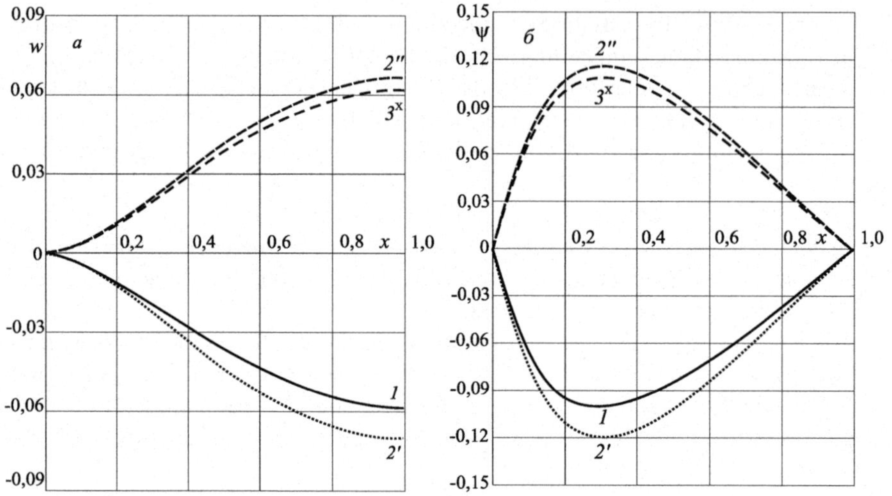

Fig. 4.8 Variation of deflection w (a) and shear ψ (b) along the axis (cantilever) under alternating loading

Figure 4.8 shows the variation of deflection w and shear ψ along the axis of a three-layer rod (cantilever clamping) under alternating loading: *1*—elastic rod, *2*—elastic–plastic, 3^X—elastic–plastic at the 10th half-cycle. As before, curves with one stroke correspond to loading from the natural state, with two stroke—repeated bending by alternating loading ($q = 2.5$ MPa).

When re-loaded, the displacement extremes decrease in approximately the same proportions as when the rod ends are embedded.

The distribution of normal stresses σ_{xx} across the cross-section in the cantilever rod embedment ($x = 0$) is illustrated in Fig. 4.9: *1*—elastic rod, *2*—elastic–plastic under loading from the natural state, *3*—elastic–plastic under reverse loading.

Accounting for nonlinearity leads to an increase in maximum stresses in the outer layers by 5%, in the filler—by 14%.

Thus, the solution (4.49) and the methodology (4.33)–(4.47) can be regarded as an extension of Moskvitin's theorem on multicycle alternating isothermal loading of elastoplastic bodies to three-layer rods.

4.4 Variable Deformation by Local Loads

Consider the deformation of a three-layer rod by a local rectangular load (see Fig. 2.13).

Loading from the natural state. Let the local surface load be uniformly distributed inside the segment [b, a] on the outer surface of a three-layer elastoplastic rod (see Fig. 2.13). In this case, the deformation of the rod will be described by a nonlinear

Fig. 4.9 Distribution of normal stresses σ_{xx} across the cross section (x = 0) (cantilever)

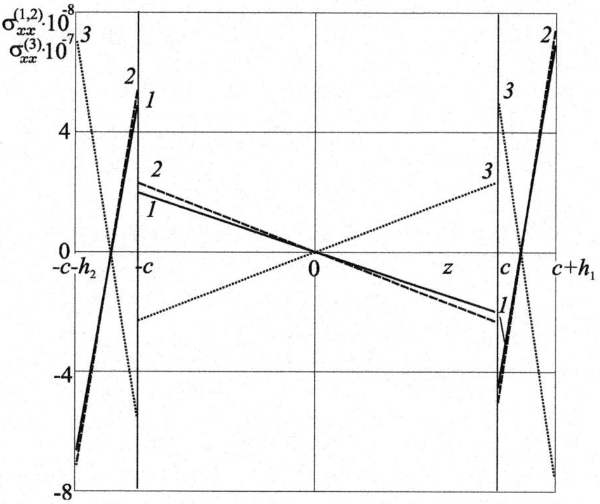

system of differential Eq. (4.9), in which the external loads are as follows:

$$q'(x) = q_0(H_0(a - x) - H_0(b - x)),$$

$$p'(x) = p_0(H_0(a - x) - H_0(b - x)). \tag{4.50}$$

At construction of analytical iteration solution of this boundary value problem we will use the solution (4.11). Substituting into it the load (4.50) and using the integrals from the Heaviside functions taken when obtaining the solution (2.32), we have the solution at the first half-cycle after some transformations (stroke above):

$$\begin{aligned}
\psi'^{(n)}(x) &= C_2'^{(n)} sh(\beta x) + C_3'^{(n)} ch(\beta x) \\
&+ \frac{1}{\beta^2}\big[\gamma_2 p_0((ch(\beta x - a\beta) - 1)H_0(a - x) \\
&- (ch(\beta x - b\beta) - 1)H_0(b - x)) \\
&- \frac{\gamma_1 q_0}{\beta}((\beta x - \beta a - \mathrm{sh}(\beta x - \beta a))H_0(a - x) \\
&-(\beta x - \beta b - sh(\beta x - \beta b))H_0(b - x)) - \gamma_1 C_1'^{(n)}\big] \\
&+ \frac{1}{\beta}\left[sh(\beta x)\int g_{\omega 1}'^{(n)} ch(\beta x)dx - ch(\beta x)\int g_{\omega 1}'^{(n)} sh(\beta x)dx\right],
\end{aligned}$$

$$u\prime^{(n)}(x) = \gamma_3\psi'^{(n)} + \frac{1}{\alpha_2}\Big[a_4L_2^{-1}\big(p_\omega'^{(n-1)}\big) - a_4(p_0(H_2(a-x) - H_2(b-x))$$
$$- a_7q_0(H_3(a-x) - H_3(b-x))$$
$$-L_3^{-1}\big(q_\omega'^{(n-1)}\big) + \frac{a_7}{2}C_1'^{(n)}x^2\Big] + C_7'^{(n)}x + C_8'^{(n)},$$

$$w\prime^{(n)}(x) = \frac{1}{\alpha_2}\Big[\alpha_1\int\psi'^{(n)}dx + a_7\big(p_0(H_3(a-x) - H_3(b-x)) + L_3^{-1}\big(p_\omega'^{(n-1)}\big)\big)$$
$$+ a_1\big(q_0(H_4(a-x) - H_4(b-x) - L_4^{-1}\big(q_\omega'^{(n-1)}\big)\big)$$
$$+\frac{1}{6}a_1C_1'^{(n)}x^3\Big] + \frac{1}{2}C_4'^{(n)}x^2 + C_5'^{(n)}x + C_6'^{(n)}, \qquad (4.51)$$

where coefficients $\alpha_1, \alpha_2, \alpha_3, \beta^2, \gamma_1, \gamma_2, \gamma_3$ and linear integral operators $L_2^{-1}, L_3^{-1}, L_4^{-1}$ are defined by formulas (2.19), (2.23); $h_\omega'^{(n-1)}$, $q_\omega'^{(n-1)}$, $q_\omega'^{(n-1)}$ are values calculated by formulas (4.10); integral of relative shear

$$\int\psi'^{(n)}dx\Big|_{x=0} = \frac{C_2^{(n)}}{\beta} + \frac{1}{\beta^2}\Big[\frac{\gamma_2 p_0}{\beta}(\mathrm{sh}(\beta b) - \mathrm{sh}(\beta a))$$
$$+\frac{\gamma_1 q_0}{\beta^2}((\mathrm{ch}(\beta a) - \mathrm{ch}(\beta b)) - \frac{1}{2}\gamma_1 q_0(a^2 - b^2)\Big]$$
$$+\frac{1}{\beta}\int sh(\beta x)\int g_{\omega 1}'^{(n)}ch(\beta x)dx - ch(\beta x)\int g_{\omega 1}'^{(n)}sh(\beta x)\mathrm{d}x\mathrm{d}x\Big|_{x=0}.$$

The integrand function differs from the one introduced in (4.11):

$$g_{\omega 1}'^{(n)}(x) = \frac{\alpha_2}{\alpha_1}\gamma_1 h_\omega'^{(n-1)} - \gamma_2 p_\omega'^{(n-1)} - \gamma_1\Big(\int q_\omega'^{(n-1)}dx + C_1^{(n)}\Big). \qquad (4.52)$$

Satisfying by the solution (4.51) at each iteration step *the boundary conditions for the left cantilever* (2.24), we obtain the integration constants $C\prime_1^{(n)}, \ldots, C\prime_8^{(n)}$: $C\prime_1^{(n)}, \ldots, C\prime_8^{(n)}$:

$$C\prime_1^{(n)} = L_1^{-1}(q_\omega^{(n-1)}|_{x=l},$$

$$C_2'^{(n)} = \frac{1}{\beta}\Big[\frac{ch(\beta l)}{sh(\beta l)}\Big(\int g_{\omega 1}'^{(n)}sh(\beta x)dx|_{x=l} - \int g_{\omega 1}'^{(n)}sh(\beta x)dx|_{x=0}$$
$$-\frac{\gamma_2 p_0}{\beta}(ch(\beta b) - ch(\beta a)) + \frac{\gamma_1 q_0}{\beta^2}(a\beta - b\beta + sh(\beta b) - sh(\beta a))\Big)$$
$$+\frac{\gamma_1 C_1'}{\beta}(1 - ch(\beta l))\Big) - \int g_{\omega 1}'^{(n)}ch(\beta x)\mathrm{d}x\Big|_{x=l} + \gamma_1 C_1'\frac{sh(\beta l)}{\beta}\Big],$$

$$C_3^{\prime(n)} = \frac{1}{\beta}\int g_{\omega 1}^{\prime(n)}(x)sh(\beta x)dx|_{x=0}$$
$$+\frac{1}{\beta^2}(\gamma_2 p_0(ch(\beta b) - ch(\beta a)),$$
$$-\frac{1}{\beta}\gamma_1 q_0(a\beta - b\beta + sh(\beta b) - sh(\beta a)) + \gamma_1 C_1\Big)$$

$$C_4^{(n)} = -\frac{a_7}{\alpha_2}L_1^{-1}(p_\omega^{\prime(n-1)})|_{x=l} + \frac{a_1}{\alpha_2}L_2^{-1}(q_\omega^{\prime(n-1)})|_{x=l} - \frac{a_1}{\alpha_2}C_1^{\prime(n)}l,$$

$$C_5^{\prime(n)} = \frac{1}{\alpha_2}\left(a_7\left(p_0\frac{a^2-b^2}{2} - L_2^{-1}(p_\omega^{\prime(n-1)})\right)|_{x=0}\right.$$
$$\left.+a_1\left(q_0\frac{a^3-b^3}{6} - L_3^{-1}(q_\omega^{\prime(n-1)})\right)|_{x=0}\right),$$

$$C\prime_6^{(n)} = \frac{1}{\alpha_2}\left(a_7\left(p_0\frac{b^3-a^3}{6} - L_3^{-1}(p_\omega^{\prime(n-1)})\right|_{x=0}\right)$$
$$-a_1\left(q_0\frac{a^4-b^4}{24} - L_4^{-1}(q_\omega^{\prime(n-1)})|_{x=0}\right) - \alpha_1\int \psi^{\prime(n)}dx\Big|_{x=0}\Big)$$

$$C\prime_7^{(n)} = -\frac{1}{\alpha_2}\left[a_4 L_1^{-1}(p\prime_\omega^{(n-1)})|_{x=l} - a_7(L_2^{-1}(q\prime_\omega^{(n-1)})|_{x=l} + C\prime_1^{(n)}l)\right], \tag{4.53}$$

$$C_8^{\prime(n)} = \frac{1}{\alpha_2}\left(a_4\left(p_0\frac{a^2-b^2}{2} - L_2^{-1}(p_\omega^{\prime(n-1)})|_{x=0}\right)\right.$$
$$\left.-a_7\left(q_0\frac{b^3-a^3}{6} - L_3^{-1}(q\prime_\omega^{(n-1)})|_{x=0}\right)\right)$$

Numerical results. When numerically realising the solution (4.51), the following values of relative layer thicknesses were assumed $h_1 = 0.02$, $h_2 = 0.04$, $c = 0.09$. The load with intensity $q_0 = -3.5$ MPa ($p_0 = 0$) is uniformly distributed on the segment $[0, b]$. Layer materials: D16T alloy, fluoroplastic-4.

Figures 4.10, 4.11 show the variation of shear in the filler and deflection along the axis of a three-layer rod as a function of the length of the section on which a rectangular distributed load is applied (without a streak—elastic rod, with a streak—elastic–plastic): *1*—$b = 0.50\ l$, *2*—$b = 0.75\ l$, *3*—$b = l$. At $b = 0.5\ l$ the areas of plastic deformation are small and the elastic–plastic deflection increases by 3–4%; at $b = 0.75\ l$ the increase is 6–7%, if the load acts over the entire surface of the rod ($b = l$), the increase is up to 15%. For shear, the corresponding increases are as follows: 5%, 9–10%; 18–19%. In the latter case ($b = l$), small decreases in deflection and shear are observed compared to the thickness symmetric rod (see Fig. 4.4).

Figures 4.12, 4.13 illustrate the distribution of areas of physical nonlinearity of the layer materials in the longitudinal section of the rod (gray fill) at $b = 0.75l$ and $b =$

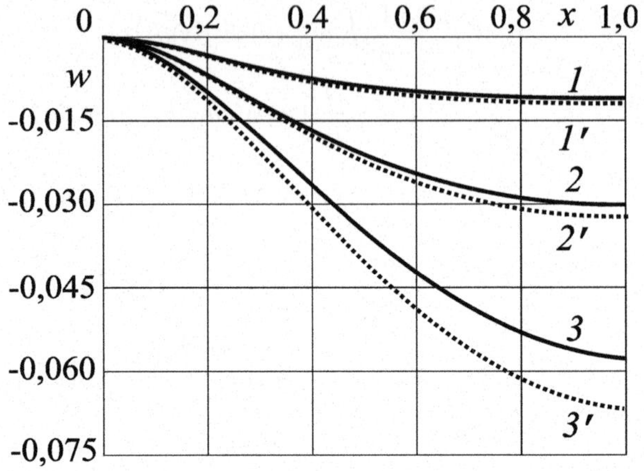

Fig. 4.10 Variation of shear along the axis of a three-layer rod as a function of the length of the loading section

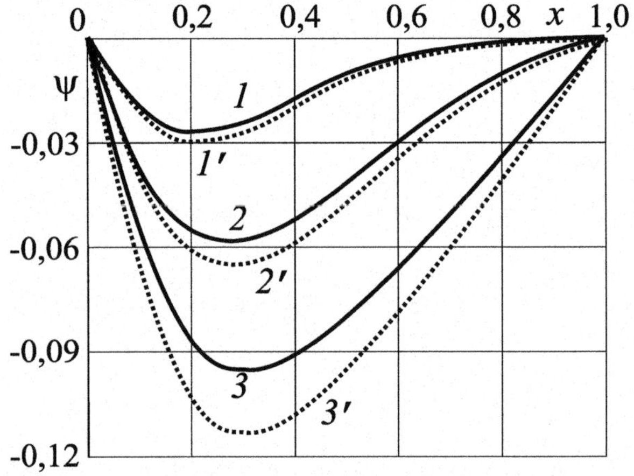

Fig. 4.11 Variation of deflection along the axis of a three-layer rod as a function of the length of the loading section

l, respectively. As one would expect, the areas of nonlinearity increase significantly in the second case.

Compared to a thickness symmetric rod (see Fig. 4.5), the nonlinearity areas decrease in the second layer and increase in the first layer.

Repeated alternating loading. Let, starting from the moment $t = t_1$, there is an instantaneous unloading and repeated loading with forces of opposite sign p'', q'', changing according to the same law as at loading from a natural state:

$$q''(x) = -q_0(H_0(a - x) - H_0(b - x)),$$

$$p''(x) = -p_0(H_0(a - x) - H_0(b - x)). \tag{4.54}$$

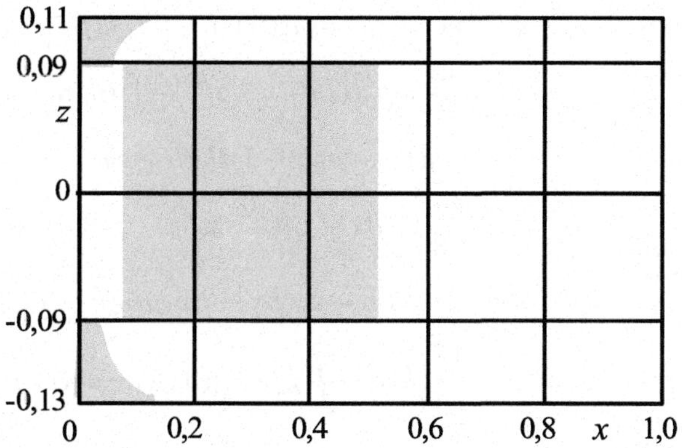

Fig. 4.12 Distribution of areas of physical nonlinearity of layer materials at $b = 0{,}75l$

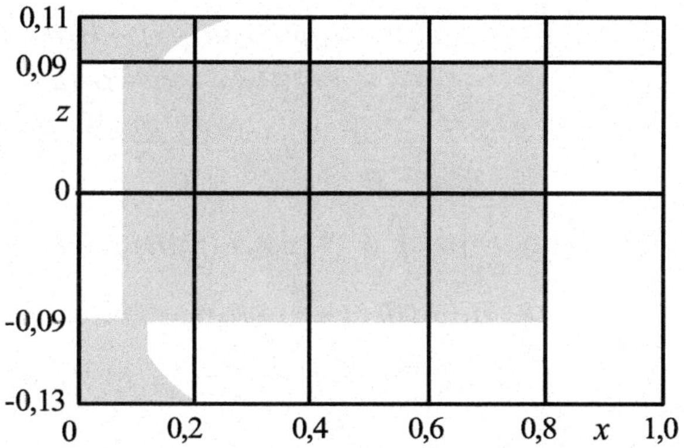

Fig. 4.13 Distribution of areas of physical nonlinearity of layer materials at $b = l$

These loads will create in the kth layer of the rod a field of displacements u'', ψ'', w'', strains $\varepsilon_x''^{(k)}$, $\varepsilon''^{(k)}$, ψ'' and stresses $\sigma_x''^{(k)}$, $s_x''^{(k)}$, $\sigma''^{(k)}$, $\sigma_{xz}''^{(3)}$. Further we proceed according to the scheme considered in Sect. 4.3.

Differences for the values of stress–strain state and load (4.15), in which the values with one stroke are stresses, strains, external loads and displacements in the rod before unloading, two streaks label similar parameters during the second half-cycle, will be the same. The relation between strains and displacements with asterisks retain the form (4.16). Let us take the nonlinear physical constitutive equations of state in the form (4.17). According to Moskvitin's hypothesis [1], we assume the universal plasticity functions $\omega^{(k)*}$ in bearing layers to be expressed through the plasticity functions under loading from the natural state—(4.18), (4.19).

As a result, the iteration equations of equilibrium (4.23) will be valid for the displacements with asterisks, taking into account the load (4.54) and boundary conditions (4.25)–(4.27). Therefore, when constructing the analytical iteration solution of the boundary value problem, we will use the solution (4.50). After some transformations we have:

$$\psi^{(n)*}(x) = C_2^{(n)*}sh(\beta x) + C_3^{(n)*}ch(\beta x)$$
$$+\frac{1}{\beta^2}\big[\gamma_2 p_0^*((ch(\beta x- a\beta) - 1)H_0(a-x)$$
$$-(ch(\beta x - b\beta) - 1)H_0(b-x))$$
$$-\frac{\gamma_1 q_0^*}{\beta}((\beta x - \beta a - \mathrm{sh}(\beta x - \beta a))H_0(a-x)$$
$$-(\beta x - \beta b - \mathrm{sh}(\beta x - \beta b))H_0(b-x)) - \gamma_1 C_1^{(n)*}\Big]$$
$$+\frac{1}{\beta}\left[sh(\beta x)\int g_{\omega 1}^{(n)*}ch(\beta x)\mathrm{d}x - ch(\beta x)\int g_{\omega 1}^{(n)*}sh(\beta x)\mathrm{d}x\right],$$

$$u^{(n)*}(x) = \gamma_3\psi^{(n)*} + \frac{1}{\alpha_2}\big[a_4 \mathrm{L}_2^{-1}(p_\omega^{(n-1)*})$$
$$-a_4(p_0^*(H_2(a-x) - H_2(b-x))$$
$$-a_7 q_0^*(H_3(a-x) - H_3(b-x))$$
$$-\mathrm{L}_3^{-1}(q_\omega^{(n-1)*}) + \frac{a_7}{2}C_1^{(n)*}x^2\Big] + C_7^{(n)*}x + C_8^{(n)*},$$

$$w^{(n)*}(x) = \frac{1}{\alpha_2}\Big[\alpha_1\int\psi^{(n)*}\mathrm{d}x + a_7(p_0^*(H_3(a-x) - H_3(b-x)) + \mathrm{L}_3^{-1}(p_\omega^{(n-1)*}))$$
$$+a_1(q_0^*(H_4(a-x) - H_4(b-x) - \mathrm{L}_4^{-1}(q_\omega^{(n-1)*}))$$
$$+\frac{1}{6}a_1 C_1^{(n)*}x^3\Big] + \frac{1}{2}C_4^{(n)*}x^2 + C_5^{(n)*}x + C_6^{(n)*}. \tag{4.55}$$

Here $h_\omega^{(n-1)*}$, $p_\omega^{(n-1)*}$, $q_\omega^{(n-1)*}$ are values calculated by the formulas like (4.10); The integrand function

$$g_{\omega 1}^{(n)*}(x) = \frac{\alpha_2}{\alpha_1}\gamma_1 h_\omega^{(n-1)*} - \gamma_2 p_\omega^{(n-1)*} - \gamma_1\left(\int q_\omega^{(n-1)*}\mathrm{d}x + C_1^{(n)*}\right), \tag{4.56}$$

shear integral with asterisk

$$\int\psi^{(n)*}\mathrm{d}x|_{x=0} = \frac{C_2^{(n)*}}{\beta} + \frac{1}{\beta^2}\left[\frac{\gamma_2 p_0^*}{\beta}(\mathrm{sh}(\beta b) - \mathrm{sh}(\beta a)\right.$$
$$\left.+\frac{\gamma_1 q_0^*}{\beta^2}((\mathrm{ch}(\beta a) - \mathrm{ch}(\beta b)) - \frac{1}{2}\gamma_1 q_0^*(a^2 - b^2)\right]$$
$$+\frac{1}{\beta}\int sh(\beta x)\int g_{\omega 1}^{(n)*}ch(\beta x)\mathrm{d}x - ch(\beta x)\int g_{\omega 1}^{(n)*}sh(\beta x)\mathrm{d}x\mathrm{d}x\bigg|_{x=0}.$$

Satisfying by the solution (4.55) the boundary conditions for the cantilever (4.25), we obtain the following integration constants:

$$C_1^{(n)*} = \mathrm{L}_1^{-1}(q_\omega^{(n-1)*})|_{x=l},$$

$$C_2^{(n)*} = \frac{1}{\beta}\Bigg[\frac{ch(\beta l)}{sh(\beta l)}\Bigg(\int g_{\omega 1}^{(n)*} sh(\beta x)\mathrm{d}x|_{x=l}$$
$$-\int g_{\omega 1}^{(n)*} sh(\beta x)\mathrm{d}x|_{x=0} - \frac{\gamma_2 p_0^*}{\beta}(ch(\beta b) - ch(\beta a))$$
$$+\frac{\gamma_1 q_0^*}{\beta^2}(a\beta - b\beta + sh(\beta b) - sh(\beta a)) + \frac{\gamma_1 C_1^{(n)*}}{\beta}(1 - ch(\beta l))\Bigg)$$
$$-\int g_{\omega 1}^{(n)*} ch(\beta x)\mathrm{d}x\Big|_{x=l} + \gamma_1 C_1^{(n)*}\frac{sh(\beta l)}{\beta}\Bigg],$$

$$C_3^{(n)*} = \frac{1}{\beta}\int g_{\omega 1}^{(n)*}(x) sh(\beta x)\mathrm{d}x|_{x=0}$$
$$+\frac{1}{\beta^2}(\gamma_2 p_0^*(ch(\beta b) - ch(\beta a))$$
$$-\frac{1}{\beta}\gamma_1 q_0^*(a\beta - b\beta + sh(\beta b) - sh(\beta a)), +\gamma_1 C_1^{(n)*}\Big)$$

$$C_4^{(n)*} = -\frac{a_7}{\alpha_2}\mathrm{L}_1^{-1}\Big(\Big(p_\omega^{(n-1)*})|_{x=l} + \frac{a_1}{\alpha_2}\mathrm{L}_2^{-1}(q_\omega^{(n-1)*})|_{x=l} - \frac{a_1}{\alpha_2}C_1^{(n)*}l,$$

$$C_5^{(n)*} = \frac{1}{\alpha_2}\Bigg(a_7\Bigg(p_0^*\frac{a^2 - b^2}{2} - \mathrm{L}_2^{-1}(p_\omega^{(n-1)*})\Bigg)|_{x=0}$$
$$+a_1\Bigg(q_0^*\frac{a^3 - b^3}{6} - \mathrm{L}_3^{-1}(q_\omega^{(n-1)*})\Bigg)|_{x=0}\Bigg),$$

$$C_6^{(n)*} = \frac{1}{\alpha_2}\Bigg(a_7\Bigg(p_0^*\frac{b^3 - a^3}{6} - \mathrm{L}_3^{-1}(p_\omega^{(n-1)*})\Big|_{x=0}\Bigg)$$
$$-a_1\Bigg(q_0^*\frac{a^4 - b^4}{24} - \mathrm{L}_4^{-1}(q_\omega^{(n-1)*})\Big|_{x=0}\Bigg) - \alpha_1\int \psi^{(n)*}\mathrm{d}x\Big|_{x=0},$$

$$C_7^{(n)*} = -\frac{1}{\alpha_2}\Big[a_4\mathrm{L}_1^{-1}(p_\omega^{(n-1)*})|_{x=l} - a_7(\mathrm{L}_2^{-1}(p_\omega^{(n-1)*})|_{x=l} + C_1^{(n)}l)\Big], \tag{4.57}$$

$$C_8^{(n)*} = \frac{1}{\alpha_2}\Bigg(a_4\Bigg(p_0^*\frac{a^2 - b^2}{2} - \mathrm{L}_2^{-1}(p_\omega^{(n-1)*})|_{x=0}\Bigg)$$
$$-a_7\Bigg(q_0^*\frac{b^3 - a^3}{6} - \mathrm{L}_3^{-1}(q_\omega^{(n-1)*})|_{x=0}\Bigg)\Bigg)$$

The obtained formulas (4.50)–(4.57) give in iterations the analytical solution of the boundary value problem in displacements for the values with asterisks under repeated local alternating loading for a cantilevered three-layer rod.

The corresponding parameters of the stress–strain state at the second half-cycle of loading are obtained from relations (4.32) taking into account solutions (4.51)–(4.55).

Cyclic alternating loading. Let, after (m–1) alternating half-cycles of loading, starting from the moment (t_m–1), there is an instantaneous unloading and a new loading by loads of opposite sign $p^{(m)}$, $q^{(m)}$:

$$q^{(m)}(x) = (-1)^{(m-1)} q_0 (H_0(a - x) - H_0(b - x)),$$

$$p^{(m)}(x) = (-1)^{(m-1)} p_0 (H_0(a - x) - H_0(b - x)). \tag{4.58}$$

These loads will create a displacement field in the kth layer of the rod $u^{(m)}$, $\psi^{(m)}$, $w^{(m)}$. As before, for all values of the stress–strain state and load, we keep the differences (4.33), in which the values with index (m–1) are stresses, strains, displacements and loads in the rod before unloading, index m denotes similar parameters in the process of the mth half-cycle.

Relations (4.16), relating deformations and displacements with asterisks, will be valid also in the considered situation. For the corresponding stresses and strains we assume the physical constitutive Eq. (4.34). We assume that the universal nonlinearity functions in the bearing layers at the mth half-cycle are expressed through the plasticity functions under loading from the natural state by the formulas (4.35). The physical nonlinearity of the filler under mth alternating loading is still described by relations (4.1). Taking into account the loads (4.58), the required functions $u^{(m,n)*}$, $\psi^{(m,n)*}$, $a_7 w^{(m,n)*}$ will satisfy the equilibrium Eq. (4.39), in which the values $p_\omega^{(m,n-1)*}$, $h_\omega^{(m,n-1)*}$, $q_\omega^{(m,n-1)*}$, corresponding to the inelastic components, at the first step ($n = 1$) are taken equal to zero, and further are calculated according to the results of the previous approximation by the formulas (4.40). Boundary conditions at cantilever termination of the left end of the rod for the values with asterisks are (4.41).

In this case, the considered boundary value problem with asterisks coincides with the boundary value problem under repeated local alternating loading with precision to notation. Therefore, its recurrence analytic solution is written in the following form:

$$\begin{aligned}
\psi^{(m,n)*}(x) &= C_2^{(m,n)*} sh(\beta x) + C_3^{(m,n)*} ch(\beta x) \\
&+ \frac{1}{\beta^2}\Big[\gamma_2 p_0^{(m)*}((ch(\beta x - a\beta) - 1)H_0(a - x) \\
&- (ch(\beta x - b\beta) - 1)H_0(b - x)) \\
&- \frac{\gamma_1 q_0^{(m)*}}{\beta}((\beta x - \beta a - \mathrm{sh}(\beta x - \beta a))H_0(a - x) \\
&-(\beta x - \beta b - \mathrm{sh}(\beta x - \beta b))H_0(b - x)) - \gamma_1 C_1^{(m,n)*}\Big] \\
&+ \frac{1}{\beta}\left[sh(\beta x)\int g_{\omega 1}^{(m,\,n)*} ch(\beta x)\mathrm{d}x - ch(\beta x)\int g_{\omega 1}^{(m,\,n)*} sh(\beta x)\mathrm{d}x\right],
\end{aligned}$$

$$
u^{(m,n)*}(x) = \gamma_3\, \psi^{(m,n)*} + \frac{1}{\alpha_2}\Big[a_4 \mathrm{L}_2^{-1}\big(p_\omega^{(m,n-1)*}\big)
- a_4\big(p_0^{(m)*}(H_2(a-x) - H_2(b-x)\big) - a_7 q_0^{(m)*}(H_3(a-x)
-H_3(b-x)) - \mathrm{L}_3^{-1}\big(q_\omega^{(m,n-1)*}\big) + \frac{a_7}{2} C_1^{(m,n)*} x^2\Big]
+ C_7^{(m,n)*} x + C_8^{(m,n)*}
$$

$$
w^{(m,n)*}(x) = \frac{1}{\alpha_2}\Big[\alpha_1 \int \psi^{(m,n)*}\mathrm{d}x
+ a_7\Big(p_0^{(m)*}(H_3(a-x) - H_3(b-x)) + \mathrm{L}_3^{-1}\big(p_\omega^{(m,n-1)*}\big)\Big)
+ a_1\Big(q_0^{(m)*}(H_4(a-x) - H_4(b-x) - \mathrm{L}_4^{-1}\big(q_\omega^{(m,n-1)*}\big)\Big)
+ \frac{1}{6} a_1 C_1^{(m,n)*} x^3\Big] + \frac{1}{2} C_4^{(m,n)*} x^2 + C_5^{(m,n)*} x + C_6^{(m,n)*}. \qquad (4.59)
$$

Here the coefficients $\alpha_1, \alpha_2, \alpha_3, \beta^2, \gamma_1, \gamma_2, \gamma_3$ and linear integral operators $\mathrm{L}_2^{-1}, \mathrm{L}_3^{-1}, \mathrm{L}_4^{-1}$ are defined by formulas (2.19), (2.23); integration constants $C_1^{(n)*}$, ..., $C_8^{(n)*}$ at each step of the approximation follow from the conditions of rod fixation; integrand function

$$
g_{\omega 1}^{(m,n)*}(x) = \frac{\alpha_2}{\alpha_1}\gamma_1 h_\omega^{(m,n-1)*} - \gamma_2 p_\omega^{(m,n-1)*} - \gamma_1\Big(\int q_\omega^{(m,n-1)*}\mathrm{d}x + C_1^{(m,n)*}\Big) \qquad (4.60)
$$

values $p_\omega^{(m,n-1)*}, h_\omega^{(m,n-1)*}, q_\omega^{(m,n-1)*}$ correspond to inelastic components and are calculated by formulae (4.40);

$$
\int \psi^{(m,n)*}\mathrm{d}x\Big|_{x=0} = \frac{C_2^{(m,n)*}}{\beta} + \frac{1}{\beta^2}\Big[\frac{\gamma_2 p_0^*}{\beta}(\mathrm{sh}(\beta b) - \mathrm{sh}(\beta a)
+\frac{\gamma_1 q_0^*}{\beta^2}((\mathrm{ch}(\beta a) - \mathrm{ch}(\beta b)) - \frac{1}{2}\gamma_1 q_0^*(a^2 - b^2)\Big]
+\frac{1}{\beta}\int sh(\beta x)\int g_{\omega 1}^{(m,n)*} ch(\beta x)\mathrm{d}x - ch(\beta x)\int g_{\omega 1}^{(m,n)*} sh(\beta x)\mathrm{d}x\mathrm{d}x\Big|_{x=0}
$$

Satisfying by the solution (4.59) the boundary conditions for the cantilever (4.25), we obtain the following integration constants:

$$
C_1^{(m,n)*} = \mathrm{L}_1^{-1}\big(q_\omega^{(m,n-1)*}\big)\big|_{x=l}
$$

$$
C_2^{(m,n)*} = \frac{1}{\beta}\Big[\frac{ch(\beta l)}{sh(\beta l)}\Big(\int g_{\omega 1}^{(m,\,n)*} sh(\beta x)\mathrm{d}x|_{x=l}
$$

$$-\int g_{\omega 1}^{(m,\,n)*} sh(\beta x)\mathrm{d}x|_{x=0} - \frac{\gamma_2 p_0^{(m)*}}{\beta} \times (ch(\beta b) - ch(\beta a))$$

$$+\frac{\gamma_1 q_0^{(m)*}}{\beta^2}(a\beta - b\beta + sh(\beta b) - sh(\beta a)) + \frac{\gamma_1 C_1^{(m,n)*}}{\beta}(1 - ch(\beta l))\Bigg)$$

$$-\int g_{\omega 1}^{(m,\,n)*} ch(\beta x)\mathrm{d}x\Big|_{x=l} + \gamma_1 C_1^{(m,n)*}\frac{sh(\beta l)}{\beta}\Bigg],$$

$$C_3^{(m,n)*} = \frac{1}{\beta}\int g_{\omega 1}^{(m,\,n)*}(x) sh(\beta x)\mathrm{d}x|_{x=0}$$

$$+\frac{1}{\beta^2}\Big(2p_0^{(m)*}(ch(\beta b) - ch(\beta a))$$

$$-\gamma\frac{1}{\beta}\gamma_1 q_0^{(m)*}(a\beta - b\beta + sh(\beta b) - sh(\beta a)) + \gamma_1 C_1^{(m,n)*}\Big),$$

$$C_4^{(m,n)*} = -\frac{a_7}{\alpha_2}\mathrm{L}_1^{-1}((p_\omega^{(m,n-1)*})|_{x=l} + \frac{a_1}{\alpha_2}\mathrm{L}_2^{-1}(q_\omega^{(m,n-1)*})|_{x=l} - \frac{a_1}{\alpha_2}C_1^{(m,n)*}l,$$

$$C_5^{(m,n)*} = \frac{a_7}{\alpha_2}\left(p_0^{(m)*}\frac{a^2 - b^2}{2} - \mathrm{L}_2^{-1}(p_\omega^{(m,n-1)*})\right)|_{x=0}$$

$$+\frac{a_1}{\alpha_2}\left(q_0^{(m)*}\frac{a^3 - b^3}{6} - \mathrm{L}_3^{-1}(q_\omega^{(m,n-1)*})\right)|_{x=0},$$

$$C_6^{(m,n)*} = \frac{a_7}{\alpha_2}\left(p_0^{(m)*}\frac{b^3 - a^3}{6} - \mathrm{L}_3^{-1}(p_\omega^{(m,n-1)*})\Big|_{x=0}\right)$$

$$-\frac{a_1}{\alpha_2}\left(q_0^{(m)*}\frac{a^4 - b^4}{24} - \mathrm{L}_4^{-1}(q_\omega^{(m,n-1)*})\big|_{x=0}\right) - \alpha_1\int \psi^{(m,n)*}\mathrm{d}x\Big|_{x=0},$$

$$C_7^{(m,n)*} = -\frac{1}{\alpha_2}\Big[a_4\mathrm{L}_1^{-1}(p_\omega^{(m,n-1)*})|_{x=l} - a_7(\mathrm{L}_2^{-1}(p_\omega^{(m,n-1)*})|_{x=l} + C_1^{(m,n)}l)\Big],$$

$$C_8^{(m,n)*} = \frac{a_4}{\alpha_2}\left(p_0^{(m)*}\frac{a^2 - b^2}{2} - \mathrm{L}_2^{-1}(p_\omega^{(m,n-1)*})|_{x=0}\right)$$

$$-\frac{a_7}{\alpha_2}\left(q_0^{(m)*}\frac{b^3 - a^3}{6} - \mathrm{L}_3^{-1}(q_\omega^{(m,n-1)*})|_{x=0}\right). \tag{4.61}$$

Consequently, the obtained analytical formulas (4.59)–(4.61) give the solution of the boundary value problem for the values with asterisks at the mth half-cycle of local alternating loading of a cantilevered three-layered rod. The corresponding parameters of the stress–strain state are obtained from relations (4.49) taking into account solutions (4.51), (4.59).

Numerical results are obtained under the assumptions introduced at the beginning of this section.

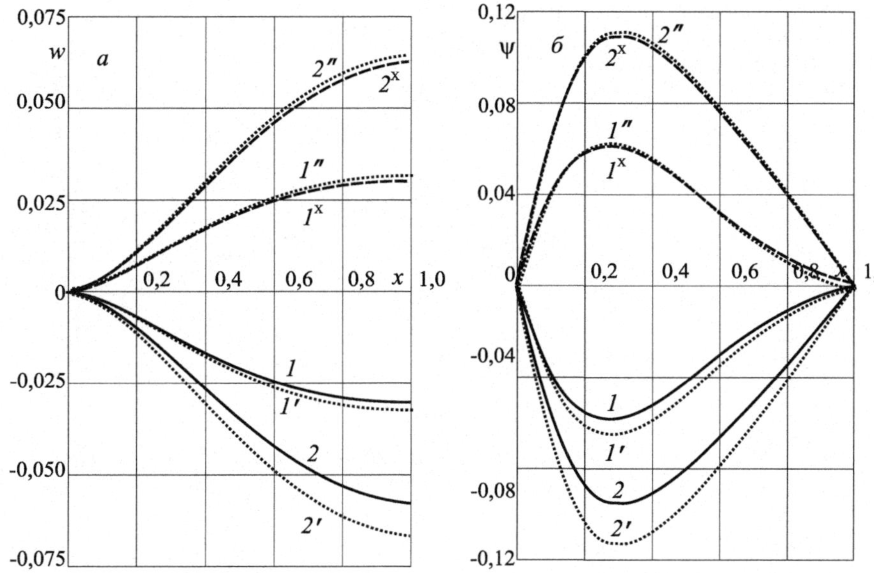

Fig. 4.14 Variation of shear (a) and deflection (b) along the axis as a function of the length of the loading section

Figure 4.14 shows the variation of shear in the filler and deflection along the axis of a three-layer rod of unit length as a function of the length of the section on which a distributed load is applied (no stroke—elastic rod, one stroke—elastic–plastic, two strokes—variable loading, index X—tenth half-cycle): *1*—$b = 0{,}75$, *2*—$b = 1$. In the case of repeated local loading, the area of plastic deformations is smaller than when the rod is loaded along its entire length, which causes a reduction in the cyclic hardening of the bearing layer materials. As a consequence, the displacement extremums also change insignificantly. At the 10th half-cycle they decrease by 3% and 5% relative to the corresponding curves under loading from the natural state.

The distribution of normal stresses σ_{xx} across the section at the embedment of a cantilever rod ($x = 0$) is illustrated in Fig. 4.15 (no stroke—elastic rod, one stroke—elastic–plastic, two strokes—variable loading): *1*—$b = 0.75$, *2*—$b = 1$.

Here the stresses in the bearing layers differ by 5–6%, which is explained by the different thickness of the layers. Taking into account the physical nonlinearity leads to an increase in the maximum stresses in the outer layers by 5% and in the filler by 14%.

Thus, the solutions (4.51), (4.55) and (4.59) allow us to study the stress–strain state of elastic–plastic three-layer rods under local multi-cycle alternating isothermal loadings with a uniformly distributed load. They can also be the basis for considering cases of variable loading of rods by concentrated forces or moments.

Consider the bending of a three-layered rod with incompressible filler by a local sinusoidal load (see Fig. 2.22).

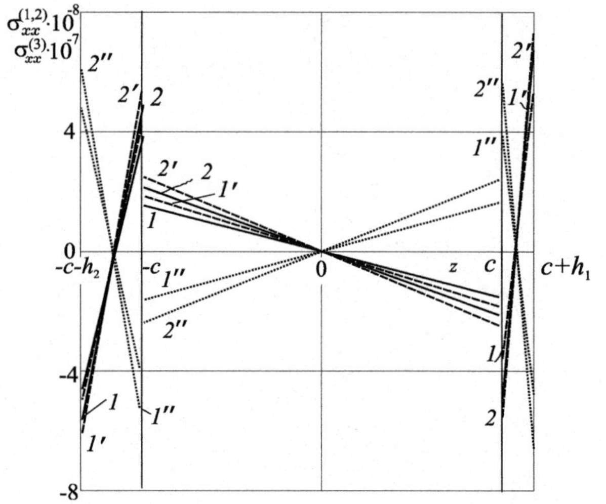

Fig. 4.15 Variation of normal stresses σ_{xx} across the section at the embedment ($x = 0$)

Loading from the natural state. The intensity of the load will be:

$$q'(x) = q_0 H_0(b - x)\sin\left(\frac{\pi x}{b}\right), p'(x) = 0. \tag{4.62}$$

We construct the analytical iteration solution of this problem on the basis of the solution (2.22). The result is

$$\psi'^{(n)}(x) = C_2'^{(n)} sh(\beta x) + C_3'^{(n)} ch(\beta x) - \frac{\gamma_1 C_1'^{(n)}}{\beta^2}$$
$$+ \frac{\gamma_1 q_0 b H_0(b - x)}{\beta^2 \pi}\left(\frac{\beta^2 b^2 \cos\left(\frac{\pi x}{b}\right) + \beta^2 b^2 ch(\beta x - \beta b)}{\pi^2 + \beta^2 b^2} - ch(\beta x - \beta b) + 1\right)$$
$$+ \frac{1}{\beta}\left[sh(\beta x)\int g_{\omega 1}''^{(n)} ch(\beta x)dx - ch(\beta x)\int g_{\omega 1}'^{(n)} sh(\beta x)dx\right],$$

$$u'^{(n)}(x) = \gamma_3 \psi'^{(n)} + \frac{1}{\alpha_2}\left(a_7 L_3^{-1}(q' - q_\omega'^{(n-1)})\right.$$
$$\left.- a_4 L_2^{-1}(p_\omega'^{(n-1)}) + \frac{a_7 C_1'^{(n)}}{2} x^2\right) + C_7'^{(n)} x + C_8'^{(n)}.$$

$$w'^{(n)}(x) = \frac{1}{\alpha_2}\left[\alpha_1 \int \psi'^{(n)} dx - a_7 L_3^{-1}(p_\omega'^{(n-1)})\right.$$
$$\left. + a_1 L_4^{-1}(q' - q_\omega'^{(n-1)}) + \frac{a_1 C_1'^{(n)}}{6} x^3\right]$$
$$+ \frac{C_4'^{(n)}}{2} x^2 + C_5'^{(n)} x + C_6'^{(n)}, \tag{4.63}$$

where the stroke at the top shows the loading at the first half-cycle; $h_{\omega}^{\prime(n-1)}$, $p_{\omega}^{\prime(n-1)}$, $q_{\omega}^{\prime(n-1)}$ are additional loads calculated by the formulas (4.10); $g_{\omega 1}^{\prime(n)}(x)$ is integrand function and linear operators from the external load

$$g_{\omega 1}^{\prime(n)}(x) = \frac{\alpha_2}{\alpha_1}\left(\gamma_1 h_{\omega}^{\prime(n-1)} - \gamma_2 p_{\omega}^{\prime(n-1)}\right) - \gamma_1\left(\int q_{\omega}^{\prime(n-1)}dx + C_1^{\prime(n)}\right),$$

$$\mathrm{L}_3^{-1}(q') = \frac{q_0 b^3}{\pi^3}\left[\cos\left(\frac{\pi x}{b}\right) + 1\right]H_0(b-x) - \frac{q_0 b}{\pi}H_2(b-x),$$

$$\mathrm{L}_4^{-1}(q') = \frac{q_0 b^4}{\pi^4}\sin\left(\frac{\pi x}{b}\right)H_0(b-x)$$
$$- \frac{q_0 b^3}{\pi^3}H_1(b-x) + \frac{q_0 b}{\pi}H_3(b-x);$$

shear integral

$$\int \psi^{\prime(n)}dx|_{x=0} = \frac{C_2}{\beta} - \frac{\gamma_1 q_0 b^3 sh(\beta b)}{\pi\beta\left(\pi^2 + \beta^2 b^2\right)}$$
$$+ \frac{\gamma_1 q_0 b sh(\beta b)}{\pi\beta^3} - \frac{\gamma_1 q_0 b^2}{\pi\beta^2}$$
$$+ \frac{1}{\beta}\int sh(\beta x)\int g_{\omega 1}^{(n)} ch(\beta x)\mathrm{d}x - ch(\beta x)\int g_{\omega 1}^{(n)} sh(\beta x)\mathrm{d}x\mathrm{d}x\bigg|_{x=0}.$$

In the case of rigid embedding of the left end of the rod, satisfying by solution (4.63) the boundary conditions (2.24), we obtain integration constants:

$$C{\prime}_1^{(n)} = L_1^{-1}(q{\prime}_{\omega}^{(n-1)})|_{x=l},$$

$$C_2^{\prime(n)} = \frac{1}{\beta}\left[\frac{ch(\beta l)}{sh(\beta l)}\left(\int g^{\prime(n)} sh(\beta x)dx|_{x=l} - \int g_{\omega 1}^{\prime(n)} sh(\beta x)dx|_{x=0}\right)\right.$$
$$\left. - \int g_{\omega 1}^{\prime(n)} ch(\beta x)dx|_{x=l} + \gamma_1 C_1^{\prime(n)}\frac{sh(\beta l)}{\beta}\right]$$
$$+ \frac{\gamma_1 q_0 b ch(\beta l)}{\pi\beta^2\left(\pi^2 + \beta^2 b^2\right)sh(\beta l)}\left(2\beta^2 b^2 + \pi^2 - \pi^2 ch(\beta x - \beta b)\right),$$

$$C_3^{\prime(n)} = \frac{1}{\beta}\int g{\prime}_{\omega 1}^{(n)}(x) sh(\beta x)dx|_{x=0}$$
$$+ \frac{\gamma_1 q_0 b}{\pi\beta^2\left(\pi^2 + \beta^2 b^2\right)}\left(\pi^2 ch(\beta x - \beta b) - 2\beta^2 b^2 - \pi^2\right),$$

$$C_4^{\prime(n)} = -\frac{a_7}{\alpha_2}L_1^{-1}(p_{\omega}^{\prime(n-1)})|_{x=l} + \frac{a_1}{\alpha_2}L_2^{-1}(q_{\omega}^{\prime(n-1)})|_{x=l} - \frac{a_1}{\alpha_2}C_1^{\prime(n)}l,$$

$$C_5^{\prime(n)} = \frac{a_7}{\alpha_2}\left[-L_2^{-1}(p_\omega^{\prime(n-1)})|_{x=0} + L_3^{-1}(q_\omega^{\prime(n-1)})|_{x=0} + \frac{q_0 b^3(\pi^2-4)}{2\pi^3}\right],$$

$$C_6^{\prime(n)} = -\frac{a_7}{\alpha_2}L_3^{-1}(p_\omega^{\prime(n-1)})\Big|_{x=0} + \frac{a_1}{\alpha_2}L_4^{-1}(q_\omega^{\prime(n-1)})\Big|_{x=0} - \frac{\alpha_1}{\alpha_2}\int \psi^{\prime(n)}dx\Big|_{x=0} + \frac{a_1 q_0 b^4(6-\pi^2)}{6\alpha_2\pi^3},$$

$$C_7^{\prime(n)} = -\frac{1}{\alpha_2}\left[a_4 L_1^{-1}(p_\omega^{\prime(n-1)})|_{x=l} - a_7(L_2^{-1}(p_\omega^{\prime(n-1)})|_{x=l} + a_7 C_1^{\prime(n)} l)\right],$$

$$C_8^{\prime(n)} = \frac{1}{\alpha_2}\left(a_7\left(\frac{a_7 q_0 b^3(\pi^2-4)}{2\pi^3} + L_3^{-1}(q_\omega^{\prime(n-1)})|_{x=0}\right) - a_4 L_2^{-1}(p_\omega^{\prime(n-1)})|_{x=0}\right). \tag{4.64}$$

Numerical results. The geometric and mechanical characteristics of the materials were assumed to be the same as in Sect. 4.4. Sinusoidal load (3.62) with amplitude $q_0 = 3.5$ MPa ($p_0 = 0$) is distributed on the segment $[0, b]$. At the amplitude $q_1 = 0.5\,\pi q_0$, according to (2.38), this load will be equivalent to the rectangular load q_0 acting on the same surface area of the rod. The displacements were calculated using the formulas (4.63), (4.64).

Figures 4.16, 4.17 show the variation of relative shear in the filler and deflection along the axis of the three-layer rod at $b = l$ as a function of the amplitude of the distributed load: *1*—sinusoidal q_0, *2*—rectangular q_0, *3*—sinusoidal q_1 (without streak– elastic rod, with streak– elastic–plastic rod).

For the same load amplitude (curves 1, 2) the displacements from a uniformly distributed load are much larger in magnitude. If the resultant forces, according to (2.38), are the same, the maximum displacement is slightly larger from the sinusoidal load, the deflections differ little. Accounting for physical nonlinearity is affected at equivalent loads (curves *1′*, *3′*).

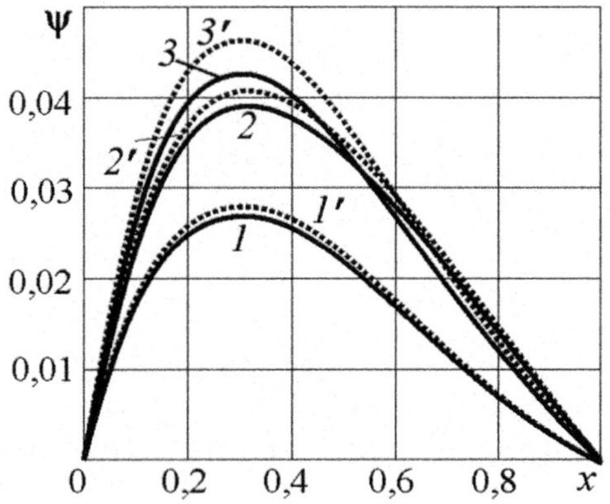

Fig. 4.16 Variation of shear along the axis under different loads

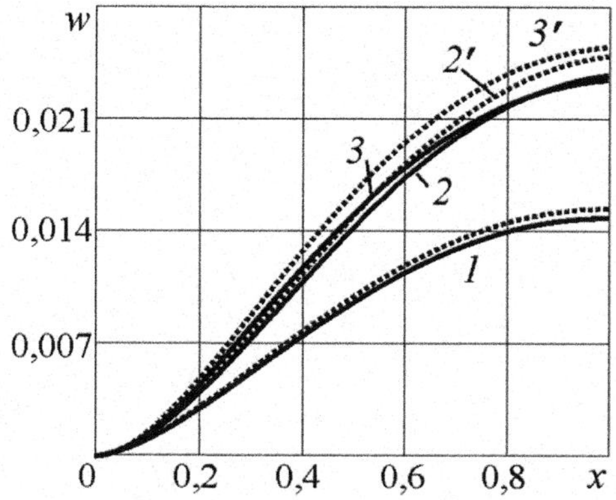

Fig. 4.17 Variation of deflection along the axis under different loads

Repeated alternating loading. Let, starting from the moment $t = t_1$, there is an instantaneous unloading and reloading with forces of the opposite sign changing according to the same law as in the loading from the natural state (4.62):

$$q''(x) = -q_0 H_0(b - x)\sin\left(\frac{\pi x}{b}\right), p''(x) = 0. \tag{4.65}$$

These loads will create in the kth layer of the rod a field of displacements u'', ψ'', w'', strains $\varepsilon_x''^{(k)}$, $\varepsilon''^{(k)}$, ψ'' and stresses $\sigma_x''^{(k)}$, $s_x''^{(k)}$, $\sigma''^{(k)}$, $\sigma_{xz}''^{(3)}$. In the following we proceed according to the scheme worked out earlier.

The differences for the values of stress–strain state and load (4.15), in which the values with one stroke are stresses, strains, external loads and displacements in the rod before unloading, with two strokes labelled similar parameters during the second half-cycle, will be the same. The relationship between strains and displacements with asterisks retain the form (4.16). We take the nonlinear physical equations of state between stresses and strains as (4.17). We assume the universal plasticity functions $\omega^{(k)*}$ to be expressed through the plasticity functions under loading from the natural state by expressions (4.18), (4.19).

As a result, the iterative equations of equilibrium (4.23), with load (4.65) and boundary conditions (4.26)–(4.27) will be valid for the displacements with asterisks. Therefore, we use the solution (4.63) to construct an analytical iterative solution of the boundary value problem. After some transformations we have:

$$\psi^{(n)*}(x) = C_2^{(n)*}\mathrm{sh}(\beta x) + C_3^{(n)*}\mathrm{ch}(\beta x) - \frac{\gamma_1 C_1^{(n)*}}{\beta^2}$$
$$+ \frac{\gamma_1 q_0^* b H_0(b - x)}{\beta^2 \pi}\left(\frac{\beta^2 b^2 \cos\left(\frac{\pi x}{b}\right) + \beta^2 b^2 ch(\beta x - \beta b)}{\pi^2 + \beta^2 b^2} - ch(\beta x - \beta b) + 1\right)$$
$$+ \frac{1}{\beta}\left[sh(\beta x)\int g_{\omega 1}^{(n-1)*} ch(\beta x)dx - ch(\beta x)\int g_{\omega 1}^{(n-1)*} sh(\beta x)dx\right],$$

$$u^{(n)*}(x) = \gamma_3\psi^{(n)*} + \frac{1}{\alpha_2}\left(a_7\left(\mathrm{L}_3^{-1}(q_0^* - q_\omega^{(n-1)*}) + \frac{C_1^{(n)*}x^2}{2}\right)\right.$$
$$\left. - a_4\mathrm{L}_2^{-1}(p_\omega^{(n-1)*})\right) + C_7^{(n)*}x + C_8^{(n)*},$$

$$w^{(n)*}(x) = \frac{1}{\alpha_2}\left[\alpha_1\int\psi^{(n)*}\mathrm{d}x - a_7\mathrm{L}_3^{-1}(p_\omega^{(n-1)*})\right.$$
$$\left. + a_1\mathrm{L}_4^{-1}(q_0^* - q_\omega^{(n-1)*}) + \frac{a_1C_1^{(n)*}}{6}x^3\right]$$
$$+ \frac{C_4^{(n)*}}{2}x^2 + C_5^{(n)*}x + C_6^{(n)*}, \qquad (4.66)$$

where $q_0^* = 2q_0$; $h_\omega^{(n-1)*}, p_\omega^{(n-1)*}, q_\omega^{(n-1)*}, g_{\omega 1}^{(n-1)*}$ are additional loads and sub-integral function calculated by formulas like (4.10), (4.56); integral of the shift function with asterisk

$$\int\psi^{(n)*}\mathrm{d}x\Big|_{x=0} = \frac{C_2^*}{\beta} - \frac{q_0^*}{\pi\beta}\left(\frac{\gamma_1 b^3 sh(\beta b)}{(\pi^2+\beta^2b^2)} + \frac{\gamma_1 b sh(\beta b)}{\beta^2} - \frac{\gamma_1 b^2}{\beta}\right)$$
$$+ \frac{1}{\beta}\int sh(\beta x)\int g_{\omega 1}^{(n)*}ch(\beta x)\mathrm{d}x - ch(\beta x)\int g_{\omega 1}^{(n)*}sh(\beta x)\mathrm{d}x\mathrm{d}x\Big|_{x=0}.$$

Substituting the solution (4.66) into the boundary conditions (4.25), we obtain the integration constants with asterisks at cantilever embedding of the left end of the rod:

$$C_1^{(n)*} = \mathrm{L}_1^{-1}(q_\omega^{(n-1)*})|_{x=l},$$

$$C_2^{(n)*} = \frac{1}{\beta}\left[\frac{ch(\beta l)}{sh(\beta l)}\left(\int g_{\omega 1}^{(n)*}sh(\beta x)\mathrm{d}x|_{x=l} - \int g_{\omega 1}^{(n)*}sh(\beta x)\mathrm{d}x|_{x=0}\right)\right.$$
$$\left. - \int g_{\omega 1}^{(n)*}ch(\beta x)\mathrm{d}x|_{x=l} + \gamma_1 C_1^{(n)*}\frac{sh(\beta l)}{\beta}\right]$$
$$+ \frac{\gamma_1 q_0^* b ch(\beta l)}{\pi\beta^2(\pi^2+\beta^2b^2)sh(\beta l)}\left(2\beta^2b^2 + \pi^2 - \pi^2 ch(\beta x - \beta b)\right),$$

$$C_3^{(n)*} = \frac{1}{\beta}\int g_{\omega 1}^{(n)*}(x)sh(\beta x)\mathrm{d}x|_{x=0}$$
$$+ \frac{\gamma_1 q_0^* b}{\pi\beta^2(\pi^2+\beta^2b^2)}\left(\pi^2 ch(\beta x - \beta b) - 2\beta^2b^2 - \pi^2\right),$$

$$C_4^{(n)*} = -\frac{a_7}{\alpha_2} \mathrm{L}_1^{-1}(p_\omega^{(n-1)*})|_{x=l} + \frac{a_1}{\alpha_2} \mathrm{L}_2^{-1}(q_\omega^{(n-1)*})|_{x=l} - \frac{a_1}{\alpha_2} C_1^{(n)*} l,$$

$$C_5^{(n)*} = \frac{a_7}{\alpha_2}\left[-L_2^{-1}(p_\omega^{(n-1)*})|_{x=0} + L_3^{-1}(q_\omega^{(n-1)*})|_{x=0} + \frac{q_0^* b^3(\pi^2-4)}{2\pi^3}\right],$$

$$C_6^{(n)*} = -\frac{a_7}{\alpha_2} L_3^{-1}(p_\omega^{(n-1)*})\Big|_{x=0} + \frac{a_1}{\alpha_2} L_4^{-1}(q_\omega^{(n-1)*})\Big|_{x=0} - \frac{\alpha_1}{\alpha_2} \int \psi^{(n)*} dx\Big|_{x=0} + \frac{a_1 q_{00}^* b^4(6-\pi^2)}{6\alpha_2 \pi^3},$$

$$C_7^{(n)*} = -\frac{1}{\alpha_2}\left[a_4 \mathrm{L}_1^{-1}(p_\omega^{(n-1)*})|_{x=l} - a_7(\mathrm{L}_2^{-1}(p_\omega^{(n-1)*})|_{x=l} + a_7 C_1^{(n)*} l)\right],$$

$$C_8^{(n)*} = \frac{1}{\alpha_2}\left(a_7\left(\frac{a_7 q_0^* b^3(\pi^2-4)}{2\pi^3} + \mathrm{L}_3^{-1}(q_\omega^{(n-1)*})|_{x=0}\right) - a_4 \mathrm{L}_2^{-1}(p_\omega^{(n-1)*})|_{x=0}\right). \tag{4.67}$$

Consequently, the obtained formulas (4.66), (4.67) give in iterations the analytical solution of the boundary value problem in displacements for the values with asterisks under repeated local alternating loading for a cantilevered three-layer rod. With their help, the parameters of the stress–strain state at the second half-cycle of loading are obtained from relations (4.32).

Cyclic alternating loading. Let, after (m–1) alternating half-cycles of loading, starting from the moment t_{m-1}, there is an instantaneous unloading and new loading by loads of the opposite sign $p^{(m)}$, $q^{(m)}$:

$$q^{(m)}(x) = (-1)^{(m-1)} q_0 H_0(b-x) \sin\left(\frac{\pi x}{b}\right), p^{(m)} = 0. \tag{4.68}$$

These loads create a displacement field $u^{(m)}$, $\psi^{(m)}$, $w^{(m)}$ in the kth layer of the rod. As before, we introduce for all values of stress–strain state and load the differences (4.33), in which the values with index (m–1) are stresses, strains, displacements and loads in the rod before unloading, index m denotes similar parameters in the process of the mth half-cycle.

Relations (4.16), (4.34), (4.35), (4.1) are also valid in this case. The universal functions of nonlinearity in the bearing layers at the mth half-cycle are assumed to be expressed through the plasticity functions under loading from the natural state by formulas (4.18), (4.19). Taking into account the loads (4.68), the required functions $u^{(m,n)*}$, $\psi^{(m,n)*}$, $a_7 w^{(m,n)*}$ will satisfy the equilibrium Eq. (4.39), in which the values $p_\omega^{(m,n-1)*}$, $h_\omega^{(m,n-1)*}$, $q_\omega^{(m,n-1)*}$ are calculated according to the results of the previous approximation by the formulas (4.40). The boundary conditions at cantilever termination of the left end of the rod for the values with asterisks are (4.41). In this case, the recurrence analytical solution of the boundary value problem for the values with asterisks at the mth half-cycle takes the following form:

$$\psi^{(m,n)*}(x) = C_2^{(m,n)*}\mathrm{sh}(\beta x) + C_3^{(m,n)*}\mathrm{ch}(\beta x)$$
$$-\frac{\gamma_1 C_1^{(m,n)*}}{\beta^2} + \frac{\gamma_1 q_0^{(m)*} b H_0(b-x)}{\beta^2 \pi}$$
$$\left(\frac{\beta^2 b^2 \cos\left(\frac{\pi x}{b}\right) + \beta^2 b^2 ch(\beta x - \beta b)}{\pi^2 + \beta^2 b^2} - ch(\beta x - \beta b) + 1\right)$$
$$+\frac{1}{\beta}\left[sh(\beta x)\int g_{\omega 1}^{(m,n)*} ch(\beta x)dx - ch(\beta x)\int g_{\omega 1}^{(m,n)*} sh(\beta x)dx\right],$$

$$u^{(m,n)*}(x) = \gamma_3 \psi^{(n)*}$$
$$+\frac{1}{\alpha_2}\left(a_7\left(\mathrm{L}_3^{-1}(q_0^{(m)*} - q_\omega^{(m,n-1)*}) + \frac{C_1^{(m,n)*}x^2}{2}\right)\right.$$
$$\left.-a_4 \mathrm{L}_2^{-1}(p_\omega^{(m,n-1)*})\right) + C_7^{(m,n)*}x + C_8^{(m,n)*},$$

$$w^{(m,n)*}(x) = \frac{1}{\alpha_2}\left[\alpha_1 \int \psi^{(m,n)*}dx - a_7 \mathrm{L}_3^{-1}(p_\omega^{(m,n-1)*})\right.$$
$$\left. a_1 \mathrm{L}_4^{-1}(q_0^{(m)*} - q_\omega^{(m,n-1)*}) + \frac{a_1 C_1^{(m,n)*}}{6}x^3\right]$$
$$+\frac{C_4^{(m,n)*}}{2}x^2 + C_5^{(m,n)*}x + C_6^{(m,n)*}, \tag{4.69}$$

where $h_\omega^{(m,n-1)*}$, $p_\omega^{(m,n-1)*}$, $q_\omega^{(m,n-1)*}$, $g_{\omega 1}^{(m,n-1)*}$ are additional loads and integrand function, calculated by formulas (4.40), (4.45); integral from the shear with asterisk at the mth half-cycle

$$\int \psi^{(m,n)*}dx|_{x=0} = \frac{C_2^*}{\beta} - \frac{q_0^*}{\pi\beta}\left(\frac{\gamma_1 b^3 sh(\beta b)}{(\pi^2 + \beta^2 b^2)} + \frac{\gamma_1 b sh(\beta b)}{\beta^2}\right.$$
$$\left.-\frac{\gamma_1 b^2}{\beta}\right) + \frac{1}{\beta}\int sh(\beta x)\int g_{\omega 1}^{(m,n)*} ch(\beta x)dx - ch(\beta x)$$
$$\left.\int g_{\omega 1}^{(m,n)*} sh(\beta x)dxdx\right|_{x=0}$$

Substituting the solution (4.69) into the boundary conditions (4.41), we obtain the integration constants with asterisks at cantilever embedding of the left end of the rod:

$$C_1^{(m,n)*} = \mathrm{L}_1^{-1}(q_\omega^{(n-1)*})|_{x=l},$$

$$C_2^{(m,n)*} = \frac{1}{\beta}\left[\frac{ch(\beta l)}{sh(\beta l)}\left(\int g_{\omega 1}^{(m,n)*} sh(\beta x)dx|_{x=l} - \int g_{\omega 1}^{(m,n)*} sh(\beta x)dx|_{x=0}\right)\right.$$

$$
\left. - \int g_{\omega 1}^{(n)*} ch(\beta x) \mathrm{d}x|_{x=l} + \gamma_1 C_1^{(m,\,n)*} \frac{sh(\beta l)}{\beta} \right]
$$

$$
+ \frac{\gamma_1 q_0^* b ch(\beta l)}{\pi \beta^2 (\pi^2 + \beta^2 b^2) sh(\beta l)} \left(2\beta^2 b^2 + \pi^2 - \pi^2 ch(\beta x - \beta b)\right),
$$

$$
C_3^{(m,\,n)*} = \frac{1}{\beta} \int g_{\omega 1}^{(m,\,n)*}(x) sh(\beta x) \mathrm{d}x|_{x=0}
$$

$$
+ \frac{\gamma_1 q_0^{(m)*} b}{\pi \beta^2 (\pi^2 + \beta^2 b^2)} \left(\pi^2 ch(\beta x - \beta b) - 2\beta^2 b^2 - \pi^2\right),
$$

$$
C_4^{(m,\,n)*} = -\frac{a_7}{\alpha_2} \mathrm{L}_1^{-1}(p_\omega^{(m,\,n-1)*})|_{x=l} + \frac{a_1}{\alpha_2} \mathrm{L}_2^{-1}(q_\omega^{(m,\,n-1)*})|_{x=l} - \frac{a_1}{\alpha_2} C_1^{(m,\,n)*} l,
$$

$$
C_5^{(m,\,n)*} = \frac{a_7}{\alpha_2} \left[-L_2^{-1}(p_\omega^{(m,\,n-1)*})|_{x=0} + L_3^{-1}(q_\omega^{(m,\,n-1)*})|_{x=0} + \frac{q_0^{(m)*} b^3 (\pi^2 - 4)}{2\pi^3} \right],
$$

$$
C_6^{(m,\,n)*} = \frac{a_1 q_0^{(m)*} b^4 (6 - \pi^2)}{6\alpha_2 \pi^3} - \frac{\alpha_1}{\alpha_2} \left. \int \psi^{(m,\,n)*} \mathrm{d}x \right|_{x=0},
$$

$$
C_6^{(m,\,n)*} = -\frac{a_7}{\alpha_2} L_3^{-1}(p_\omega^{(m,\,n-1)*})\Big|_{x=0} + \frac{a_1}{\alpha_2} L_4^{-1}(q_\omega^{(m,\,n-1)*})\Big|_{x=0}
$$

$$
- \frac{\alpha_1}{\alpha_2} \left. \int \psi^{(m,\,n)*} dx \right|_{x=0} + \frac{a_1 q_0^{(m)*} b^4 (6 - \pi^2)}{6\alpha_2 \pi^3},
$$

$$
C_7^{(m,\,n)*} = -\frac{1}{\alpha_2} \left[a_4 \mathrm{L}_1^{-1}(p_\omega^{(m,\,n-1)*})|_{x=l} - a_7 (\mathrm{L}_2^{-1}(p_\omega^{(m,\,n-1)*})|_{x=l} + a_7 C_1^{(m,\,n)*} l) \right], \tag{4.70}
$$

$$
C_8^{(m,\,n)*} = -\frac{1}{\alpha_2} \left(a_7 \left(\frac{a_7 q_0^{(m)*} b^3 (\pi^2 - 4)}{2\pi^3} + \mathrm{L}_3^{-1}(q_\omega^{(m,\,n-1)*})|_{x=0} \right) - a_4 (\mathrm{L}_2^{-1}(p_\omega^{(m,\,n-1)*})|_{x=o} \right)
$$

The parameters of the stress–strain state of the cantilevered three-layer rod at the mth half-cycle of local sinusoidal alternating loading are obtained from relations (4.49) using formulas (4.69), (4.70).

Numerical results. The geometric and mechanical characteristics of the materials were assumed to be the same as at the beginning of this section.

Figures 4.18, 4.19 show the variation of deflection and shear in the filler along the axis of the rod at $b = l$ (no stroke—elastic rod, one stroke—elastic–plastic deformation, two strokes—variable loading, index X—tenth half cycle) under the following loads: *1*—sinusoidal q_0, 2—rectangular q_0, *3*—sinusoidal q_1. In this case, as before, the displacement extrema at repeated loading are reduced compared to loading at the first half-cycle due to cyclic hardening of the material of bearing layers. As the amplitude of the sinusoidal load increases, the plastic strain region

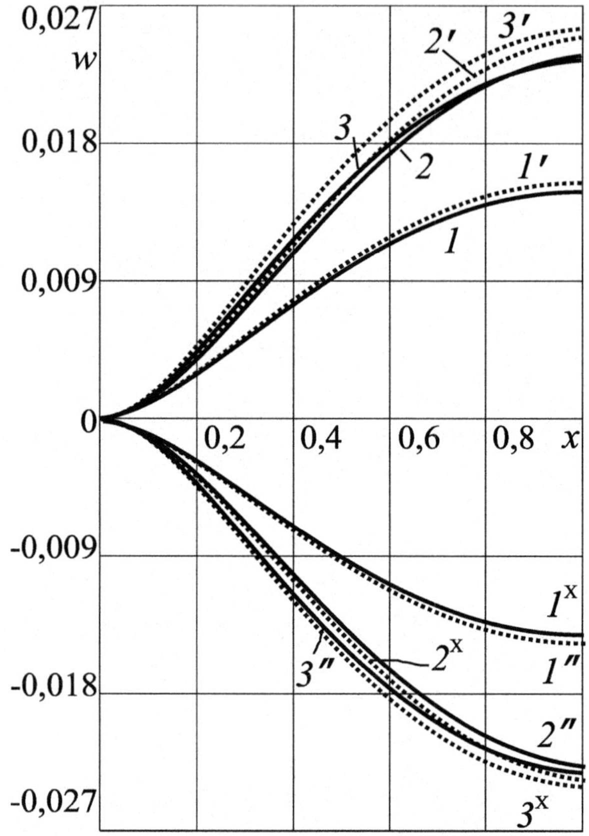

Fig. 4.18 Variation of deflection along the axis under different loads

increases and the hardening intensifies. At the 10th half-cycle, the extrema decrease by 4% and 5%, respectively, relative to the curves under loading from the natural state.

The distribution of normal stresses σ_{xx} across the section at the embedment ($x = 0$) of a cantilevered rod is illustrated in Fig. 4.20 (no stroke—elastic rod, one stroke—elastic–plastic, two strokes—alternating loading) at different amplitudes of sinusoidal load: *1*—q_0, *2*—q_1.

In this case, also due to the different thicknesses, the stresses in the bearing layers differ by 5–6%. Taking into account physical nonlinearity leads to an increase of maximum stresses in the outer layers by 5–6%, in the filler—by 12–14%. After changing the sign of the load, the stresses are almost unchanged.

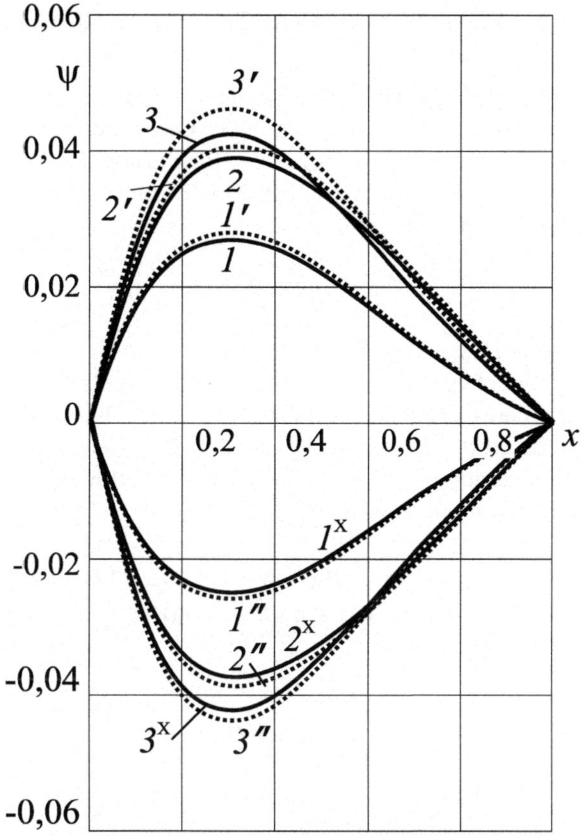

Fig. 4.19 Variation of shear along the axis under different loads

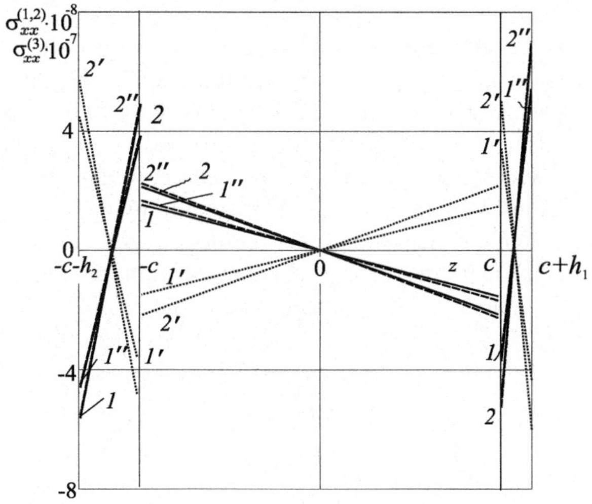

Fig. 4.20 Distribution of normal stresses σ_{xx} across the section at the embedment ($x = 0$)

4.5 Physics Nonlinear Deformation of a Stepped Rod

We consider a thickness asymmetric three-layered rod with an irregular boundary (see Fig. 2.31), the outer bearing layers of which are made of elastic–plastic material, and the thickness incompressible filler is nonlinearly elastic. The package kinematics corresponds to the previously adopted model (see Sect. 2.5). The thickness of layer 2 is step-variable:

$$h_3 = 2c, h_2 = h_{20} + h_{21} H_0(x - x_0), \tag{4.71}$$

where $H_0(x)$ is Heaviside function.

The physical equations of state of Ilyushin's theory of small elastoplastic deformations are used in the layers of the rod (4.1). The methodology of solving the problem is based on the approaches described in Sects. 2.5, 4.1. The system of equations for determining the required functions follows from (4.7), taking into account the irregularity of the rod boundary (4.71):

$$\begin{aligned}&(a_1 + a_{10} H(x - x_0))u,_{xx} + (a_6 + a_{60} H(x - x_0))\psi,_{xx}\\&- (a_7 + a_{70} H(x - x_0))w,_{xxx} = -p + p_\omega,\end{aligned}$$

$$\begin{aligned}&(a_6 + a_{60} H(x - x_0))u,_{xx} + (a_2 + a_{20} H(x - x_0))\psi,_{xx}\\&- (a_3 + a_{30} H(x - x_0))w,_{xxx} - a_5\psi = h_\omega,\end{aligned}$$

$$\begin{aligned}&(a_7 + a_{70} H(x - x_0))u,_{xxx} + (a_3 + a_{30} H(x - x_0))\psi,_{xxx}\\&- (a_4 + a_{40} H(x - x_0))w,_{xxxx} = -q + q_\omega.\end{aligned} \tag{4.72}$$

Here coefficients $a_1, \ldots, a_7$ and $a_{10}, \ldots, a_{70}$ are determined by formulas (2.17), (2.44); the terms p_ω, h_ω, q_ω take into account the physical nonlinearity of layer materials and are calculated by formulas (4.8), (4.6), taking into account the condition (4.71).

According to the solution procedure, we consider the system of Eq. (4.72) in two regions: I—before the point of irregularity ($x < x_0$); II—after it ($x \geq x_0$). In the first of them the Heaviside function turns to zero and according to (4.71) $h_2 = h_{20}$. As before, mark the required displacements in this area with the index "1" at the bottom. The system (4.72) is transformed to the form:

$$\begin{aligned}&a_1 u_1,_{xx} + a_6\psi_1,_{xx} - a_7 w_1,_{xxx} = -p + p_\omega,\\&a_6 u_1,_{xx} + a_2\psi_1,_{xx} - a_3 w_1,_{xxx} - a_5\psi_1 = h_\omega,\\&a_7 u_1,_{xxx} + a_3\psi_1,_{xxx} - a_4 w_1,_{xxxx} = -q + q_\omega.\end{aligned} \tag{4.73}$$

Now consider the system of Eq. (4.72) in the second region where the Heaviside function is equal to one ($x \geq x_0$, after stepping) and $h_2 = h_{20} + h_{21}$. The displacements in this area are marked with the index "2" at the bottom. The result is:

$$a_{11}u_{2},_{xx} + a_{61}\psi_{2},_{xx} - a_{71}w_{2},_{xxx} = -p + p_{\omega},$$
$$a_{61}u_{2},_{xx} + a_{21}\psi_{2},_{xx} - a_{31}w_{2},_{xxx} - a_{5}\psi_{2} = h_{\omega},$$
$$a_{71}u_{2},_{xxx} + a_{31}\psi_{2},_{xxx} - a_{41}w_{2},_{xxxx} = -q + q_{\omega}, \quad (4.74)$$

where

$$a_{11} = a_1 + a_{10},\ a_{21} = a_2 + a_{20},\ a_{31} = a_3 + a_{30},$$
$$a_{41} = a_4 + a_{40},\ a_{61} = a_6 + a_{60},\ a_{71} = a_7 + a_{70}.$$

In the case of *rigid embedding of the left end of the rod with a free right end*, the following boundary conditions must be fulfilled:

$$x = 0 : w_1 = w_{1,x} = u_1 = \psi_1 = 0;$$
$$x = l : \psi_2 = 0, N_2 = 0, M_2 = 0, M_{2},_{x} = 0. \quad (4.75)$$

For a rod *hinged at both ends*, the following boundary conditions should be used instead of (4.75):

$$x = 0 : w_1 = u_1 = \psi_1 = 0, M_1 = 0;$$
$$x = l : w_2 = u_2 = \psi_2 = 0, M_2 = 0, \quad (4.76)$$

where the lower index indicates the region of the required solution.

If the *rod is pinched at both ends*, the following conditions must be fulfilled:

$$x = 0 : w_1 = w_{1,x} = u_1 = \psi_1 = 0;$$
$$x = l : w_2 = w_{2},_{x} = u_2 = \psi_2 = 0. \quad (4.77)$$

Let's write out an algebraic system of eight equations that connect the solutions of the systems (4.73), (4.74) at the point of irregularity $x = x_0$ in the considered case of a physically nonlinear rod. For this purpose, we require fulfilment of eight conditions (2.28). The result is:

$$\langle \psi_2(x) - \psi_1(x) \rangle|_{x=x_0} = 0,\ \langle \psi_{2},_{x}(x) - \psi_{1},_{x}(x) \rangle|_{x=x_0} = 0,$$
$$\langle \psi_{2},_{xx}(x) - \psi_{1},_{xx}(x) \rangle|_{x=x_0} = 0,\ \langle w_2(x) - w_1(x) \rangle|_{x=x_0} = 0,$$
$$\langle w_{2},_{x}(x) - w_{1},_{x}(x) \rangle|_{x=x_0} = 0,\ \langle w_{2},_{xx}(x) - w_{1},_{xx}(x) \rangle|_{x=x_0} = 0,$$
$$\langle u_2(x) - u_1(x) \rangle|_{x=x_0} = 0,\ \langle u_{2},_{x}(x) - u_{1},_{x}(x) \rangle|_{x=x_0} = 0. \quad (4.78)$$

Thus, the boundary value problem for the determination of displacements in a three-layer rod with a stepped boundary is closed by joining to the system of differential Eq. (4.72) the boundary conditions (4.75)–(4.77) and the conjugation conditions (4.78). To solve the obtained nonlinear system of differential equations we again use the method of "elastic" Ilyushin solutions. Suppose that (4.72)–(4.78) contains a small parameter, e.g., all physical nonlinearity functions are small ($\omega < 1$). Then, on the basis of the method of elastic solutions, the method of iterations is possible, at which for any nth approximation the system of Eq. (4.72) is transformed to the form:

$$
\begin{aligned}
&(a_1 + a_{10}H(x - x_0))u,_{xx}^{(n)} + (a_6 + a_{60}H(x - x_0))\psi,_{xx}^{(n)} \\
&- (a_7 + a_{70}H(x - x_0))w,_{xxx}^{(n)} = -p + p_\omega^{(n-1)}, \\
&(a_6 + a_{60}H(x - x_0))u,_{xx}^{(n)} + (a_2 + a_{20}H(x - x_0))\psi,_{xx}^{(n)} \\
&- (a_3 + a_{30}H(x - x_0))w,_{xxx}^{(n)} - a_5\psi = h_\omega^{(n-1)}, \\
&(a_7 + a_{70}H(x - x_0))u,_{xxx}^{(n)} + (a_3 + a_{30}H(x - x_0))\psi,_{xxx}^{(n)} \\
&- (a_4 + a_{40}H(x - x_0))w,_{xxxx}^{(n)} = -q + q_\omega^{(n-1)}.
\end{aligned}
\tag{4.79}
$$

The force boundary conditions in (2.12) at the rod ends are transformed as follows:

$$
\begin{aligned}
&x = 0 : N^{(n)} = N_0 + N^{\omega(n-1)}, M^{(n)} = M_0 + M^{\omega(n-1)}, \\
&M^{(n)},_x = Q_0 + M^{\omega(n-1)},_x, H^{(n)} = H^{\omega(n-1)}; \\
&x = l : N^{(n)} = N_l + N^{\omega(n-1)}, M^{(n)} = M_l + M^{\omega(n-1)}, \\
&M^{(n)},_x = Q_l + M^{\omega(n-1)},_x, H^{(n)} = H^{\omega(n-1)}.
\end{aligned}
\tag{4.80}
$$

The values $p_\omega^{(n-1)}$, $h_\omega^{(n-1)}$, $q_\omega^{(n-1)}$, corresponding to the inelastic components in (4.79), (4.80), are assumed to be zero at the first step, and are further calculated from the results of the previous approximation by formulae (4.10).

Application of the proposed method of successive approximations allows at each approximation step to reduce the considered problem to a linear problem of the theory of elasticity with additional "external" loads.

Let us write out the solution of the system of Eq. (4.79) according to the procedure described in Sect. 2.5. As a result:

$$
\begin{aligned}
&\psi^{(n)}(x) = \psi_1^{(n)}(x) + (\psi_2^{(n)}(x) - \psi_1^{(n)}(x))H(x - x_0), \\
&u^{(n)}(x) = u_1^{(n)}(x) + (u_2^{(n)}(x) - u_1^{(n)}(x))H(x - x_0), \\
&w^{(n)}(x) = w_1^{(n)}(x) + (w_2^{(n)}(x) - w_1^{(n)}(x))H(x - x_0).
\end{aligned}
\tag{4.81}
$$

Here, the displacements with lower index 1 corresponding to the solution in the first region, i.e., up to the point of irregularity of the bottom layer, are defined as follows:

$$
\begin{aligned}
\psi_1^{(n)}(x) = {} & C_{21}^{(n)} sh(\beta_1 x) + C_{31}^{(n)} ch(\beta_1 x) \\
& + \frac{1}{\beta_1}\left[sh(\beta_1 x) \int g_1^{(n)}(x) ch(\beta_1 x) dx \right. \\
& \left. - ch(\beta_1 x) \int g_1^{(n)}(x) sh(\beta_1 x) dx \right],
\end{aligned}
$$

$$u_1^{(n)}(x) = \gamma_{31}\psi_1^{(n)} + \frac{1}{\alpha_{21}}\left[-a_4 L_2^{-1}(p - p_\omega^{(n-1)}) + a_7 L_3^{-1}(q - q_\omega^{(n-1)}) + \frac{a_7}{2}C_{11}^{(n)}x^2\right] + C_{71}^{(n)}x + C_{81}^{(n)},$$

$$w_1^{(n)}(x) = \frac{1}{\alpha_{21}}\left[\alpha_{11}\int \psi_1^{(n)}\mathrm{d}x - a_7 L_3^{-1}(p - p_\omega^{(n-1)}) + a_1 L_4^{-1}(q - q_\omega^{(n-1)}) + \frac{a_1 C_{11}^{(n)}}{6}x^3\right] + \frac{C_{41}^{(n)}}{2}x^2 + C_{51}^{(n)}x + C_{61}^{(n)}, \tag{4.82}$$

where

$$g_1^{(n)}(x) = \frac{\alpha_{21}}{\alpha_{11}}\gamma_{11}h_\omega^{(n-1)} + \gamma_{21}\,(p - p_\omega^{(n-1)}) + \gamma_{11}\left(\int (q - q_\omega^{(n-1)})\mathrm{d}x + C_{11}^{(n)}\right)$$

Displacements with index 2 are the next solution in the second region:

$$\psi_2^{(n)}(x) = C_{22}^{(n)}sh(\beta_2 x) + C_{32}^{(n)}ch(\beta_2 x) + \frac{1}{\beta_2}\left[sh(\beta_2 x)\int g_2^{(n)}(x)ch(\beta_2 x)\mathrm{d}x - ch(\beta_2 x)\int g_2^{(n)}(x)sh(\beta_2 x)\mathrm{d}x\right],$$

$$u_2^{(n)}(x) = \gamma_{32}\psi_2^{(n)} + \frac{1}{\alpha_{22}}\left[-a_{41}L_2^{-1}(p - p_\omega^{(n-1)}) + a_{71} L_3^{-1}(q - q_\omega^{(n-1)}) + \frac{a_{71}}{2}C_{12}^{(n)}x^2\right] + C_{72}^{(n)}x + C_{82}^{(n)},$$

$$w_2^{(n)}(x) = \frac{1}{\alpha_{22}}\left[\alpha_{12}\int \psi_2^{(n)}\mathrm{d}x - a_{71}L_3^{-1}(p - p_\omega^{(n-1)}) + a_{11}L_4^{-1}(q - q_\omega^{(n-1)}) + \frac{a_{11}C_{12}^{(n)}}{6}x^3\right] + \frac{C_{42}^{(n)}}{2}x^2 + C_{52}^{(n)}x + C_{62}^{(n)}, \tag{4.83}$$

where $\alpha_{11}, \alpha_{21}, \alpha_{31}, \beta_1^2, \gamma_{11}, \gamma_{21}, \alpha_{12}, \alpha_{22}, \alpha_{32}, \beta_2^2, \gamma_{12}, \gamma_{22}, \gamma_{23}$ are parameters introduced in (2.45), (2.47); L_2^{-1}, L_3^{-1}, L_4^{-1} are linear integral operators, are defined in (2.23); function

$$g_2^{(n)}(x) = \frac{\alpha_{22}}{\alpha_{12}}\gamma_{12}\, h_\omega^{(n-1)} + \gamma_{22}\,(p - p_\omega^{(n-1)}) + \gamma_{12}\left(\int (q - q_\omega^{(n-1)})\mathrm{d}x + C_{12}^{(n)}\right).$$

The integration constants $C_{11}^{(n)}, \ldots, C_{81}^{(n)}; C_{12}^{(n)}, \ldots, C_{82}^{(n)}$ follow from the 8 conditions for fixation of the rod at the ends and the additional 8 requirements for conjugation of the solutions at the point of irregularity.

In the case of *rigid embedding of the left end of the rod with a free right end*, the boundary conditions follow from (4.75):

$$\begin{aligned}
&x = 0 : w_1^{(n)} = w_1^{(n)}{}_{,x} = u_1^{(n)} = \psi_1^{(n)} = 0;\\
&x = l : \psi_2^{(n)} = 0, N_2^{(n)} = N_2^{\omega(n-1)}, M_2^{(n)} = M_2^{\omega(n-1)}, M_2^{(n)}{}_{,x} = M_2^{\omega(n-1)}{}_{,x}.
\end{aligned} \tag{4.84}$$

For a *rod hinged at both ends*, the boundary conditions are obtained from Eq. (4.76):

$$\begin{aligned}
&x = 0 : w_1^{(n)} = u_1^{(n)} = \psi_1^{(n)} = 0, M_1^{(n)} = M_1^{\omega(n-1)};\\
&x = l : w_2^{(n)} = u_2^{(n)} = \psi_2^{(n)} = 0, M_2^{(n)} = M_2^{\omega(n-1)}.
\end{aligned} \tag{4.85}$$

If the rod is *clamped at both ends*, the conditions like (4.77) must be satisfied:

$$\begin{aligned}
&x = 0 : w_1^{(n)} = w_1^{(n)}{}_{,x} = u_1^{(n)} = \psi_1^{(n)} = 0;\\
&x = l : w_2^{(n)} = w_2^{(n)}{}_{,x} = u_2^{(n)} = \psi_2^{(n)} = 0.
\end{aligned} \tag{4.86}$$

Let's write out an algebraic system of eight equations that cross-link the solutions of (4.82), (4.83) at the point $x = x_0$. For this purpose we require fulfillment of eight conditions (4.78). As a result:

$$\begin{aligned}
&C_{21}^{(n)} sh(\beta_1 x_0) + C_{31}^{(n)} ch(\beta_1 x_0) - C_{22}^{(n)} sh(\beta_2 x_0) - C_{32}^{(n)} ch(\beta_2 x_0)\\
&= \frac{1}{\beta_2}\left[sh(\beta_2 x) \int g_2^{(n)}(x) ch(\beta_2 x)\mathrm{d}x - ch(\beta_2 x) \int g_2^{(n)}(x) sh(\beta_2 x)\mathrm{d}x \right]\Bigg|_{x=x_0}\\
&- \frac{1}{\beta_1}\left[sh(\beta_1 x) \int g_1^{(n)}(x) ch(\beta_1 x)\mathrm{d}x - ch(\beta_1 x) \int g_1^{(n)}(x) sh(\beta_1 x)\mathrm{d}x \right]\Bigg|_{x=x_0},
\end{aligned}$$

$$\begin{aligned}
&C_{21}^{(n)} \beta_1 ch(\beta_1 x_0) + C_{31}^{(n)} \beta_1 sh(\beta_1 x_0)\\
&- C_{22}^{(n)} \beta_2 sh(\beta_2 x_0) - C_{32}^{(n)} \beta_2 ch(\beta_2 x_0)\\
&= \frac{1}{\beta_2}\Bigg[sh(\beta_2 x) \int g_2^{(n)}(x) ch(\beta_2 x)\mathrm{d}x\\
&- ch(\beta_2 x) \int g_2^{(n)}(x) sh(\beta_2 x)\mathrm{d}x \Bigg]_{,x} \Big|_{x=x_0}\\
&- \frac{1}{\beta_1}\Bigg[sh(\beta_1 x) \int g_1^{(n)}(x) ch(\beta_1 x)\mathrm{d}x
\end{aligned}$$

$$-ch(\beta_1 x)\int g_1^{(n)}(x)sh(\beta_1 x)\mathrm{d}x\Big],_x|_{x=x_0},$$

$$\begin{aligned}
&C_{21}^{(n)}\beta_1^2 sh(\beta_1 x_0)+C_{31}^{(n)}\beta_1^2 ch(\beta_1 x_0)\\
&-C_{22}^{(n)}\beta_2^2 sh(\beta_2 x_0)-C_{32}^{(n)}\beta_2^2 ch(\beta_2 x_0)\\
&=\frac{1}{\beta_2}\Big[sh(\beta_2 x)\int g_2^{(n)}(x)ch(\beta_2 x)\mathrm{d}x\\
&-ch(\beta_2 x)\int g_2^{(n)}(x)sh(\beta_2 x)\mathrm{d}x\Big],_{xx}|_{x=x_0}\\
&-\frac{1}{\beta_1}\Big[sh(\beta_1 x)\int g_1^{(n)}(x)ch(\beta_1 x)\mathrm{d}x\\
&-ch(\beta_1 x)\int g_1^{(n)}(x)sh(\beta_1 x)\mathrm{d}x\Big],_{xx}|_{x=x_0},
\end{aligned}$$

$$\begin{aligned}
&\frac{1}{\alpha_{21}}\Big[\alpha_{11}\int \psi_1^{(n)}\mathrm{d}x-a_7L_3^{-1}(p-p_\omega^{(n-1)})\\
&+a_1L_4^{-1}(q-q_\omega^{(n-1)})+\frac{a_1C_{11}^{(n)}}{6}x^3\Big]|_{x=x_0}\\
&+\frac{C_{41}^{(n)}}{2}x_0^2+C_{51}^{(n)}x_0+C_{61}^{(n)}\\
&-\frac{1}{\alpha_{22}}\Big[\alpha_{12}\int \psi_2^{(n)}\mathrm{d}x-a_{71}L_3^{-1}(p-p_\omega^{(n-1)})\\
&+a_{11}L_4^{-1}(q-q_\omega^{(n-1)})+\frac{a_{11}C_{12}^{(n)}}{6}x^3\Big]|_{x=x_0}\\
&-\frac{C_{42}^{(n)}}{2}x_0^2-C_{52}^{(n)}x_0-C_{62}^{(n)}=0,
\end{aligned}$$

$$\begin{aligned}
&\frac{1}{\alpha_{21}}\Big[\alpha_{11}\psi_1^{(n)}-a_7L_2^{-1}(p-p_\omega^{(n-1)})+a_1L_3^{-1}(q-q_\omega^{(n-1)})+\frac{a_1C_{11}^{(n)}}{2}x^2\Big]\Big|_{x=x_0}\\
&-\frac{1}{\alpha_{22}}\Big[\alpha_{12}\psi_2^{(n)}-a_{71}L_2^{-1}(p-p_\omega^{(n-1)})\\
&+a_{11}L_3^{-1}(q-q_\omega^{(n-1)})+\frac{a_{11}C_{12}^{(n)}}{2}x^2\Big]|_{x=x_0}\\
&+C_{41}^{(n)}x_0+C_{51}^{(n)}-C_{42}^{(n)}x_0-C_{52}^{(n)}=0,
\end{aligned}$$

$$\frac{1}{\alpha_{21}}\Big[\alpha_{11}\psi_1^{(n)},_x-a_7L_1^{-1}(p-p_\omega^{(n-1)})$$

$$+a_1L_2^{-1}(q-q_\omega^{(n-1)})+a_1C_{11}^{(n)}x\Big]\Big|_{x=x_0}$$
$$+C_{41}^{(n)}-\frac{1}{\alpha_{22}}\Big[\alpha_{12}\,\psi_2^{(n)},_x-a_{71}L_1^{-1}(p-p_\omega^{(n-1)})$$
$$+a_{11}L_2^{-1}(q-q_\omega^{(n-1)})+a_{11}C_{12}^{(n)}x\Big]\Big|_{x=x_0}$$
$$-C_{42}^{(n)}=0,$$

$$\gamma_{31}\psi_1^{(n)}\Big|_{x=x_0}+\frac{1}{\alpha_{21}}\Big[-a_4L_2^{-1}(p-p_\omega^{(n-1)})$$
$$+a_7L_3^{-1}(q-q_\omega^{(n-1)})+\frac{a_7}{2}C_{11}^{(n)}x^2\Big]\Big|_{x=x_0}$$
$$-\frac{1}{\alpha_{22}}\Big[-a_{41}L_2^{-1}(p-p_\omega^{(n-1)})$$
$$+a_{71}L_3^{-1}(q-q_\omega^{(n-1)})+\frac{a_{71}}{2}C_{12}^{(n)}x^2\Big]\Big|_{x=x_0}$$
$$+C_{71}^{(n)}x_0+C_{81}^{(n)}-\gamma_{32}\psi_2^{(n)}|_{x=x_0}-C_{72}^{(n)}x_0-C_{82}^{(n)}=0,$$

$$\gamma_{31}\psi_1^{(n)},_x\Big|_{x=x_0}+\frac{1}{\alpha_{21}}\Big[-a_4L_1^{-1}(p-p_\omega^{(n-1)})$$
$$+a_7L_2^{-1}(q-q_\omega^{(n-1)})+a_7C_{11}^{(n)}x\Big]\Big|_{x=x_0}$$
$$-\frac{1}{\alpha_{22}}\Big[-a_{41}L_1^{-1}(p-p_\omega^{(n-1)})$$
$$+a_{71}L_2^{-1}(q-q_\omega^{(n-1)})+a_{71}C_{12}^{(n)}x\Big]\Big|_{x=x_0}$$
$$+C_{71}^{(n)}-\gamma_{32}\,\psi_2^{(n)},_x\Big|_{x=x_0}-C_{72}^{(n)}=0. \tag{4.87}$$

Thus, the general recurrence solution of the problem on deformation of an elastic–plastic three-layer rod with an irregular lower boundary is represented by Eq. (4.81) taking into account (4.82), (4.83). The integration constants are found from system (4.84)–(4.87).

Numerical results are obtained in the case of boundary conditions (4.86), (4.87). Three-layer rod package: D16T-Fluoroplastic-D16T, geometric characteristics $c=0.09$, $h_1=h_2=0.03$, $h_{20}=0.02$, $h_{21}=0.01$. Coordinate of the step start is: $x_0=0.5$. Load intensities are: $p=0$, $q=-10$ MPa. The plasticity function for the bearing layers ($k=1,2$) and the physical nonlinearity function for the filler are taken in the form (1.38).

Figure 4.21 illustrates the convergence of the successive approximation method process. The curve number corresponds to the iteration number. Here, the second approximation differs from the first approximation by about 10% for deflection and shear. The 6th approximation, which differs from the previous one by less than 1%, is taken as the desired solution. If we continue the iterative process, the difference between the subsequent and previous approximations decreases and tends to zero, which is confirmed in Fig. 4.10. Therefore, we stop the iterative process when the difference of less than 1% between the subsequent and previous maximum deflection values is reached. The resulting elastic–plastic deflection and shear of the rod exceed the elastic deflection and shear by 23% and 25%, respectively. This exceeds the corresponding values of the smooth rod, due to its thinning and larger plastic deformation area.

Figure 4.22 shows the areas of physical nonlinearity in the rod layers (dark fill). The first plastic deformations appear on the outer surfaces of the left end face of the second load-bearing layer, then in the same section of the first layer. As the load increases, similar regions are formed first symmetrically on the right end and then in the middle of the rod on the outer planes. The nonlinearity of the filler deformation is due to both deflection and relative shear. As a result, plasticity regions occupy about 45% of the volume in the load-bearing layers and 76% in the filler. All material nonlinearity regions are slightly shifted to the left.

Variation of normal $\sigma_x^{(3)}$ and tangential $\sigma_{xz}^{(3)}$ stresses in the filler are shown in Fig. 4.23: 1, $1'$—$\sigma_{xz}^{(3)}$; 2, $2'$—$\sigma_{xx}^{(3)}$ $(z = c)$; 3, $3'$—$\sigma_x^{(3)}$ $(z = -c)$. Tangential stresses

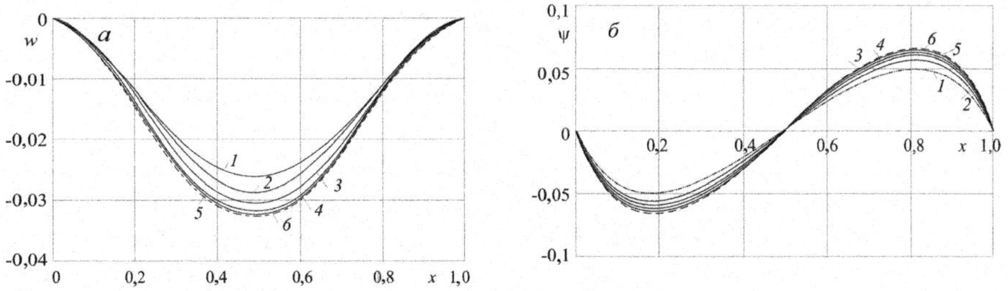

Fig. 4.21 Convergence of the method of successive approximations

Fig. 4.22 Areas of physical nonlinearity in the layers of the rod

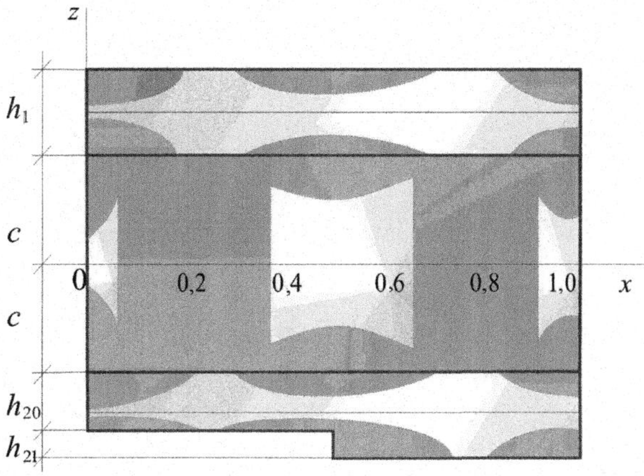

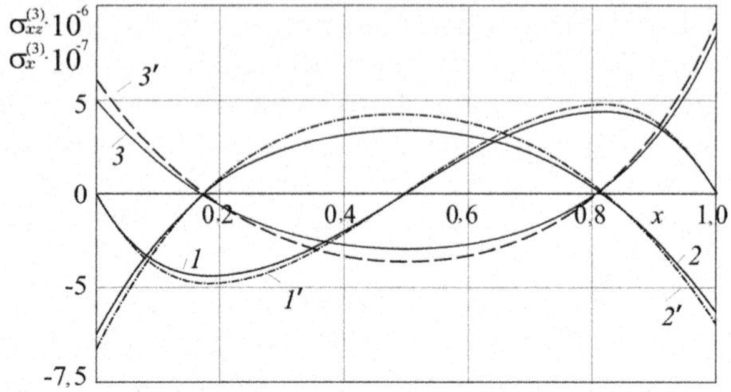

Fig. 4.23 Variation of normal $\sigma_x^{(3)}$ and tangential $\sigma_{xz}^{(3)}$ stresses in the filler

along the thickness of the plate do not change. Normal stresses exceed tangential stresses by an order of magnitude.

The distribution of normal stresses $\sigma_x^{(1)}$ along the axis of the rod on the surfaces of the first bearing layer is illustrated in Fig. 4.24: 1, 1′—at $z = c$; 2, 2′—at $z = c + h_2$. The maximum increase of stresses due to nonlinearity of materials in the considered case is about 12–14%.

Variation of normal stresses $\sigma_x^{(2)}$ along the axis of the rod on the boundary surfaces of the second bearing layer is shown in Fig. 4.25: 1, 1′—$\sigma_x^{(2)}$ at $z = -h_2 - c$; 2, 2′—$\sigma_x^{(2)}$ at $z = -c$.

The maximum stress increase due to physical nonlinearity of materials in this case is also of the order of 12–13%.

Figure 4.26 illustrates the change of normal stresses $\sigma_x^{(k)}$ along the thickness of the rod at the left end under elastic and physically nonlinear deformation: *1*—elastic rod; 2—elastoplastic rod. Here the stresses in the outer layers are referred to $q_1 = 10^8$ Pa, in the filler to $q_2 = 10^7$ Pa. Their distribution over the height of the rod is

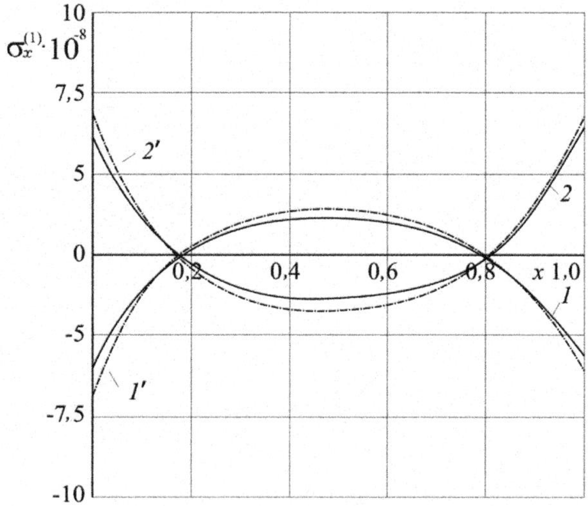

Fig. 4.24 Distribution of normal stresses $\sigma_x^{(1)}$ along the axis of the rod on the surfaces of the first bearing layer

Fig. 4.25 Variation of normal stresses $\sigma_x^{(2)}$ along the axis of the rod on the boundary surfaces of the second bearing layer

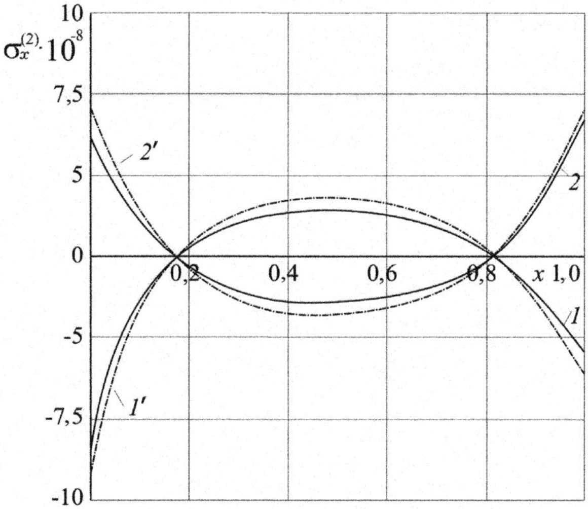

Fig. 4.26 Variation of normal stresses $\sigma_x^{(k)}$ along the thickness of the rod at the left end under elastic and physically nonlinear deformation

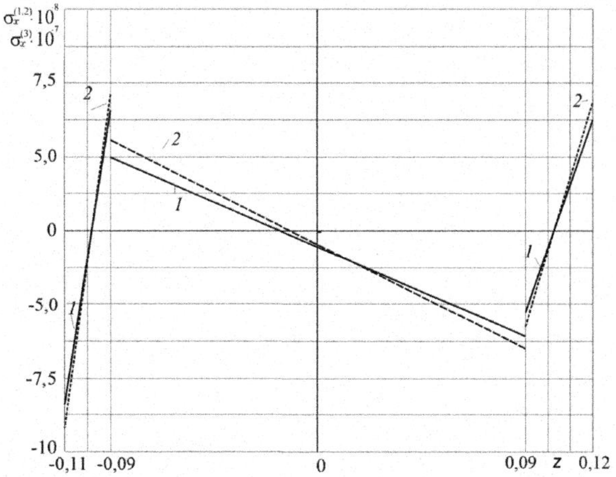

linear. Due to the presence of the step, the maximum stresses in the left end on the bottom surface of layer 2 exceed those of layer 1 by almost 33%.

The parametric numerical study of the stress–strain state of asymmetric in thickness elastic–plastic three-layer rods with continuous and stepped surfaces has shown the necessity of taking into account physically nonlinear properties of materials of layers, both at single and variable isothermal deformation. It should be emphasized that the above results are valid only in the region of small elastoplastic deformations and under simple cyclic loadings.

Reference

1. Moskvitin, V. V. 1981. *Cyclic loading of structural elements*, 344. Moscow: Nauka. (in Russian).

Chapter 5
Three-Layer Elastic–Plastic Rods in a Temperature Field

Studies of single and variable deformation of asymmetric in thickness elastic–plastic three-layer rods in a temperature field are presented. The load is uniformly distributed. The kinematic hypotheses are based on the "broken" normal hypothesis with an incompressible filler. The work of tangential stresses in the filler is taken into account. Analytical solutions are obtained and numerical analysis of thermoelasticity and thermoelastic-plasticity problems is performed.

5.1 Statement of the Boundary Value Problem of Thermoelastic-Plasticity

Consider the deformation of a three-layer elastic–plastic rod in the temperature field when loaded from the natural state. Let to the external surface $z = c + h_1$ of the rod under consideration, in addition to the distributed power load $p'(x)$, $q'(x)$, be also supplied with a heat flux of density q_t in the direction opposite to the external normal (Fig. 5.1). The surface $z = -c - h_2$ and the contour of the rod are thermally isolated. This allows the inhomogeneous temperature field in the rod $T(z)$, counted from some initial temperature T_0, to be calculated with sufficient accuracy by formula (1.55):

$$T = \frac{q_t H}{\lambda}\left\{\tau + \frac{1}{2}\left(\frac{z + c + h_2}{H}\right)^2, \right.$$

$$\left. -\frac{1}{6} - \frac{2}{\pi^2}\sum_{n=1}^{\infty}\frac{(-1)^n}{n^2}\cos\left(\pi n\frac{z + c + h_2}{H}\right)e^{-n^2\pi^2\tau}\right\},$$

$$\tau = \frac{at}{H^2}, a = \sum_{k=1}^{3}\lambda_{tk}h_k \Big/ \sum_{k=1}^{3}\rho_k C_k h_k, H = \sum_{k=1}^{3}h_k \tag{5.1}$$

S. Eduard et al., *Deformation of Three-layer Structural Elements in Thermal Radiation Fields*, https://doi.org/10.1007/978-981-97-7217-9_5

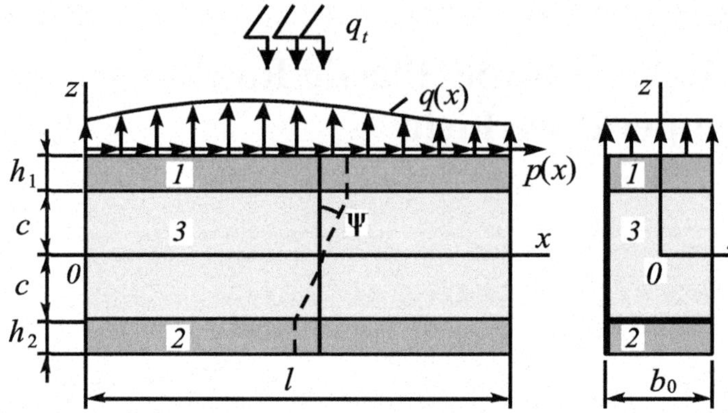

Fig. 5.1 Design diagram of the thermal force loading of the rod

where λ_{tk}, C_k—thermal conductivity and heat capacity of the material of the kth layer.

It is assumed that the load-bearing layers of the rod are made of elastic–plastic material, and the incompressible filler in its thickness is nonlinear-elastic. To describe the kinematics of the package, the previously adopted broken-line hypotheses are valid. The loading trajectories are considered to belong to the class of simple trajectories (according to Ilyushin [1]). Values with one streak are stresses, strains, external loads and displacements in the rod before unloading. Physical equations of state of the theory of small elastic–plastic deformations with regard to temperature are used in the bearing layers of the rod:

$$s_x^{\prime(k)} = 2G_k(T) f^{\prime(k)}(\varepsilon_u^{(k)}, T)\ \text{э}_x^{\prime(k)},$$
$$s_{xz}^{(3)} = 2G_3(1 - \omega^{(k)}(\varepsilon_u^{(k)}, T_k))\ \text{э}_{xz}^{(3)},$$

$$\sigma^{\prime(k)} = 3K_k(T)(\varepsilon^{\prime(k)} - \alpha_{0k}T)(k = 1, 2). \tag{5.2}$$

where $s_x^{\prime(k)}$, $\sigma^{\prime(k)}$ и $\text{э}_{xx}^{\prime(k)}$, $\varepsilon^{\prime(k)}$ are deviator and spherical parts of the stress and strain tensors of the kth layer; $s_{xz}^{(3)}$, $\text{э}_{xz}^{(3)}$ are tangential stress and shear strain in the filler; $G_k(T)$, $K_k(T)$ are temperature-dependent shear and volume deformation modules; α_{0k} is material linear thermal expansion coefficient; $\varepsilon_u^{\prime(k)}$ is strain intensity (rate); $\varepsilon_y^{\prime(k)}$ is the deformation yield strength of the bearing layer material; $f\prime^{(k)}$ is temperature-dependent nonlinearity function (1.38), (1.39) under loading from the natural state:

$$f^{\prime(k)}(\varepsilon_u^{\prime(k)}, T) = \begin{cases} 1, & \varepsilon_u^{\prime(k)} \le \varepsilon_y^{\prime(k)}(T), \\ 1 - \omega^{\prime(k)}(\varepsilon\prime_u^{(k)}, T), & \varepsilon_u^{\prime(k)} > \varepsilon_y^{\prime(k)}(T). \end{cases}$$

The dependence of elastic modulus on temperature is described by Bell's formula (1.37). The values of the deformation yield strength $\varepsilon_y^{\prime(k)}(T)$ follow from the formula for the dependence of the yield strength of metals $\sigma_y'^{(k)}(T)$ on the temperature, proposed by Makhutov [2]:

$$\sigma_y' = \sigma_{y0}' \exp\left\{\kappa\left(\frac{1}{T} - \frac{1}{T_0}\right)\right\}, \varepsilon_y'^{(k)}(T) = \frac{\sigma_y'^{(k)}(T)}{E(T)}. \tag{5.3}$$

where κ is material constant, obtained experimentally.

Similar equations of state are used for a physically nonlinear filler material:

$$s_x'^{(3)} = 2G_3(T)f'^{(3)}(\varepsilon_u^{(3)})\ э_x'^{(3)},$$
$$s_{xz}'^{(3)} = 2G_3(T)f'^{(3)}(\varepsilon_u^{(3)})\ э_{xz}'^{(3)},$$

$$\sigma'^{(3)} = 3K_3(T)(\varepsilon'^{(3)} - \alpha_{03}T). \tag{5.4}$$

Here $f'^{(3)}$ is universal function of physical nonlinearity of the filler ($\varepsilon_s'^{(3)}$ is nonlinearity limit of the filler material)

$$f'^{(3)}(\varepsilon_u'^{(3)}) = \begin{cases} 1, \varepsilon_u'^{(3)} \le \varepsilon_s'^{(3)}(T), \\ 1 - \omega'^{(3)}(\varepsilon_u'^{(3)}), \varepsilon_u'^{(3)} > \varepsilon_s'^{(3)}(T). \end{cases} \tag{5.5}$$

The dependence of elasticity parameters on temperature is taken as follows:

$$G_3(T) = \frac{G_3(T_0)}{\varphi_3(T)}, K_3(T) = \frac{K_3(T_0)}{\varphi_3(T)},$$
$$K_3(T) = \frac{K_3(T_0)}{\varphi_3(T)} \varphi_3(T) = 1 + B\left(\frac{T - T_0}{T_m}\right)^\gamma \tag{5.6}$$

where *B*, γ are material constants obtained experimentally.

Using (5.2), (5.4), let us represent the components of the stress tensor in the layers through the deviator and the spherical part of the strain tensor

$$\sigma_x'^{(k)} = s_x'^{(k)} + \sigma'^{(k)} = 2G_k\ э_x'^{(k)}$$
$$+ 3K_k\left(\varepsilon'^{(k)} - \alpha_k T\right) - 2G_k\ э_x'^{(k)}\ \omega'^{(k)},$$

$$\sigma_{xz}'^{(3)} = s_{xz}'^{(3)} = 2G_3\ э_{xz}'^{(3)} - 2G_3\ э_{xz}'^{(3)}\ \omega'^{(3)} (k = 1, 2, 3). \tag{5.7}$$

As in the isothermal problem, let us distinguish in the stress tensor the elastic (index "*e*") and inelastic (index "ω") terms:

$$\sigma_x'^{(k)} = \sigma_{xe}'^{(k)} - \sigma_{x\omega}'^{(k)}, \sigma_{xz}'^{(3)} = \sigma_{xze}'^{(3)} - \sigma_{xz\omega}'^{(3)},$$

$$\sigma_{xe}'^{(k)} = 2G_k\ э_x'^{(k)} + 3K_k(\varepsilon'^{(k)} - \alpha_{0k}T), \sigma_{x\omega}'^{(k)} = 2G_k\ э_x'^{(k)}\ \varepsilon'^{(k)},$$

$$\sigma_{xze}^{\prime(3)} = 2G_3\ \ni_{xz}^{\prime(3)},\ \sigma_{xz\omega}^{\prime(3)} = 2G_3\ \ni_{xz}^{\prime(3)}\ \omega^{\prime(3)}. \tag{5.8}$$

Let's do the same with the internal forces (efforts) (2.4):

$$N^{\prime(k)} = N_e^{\prime(k)} - N_\omega^{\prime(k)},\ M^{\prime(k)} = M_e^{\prime(k)} - M_\omega^{\prime(k)},\ Q^{\prime(3)} = Q_e^{\prime(3)} - Q_e^{\prime(3)} \tag{5.9}$$

The elastic (index "e") and inelastic (index "ω") components in (5.9) are calculated by the expressions:

$$N_e^{\prime(k)} = b_0 \int \sigma_{xe}^{\prime(k)} dz,\ M_e^{\prime(k)} = b_0 \int\limits_{h_k} \sigma_{xe}^{\prime(k)} z dz,$$

$$M_\omega^{\prime(k)} = b_0 \int\limits_{h_k} \sigma_{x\omega}^{\prime(k)} z dz,\ Q_e^{\prime(3)} = b_0 \int\limits_{-c}^{c} \sigma_{xze}^{\prime(3)} dz,\ Q_\omega^{\prime(3)} = b_0 \int\limits_{-c}^{c} \sigma_{xz\omega}^{\prime(3)} dz. \tag{5.10}$$

Since the equations of equilibrium of the three-layer rod in forces (2.11) and the boundary conditions (2.12) were obtained independently of the physical equations of state, we can use them in the case under consideration. By substituting the internal forces (5.9) into the equations of equilibrium (2.11), we reduce the system to the form:

$$N_{e,x}^{\prime} = -b_0 p' + \sigma_{\omega,x}^{\prime},$$

$$H_{e,x}^{\prime} - Q_e^{\prime} = H_{\omega,x}^{\prime} - Q_\omega^{\prime},$$

$$M_{e,xx}^{\prime} = -b_0 q' + M_{\omega,xx}^{\prime}, \tag{5.11}$$

where

$$N_e^{\prime} = \sum_{k=1}^{3} N_E^{\prime(k)},\ M_e^{\prime} = \sum_{k=1}^{3} M_E^{\prime(k)},\ Q_e^{\prime} = Q_E^{\prime(3)},$$

$$H_e^{\prime} = c(N_e^{\prime(1)} - N_e^{\prime(2)}) + M_e^{\prime(3)},$$

$$N_\omega^{\prime} = \sum_{k=1}^{3} N_\omega^{\prime(k)},\ M_\omega^{\prime} = \sum_{k=1}^{3} M_\omega^{\prime(k)},$$

$$Q_\omega^{\prime} = Q_\omega^{\prime(3)},\ H_\omega^{\prime} = c(N_\omega^{\prime(1)} - N_\omega^{\prime(2)}) + M_\omega^{\prime(3)}.$$

The main difference between system (5.11) and Eq. (5.5) is in the consideration of temperature through stresses (5.8), (5.10) and elastic modules. The formulas for calculating the inelastic components in (5.11) follow from relations (5.8) and (5.10):

$$N'_\omega = \frac{4}{3}b_0 \sum_{k=1}^{3} \int_{h_k} G_k \omega'_k \varepsilon_x'^{(k)} \mathrm{d}z,\; M'_\omega = \frac{4}{3}b_0 \sum_{k=1}^{3} \int_{h_k} G_k \omega'_k \varepsilon_x'^{(k)} z\mathrm{d}z,$$

$$Q'_\omega = 2b_0 \int_{-c}^{c} G_3 \omega'_3 \Psi' \mathrm{d}z. \tag{5.12}$$

Using relations (5.8), (2.2), express the elastic components of the generalized forces (5.10) through the required functions $u'(x)$, $\Psi'(x)$, $w'(x)$ and substitute the result in Eq. (5.11). As a result, we obtain a system of nonlinear differential equations of equilibrium of a three-layer thermoelastic-plastic rod in displacements:

$$a_1 u'_{xxx}, + a_6 \psi'{,}_{xx} - a_7 w'{,}_{xxx} = -p' + p'_\omega,$$

$$a_6 u'{,}_{xx} + a_2 \psi'{,}_{xx} - a_3 w'{,}_{xxx} - a_5 \psi' = h'_\omega,$$

$$a_7 u'{,}_{xxx} + a_3 \Psi'{,}_{xxx} - a_4 w'{,}_{xxxx} = -q' + q'_\omega. \tag{5.13}$$

Here, the coefficients a_i are determined by integral relations, since the elastic modules of the materials in the layers change along their thickness together with the temperature (5.1):

$$a_1 = \sum_{k=1}^{3} K_{k0},\; a_2 = K_{32} + c^2(K_{10} + K_{20}),\; a_3 = K_{32} + c(K_{11} + K_{21}),$$

$$a_4 = \sum_{k=1}^{3} K_{k2},\; a_5 = \int_{h_3} G_3 dz,\; a_6 = c(K_{10} - K_{20}),\; a_7 = \sum_{k=1}^{3} K_{k1},$$

$$K_{km} = \int_{h_k} \left[K_k(T) + \tfrac{4}{3}G_k(T)\right] z^m \mathrm{d}z,\; (m = 0, 1, 2). \tag{5.14}$$

The quantities p'_ω, h'_ω, q'_ω included in the right-hand sides of Eq. (5.13) take into account the physical nonlinearity of the layer materials and are calculated using relations (5.12):

$$p'_\omega = \frac{1}{b_0} N'_{\omega,x},\; h'_\omega = \frac{1}{b_0}(H'_{\omega,x} - Q'_\omega),\; q'_\omega = \frac{1}{b_0} M'_{\omega,xx}. \tag{5.15}$$

If the temperature field is averaged over the thickness of the kth layer of the rod, the values of elastic modules are constant within the layer

$$T_k(t) = \frac{1}{h_k}\int\limits_{h_k} T(z,t)dz,\ G_k(T) \equiv G_k(T_k),\ K_k(T) \equiv K_k(T_k). \tag{5.16}$$

For the coefficients a_i (5.14) the formulas (2.17) will be true.

The elastic components of the internal forces in (5.11) can be expressed as displacement functions using (2.2), (5.8) and (5.10):

$$N'^{(1)} = b_0 \int\limits_{c}^{c+h_1} \sigma_x'^{(1)} \mathrm{d}z = b_0 h_1 \left[K_1^+ \left((u'_{,x} + c\psi_{,x}) - \left(c + \frac{1}{2}h_1\right) w'_{,xx}\right) - 3\alpha_{01} K_1 T_1\right],$$

$$N'^{(2)} = b_0 h_2 \left[\left(K_2^+ (u'_{,x} - c\Psi'_{,x}) + \left(c + \frac{1}{2}h_2\right) h_2 w'_{,xx}\right) - 3\alpha_{02} K_2 T_2\right],$$

$$N'^{(3)} = b_0 \int\limits_{-c}^{c} \sigma_x'^{(3)} dz = 2b_0 c\left(K_3^+ u'_{,x} - 3K_3 T_3 \alpha_{03}\right),$$

$$M'^{(1)} = b_0 h_1 \left[K_1^+ \left(\left(c + \frac{1}{2}h_1\right)(u_{,x} + c\psi_{,x})\right.\right.$$
$$\left.\left.\left(c^2 + ch_1 + \frac{1}{3}h_1^2\right) w_{,xx}\right) - 3K_1 T_1 \alpha_{01}\left(c + \frac{1}{2}h_1\right)\right],$$

$$M'^{(2)} = -b_0 h_2 \left[K_2^+ \left(\left(c + \frac{1}{2}h_2\right)(u_{,x} - c\psi_{,x})\right.\right.$$
$$\left.\left. + \left(c^2 + ch_2 + \frac{1}{3}h_2^2\right) w_{,xx}\right)\right] - 3K_2 T_2 \alpha_{02}\left(c + \frac{1}{2}h_2\right)\Bigg],$$

$$M'^{(3)} = b_0 \int\limits_{-c}^{c} \sigma_x^{(3)} z\mathrm{d}z = \frac{2}{3} b_0 c^3 \left[K_3^+ (\psi_{,x} - w_{,xx}) - 3K_3 T_3 \alpha_{03}\right],$$

$$Q'^{(3)} = b_0 \int\limits_{-c}^{c} \sigma_{xz}^{(3)} \mathrm{d}z = 2b_0 c G_3 \psi. \tag{5.17}$$

The coefficients a_i after integration in (5.14) will formally take the form (2.17), but will take into account the temperature in the rod layers through elasticity parameters using relations (5.16).

5.2 Solution of the Boundary Value Problem of Thermoelastic-Plasticity

It should be noted that the temperature is not explicitly included in the equilibrium Eq. (5.13). This system of differential equations is nonlinear, so assume, as before, that (5.13), (5.15) contains a small parameter, for example, all $\omega_k < 1$. Then the method of elastic solutions reduces the system of Eqs. (5.13) at the nth step of approximation to the following iterative form:

$$a_1 u,_{xx}^{\prime(n)} + a_6 \psi,_{xx}^{\prime(n)} - a_7 w,_{xxx}^{\prime(n)} = -p' + p_\omega^{\prime(n-1)},$$

$$a_6 u,_{xx}^{\prime(n)} + a_2 \psi,_{xx}^{\prime(n)} - a_3 w,_{xxx}^{\prime(n)} - a_5 \psi^{\prime(n)} = h_\omega^{\prime(n-1)},$$

$$a_7 u,_{xxx}^{\prime(n)} + a_3 \psi,_{xxx}^{\prime(n)} - a_4 w,_{xxxx}^{\prime(n)} = -q' + q_\omega^{\prime(n-1)}. \tag{5.18}$$

The values $p_\omega^{\prime(n-1)}, h_\omega^{\prime(n-1)}, q_\omega^{\prime(n-1)}$ corresponding to the inelastic components of (5.15) are taken to be zero at the first step ($n = 1$), and then are calculated from the results of the previous approximation and are called additional "external" loads:

$$p_\omega^{\prime(n-1)} = \frac{1}{b_0} N_\omega^{(n-1)},_x,\; h_\omega^{\prime(n-1)} = \frac{1}{b_0}(H_\omega^{\prime(n-1)},_x - Q_\omega^{\prime(n-1)}),\; q_\omega^{\prime(n-1)} = \frac{1}{b_0} M_\omega^{\prime(n-1)},_{xx},$$

$$N_\omega^{\prime(n-1)} = \sum_{k=1}^{3} N_\omega^{\prime(k)(n-1)} = \frac{4}{3} b_0 \sum_{k=1}^{3} \int_{h_k} G_k \omega^{\prime(k)} (\varepsilon_u^{\prime(k)(n-1)}) \varepsilon_x^{\prime(k)(n-1)} dz,$$

$$H_\omega^{\prime(n-1)} = c(N_\omega^{\prime(1)(n-1)} - N_\omega^{\prime(2)(n-1)}) + M_\omega^{\prime(3)(n-1)},$$

$$Q_\omega^{\prime(k)(n-1)} = 2b_0 \int_{-c}^{c} G_3 \omega^{\prime(3)} (\varepsilon_u^{\prime(3)(n-1)}) \psi^{\prime(n-1)} dz,$$

$$M_\omega^{\prime(n-1)} = \sum_{k=1}^{3} M_\omega^{\prime(k)(n-1)} = \frac{4}{3} b_0 \sum_{k=1}^{3} \int_{h_k} G_k \omega^{\prime(k)} (\varepsilon_u^{\prime(k)(n-1)}) \varepsilon_x^{\prime(k)(n-1)} z dz. \tag{5.19}$$

Application of the method of successive approximations allows to reduce the considered problem to a linear problem of thermoelasticity with additional "external" loads at each step of approximation. In this case the system of Eq. (5.18) does not differ formally at each step of approximation from the corresponding system (3.9) in the problem of isothermal elastic-plasticity. Therefore, the recurrence solution of the thermoelastic-plasticity problem will be:

$$\psi'^{(n)}(x) = C_2^{(n)}\mathrm{sh}(\beta x) + C_3^{(n)}\mathrm{ch}(\beta x)$$
$$+\frac{1}{\beta}\left[\mathrm{sh}(\beta x)\int g'^{(n)}\mathrm{ch}(\beta x)dx - \mathrm{ch}(\beta x)\int g'^{(n)}\mathrm{sh}(\beta x)dx\right].$$
$$u'^{(n)}(x) = \frac{1}{\alpha_2}\Big[-a_4L_2^{-1}(p' - p_\omega'^{(n-1)}) + a_7L_3^{-1}(q' - q_\omega'^{(n-1)})$$
$$+\frac{a_7}{2}C_1^{(n)}x^2\Big] + \gamma_3\psi'^{(n)} + C_7^{(n)}x + C_8^{(n)},$$

$$w'^{(n)}(x) = \frac{1}{\alpha_2}\left[\alpha_1\int\psi'^{(n)}\mathrm{d}x - a_7L_3^{-1}(p' - p_\omega'^{(n-1)})\right.$$
$$\left. + a_1\mathrm{L}_4^{-1}(q' - q_\omega'^{(n-1)}) + \frac{1}{6}a_1C_1^{(n)}x^3\right]$$
$$+\frac{1}{2}C_4^{(n)}x^2 + C_5^{(n)}x + C_6^{(n)}. \tag{5.20}$$

Here the coefficients $\alpha_1, \alpha_2, \alpha_3, \beta^2, \gamma_1, \gamma_2, \gamma_3$ and linear integral operators L_2^{-1}, L_3^{-1}, L_4^{-1} are defined by formulas (2.19), (2.23), the function $g'^{(n)}(x)$ is introduced in (3.11). Integration constants $C_1^{(n)}, \ldots, C_8^{(n)}$ take into account the effect of the temperature field and at each approximation step follow from the rod anchoring conditions.

In the case of *rigid cantilever embedding* of the left end of the rod, the boundary conditions (2.24) must be satisfied. In this case:

$$C_1^{(n)} = -\mathrm{L}_1^{-1}(q' - q_\omega'^{(n-1)})|_{x=l},$$

$$C_2^{(n)} = \frac{1}{\beta}\left[\frac{ch(\beta l)}{sh(\beta l)}\left(\int g'^{(n)}sh(\beta x)\mathrm{d}x|_{x=l} - \int g'^{(n)}sh(\beta x)\mathrm{d}x|_{x=0}\right) - \int g'^{(n)}ch(\beta x)\mathrm{d}x|_{x=l},\right.$$

$$C_3^{(n)} = \frac{1}{\beta}\int g'^{(n)}(x)sh(\beta x)\mathrm{d}x|_{x=0},$$

$$C_4^{(n)} = \frac{a_7}{\alpha_2}\left(L_1^{-1}(p' - p_\omega'^{(n-1)}) + 3\sum_{k=1}^{3}\alpha_{0k}\int_{h_k}K_kTdz\right)|_{x=l}$$
$$-\frac{a_1}{\alpha_2}\left(L_2^{-1}(q' - q_\omega'^{(n-1)}) + 3\sum_{k=1}^{3}\alpha_{0k}\int_{h_k}K_kTzdz\right)|_{x=l} - \frac{a_1}{\alpha_2}C_1^{(n)}l,$$

$$C_5^{(n)} = \frac{a_7}{\alpha_2}L_2^{-1}(p' - p_\omega'^{(n-1)})|_{x=0} - \frac{a_1}{\alpha_2}L_3^{-1}(q' - q_\omega'^{(n-1)})|_{x=0},$$

$$C_6^{(n)} = \frac{a_7}{\alpha_2}L_3^{-1}(p' - p_\omega'^{(n-1)})\Big|_{x=0} - \frac{a_1}{\alpha_2}L_4^{-1}(q' - q_\omega'^{(n-1)})\Big|_{x=0} - \frac{\alpha_1}{\alpha_2}\int\psi'^{(n)}dx\Big|_{x=0},$$

$$C_7^{(n)} = \frac{a_4}{\alpha_2}\left(L_1^{-1}(p' - p_\omega'^{(n-1)})|_{x=l} + 3\sum_{k=1}^{3}\alpha_{0k}\int_{h_k} K_k T dz\right)$$
$$- \frac{a_7}{\alpha_2}\left(\mathrm{L}_2^{-1}(q' - q_\omega'^{(n-1)})|_{x=l} + 3\sum_{k=1}^{3}\alpha_{0k}\int_{h_k} K_k Tz\mathrm{d}z\right) - \frac{a_7}{\alpha_2}C_1^{(n)}l,$$
$$C_8^{(n)} = \frac{a_4}{\alpha_2}L_2^{-1}(p' - p_\omega'^{(n-1)})|_{x=0} - \frac{a_7}{\alpha_2}L_3^{-1}(q' - q_\omega'^{(n-1)})|_{x=0}, \quad (5.21)$$

For a *hinged rod* from the conditions (2.26) we have:

$$C_1^{(n)} = \frac{1}{l}(L_2^{-1}(q' - q_\omega'^{(n-1)})|_{x=0} - L_2^{-1}(q' - q_\omega'^{(n-1)})|_{x=l}),$$
$$C_4^{(n)} = \frac{a_7}{a_4}\left(D_1^{(n)}|_{x=0} - D_1^{(n)}|_{x=l}\right)$$
$$- \frac{1}{a_4}\left(L_2^{-1}(q' - q_\omega'^{(n-1)})|_{x=0} + 3\sum_{k=1}^{3}\alpha_{0k}\int_{h_k} K_k Tzdz\right),$$
$$C_5^{(n)} = D_2^{(n)}|_{x=0} - D_2^{(n)}|_{x=l} - \frac{a_7 l}{2a_4}\left(D_1^{(n)}|_{x=0} - D_1^{(n)}|_{x=l}\right)$$
$$- \frac{l}{2a_4}\left(\mathrm{L}_2^{-1}(q' - q_\omega'^{(n-1)})|_{x=0} + 3\sum_{k=1}^{3}\alpha_{0k}\int_{h_k} K_k Tzdz\right),$$
$$C_7^{(n)} = D_1^{(n)}|_{x=0} - D_1^{(n)}|_{x=l}; \quad (5.22)$$

$$D_1^{(n)} = \frac{a_7}{l\alpha_2}L_3^{-1}(q' - q_\omega'^{(n-1)})$$
$$- \frac{a_7 l}{2\alpha_2}\left(\mathrm{L}_2^{-1}(q' - q_\omega'^{(n-1)}) + 3\sum_{k=1}^{3}\alpha_{0k}\int_{h_k} K_k Tzdz\right)$$
$$- \frac{a_4}{l\alpha_2}L_2^{-1}(p' - p_\omega'^{(n-1)}),$$
$$D_2^{(n)} = \frac{a_1}{l\alpha_2}\left(\int \psi'^{(n)}\mathrm{d}x + L_4^{-1}(q' - q_\omega'^{(n-1)})\right)$$
$$- \frac{a_1 l}{6\alpha_2}\left(L_2^{-1}(q' - q_\omega'^{(n-1)}) + 3\sum_{k=1}^{3}\alpha_{0k}\int_{h_k} K_k Tzdz\right)$$
$$- \frac{a_7}{l\alpha_2}\mathrm{L}_3^{-1}(p' - p_\omega'^{(n-1)}).$$

The integration constants $C_2^{(n)}$, $C_3^{(n)}$, $C_6^{(n)}$, $C_8^{(n)}$ retain the form (5.21).

In the case of *rigid embedding of the rod's ends* the constant $C_1^{(n)}$ is the same as in (5.22). The integration constants $C_2^{(n)}$, $C_3^{(n)}$, $C_5^{(n)}$, $C_6^{(n)}$, $C_8^{(n)}$ retain the form (5.21).

$$C_4^{(n)} = \frac{1}{\alpha_2 l}\left[a_7 L_2^{-1}\left(p' - p_\omega'^{(n-1)}\right)\Big|_{x=l} - a_1\left(\mathrm{L}_3^{-1}\left(q' - q_\omega'^{(n-1)}\right)\right)\Big|_{x=l} - \frac{a_1 C_1^{(n)} l^2}{2}\right] - \frac{C_5^{(n)}}{l},$$

$$C_7^{(n)} = \frac{1}{\alpha_2 l}\left[a_4\left(L_2^{-1}\left(p' - p_\omega'^{(n-1)}\right)\right)\Big|_{x=l} - a_7\left(L_3^{-1}\left(q' - q_\omega'^{(n-1)}\right)\right)\Big|_{x=l} - \frac{a_7 C_1^{(n)} l^2}{2}\right] - \frac{C_8^{(n)}}{l}. \quad (5.23)$$

Thus, formulas (5.20)–(5.23) give, in iterations, the analytical solution of the problem on deformation of a three-layer elastic–plastic rod under thermoloading from the natural state in the case of three kinds of boundary conditions. In contrast to the integration constants in the isothermal problem (2.25), (2.27), (2.29), here the temperature enters explicitly.

5.3 Analytical Solution of the Thermoelasticity Problem

As a consequence, we obtain from the solution (5.20)–(5.22) at $n = 1$ the exact solution of the thermoelasticity problem for the three-layer rod in question, which also has a certain novelty and has an independent interest (the stroke at the top is omitted, since there is no plasticity):

$$\psi(x) = C_2 \mathrm{sh}(\beta x) + C_3 \mathrm{ch}(\beta x) + \frac{1}{\beta}\left[\mathrm{sh}(\beta x)\int g(x)\mathrm{ch}(\beta x)\mathrm{d}x - \mathrm{ch}(\beta x)\int g(x)\mathrm{sh}(\beta x)dx\right],$$

$$w(x) = \frac{1}{\alpha_2}\left[\alpha_1\int \psi dx - a_7 L_3^{-1}(p) + a_1 L_4^{-1}(q) + \frac{a_1 C_1}{6}x^3\right] + \frac{C_4}{2}x^2 + C_5 x + C_6,$$

$$u(x) = \gamma_3\psi + \frac{1}{\alpha_2}\left[-a_4 L_2^{-1}(p) + a_7 L_3^{-1}(q) + \frac{a_7}{2}C_1 x^2\right] + C_7 x + C_8, \quad (5.24)$$

where the coefficients $\alpha_1, \alpha_2, \alpha_3, \beta^2, \gamma_1, \gamma_2, \gamma_3$ and linear integral operators L_2^{-1}, L_3^{-1}, L_4^{-1} are defined by formulas (2.17), (2.23), the coefficients a_i are defined by (5.14), the function $g(x)$ follows from (2.19), i.e.

$$g^{(n)}(x) = \gamma_2 p + \gamma_1\left(\int q\mathrm{d}x + C_1^{(n)}\right).$$

The temperature in the solution (5.24) is taken into account by the elasticity parameters of the layer materials (5.14) and the boundary conditions through internal forces (5.17).

Using relations (5.10), write the final expressions of the generalized internal longitudinal force and bending moment through displacements (5.24) and temperature. As a result of this:

$$N = b_0\left(a_1 u,_x + a_6\psi,_x - a_7 w,_{xx} - 3\sum_{k=1}^{3}\alpha_{0k}\int_{h_k} K_k T\mathrm{d}z\right),$$

$$M = b_0\left(a_7 u,_x + a_3\psi,_x - a_4 w,_{xx} - 3\sum_{k=1}^{3}\alpha_{0k}\int_{h_k} K_k Tz\mathrm{d}z\right). \tag{5.25}$$

At cantilever fixing of the end of the three-layer rod, we obtain the integration constants by substituting the solution (5.24) into the boundary conditions (5.24) and considering (5.25):

$$C_1 = -\mathrm{L}_1^{-1}(q)|_{x=l},\ C_3 = \frac{1}{\beta}\int g(x)\mathrm{sh}(\beta x)dx|_{x=0},$$

$$C_2 = \frac{1}{\beta}\left[\frac{\mathrm{ch}(\beta l)}{\mathrm{sh}(\beta l)}\left(\int g\ \mathrm{sh}(\beta x)\mathrm{d}x|_{x=l} - \int g\ \mathrm{sh}(\beta x)\mathrm{d}x|_{x=0}\right) - \int g\ \mathrm{ch}(\beta x)\mathrm{d}x|_{x=l}\right]$$

$$C_4 = \frac{a_7}{\alpha_2}\left(\mathrm{L}_1^{-1}(p)|_{x=l} + 3\sum_{k=1}^{3}\alpha_{0k}\int_{h_k} K_k T dz\right)$$
$$- \frac{a_1}{\alpha_2}\left(\mathrm{L}_2^{-1}(q)|_{x=l} + 3\sum_{k=1}^{3}\alpha_{0k}\int_{h_k} K_k T z dz\right) - \frac{a_1}{\alpha_2}C_1 l,$$

$$C_5 = \frac{a_7}{\alpha_2}\mathrm{L}_2^{-1}(p)|_{x=0} - \frac{a_1}{\alpha_2}\mathrm{L}_3^{-1}(q)|_{x=0},$$

$$C_6 = \frac{a_7}{\alpha_2}\mathrm{L}_3^{-1}(p)\Big|_{x=0} - \frac{a_1}{\alpha_2}\mathrm{L}_4^{-1}(q)\Big|_{x=0} - \frac{\alpha_1}{\alpha_2}\int \psi\mathrm{d}x\Big|_{x=0},$$

$$C_7 = \frac{a_4}{\alpha_2}\left(\mathrm{L}_1^{-1}(p)|_{x=l} + 3\sum_{k=1}^{3}K_k\alpha_{0k}\int_{h_k} T_k\mathrm{d}z\right)$$
$$- \frac{a_7}{\alpha_2}\left(\mathrm{L}_2^{-1}(q)|_{x=l} + 3\sum_{k=1}^{3}K_k\alpha_{0k}\int_{h_k} T_k z\mathrm{d}z\right) - \frac{a_7}{\alpha_2}C_1 l,$$

$$C_8 = \frac{a_4}{\alpha_2}\mathrm{L}_2^{-1}(p)|_{x=0} - \frac{a_7}{\alpha_2}\mathrm{L}_3^{-1}(q)|_{x=0}. \tag{5.26}$$

In the case of a rod hinged at both ends, we obtain the integration constants from the boundary conditions (5.26):

$$C_1 = \frac{1}{l}\left(\mathrm{L}_2^{-1}(q)|_{x=0} - \mathrm{L}_2^{-1}(q)|_{x=l}\right),$$

$$C_4 = \frac{a_7}{a_4}(D_1|_{x=0} - D_1|_{x=l}) - \frac{1}{a_4}\left(\mathrm{L}_2^{-1}(q)|_{x=0} + 3\sum_{k=1}^{3} K_k\alpha_{0k}\int_{h_k} T_k z\mathrm{dz}\right),$$

$$C_5 = D_2|_{x=0} - D_2|_{x=l} - \frac{a_7 l}{2a_4}(D_1|_{x=0} - D_1|_{x=l})$$
$$- \frac{l}{2a_4}\left(\mathrm{L}_2^{-1}(q)|_{x=0} + 3\sum_{k=1}^{3} K_k\alpha_{0k}\int_{h_k} T_k z\mathrm{dz}\right),$$

$$C_7 = D_1|_{x=0} - D_1|_{x=l},\ D_1$$
$$= \frac{a_7}{l\alpha_2}\mathrm{L}_3^{-1}(q) - \frac{a_7 l}{2\alpha_2}\mathrm{L}_2^{-1}(q) - \frac{a_4}{l\alpha_2}\mathrm{L}_2^{-1}(p),$$

$$D_2 = \frac{a_1}{l\alpha_2}\left(\int \psi\mathrm{dx} + \mathrm{L}_4^{-1}(q)\right) - \frac{a_1 l}{6\alpha_2}\mathrm{L}_2^{-1}(q) - \frac{a_7}{l\alpha_2}\mathrm{L}_3^{-1}(p). \tag{5.27}$$

The constants C_2, C_3, C_6, C_8 retain their previous form (5.26).

When both ends of the rod are *rigidly embedded*, the solution must satisfy the boundary conditions (5.28). From here:

$$C_4 = \frac{1}{\alpha_2 l}\left[a_7\left(\mathrm{L}_2^{-1}(p)\right)\Big|_{x=l} - a_1\left(\mathrm{L}_2^{-1}(p)\right)\Big|_{x=l}\frac{a_1 C_1 l^2}{2}\right] - \frac{C_5}{l},$$

$$C_7 = \frac{1}{\alpha_2 l}\left[a_4\left(\mathrm{L}_2^{-1}(p)\right)\Big|_{x=l} - a_7\left(L_3^{-1}(q)\right)\Big|_{x=l} - \frac{a_7 C_1 l^2}{2}\right] - \frac{C_8}{l}. \tag{5.28}$$

The constant $C_1^{(n)}$ is the same as in (5.27); $C_2^{(n)}$, $C_3^{(n)}$, $C_5^{(n)}$, $C_6^{(n)}$, $C_8^{(n)}$ retain the form (5.26).

Next, we need to solve the thermoelasticity problem for a cantilevered elastic three-layer rod under constant loads $q = \text{const}$, $p = \text{const}$. In this case, from (5.24), (5.26) we obtain:

$$\psi(x) = C_2 sh(\beta x) + C_3 ch(\beta x) - \frac{1}{\beta^2}\left[\gamma_2 p + \gamma_1(qx + C_1)\right],$$

$$u(x) = \gamma_3\psi + \frac{1}{\alpha_2}\left[-\frac{1}{2}a_4px^2 + \frac{1}{6}a_7qx^3 + \frac{a_7}{2}C_1x^2\right] + C_7x,$$

$$w = \frac{1}{\alpha_2}\left[\alpha_1\int\psi\mathrm{d}x - \frac{1}{6}a_7px^3 + \frac{1}{24}a_1qx^4 + \frac{1}{6}a_1C_1x^3\right] + \frac{1}{2}C_4x^2 + C_6, \quad (5.29)$$

where

$$\int\psi\mathrm{d}x = \frac{1}{\beta}(C_2\mathrm{ch}(\beta x) + C_3\mathrm{sh}(\beta x)) - \frac{x}{\beta^2}\left(\gamma_2p + \gamma_1(\frac{qx}{2} + C_1)\right),$$

$$C_1 = -ql,\ C_6 = -\frac{\alpha_1}{\beta\alpha_2}C_2,$$

$$C_2 = \frac{1}{2\beta^2sh(\beta l)}\left[2\gamma_2p(1 - ch(\beta l)) + \gamma_1qlch(\beta l)\right],$$

$$C_3 = \frac{1}{2\beta^2}\left[2\gamma_2p - \gamma_1ql\right],$$

$$C_4 = \frac{1}{\alpha_2}\left[a_7\left(pl + 3\sum_{k=1}^{3}\alpha_{0k}\int_{h_k}K_kTdz\right) + a_1\left(\frac{ql^2}{2} - 3\sum_{k=1}^{3}\alpha_{0k}\int_{h_k}K_kTzdz\right)\right],$$

$$C_7 = \frac{1}{\alpha_2}\left[a_4\left(pl + 3\sum_{k=1}^{3}\alpha_{0k}\int_{h_k}K_kT\mathrm{d}z\right) + a_7\left(\frac{ql^2}{2} - 3\sum_{k=1}^{3}\alpha_{0k}\int_{h_k}K_kTz\mathrm{d}z\right)\right].$$

If both ends of the rod are embedded, the following integration constants must be used in the solution of (5.29):

$$C_1 = -\frac{ql}{2},\ C_2 = \frac{1}{2\beta^2sh(\beta l)}\left[2\gamma_2p(1 - ch(\beta l)) + \gamma_1ql(1 + ch(\beta l))\right],$$

$$C_4 = \frac{l}{2\alpha_2}\left(a_7p + \frac{1}{6}a_1ql\right),\ C_7 = \frac{l}{2\alpha_2}\left[a_4p + a_7\frac{ql}{6}\right]. \quad (5.30)$$

The constants C_6, C_6 retain the form (5.29).

Thus, expressions (5.24)–(5.30) give an analytical solution to the problem of deformation of a thermoelastic three-layer rod under different types of boundary conditions.

Numerical results were obtained for a three-layer elastic rod with the layers: “D16T-fluoroplastic-D16T”, layer thickness*:* $h_1 = h_2 = 0.03$, $c = 0.09$.

Figures 5.2, 5.3, 5.4, 5.5, 5.6and5.7 show the effect of temperature on the displacements, strains, and stresses of an elastic rod: *1*—$T_1 = 293$ K, *2*—$T_2 = 343$ K, *3*—$T_3 = 393$ K. Boundary conditions: embedding–embedding and cantilevered anchoring. The value of the distributed load intensity $q' = -0.15$ MPa, $p' = 0$, which allows the

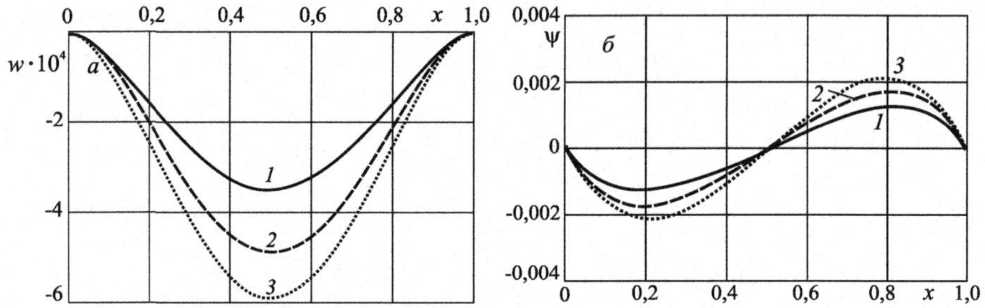

Fig. 5.2 Changes in deflection w (a) and relative shear ψ (b) along the axis at different temperatures (embedding–embedding)

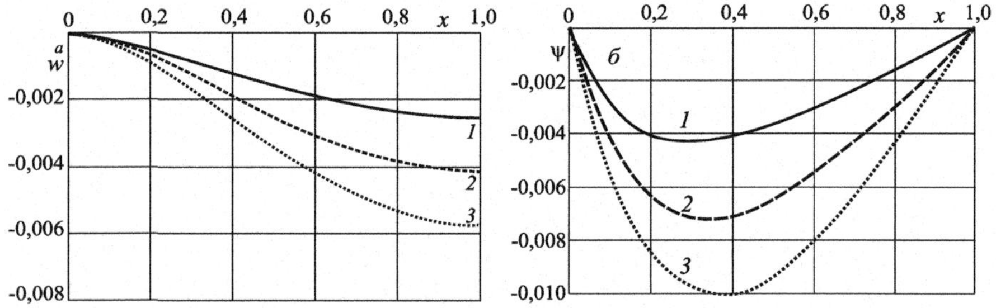

Fig. 5.3 Changes in deflection w (a) and relative shear ψ (b) along the axis at various temperatures (cantilever)

rod to deform in the elastic region as the temperature increases. The temperature is constant across the thickness of the rod.

Figure 5.2 shows: a—change in deflection w, b—change in relative shear ψ along the axis of a three-layer elastic rod with rigidly fixed ends at different temperatures. Heating the rod by 50 K leads to an increase in deflection by 42% and shear by 12% compared to isothermal deformation. When heated at 100 K, a similar increase reaches 71 and 31%.

It should be noted that for a rod terminated at the ends, the thermal effect does not lead to volume deformation of the material and shear. The increase in shear occurs only due to the temperature decrease of the elasticity parameters of the materials.

When a rod is cantilevered, the heating has a stronger effect on its deformation (Fig. 5.3). In this case, an increase in temperature by 50 K leads to an increase in deflection by 70% and shear by 64% compared to isothermal deformation. When heated at 100 K, a similar increase reaches 141 and 127%. This is due to thermal linear and volumetric deformation, a decrease in the elastic modulus of materials.

The distribution of linear strains ε_x^k in the embedment ($x = 0$) along the height of the cross-section of the rod with rigidly fixed ends is shown in Fig. 5.4. As you can see, the graphs rotate relative to the middle of the layers, increasing extremums.

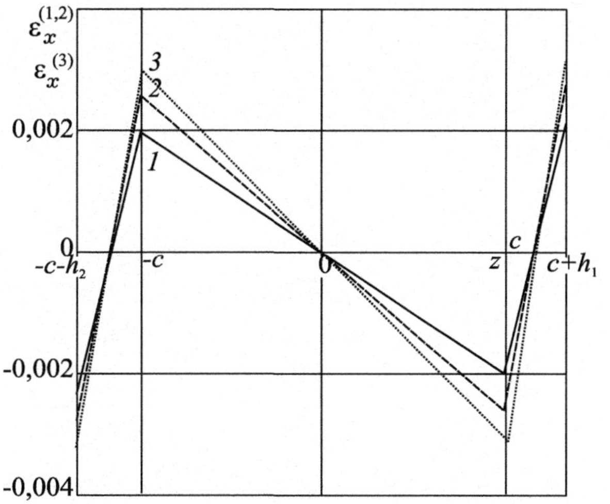

Fig. 5.4 Distribution of linear strains ε_x^k in the termination ($x = 0$) over the height of the cross section (embedding–embedding)

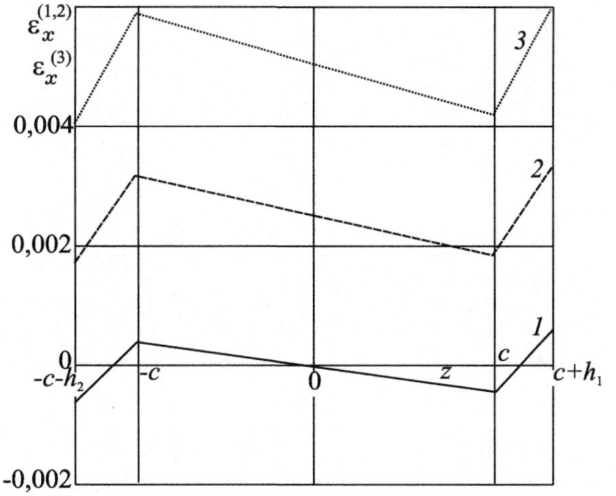

Fig. 5.5 Distribution of linear strains ε_x^k in the termination ($x = 0$) over the height of the cross section (console)

Similar graphs at the cantilever rod (Fig. 5.5) shift almost parallel to themselves in the positive area. The strains on the adhesions of the layers are continuous.

The distribution of normal stresses σ_x at the ends ($x = 0$) of an elastic three-layer rod, fixed at the ends, at different temperatures is shown in Fig. 5.6. When heated, the stresses shift to the negative area parallel to themselves, increasing modulus many times over.

When cantilevered (Fig. 5.7), stresses are extreme on the outer planes. When heated to 50 and 100 degrees, they increase in the bearing layers by 14% and 43%, respectively, and in the filler—by 20 and 60%. At the boundaries of the layers, stresses, unlike strains, are discontinuous. This effect is explained by the different elastic modules of the materials of the layers.

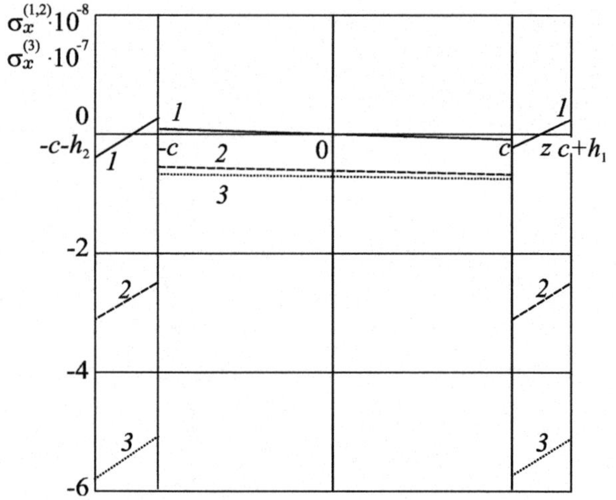

Fig. 5.6 Distribution of normal stresses σ_x at the ends ($x = 0$) at different temperatures (embedding–embedding)

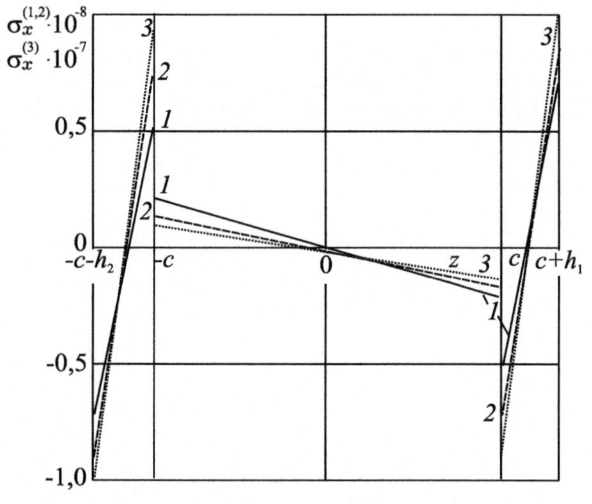

Fig. 5.7 Distribution of normal stresses σ_x at the ends ($x = 0$) at different temperatures (console)

5.4 Repeated Thermal Force Deformation of an Elastic–plastic Rod

Let, starting from the moment $t = t_1$, instantaneous unloading and repeated loading with forces of the opposite sign p'', q'', changing by the same law as during loading from the natural state. These forces will create in the kth layer of the rod the displacement field $u''(x), w''(x)$, strains $\varepsilon_x''^{(k)}(x, z)$, $\text{э}_x''^{(k)}\ (x, z)$, $\varepsilon''^{(k)}(x, z)$, $\psi''(x, z)$, and stresses $\sigma_x''^{(k)}(x, z)$, $s_x''^{(k)}(x, z)$, $\sigma''^{(k)}(x, z)$, $\sigma_{xz}''^{(3)}(x, z)$. We will assume that during the unloading and subsequent alternating loading the temperature at all points of the body remains unchanged. It coincides with the temperature field by the time the unloading starts, i.e. $T\prime(z) = T(z, t_1)$. In addition, the elastic modules of the layers materials took values $G_k(z) \equiv G_k(T's(z))$, $K_k(z) \equiv K_k(T'(z))$, with the corresponding loading trajectories still belonging to the class of simple trajectories.

As in the isothermal problem, we introduce differences for all values of the stress–strain state and load, in which the values with one stroke are the stresses, strains and displacements in the rod before unloading, with two strokes we mark similar parameters during the second half-cycle:

$$\sigma_x^{(k)*} = \sigma_x^{\prime(k)} - \sigma_x^{\prime\prime(k)},\ s_x^{(k)*} = s_x^{\prime(k)} - s_x^{\prime\prime(k)},\ \sigma^{(k)*} = \sigma^{\prime(k)} - \sigma^{\prime\prime(k)},$$

$$\sigma_{xz}^{(3)*} = \sigma_{xz}^{\prime(3)} - \sigma_{xz}^{\prime\prime(3)},\ \psi^* = \psi' - \psi'',$$

$$\varepsilon_x^{(k)*} = \varepsilon_x^{\prime(k)} - \varepsilon_x^{\prime\prime(k)},\ э_x^{(k)*} = э_x^{\prime\prime(k)} - э_x^{\prime\prime(k)},\ \varepsilon^{(k)*} = \varepsilon^{\prime(k)} - \varepsilon^{\prime\prime(k)},$$

$$q^* = q' - q'',\ p^* = p' - p'',\ u^* = u' - u'',\ w^* = w' - w''. \tag{5.31}$$

Because of linearity of connection of deformations with displacements in the rod layers (relations (2.2)), similar relations will be true for corresponding magnitudes with asterisks in the temperature field:

$$\varepsilon_x^{(1)*} = u,_x^* + c\psi,_x^* - zw,_{xx}^*,\ \varepsilon_{xz}^{(1)*} = 0 (c \le z \le c + h_1),$$

$$\varepsilon_x^{(3)*} = u,_x^* + z\psi,_x^* - zw,_{xx}^*,\ \varepsilon_{xz}^{(3)*} = \frac{1}{2}\psi^* (- c \le z \le c),$$

$$\varepsilon_x^{(2)*} = u,_x^* - c\psi,_x^* - zw,_{xx}^*,\ \varepsilon_{xz}^{(2)*} = 0 (- c - h_2 \le z \le -c). \tag{5.32}$$

For the stresses and strains (5.31), (5.32), marked with an asterisk, let us assume physical equations of state of the type (5.2):

$$s_x^{(k)*} = 2G_k f^{(k)*}\ э_x^{(k)*},\ s_{xz}^{(3)*} = 2G_3 f^{(3)*}\ э_{xz}^{(3)*},$$

$$\sigma^{(k)*} = 3K(\varepsilon^{(k)} - \alpha_{0k} T_k)\varepsilon^{(k)*} (k = 1, 2, 3). \tag{5.33}$$

The corresponding universal nonlinearity functions in the bearing layers are assumed, according to Moskvitin's hypothesis, to be expressed in terms of plasticity functions under loading from the natural state:

$$f^{(k)*}(\varepsilon_u^{(k)}, \varepsilon_y^{(k)*}, T\prime) = \begin{cases} 1, & \varepsilon_u^{(k)*} \le \varepsilon_y^{(k)*}(T') \\ 1 - \omega^{(k)*}, & \varepsilon_u^{(k)*} > \varepsilon_y^{(k)*}(T') \end{cases},$$

$$\omega^{(k)*} = \omega^{\prime(k)}(\varepsilon_u^{(k)*}, \varepsilon_y^{(k)*}, T', d_l^*). \tag{5.34}$$

Here d_l^* are the experimental material parameters included in the approximation formula for the nonlinearity function of the type (2.39):

$$\omega^{(k)*} = \begin{cases} 0, & \varepsilon_u^{(k)*} \le \varepsilon_y^{(k)*}(T'), \\ A_k^*\left(1 - \frac{\varepsilon_y^{(k)*}(T')}{\varepsilon_u^{(k)*}}\right)^{\alpha_k^*}, & \varepsilon_u^{(k)*} > \varepsilon_y^{(k)*}(T'). \end{cases} \tag{5.35}$$

The physical nonlinearity of the filler under repeated alternating loading is still described by relations (5.5).

The stress and strain components with asterisks, using the physical relations (5.33), will be represented in this form

$$\sigma_x^{(k)*} = \sigma_{xe}^{(k)*} - \sigma_{x\omega}^{(k)*}, \sigma_{xz}^{(3)*} = \sigma_{xze}^{(3)*} - \sigma_{xz\omega}^{(3)*},$$

$$\sigma_{xe}^{(k)*} = 2G_k\, \text{э}_x^{(k)*} + 3K_k \varepsilon^{(k)*}, \sigma_{x\omega}^{(k)*}$$
$$= 2G_k\, \text{э}_x^{(k)*}\, \omega^{(k)*}, \sigma_{xze}^{(3)*} = 2G_3\, \text{э}_{xz}^{(3)*},$$

$$\sigma_{xz\omega}^{(k)*} = 2G_3\, \text{э}_{xz}^{(3)*}\, \omega^{(3)*}, G_k(z) \equiv G_k(T'(z)), K_k(z) \equiv K_k(T'(z)). \tag{5.36}$$

Here, in contrast to similar values with the stroke (5.7), the temperature is not contained explicitly, and the nonlinearity function is calculated by formula (5.35). By performing a similar operation with values of the type of internal forces (2.4), we obtain:

$$N^{(k)*} = N_e^{(k)*} - N_\omega^{(k)*}, M^{(k)*} = M_e^{(k)*} - M_\omega^{(k)*}, Q^* = Q_e^* - Q_\omega^*. \tag{5.37}$$

The elastic (index "*e*") and inelastic (index "ω") components in (5.37) are calculated through stresses (5.36) by the formulas:

$$N_e^{(k)*} = b_0 \int_{h_k} \sigma_{xe}^{(k)*} dz, N_\omega^{(k)*} = b_0 \int_{h_k} \sigma_{x\omega}^{(k)*} dz, M_e^{(k)*} = b_0 \int_{h_k} \sigma_{xe}^{(k)*} z dz,$$

$$M_\omega^{(k)*} = b_0 \int_{h_k} \sigma_{x\omega}^{(k)*} z dz, Q_e^* = b_0 \int_{-c}^{c} \sigma_{xze}^{(3)*} dz, Q_\omega^* = b_0 \int_{-c}^{c} \sigma_{xz\omega}^{(3)*} dz. \tag{5.38}$$

Iterative equilibrium Eq. (5.18), due to their linearity, for the magnitudes with asterisks we write in the form

$$a_1 u,_{xx}^{(n)*} + a_6 \psi,_{xx}^{(n)*} - a_7 w,_{xxx}^{(n)*} = -p^* + p_\omega^{(n-1)*},$$

$$a_6 u,_{xx}^{(n)*} + a_2 \psi,_{xx}^{(n)*} - a_3 w,_{xxx}^{(n)*} - a_5 \psi^{(n)*} = h_\omega^{(n-1)*},$$

$$a_7 u,_{xxx}^{(n)*} + a_3 \psi,_{xxx}^{(n)*} - a_4 w,_{xxxx}^{(n)*} = -q^* + q_\omega^{(n-1)*}. \tag{5.39}$$

The values $p_\omega^{(n-1)*}$, $h_\omega^{(n-1)*}$, $q_\omega^{(n-1)*}$ corresponding to the inelastic components in (5.39) are taken to be zero at the first step ($n = 1$) and then, using (5.38), are calculated from the results of the previous approximation and are called additional "external" loads.

They act as corrections for the physical nonlinearity of the materials of the layers under repeated loading and take into account the effects of the temperature field:

$$p_\omega^{(n-1)*} = \frac{1}{b_0} N_\omega^{(n-1)*},_x,\ h_\omega^{(n-1)*} = \frac{1}{b_0}(H_\omega^{(n-1)*},_x - Q_\omega^{(n-1)*}),\ q_\omega^{(n-1)*} = \frac{1}{b_0} M_\omega^{(n-1)*},_{xx},$$

$$N_\omega^{(n-1)*} = \sum_{k=1}^{3} N_\omega^{(k,\,n-1)*} = \frac{4}{3} b_0 \sum_{k=1}^{3} \int_{h_k} G_k \omega^{(k)*}(\varepsilon_u^{(k,\,n-1)*}) \varepsilon_x^{(k,\,n-1)*} dz,$$

$$M_\omega^{(n-1)*} = \sum_{k=1}^{3} M_\omega^{(k,\,n-1)*} = \frac{4}{3} b_0 \sum_{k=1}^{3} \int_{h_k} G_k \omega^{(k)*}(\varepsilon_u^{(k,\,n-1)*}) \varepsilon_x^{(k,\,n-1)*} z dz,$$

$$H_\omega^{(n-1)*} = c(N_\omega^{(1,\,n-1)*} - N_\omega^{(2,\,n-1)*}) + M_\omega^{(3,\,n-1)*},$$

$$Q_\omega^{(n-1)*} = 2b_0 \int_{-c}^{c} G_3 \omega^{(3)*}(\varepsilon_u^{(3,\,n-1)*}) \psi^{(n-1)*} dz. \tag{5.40}$$

The boundary conditions for the values with asterisks in the cantilever termination will be of the type (5.25), in the hinge-hinge case—(5.26), in the termination-buttons case—(5.27). Thus, the boundary problem for values with asterisks (5.39), (5.40), (5.25)–(5.27) is exactly the same as the boundary problem (5.18), (5.19), (2.24), (2.26), (2.28). Therefore, her analytical solution will be:

$$\psi^{(n)*}(x) = C_2^{(n)*} sh(\beta x) + C_3^{(n)*} ch(\beta x)$$
$$+ \frac{1}{\beta}\left[sh(\beta x) \int g^{(n)*} ch(\beta x) \mathrm{d}x - ch(\beta x) \int g^{(n)*} sh(\beta x) \mathrm{d}x \right].$$

$$u^{(n)*}(x) = \gamma_3 \psi^{(n)*} + \frac{1}{\alpha_2}[-a_4 \mathrm{L}_2^{-1}(p^* - p_\omega^{(n-1)*}) + a_7 \mathrm{L}_3^{-1}(q - q_\omega^{(n-1)*})]$$
$$+ \frac{a_7}{2\alpha_2} C_1^{(n)*} x^2 + C_7^{(n)*} x + C_8^{(n)*},$$

$$w^{(n)*}(x) = \frac{1}{\alpha_2}\Big[\alpha_1 \int \psi^{(n)*} \mathrm{d}x - a_7 \mathrm{L}_3^{-1}(p^* - p_\omega^{(n-1)*}) + a_1 \mathrm{L}_4^{-1}(q^* - q_\omega^{(n-1)*})$$

$$+\frac{1}{6}a_1C_1^{(n)*}x^3\Big]+\frac{1}{2}C_4^{(n)*}x^2+C_5^{(n)*}x+C_6^{(n)*}. \tag{5.41}$$

The following function is introduced here:
$g^{(n)*}(x)=\frac{\alpha_2}{\alpha_1}\gamma_1 h_\omega^{(n-1)*}+\gamma_2(p^*-p_\omega^{(n-1)*})+\gamma_1\left(\int(q^*-q_\omega^{(n-1)*})\mathrm{d}x+C_1^{(n)*}\right)$;
coefficients $\alpha_1,\alpha_2,\alpha_3,\beta^2,\gamma_1,\gamma_2,\gamma_3$ and linear integral operators $\mathrm{L}_2^{-1}, L_3^{-1}, L_4^{-1}$ are defined by formulas (2.23) taking into account the fixation of elastic modules at the moment of unloading; the integration constants $C_1^{(n)*}$, ..., $C_8^{(n)*}$ at each approximation step follow from the rod anchoring conditions.

We obtain the integration constants $C_1^{(n)*}$, ..., $C_8^{(n)*}$ in the form (5.28)–(5.30) by satisfying for the solution (5.41) the boundary conditions for the console (5.25), the ball-supported rod (5.26), and the rigid termination of both ends (5.27). But, in this case the elasticity parameters have to be calculated at the unloading temperature.

Thus, the obtained formulas (5.41) with integration constants (5.28)–(5.30) give in iterations the analytical solution of the problem for the values with asterisks under repeated alternating loading in the case of different types of boundary conditions.

The corresponding solution of the problem in the second half-cycle is obtained from the relations (5.31), taking into account the solutions (5.20)–(5.22), (5.41):

$$\sigma_x^{\prime\prime(k)}(x,z)=\sigma_x^{\prime(k)}(x,z)-\sigma_x^{(k)*}(x,z),\ \sigma_{xz}^{\prime\prime(3)}(x,z)=\sigma_{xz}^{\prime(3)}(x,z)-\sigma_{xz}^{(3)*}(x,z),$$

$$\varepsilon_x^{\prime\prime(k)}(x,z)=\varepsilon_x^{\prime(k)}(x,z)-\varepsilon_x^{(k)*}(x,z),$$

$$u''(x)=u'(x)-u^*(x),\ \psi''(x)=\psi'(x)-\psi^*(x),\ w''(x)=w'(x)-w^*(x),$$

$$\psi''(x)=\psi'(x)-\psi^*(x),\ w''(x)=w'(x)-w^*(x), \tag{5.42}$$

where the values with one stroke are the stresses, strains and displacements in the rod before unloading.

5.5 Cyclic Bending in the Temperature Field

By a method quite similar to that described in the previous sections, we obtain the solution for any mth load variable. Let, after (m–1) sign-variable half-cycles of loading, starting from the moment $t_{(m-1)}$, instantaneous unloading and new loading are carried out by the forces of the reverse sign $p^{(m)}$, $q^{(m)}$ (upper index "m" is loading number). These forces will create in the kth layer of the rod a field of displacements $u^{(m)}$, $\psi^{(m)}$, $w^{(m)}$, strains $\varepsilon_{ij}^{(m,k)}$ and stresses $\sigma_{ij}^{(m,k)}$.

Let's assume that during (m–1) sign-variable half-cycles the temperature at all points of the body remains unchanged after the first unloading begins, i.e. $T^{(m)}(z)$

$= T'(z, t_1)$. At the same time, the elastic modules retained the values of: $G_k(z) \equiv G_k(T'(z, t_1))$, $K_k(z) \equiv K_k(T'(z, t_1))$.

As before, let's introduce for all values of the stress–strain state and load differences, in which the values with index $(m–1)$ are stresses, strains, displacements and loads in the rod before unloading, the index m denotes similar parameters in the process of the mth half-cycle:

$$\sigma_x^{(m,k)*} = (-1)^n(\sigma_x^{(m-1,k)} - \sigma_x^{(m,k)}),\ \varepsilon_x^{(m,k)*} = (-1)^n(\varepsilon_x^{(m-1,k)} - \varepsilon_x^{(m,k)}),$$

$$u^{(m)*} = (-1)^n(u^{(m-1)} - u^{(m)}),\ \psi^{(m)*} = (-1)^n(\psi^{(m-1)} - \psi^{(m)}),$$

$$w^{(m)*} = (-1)^n(w^{(m-1)} - w^{(m)}),$$

$$q^{(m)*} = (-1)^n(q^{(m-1)} - q^{(m)}), p^{(m)*} = (-1)^n(p^{(m-1)} - p^{(m)}). \tag{5.43}$$

The relations (5.32) linking the deformations and displacements with the asterisks in (5.43) will also be true here. For the stresses and strains marked with an asterisk, we will accept the physical equations of state (5.33):

$$s_x^{(m,k)*} = 2G_k\ \ni_x^{(m,k)*} f^{(m,k)*}(\varepsilon_u^{(m,k)*}, T'), \sigma^{(m,k)*} = 3K_k\varepsilon^{(m,k)*}(k = 1, 2, 3),$$

$$s_{xz}^{(m,3)*} = 2G_3\ \ni_{xz}^{(m,3)*} f^{(m,3)*}(\varepsilon_u^{(m,3)*}, T'). \tag{5.44}$$

The universal nonlinearity functions in the bearing layers are assumed to be expressed through the plasticity functions under loading from the natural state:

$$f^{(m,k)*}(\varepsilon_u^{(m,k)*}, T') = \begin{cases} 1, & \varepsilon_u^{(m,k)*} \le \varepsilon_y^{(m,k)*}(T'), \\ 1 - \omega^{(m,k)*}(\varepsilon_u^{(m,k)*}, T'), & \varepsilon_u^{(m,k)*} > \varepsilon_y^{(m,k)*}(T'), \end{cases}$$

$$\omega^{(m,k)*} = \omega'^{(k)}(\varepsilon_u^{(m,k)*}, \varepsilon_y^{(m,k)*}, T', d_l^{(m)*}), \tag{5.45}$$

where $\varepsilon_y^{(m,k)*}(T')$ is the value, corresponding to the yield strength of the material at the mth half-cycle, $d_l^{(m)*}$ are experimental material parameters included in the approximation formula of the type (5.35).

The physical nonlinearity of the filler under the m sign-variable loading is still described by relations (5.5). In the stress and strain components with asterisks, using the physical relations (5.44), as before, let us select the nonlinear components:

$$\sigma_x^{(m,k)*} = \sigma_{xe}^{(m,k)*} - \sigma_{x\omega}^{(m,k)*}, \sigma_{xz}^{(m,3)*} = \sigma_{xze}^{(m,3)*} - \sigma_{xz\omega}^{(m,k)*}, \tag{5.46}$$

where

$$\sigma_{xe}^{(m,k)*} = 2G_k \ \text{э}_x^{(m,k)*} + 3K_k\varepsilon^{(m,k)*}, \sigma_{x\omega}^{(m,k)*} = 2G_k \ \text{э}_x^{(m,k)*} \ \omega^{(m,k)*},$$

$$\sigma_{xze}^{(m,3)*} = 2G_3 \ \text{э}_{xz}^{(m,3)*}, \sigma_{xz\omega}^{(m,3)*} = 2G_3 \ \text{э}_{xz}^{(m,k)*} \ \omega^{(m,3)*}.$$

Here the nonlinearity function is calculated by formula (5.45).

By performing a similar operation with values of the type of internal forces (2.8), we obtain:

$$N^{(m,k)*} = N_e^{(m,k)*} - N_\omega^{(m,k)*}, M^{(m,k)*} = M_e^{(m,k)*} - M_\omega^{(m,k)*},$$

$$Q^{(m,3)*} = Q_e^{(m,3)*} - Q_\omega^{(m,3)*}. \tag{5.47}$$

The elastic (index "*e*") and inelastic (index "ω") components in (5.47) are calculated through the values in (5.46) by the formulas:

$$N_e^{(m,k)*} = b_0 \int \sigma_{xe}^{(m,k)*} dz, N_\omega^{(m,k)*} = b_0 \int_{h_k} \sigma_{x\omega}^{(m,k)*} dz,$$

$$M_e^{(m,k)*} = b_0 \int_{h_k} \sigma_{xe}^{(m,k)*} zdz, M_\omega^{(m,k)*} = b_0 \int_{h_k} \sigma_{x\omega}^{(m,k)*} zdz,$$

$$Q_e^{(m,3)*} = b_0 \int_{-c}^{c} \sigma_{xze}^{(m,3)*} dz, Q_\omega^{(m,k)*} = b_0 \int_{-c}^{c} \sigma_{xz\omega}^{(m,3)*} dz. \tag{5.48}$$

The corresponding iterative equilibrium Eq. (5.18) at the *n*th step for the values with asterisks at the *m*th half-cycle are written in the form

$$a_1 u,_{xx}^{(m,n)*} + a_6 \psi,_{xx}^{(m,n)*} - a_7 w,_{xxx}^{(m,n)*} = -p^{(m)*} + p_\omega^{(m,n-1)*},$$

$$a_6 u,_{xx}^{(m,n)*} + a_2 \psi,_{xx}^{(m,n)*} - a_3 w,_{xxx}^{(m,n)*} - a_5 \psi^{(m,n)*} = h_\omega^{(m,n-1)*},$$

$$a_7 u,_{xxx}^{(m,n)*} + a_3 \psi,_{xxx}^{(m,n)*} - a_4 w,_{xxxx}^{(m,n)*} = -q^{(m)*} + q_\omega^{(m,n-1)*}. \tag{5.49}$$

The values $p_\omega^{(m,n-1)*}$, $h_\omega^{(m,n-1)*}$, $q_\omega^{(m,n-1)*}$ corresponding to the inelastic components in (5.48) are taken to be zero at the first step ($n = 1$), and then are calculated from the results of the previous approximation and are called additional "external" loads:

$$p_\omega^{(m,n-1)*} = \frac{1}{b_0} N_\omega^{(m,n-1)*},_x, h_\omega^{(m,n-1)*} = \frac{1}{b_0}(H_\omega^{(m,n-1)*},_x - Q_\omega^{(m,n-1)*}),$$

$$q_{\omega}^{(m,n-1)*} = \frac{1}{b_0} M_{\omega}^{(m,n-1)*}{}_{,xx},$$

$$N_{\omega}^{(m,n-1)*} = \sum_{k=1}^{3} N_{\omega}^{(m,n-1,k)*} = \frac{4}{3} b_0 \sum_{k=1}^{3} \int_{h_k} G_k \omega^{(k)*} (\varepsilon_u^{(m,n-1,k)*}) \varepsilon_x^{(m,n-1,k)*} dz,$$

$$M_{\omega}^{(m,n-1)*} = \sum_{k=1}^{3} M_{\omega}^{(m,n-1,k)*} = \frac{4}{3} b_0 \sum_{k=1}^{3} \int_{h_k} G_k \omega^{(k)*} (\varepsilon_u^{(m,n-1,k)*}) \varepsilon_x^{(m,n-1,k)*} z dz,$$

$$H_{\omega}^{(m,n-1)*} = c(N_{\omega}^{(m,n-1,1)*} - N_{\omega}^{(m,n-1,2)*}) + M_{\omega}^{(m,n-1,3)*},$$

$$Q_{\omega}^{(m,n-1)*} = 2b_0 \int_{-c}^{c} G_3 \omega^{(3)*} (\varepsilon_u^{(m,n-1,3)*}) \psi dz. \tag{5.50}$$

The boundary conditions of the "left-hand cantilever", "hinge-hinge", "rigid embedding-rigid embedding" type for values with asterisks will be determined by requirements formally coinciding with (5.41)–(5.43).

Thus, the boundary value problem (5.49), (5.50), (5.45) for values with asterisks coincides with the boundary value problem for a three-layer elastic–plastic rod from the natural state (5.39), (5.40), (5.34) with the accuracy of notation. Therefore, its recurrence analytic solution is as follows:

$$\psi^{(m,n)*}(x) = C_2^{(m,n)*} sh(\beta x) + C_3^{(m,n)*} ch(\beta x)$$

$$+\frac{1}{\beta}\left[sh(\beta x) \int g^{(m,n)*} ch(\beta x) \mathrm{d}x - ch(\beta x) \int g^{(m,n)*} sh(\beta x) dx\right],$$

$$u^{(m,n)*}(x) = \frac{1}{\alpha_2}\Big[a_7 L_3^{-1}(q^{(m)*} - q_{\omega}^{(m,n)*}) - a_4 \mathrm{L}_2^{-1}(p^{(m)*} - p_{\omega}^{(m,n)*})$$
$$+\frac{a_7}{2} C_1^{(m,n)*} x^2\Big] + C_7^{(m,n)*} x + C_8^{(m,n)*} + \gamma_3 \psi^{(m,n)*},$$

$$w^{(m,n)*}(x) = \frac{1}{\alpha_2}\Big[\alpha_1 \int \psi^{(m,n)*} \mathrm{d}x - a_7 L_3^{-1}(p^{(m)*} - p_{\omega}^{(m,n)*})$$
$$+ a_1 L_4^{-1}(q^{(m)*} - q_{\omega}^{(m,n)*}) + \frac{1}{6} a_1 C_1^{(m,n)*} x^3\Big]$$
$$+ \frac{1}{2} C_4^{(m,n)*} x^2 + C_5^{(m,n)*} x + C_6^{(m,n)*}. \tag{5.51}$$

Here the coefficients $\alpha_1, \alpha_2, \alpha_3, \beta^2, \gamma_1, \gamma_2, \gamma_3$ and linear integral operators $\mathrm{L}_2^{-1}, L_3^{-1}, L_4^{-1}$ are defined by formulas (2.17), (2.19), (2.23) and are calculated at the temperature in the case of first unloading. The function $g^{(m,n)*}$ is defined by the formula:

$$g^{(m,n)*}(x) = \frac{\alpha_2}{\alpha_1}\gamma_1 h_\omega^{(m,n-1)*} + \gamma_2\left(p^{(m)*} - p_\omega^{(m,n-1)*}\right)$$
$$+ \gamma_1\left(\int (q^{(m)*} - q_\omega^{(m,n-1)*})\mathrm{d}x + C_1^{(m,n)*}\right)$$

The integration constants $C_1^{(m,n)*}$, ..., $C_8^{(m,n)*}$ follow from the conditions of the rod fastening (5.41)–(5.43) and are defined by formulas (5.46)–(5.48), in which the elasticity parameters must be calculated at the temperature of the first unloading.

Thus, formulas (5.51), (5.46)–(5.48) give in iterations the analytical solution of the problem in displacements for quantities with asterisks at arbitrary m half-cycle of sign-variable loading in case of different kinds of boundary conditions. We obtain the solution of the problem from relations (5.43), taking into account solutions (5.20)–(5.22) and (5.51):

$$\sigma_x^{(m,k)}(x,z) = \sigma_x' k(x,z) - \sum_{l=2}^{m} (-1)^k \sigma_x^{(l,k)*}(x,z),$$

$$\sigma_{xz}^{(m,3)}(x,z) = \sigma_x'^{(3)}(x,z) - \sum_{l=2}^{m} (-1)^k \sigma_x^{(l,3)*}(x,z),$$

$$\varepsilon_x^{(m,k)}(x,z) = \varepsilon_{ij}' k(x,z) - \sum_{l=2}^{m} (-1)^k \varepsilon_x^{(l,k)*}(x,z),$$

$$u^{(m)}(x) = u'(x) - \sum_{l=2}^{m} (-1)^l u^{(l)*}(x),$$

$$\psi^{(m)}(x) = \psi'(x) - \sum_{l=2}^{m} (-1)^l \psi^{(l)*}(x),$$

$$w^{(m)}(x) = w'(x) - \sum_{l=2}^{m} (-1)^l w^{(l)*}(x), \tag{5.52}$$

where the values with one stroke are the stresses, strains and displacements in the rod before the first unloading.

5.6 Parametric Analysis of Solutions

The figures further illustrate the effect of increasing the temperature by 50° on the stress–strain state of an elastic–plastic rod under single loads. For the cantilever rod, the load was taken $q' = -0.75$ MPa, with rigid embedding of both ends $q' = -10$

MPa, the yield strength was calculated by the formula (5.3). The plasticity function for the load-bearing layers was calculated by formula (5.2), the nonlinearity function of the filler—by formula (5.5).

Numerical analysis confirmed the practical convergence of the method of elastic solutions. At the 4th step of iteration at temperature $T_2 = 343$ K, both at cantilever fastening (Fig. 5.8) of the rod and at embedded fastening (Fig. 5.9), the displacements differed from the previous values by no more than 0.2%. At increased temperature, the physical nonlinearity leads to an increase in calculated displacements at the cantilever rod by 14 and 20%, with both ends embedded by 10 and 16%, respectively.

Figure 5.10 shows: *a*—deflection, *b*—relative shear in the filler of a three-layer elastic–plastic rod when the left end is cantilevered; *1*, *2*—elastic displacements at $T_1 = 293$ K and $T_2 = 343$ K; *3*, *4*—elastic–plastic at $T_1 = 293$ K and $T_2 = 343$ K. Taking into account the temperature effects on the physical nonlinearity of the layer materials increases the calculated elastic displacements by 84% and 100%, respectively.

Here, the temperature dependence of the elasticity parameters and the physical nonlinearity limit of the material of the layers, as well as their volumetric thermal expansion, were taken into account.

Figure 5.11 shows: *a*—deflection, *b*—relative shear in the filler of a three-layer elastic–plastic rod when there is rigid embedding of both ends of rod; *1*, *2*—elastic displacements at $T_1 = 293$ K and $T_2 = 343$ K; *3*, *4*—elastic–plastic at $T_1 = 293$

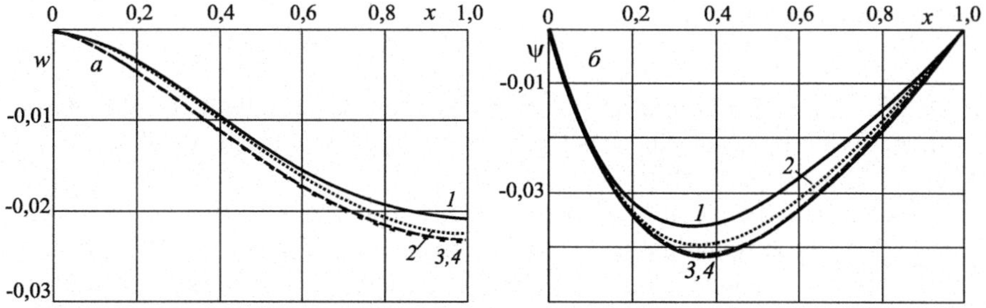

Fig. 5.8 Convergence of the method of elastic solutions (cantilever)

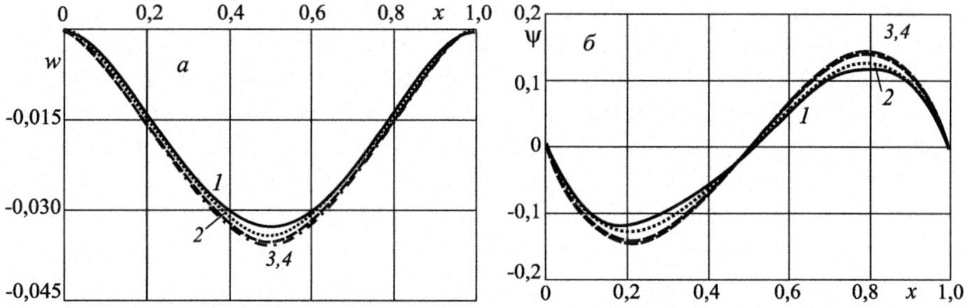

Fig. 5.9 Convergence of the method of elastic solutions (embedding-embedding)

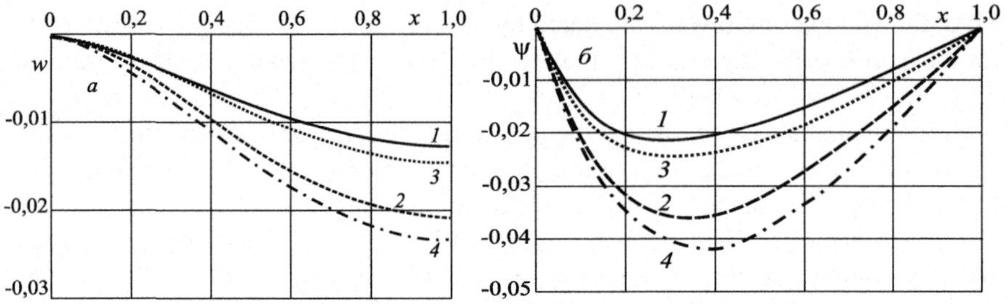

Fig. 5.10 Changing of deflection w (a) and relative shear ψ (b) along the axis at various temperatures (cantilever)

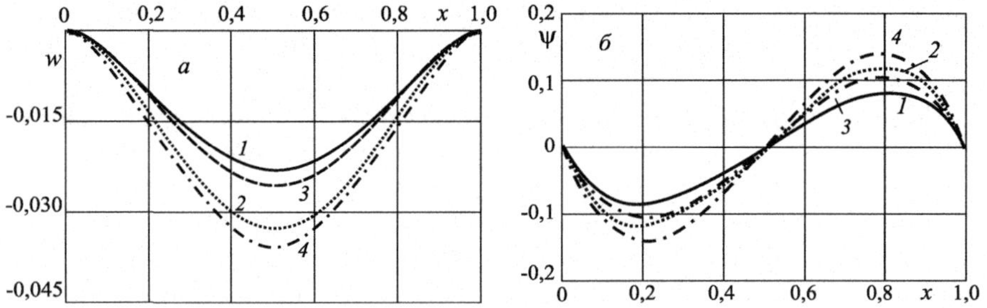

Fig. 5.11 Changing of deflection w (a) and relative shear ψ (b) along the axis at various temperatures (embedding)

K and T_2 = 343 K. Taking into account the temperature effects on the physical nonlinearity of the layer materials increases the calculated elastic displacements by 55% and 75%, respectively.

The distribution of areas of plasticity and physical nonlinearity of materials (gray filling) along the longitudinal section of the three-layer rod at the left end is illustrated in Fig. 5.12 (a—cantilever, b—embedding- embedding): *1*—isothermal loading, *2*—additional areas under thermal-force loading (T_2 = 343 K, darker fill).

The nonlinearity regions, at a 50° temperature rise, occupy a slightly larger volume (by about 8–10%) than during isothermal loading.

Figures 5.13 through 5.16 illustrate the effect of temperature (T_2 = 343 K) on the parameters of the stress–strain state of a three-layer rod under low-cycle loading.

Figures 5.13, 5.14 show the change in deflection w (a) and shear ψ (b) along the axis of the three-layer rod after 8 half-cycles of loading, calculated according to different physical equations of state for cantilevering the left end and for embedding both ends, respectively. Curves with one stroke represent loading from the natural state, two strokes represent repeated alternating bending: *1*—elastic rod, *2*—isothermal elastoplasticity, *3*—thermoelasticity at T_2 = 343 K, 3^8—bending at the 8th loading half-cycle (T_2 = 343 K).

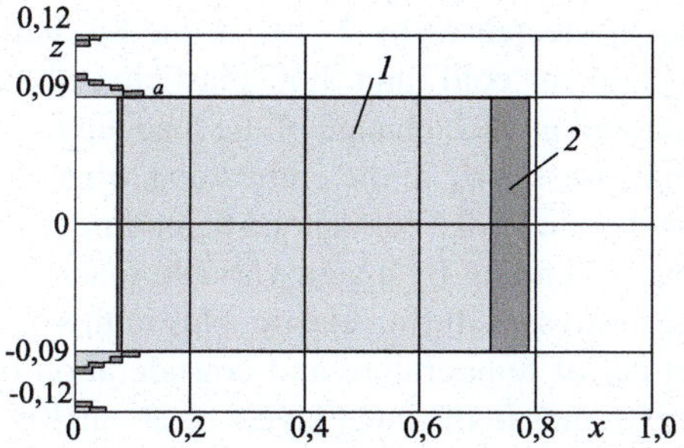

Fig. 5.12 Distribution of areas of plasticity and physical nonlinearity in the cantilevered (*a*) and rigidly clamped (*b*) rod

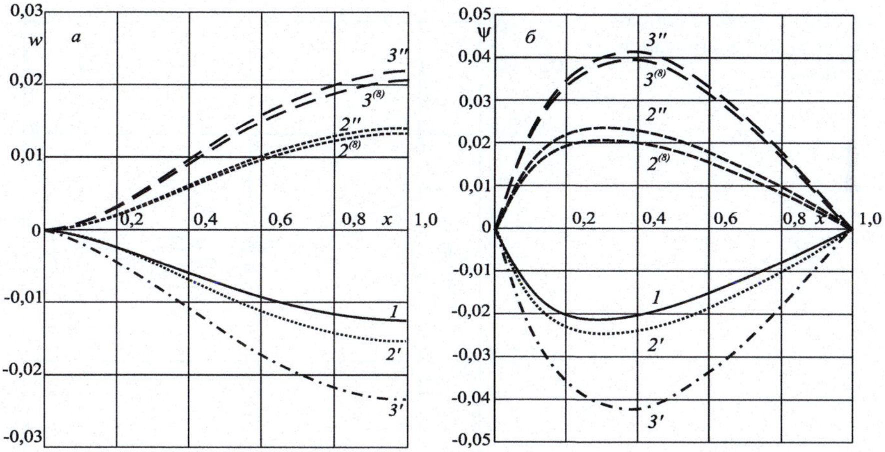

Fig. 5.13 Change of deflection *w* (*a*) and relative shear ψ (*b*) along the axis under alternating loading (cantilever)

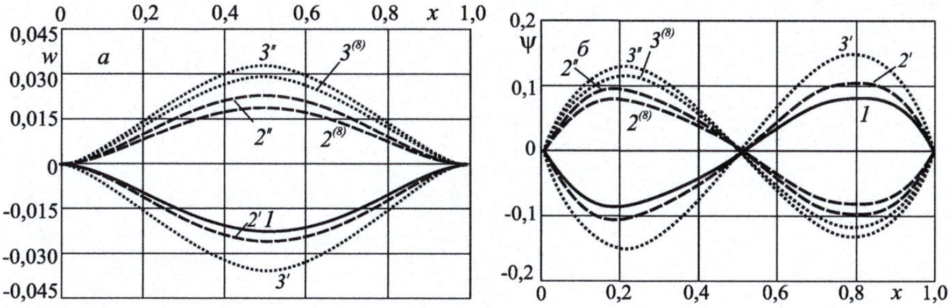

Fig. 5.14 Change of deflection *w* (*a*) and relative shear ψ (*b*) along the axis under sign-variable loading (embedding- embedding)

During repeated loading the displacement decreases by 2–3%, at the 8th half-cycle—by $\sigma_{xz}^{(m,3)*} = \sigma_{xze}^{(m,3)*} - \sigma_{xz\omega}^{(m,k)*}$–5%, both at "cold" and "hot" plasticity, which is explained by cyclic hardening of the material at each change of the load sign.

Figure 5.15 shows the change in normal stresses σ_x in the embedding when the sign of the external load changes. Graphs with one stroke correspond to loading from the natural state, with two strokes—repeated bending by a sign-variable load: *1*—stresses in the elastic rod ($T_1 = 293$ K), *2*—stresses during elastic–plastic thermo-force deformation ($T_2 = 343$ K). The effect of temperature and consideration of nonlinearity leads to an increase of stress extremes in the outer layers by about 30%, in the filler they decrease by 40%. In the second half-cycle the stresses change sign, their extremums change insignificantly.

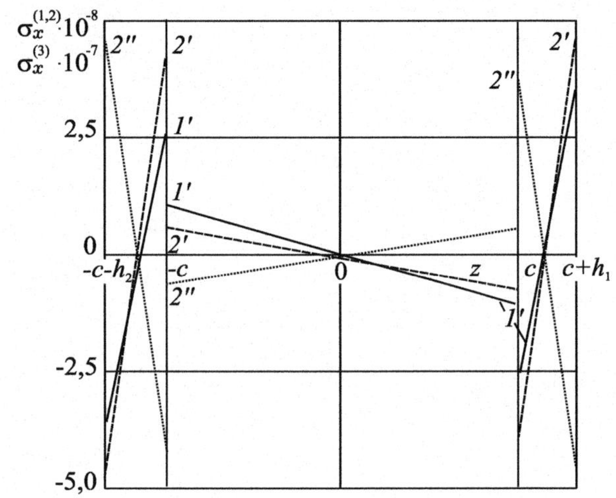

Fig. 5.15 Change of normal stresses σ_x in the termination (cantilever)

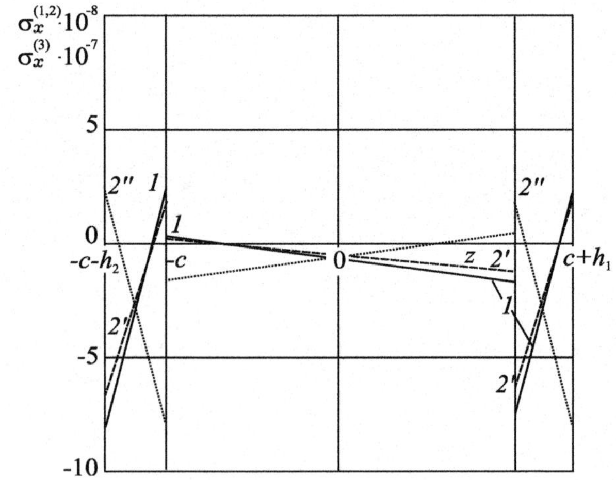

Fig. 5.16 Change of normal stresses σx in the left buttress (embedding- embedding)

Similar graphs of stresses at the left end when both ends of the rod are embedded are shown in Fig. 5.16. In this case, there is a shift of the curves to the negative region due to the thermal effect, with stresses in the load-bearing layers increasing, and in the filler decreasing.

This study has shown that the stress–strain state of asymmetric in thickness elastic–plastic three-layer rods depends significantly on the effect of the temperature field, which must be taken into account both during single and alternating thermo-force deformation. The results obtained in this section can be considered as an extension of the theory of V. V. Moskvitin on cyclic sign-variable loading of three-layer rods with incompressible filler in the temperature field.

References

1. Ilyushin, A. A. 1948. *Plasticity. Part 1. Elastic-plastic deformations*, 392. Moscow: Gostekhizdat. (in Russian).
2. Makhutov N. A., et al. 1986. *Mechanics of low-cycle fracture*s, 264. Moscow: Nauka. (in Russian).

Chapter 6
Deformation of a Three-Layer Rod in Neutron Flux

Single and alternating bending of elastic and elastoplastic three-layered rods with rigid filler in neutron flux is investigated. Kinematic hypotheses are based on the "broken" normal hypothesis, the filler is incompressible. The work of tangential stresses in the filler is taken into account. Analytical solutions of the problems of radiation elasticity and elastoplasticity are obtained. Numerical analysis of solutions is performed.

6.1 Elastoplastic Three-Layer Rod Under Neutron Irradiation

Let a neutron flux of density φ_0 in the direction opposite to the external normal be applied to the external surface $z = c + h_1$ of the considered three-layered rod in addition to the distributed force load $p'(x),\ q'(x)$ (Fig. 6.1).

Radiation irradiation of solids is accompanied by numerous effects. As a result, in solids volumetric deformation θ_I occurs, elastic and especially plastic characteristics of the material change [1, 2].

Assume that the materials of the bearing layers of the rod have elastic–plastic properties and the filler is nonlinearly elastic. To describe the deformation of layers in neutron flux we use the following physical equations of state:

$$
\begin{aligned}
s_x^{\prime(k)} &= 2G_k(1 - \omega^{\prime(k)}(\varepsilon_u^{\prime(k)}, I))\ \text{э}_x^{\prime(k)},\\
s_{xz}^{\prime(3)} &= 2G_3(1 - \omega^{\prime(3)}(\varepsilon_u^{(3)}, I))\ \text{э}_{xz}^{\prime(3)},\\
\sigma^{\prime(k)} &= K_k\ (3\varepsilon^{\prime(k)} - B_k I) \quad (k = 1,\ 2,\ 3).
\end{aligned} \tag{6.1}
$$

Here $\omega^{(k)}$ are universal functions of nonlinearity of materials of bearing layers ($k =$ 1, 2) under loading from the natural state and for the filler (for example, for D16T

S. Eduard et al., *Deformation of Three-layer Structural Elements in Thermal Radiation Fields*, https://doi.org/10.1007/978-981-97-7217-9_6

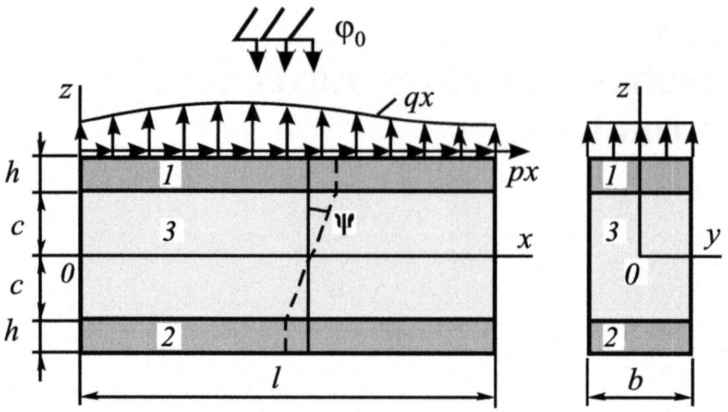

Fig. 6.1 Calculation diagram of a rod in neutron flux

alloy) have the following form:

$$\omega^{\prime(k)}(\varepsilon_u^{\prime(k)}, I) = \begin{cases} 0, & \varepsilon_u^{\prime(k)} \le \varepsilon_y^{\prime(k)}(I), \\ A_k\left(1 - \frac{\varepsilon_y^{\prime(k)}(I)}{\varepsilon_u^{\prime(k)}}\right)^{\alpha_k}, & \varepsilon_u^{\prime(k)} > \varepsilon_y^{\prime(k)}(I), \end{cases}$$

$$\varepsilon_y^{\prime(k)}(I) = \varepsilon_{y0}^{\prime}\left[1 + A(1 - \exp(-\xi I))^{1/2}\right], \tag{6.2}$$

$\varepsilon_u^{\prime(k)}$ is strain intensity; $\varepsilon_y^{\prime(k)}$ is deformation yield strength of the bearing layer material; I is integral neutron flux; B_k are material constants obtained experimentally.

In the following, we will assume that time variations of external loads and boundary displacements occur in such a way that the corresponding loading trajectories do not belong to the class of complex loading, and the radiation growth of the plasticity limit does not exceed the growth of strain intensity in the irradiated points of the solid body, which would prevent the formation of plastic deformations.

The components of the stress tensor in the layers, using (6.1), are represented through the deviator and the spherical part of the strain tensor

$$\sigma_x^{\prime(k)} = s_x^{\prime(k)} + \sigma^{\prime(k)} = 2G_k\, \text{э}_x^{\prime(k)} + K_k(3\varepsilon^{\prime(k)} - B_k I) - 2G_k\, \text{э}_x^{\prime(k)}\, \omega^{\prime(k)} (k = 1, 2, 3)$$

$$\sigma_{xz}^{\prime(3)} = s_{xz}^{\prime(3)} = 2G_3\, \text{э}_{xz}^{\prime(3)} - 2G_3\, \text{э}_{xz}^{\prime(3)}\, \omega^{\prime(3)}. \tag{6.3}$$

As before we identify elastic (index "e") and inelastic (index "ω") summands in the stress tensor:

$$\sigma_x^{\prime(k)} = \sigma_{xe}^{\prime(k)} - \sigma_{x\omega}^{\prime(k)}, \sigma_{xz}^{\prime(3)} = \sigma_{xze}^{\prime(3)} - \sigma_{xz\omega}^{\prime(3)}, \tag{6.4}$$

where

$$\sigma_{xe}^{\prime(k)} = 2G_k\, \text{э}_x^{\prime(k)} + K_k(3\varepsilon^{\prime(k)} - B_k I), \quad \sigma_{x\omega}^{\prime(k)} = 2G_k\, \text{э}_x^{\prime(k)}\, \omega^{\prime(k)},$$

$$\sigma'^{(3)}_{xze} = 2G_3\, \ni'^{(3)}_{xz}, \quad \sigma'^{(3)}_{xz\omega} = 2G_3\, \ni'^{(3)}_{xz}\, \omega'^{(3)}.$$

Carrying out a similar operation with the internal forces (3.4), we obtain:

$$N'^{(k)} = N'^{(k)}_e - N'^{(k)}_\omega, \quad M'^{(k)} = M'^{(k)}_e - M'^{(k)}_\omega, Q'^{(3)} = Q'^{(3)}_e - Q'^{(3)}_\omega,$$

$$N'^{(k)}_e = b_0 \int \sigma'^{(k)}_{xe} \mathrm{d}z, N'^{(k)}_\omega = b_0 \int\limits_{h_k} \sigma'^{(k)}_{x\omega} \mathrm{d}z, M'^{(k)}_e = b_0 \int\limits_{h_k} \sigma'^{(k)}_{xe} z \mathrm{d}z,$$

$$M'^{(k)}_\omega = b_0 \int\limits_{h_k} \sigma'^{(k)}_{x\omega} z\mathrm{d}z, Q'^{(3)}_e = b_0 \int\limits_{-c}^{c} \sigma'^{(3)}_{xze} \mathrm{d}z, Q'^{(3)}_\omega = b_0 \int\limits_{-c}^{c} \sigma'^{(3)}_{xz\omega} \mathrm{d}z. \tag{6.5}$$

Expressions of elastic components of internal forces in (6.5) through required displacements are obtained using (6.3, 6.4):

$$N'^{(1)} = b_0 \int\limits_{c}^{c+h} \sigma'^{(1)}_x \,\mathrm{d}z = b_0 h_1 \left[\left(K_1^+ (u'_{,x} + c\psi'_{,x}) - \left(c + \frac{1}{2} h_1 \right) w'_{,xx} \right) - K_1 B_1 I_1 \right],$$

$$N'^{(2)} = b_0 \int\limits_{-c-h_2}^{-c} \sigma'^{(2)}_x \,\mathrm{d}z = b_0 h_2 \left[\left(K_2^+ (u'_{,x} - c\psi'_{,x}) + \left(c + \frac{1}{2} h_2 \right) w'_{,xx} \right) - K_2 B_2 I_2 \right],$$

$$N'^{(3)} = b_0 \int\limits_{-c}^{c} \sigma'^{(3)}_x \mathrm{d}z = 2 b_0 c \left(K_3^+ u'_{,x} - K_3 B_3 I_3 \right),$$

$$M'^{(1)} = b_0 h_1 \left[K_1^+ \left(\left(c + \frac{1}{2} h_1 \right) (u'_{,x} + c\psi'_{,x}) - \left(c^2 + \mathrm{ch}_1 + \frac{1}{3} h_1^2 \right) w'_{,xx} \right), \right.$$
$$\left. - K_1 B_1 I_1 (c + \frac{1}{2} h_1) \right]$$

$$M'^{(2)} = -b_0 h_2 \left[K_2^+ \left(\left(c + \frac{1}{2} h_2 \right) (u'_{,x} - c\psi'_{,x}) + \left(c^2 + \mathrm{ch}_2 + \frac{1}{3} h_2^2 \right) w'_{,xx} \right), \right.$$
$$\left. - K_2 B_2 I_2 \left(c + \frac{1}{2} h_2 \right) \right]$$

$$M'^{(3)} = b_0 \int\limits_{-c}^{c} \sigma^{(3)}_x z \,\mathrm{d}z = \frac{2}{3} b_0 c^3 \left[K_3^+ (\psi'_{,x} - w'_{,xx}) - K_3 B_3 I_3 \right],$$

$$Q'^{(3)} = b_0 \int\limits_{-c}^{c} \sigma^{(3)}_{xz} \mathrm{d}z = 2 b_0 c G_3 \psi'. \tag{6.6}$$

Here $I_k(t)$ is the value of the integral neutron flux averaged over the thickness of the kth layer, obtained using the relations of flux values at the boundaries of layers $I^{(k)}$:

$$\varphi^{(1)}(z,t) = \varphi_0 t e^{\mu_1(z-c-h_1)}, \varphi^{(3)}(z,t) = \varphi_0 t e^{-\mu_1 h_1} e^{\mu_3(z-c)},$$
$$\varphi^{(2)}(z,t) = \varphi_0 t e^{-\mu_1 h_1 - 2\mu_3 c} e^{\mu_2(z+c)}$$

$$I_1(t) = \frac{1}{h_1}\int\limits_{h_1} \varphi_0 t e^{\mu_1(z-c-h_1)}dz = \frac{\varphi_0 t}{h_1\mu_1}\left(1 - e^{-\mu_1 h_1)}\right)$$

$$I_3(t) = \frac{\varphi_0 t e^{-\mu_1 h_1}}{2c\mu_3}\left(1 - e^{-2\mu_3 c}\right), I_2(t) = \frac{\varphi_0 t e^{-\mu_1 h_1 - 2\mu_3 c}}{h_2\mu_2}\left(1 - e^{-\mu_2 h_2}\right) \tag{6.7}$$

The equations of equilibrium of the rod under consideration and the boundary conditions in the forces formally retain the form (5.11, 3.12). However, in the considered case the influence of neutron irradiation is taken into account by expressions (6.6, 6.7) and the nonlinearity function in (6.4). Substituting into these equations the expressions of internal forces through the required displacements (6.6) and following the previously discussed derivation procedure, we obtain a system of nonlinear differential equations of equilibrium in displacements of the three-layer irradiated rod:

$$\begin{aligned} a_1 u'_{,xx} + a_6\psi'_{,xx} - a_7 w'_{,xxx} &= -p' + p'_\omega, \\ a_6 u'_{,xx} + a_2\psi'_{,xx} - a_3 w'_{,xxx} - a_5\psi' &= h'_\omega, \\ a_7 u'_{,xxx} + a_3\psi'_{,xxx} - a_4 w'_{,xxxx} &= -q' + q'_\omega. \end{aligned} \tag{6.8}$$

Here the coefficients a_i are determined by formula (3.17); the values $p'_\omega,\ h'_\omega,\ q'_\omega$ take into account the physical nonlinearity of layer materials and irradiation level, they are calculated using formulas (4.8, 6.4, 6.5). It should be noted that the integral neutron flux is not explicitly included in the equilibrium Eq. (6.8).

The system of differential Eq. (6.8) is nonlinear, so for its solution, as before, we use the method of elastic solutions. In this case it takes the following iterative form:

$$\begin{aligned} a_1 u'^{(n)}_{,xx} + a_6\psi'^{(n)}_{,xx} - a_7 w'^{(n)}_{,xxx} &= -p' + p'^{(n-1)}_\omega, \\ a_6 u'^{(n)}_{,xx} + a_2\psi'^{(n)}_{,xx} - a_3 w'^{(n)}_{,xxx} - a_5\psi'^{(n)} &= h'^{(n-1)}_\omega, \\ a_7 u'^{(n)}_{,xxx} + a_3\psi'^{(n)}_{,xxx} - a_4 w'^{(n)}_{,xxxx} &= -q' + q'^{(n-1)}_\omega. \end{aligned} \tag{6.9}$$

The values $p'^{(n-1)}_\omega,\ h'^{(n-1)}_\omega,\ q'^{(n-1)}_\omega$ at the first step ($n = 1$) are assumed to be equal to zero, and further they are calculated by the results of the previous approximation. They are determined by formulas(4.10), in which the nonlinearity functions depend on the irradiation level.

Application of the method of successive approximations allows at each step to reduce the considered problem to a linear elasticity problem with additional "external" loads. In this case the system of Eq. (6.9) formally at each step of approximation does not differ from the corresponding system (4.9) in the problem of isothermal elastoplasticity. Therefore, the recurrence solution of the problem of radiation elastic plasticity has the form:

$$\psi'^{(n)}(x) = C_2^{(n)}\mathrm{sh}(\beta x) + C_3^{(n)}\mathrm{ch}(\beta x)$$

$$
\begin{aligned}
&+\frac{1}{\beta}\left[\operatorname{sh}(\beta x)\int g'^{(n)}\operatorname{ch}(\beta x)\mathrm{d}x-\operatorname{ch}(\beta x)\int g'^{(n)}\operatorname{sh}(\beta x)\mathrm{d}x\right],\\
u'^{(n)}(x) &= \gamma_3\,\psi'^{(n)}+C_7^{(n)}x+C_8^{(n)}\\
&+\frac{1}{\alpha_2}\left[-a_4L_2^{-1}(p'-p'^{(n-1)}_\omega)+a_7L_3^{-1}(q'-q'^{(n-1)}_\omega)+\frac{a_7}{2}C_1^{(n)}x^2\right],\\
w'^{(n)}(x) &= \frac{1}{\alpha_2}\left[\alpha_1\int\psi'^{(n)}\mathrm{d}x-a_7L_3^{-1}(p'-p'^{(n-1)}_\omega)+a_1L_4^{-1}(q'-q'^{(n-1)}_\omega)\right.\\
&\left.+\frac{1}{6}a_1C_1^{(n)}x^3\right]+\frac{1}{2}C_4^{(n)}x^2+C_5^{(n)}x+C_6^{(n)}.
\end{aligned}
\tag{6.10}
$$

Here $g'^{(n)}(x)$ formally coincides with the function introduced in (4.11); the coefficients $\alpha_1,\ \alpha_2,\ \alpha_3,\ \beta^2,\ \gamma_1,\ \gamma_2,\ \gamma_3$ and linear integral operators $\mathrm{L}_2^{-1},\ \mathrm{L}_3^{-1},\ L_4^{-1}$ are defined by formula (3.17, 3.23). The integration constants $C_1^{(n)},\ldots,C_8^{(n)}$ take into account the impact of neutron flux and at each step follow from the rod fixation conditions.

In the case of cantilever embedding of the left end of the rod at the free right end, the boundary conditions (3.24) must be fulfilled, so we obtain:

$$
\begin{aligned}
C_1^{(n)} &= -L_1^{-1}(q'-q'^{(n-1)}_\omega)|_{x=l}\,,\\
C_2^{(n)} &= \frac{1}{\beta}\frac{\operatorname{ch}(\beta l)}{\operatorname{sh}(\beta l)}\left(\int g'^{(n)}\operatorname{sh}(\beta x)\mathrm{d}x|_{x=l}-\int g'^{(n)}\operatorname{sh}(\beta x)\mathrm{d}x|_{x=0}\right)\\
&-\frac{1}{\beta}\int g'^{(n)}\operatorname{ch}(\beta x)\mathrm{d}x|_{x=l}\,,\\
C_3^{(n)} &= \frac{1}{\beta}\int g'^{(n)}(x)\,\operatorname{sh}(\beta x)\mathrm{d}x|_{x=0}\,,\\
C_4^{(n)} &= \frac{a_7}{\alpha_2}\left(L_1^{-1}(p'-p'^{(n-1)}_\omega)+\sum_{k=1}^{3}B_kK_kI_kh_k\right)|_{x=l}\\
&-\frac{a_1}{\alpha_2}\left(L_2^{-1}(q'-q'^{(n-1)}_\omega)+B_1K_1I_1h_1\left(c+\frac{h_1}{2}\right)-B_2K_2I_2h_2\left(c+\frac{h_2}{2}\right)\right)|_{x=l}\\
&-\frac{a_1}{\alpha_2}C_1^{(n)}l,\\
C_5^{(n)} &= \frac{a_7}{\alpha_2}L_2^{-1}(p'-p'^{(n-1)}_\omega)|_{x=0}-\frac{a_1}{\alpha_2}L_3^{-1}(q'-q'^{(n-1)}_\omega)|_{x=0}\,,\\
C_6^{(n)} &= \frac{a_7}{\alpha_2}L_3^{-1}(p'-p'^{(n-1)}_\omega)\Big|_{x=0}-\frac{a_1}{\alpha_2}L_4^{-1}(q'-q'^{(n-1)}_\omega)\Big|_{x=0}-\frac{\alpha_1}{\alpha_2}\int\psi'^{(n)}\mathrm{d}x\Big|_{x=0},\\
C_7^{(n)} &= \frac{a_4}{\alpha_2}\left(L_1^{-1}(p'-p'^{(n-1)}_\omega)|_{x=l}+\sum_{k=1}^{3}B_kK_kI_kh_k\right)\\
&-\frac{a_7}{\alpha_2}\left(L_2^{-1}(q'-q'^{(n-1)}_\omega)|_{x=l}+B_1K_1I_1h_1\left(c+\frac{h_1}{2}\right)-B_2K_2I_2h_2\left(c+\frac{h_2}{2}\right)-C_1^{(n)}l\right),
\end{aligned}
$$

$$C_8^{(n)} = \frac{a_4}{\alpha_2} L_2^{-1}(p' - p_\omega'^{(n-1)})|_{x=0} - \frac{a_7}{\alpha_2} L_3^{-1}(q' - q_\omega'^{(n-1)})|_{x=0}. \tag{6.11}$$

For a rod hinged at the ends, from (3.26) we have:

$$\begin{aligned}
C_1^{(n)} &= \frac{1}{l}\left(L_2^{-1}(q' - q_\omega'^{(n-1)})|_{x=0} - L_2^{-1}(q' - q_\omega'^{(n-1)})|_{x=l}\right),\\
C_4^{(n)} &= \frac{a_7}{a_4}\left(D_1^{(n)}|_{x=0} - D_1^{(n)}|_{x=l}\right)\\
&\quad - \frac{1}{a_4}\left(L_2^{-1}(q' - q_\omega'^{(n-1)})|_{x=0} + B_1K_1I_1h_1\left(c + \frac{h_1}{2}\right) - B_2K_2I_2h_2\left(c + \frac{h_2}{2}\right)\right),\\
C_5^{(n)} &= D_2^{(n)}|_{x=0} - D_2^{(n)}|_{x=l} - \frac{a_7 l}{2a_4}\left(D_1^{(n)}|_{x=0} - D_1^{(n)}|_{x=l}\right)\\
&\quad - \frac{l}{2a_4}\left(L_2^{-1}(q' - q_\omega'^{(n-1)})|_{x=0} + B_1K_1I_1h_1\left(c + \frac{h_1}{2}\right) - B_2K_2I_2h_2\left(c + \frac{h_2}{2}\right)\right),\\
C_7^{(n)} &= D_1^{(n)}|_{x=0} - D_1^{(n)}|_{x=l},\\
D_1^{(n)} &= \frac{a_7}{l\alpha_2} L_3^{-1}(q - q_\omega^{(n-1)}) - \frac{a_7 l}{2\alpha_2} L_2^{-1}(q - q_\omega^{(n-1)}) - \frac{a_4}{l\alpha_2} L_2^{-1}(p - p_\omega^{(n-1)})\\
&\quad + B_1K_1I_1h_1\left(c + \frac{h_1}{2}\right) - B_2K_2I_2h_2\left(c + \frac{h_2}{2}\right),\\
D_2^{(n)} &= \frac{a_1}{l\alpha_2}\left(\int \psi'^{(n)} dx + L_4^{-1}(q' - q_\omega'^{(n-1)})\right)\\
&\quad - \frac{a_1 l}{6\alpha_2}\left(L_2^{-1}(q' - q_\omega'^{(n-1)}) + B_1K_1I_1h_1\left(c + \frac{h_1}{2}\right) - B_2K_2I_2h_2\left(c + \frac{h_2}{2}\right)\right)\\
&\quad - \frac{a_7}{l\alpha_2} L_3^{-1}(p' - p_\omega'^{(n-1)})
\end{aligned} \tag{6.12}$$

The integration constants $C_2^{(n)}$, $C_3^{(n)}$, $C_6^{(n)}$, $C_8^{(n)}$ keep the form (6.11).

In the case of rigid embedding of both ends of the rod, the conditions (3.28) must be fulfilled. Then the integration constant $C_1^{(n)}$ is described by formula (6.12), and the constants $C_2^{(n)}$, $C_3^{(n)}$, $C_5^{(n)}$, $C_6^{(n)}$, $C_8^{(n)}$ are described by formula (6.11):

$$\begin{aligned}
C_4^{(n)} &= \frac{1}{\alpha_2 l}\left[a_7 L_2^{-1}\left(p' - p_\omega'^{(n-1)}\right)\Big|_{x=l} - a_1 L_3^{-1}\left(q' - q_\omega'^{(n-1)}\right)\Big|_{x=l} - \frac{a_1 C_1^{(n)} l^2}{2}\right] - \frac{C_5^{(n)}}{l},\\
C_7^{(n)} &= \frac{1}{\alpha_2 l}\left[a_4 L_2^{-1}\left(p' - p_\omega'^{(n-1)}\right)\Big|_{x=l} - a_7 L_3^{-1}\left(q' - q_\omega'^{(n-1)}\right)\Big|_{x=l} - \frac{a_7 C_1^{(n)} l^2}{2}\right] - \frac{C_8^{(n)}}{l}.
\end{aligned} \tag{6.13}$$

Consequently, formulas (6.10)–(6.13) give an iteration solution to the problem of deformation of a three-layer elastic–plastic rod under loading in neutron flux under different boundary conditions.

6.2 Analytical Solution of the Radiation Elasticity Problem

As a consequence, we obtain from the solution (6.10)–(6.13) at $n = 1$ the exact solution of the elasticity problem on deformation of a three-layer rod in neutron flux, which is of independent interest (the stroke above is omitted):

$$\begin{aligned} \psi(x) &= C_2\mathrm{sh}(\beta x) + C_3\mathrm{ch}(\beta x) \\ &\quad + \frac{1}{\beta}\left[\mathrm{sh}(\beta x)\int g(x)\mathrm{ch}(\beta x)\mathrm{d}x - \mathrm{ch}(\beta x)\int g(x)\mathrm{sh}(\beta x)\mathrm{d}x\right], \\ w(x) &= \frac{1}{\alpha_2}\left[\alpha_1\int\psi\mathrm{d}x - a_7L_3^{-1}(p) + a_1L_4^{-1}(q) + \frac{a_1C_1}{6}x^3\right] + \frac{C_4}{2}x^2 + C_5x + C_6, \\ u(x) &= \gamma_3\psi + \frac{1}{\alpha_2}\left[-a_4L_2^{-1}(p) + a_7L_3^{-1}(q) + \frac{a_7}{2}C_1x^2\right] + C_7x + C_8, \end{aligned} \tag{6.14}$$

where coefficients $\alpha_1,\ \alpha_2,\ \alpha_3,\ \beta^2,\ \gamma_1,\ \gamma_2,\ \gamma_3$ and linear integral operators $\mathrm{L}_2^{-1},\ \mathrm{L}_3^{-1},\ L_4^{-1}$ are defined by formulas (3.17, 3.23).

Using relations (6.5, 6.6), let us write out the final expressions of internal bending moment and generalised longitudinal force through displacements and integral neutron flux (6.7):

$$\begin{aligned} M &= b_0\left(a_7u_{,x} + a_3\psi_{,x} - a_4w_{,xx} - B_1K_1I_1h_1\left(c + \frac{h_1}{2}\right) + B_2K_2I_2h_2\left(c + \frac{h_2}{2}\right)\right), \\ N &= b_0\left(a_1u_{,x} + a_6\psi_{,x} - a_7w_{,xx} - \sum_{k=1}^{3}B_kK_kI_kh_k\right). \end{aligned} \tag{6.15}$$

Let us write out the integration constants for the previously considered boundary conditions at randomly distributed loads.

At cantilever fixing of the end of the three-layer rod, the integration constants C_1, …, C_8 are obtained by substituting the solution (6.14) into the boundary conditions (3.24) and considering (6.15). As a result, we have:

$$\begin{aligned} C_1 &= -L_1^{-1}(q)|_{x=l}, \\ C_2 &= \frac{1}{\beta} - \int g(x)\ \mathrm{ch}(\beta x)\mathrm{d}x|_{x=l}\Big] - \int g(x)\ \mathrm{ch}(\beta x)\mathrm{d}x|_{x=l}\Big], \\ C_4 &= \frac{a_7}{\alpha_2}\left(L_1^{-1}(p)|_{x=l} + \sum_{k=1}^{3}B_kK_kI_kh_k\right) \\ &\quad - \frac{a_1}{\alpha_2}\left(L_2^{-1}(q)|_{x=l} + B_1K_1I_1h_1\left(c + \frac{h_1}{2}\right) - B_2K_2I_2h_2\left(c + \frac{h_2}{2}\right) + C_1l\right), \\ C_3 &= \frac{1}{\beta}\int g(x)\ \mathrm{sh}(\beta x)\mathrm{d}x|_{x=0}, \end{aligned}$$

$$
\begin{aligned}
C_5 &= \frac{a_7}{\alpha_2} L_2^{-1}(p)|_{x=0} - \frac{a_1}{\alpha_2} L_3^{-1}(q)|_{x=0},\\
C_6 &= \frac{a_7}{\alpha_2} L_3^{-1}(p)\Big|_{x=0} - \frac{a_1}{\alpha_2} L_4^{-1}(q)\Big|_{x=0} - \frac{\alpha_1}{\alpha_2}\int \psi\,\mathrm{d}x\Big|_{x=0},\\
C_7 &= \frac{a_4}{\alpha_2}\left(L_1^{-1}(p)|_{x=l} + \sum_{k=1}^{3} B_k K_k I_k h_k\right)\\
&\quad - \frac{a_7}{\alpha_2}\left(L_2^{-1}(q)|_{x=l} + B_1K_1I_1h_1\left(c+\frac{h_1}{2}\right) - B_2K_2I_2h_2\left(c+\frac{h_2}{2}\right)\right) - \frac{a_7}{\alpha_2}C_1 l,\\
C_8 &= \frac{a_4}{\alpha_2} L_2^{-1}(p)|_{x=0} - \frac{a_7}{\alpha_2} L_3^{-1}(q)|_{x=0}
\end{aligned}
\tag{6.16}
$$

For a rod hinged at both ends, from the boundary conditions (3.26) we obtain:

$$
\begin{aligned}
C_1 &= \frac{1}{l}\left(L_2^{-1}(q)|_{x=0} - L_2^{-1}(q)|_{x=l}\right),\\
C_4 &= \frac{a_7}{a_4}\left(D_1|_{x=0} - D_1|_{x=l}\right)\\
&\quad - \frac{1}{a_4}\left(L_2^{-1}(q)|_{x=0} + B_1K_1I_1h_1\left(c+\frac{h_1}{2}\right) - B_2K_2I_2h_2\left(c+\frac{h_2}{2}\right)\right),\\
C_5 &= D_2|_{x=0} - D_2|_{x=l} - \frac{a_7 l}{2a_4}\left(D_1|_{x=0} - D_1|_{x=l}\right)\\
&\quad - \frac{l}{2a_4}\left(L_2^{-1}(q)|_{x=0} + B_1K_1I_1h_1\left(c+\frac{h_1}{2}\right) - B_2K_2I_2h_2\left(c+\frac{h_2}{2}\right)\right),\\
C_7 &= D_1|_{x=0} - D_1|_{x=l},\\
D_1 &= \frac{a_7}{l\alpha_2} L_3^{-1}(q) - \frac{a_7 l}{2\alpha_2}\left(L_2^{-1}(q) + B_1K_1I_1h_1\left(c+\frac{h_1}{2}\right) - B_2K_2I_2h_2\left(c+\frac{h_2}{2}\right)\right)\\
&\quad - \frac{a_4}{l\alpha_2} L_2^{-1}(p),\\
D_2 &= \frac{a_1}{l\alpha_2}\left(\int \psi\,dx + L_4^{-1}(q)\right)\\
&\quad - \frac{a_1 l}{6\alpha_2}\left(L_2^{-1}(q) + B_1K_1I_1h_1\left(c+\frac{h_1}{2}\right) - B_2K_2I_2h_2\left(c+\frac{h_2}{2}\right)\right) - \frac{a_7}{l\alpha_2} L_3^{-1}(p).
\end{aligned}
\tag{6.17}
$$

The integration constants C_2, C_3, C_6, C_8 keep the form (6.16).

When both ends are rigidly embedded, the conditions (3.28) must be fulfilled, so the integration constants C_4, C_7 are described by the formula (6.13); the constant C_1—(6.17); and the constants C_2, C_3, C_5, C_6, C_8 retain the form (6.16).

In the following, we will need expressions for deflection and relative shear obtained when solving the radiation elasticity problem at constant loads $q = \text{const}$, $p = \text{const}$ in the neutron flux. Then from (6.14, 6.16) we obtain in the case of the left cantilever:

$$\psi(x) = C_2\mathrm{sh}(\beta x) + C_3\mathrm{ch}(\beta x) - \frac{1}{\beta^2}\left[\gamma_2 p + \gamma_1 q(x - \frac{l}{2})\right],$$
$$u(x) = \gamma_3\,\psi + \frac{1}{\alpha_2}\left[-\frac{1}{2}a_4 p x^2 + \frac{1}{6}a_7 q x^3 - \frac{a_7}{4} q l x^2\right] + C_7 x,$$
$$w = \frac{1}{\alpha_2}\left[\alpha_1\int\psi\,\mathrm{d}x - \frac{1}{6}a_7 p x^3 + \frac{1}{12}a_1 q x^3\left(\frac{x}{2} - l\right)\right] + \frac{1}{2}C_4 x^2 + C_6, \quad (6.18)$$

where

$$C_2 = \frac{1}{2\beta^2\mathrm{sh}(\beta l)}\left[2\gamma_2 p(1 - \mathrm{ch}(\beta l)) + \gamma_1 q l(1 + \mathrm{ch}(\beta l))\right],$$
$$C_3 = \frac{1}{2\beta^2}\left[2\gamma_2 p - \gamma_1 q l\right],$$
$$C_4 = \frac{1}{\alpha_2}\left[a_7\left(pl + \sum_{k=1}^{3} B_k K_k I_k h_k\right)\right.$$
$$\left.+a_1\left(\frac{ql^2}{2} - B_1 K_1 I_1 h_1\left(c + \frac{h_1}{2}\right) + B_2 K_2 I_2 h_2\left(c + \frac{h_2}{2}\right)\right)\right],$$
$$C_6 = -\frac{\alpha_1}{\beta\alpha_2}C_2,$$
$$C_7 = \frac{a_4}{\alpha_2}\left(pl + \sum_{k=1}^{3} B_k K_k I_k h_k\right)$$
$$+\frac{a_7}{\alpha_2}\left(\frac{ql^2}{2} - B_1 K_1 I_1 h_1\left(c + \frac{h_1}{2}\right) + B_2 K_2 I_2 h_2\left(c + \frac{h_2}{2}\right)\right),$$
$$\int\psi\,\mathrm{d}x = \frac{1}{\beta}(C_2\mathrm{ch}(\beta x) + C_3\mathrm{sh}(\beta x)) - \frac{x}{2\beta^2}(2\gamma_2 p + \gamma_1 q(x - l)).$$

When both ends of the rod are embedded and uniformly distributed loads are applied, in solution (6.18) it is necessary to replace constants C_4, C_7 and take their representations from (6.13). In this case displacements and strains do not depend on the neutron flux. It only affects stresses through volumetric strain.

Thus, expressions (6.14, 6.15)–(6.18) give an analytical solution to the problem of deformation of an elastic three-layer rod in neutron flux under three types of boundary conditions.

Numerical results are obtained for the D16T–fluoroplastic–D16T rod package. Relative thicknesses of layers $h_1 = h_2 = 0{,}03$, $c = 0{,}09$; elastic and radiation parameters of materials are given in Tables 2.1–2.3 and in Sect. 6.1. Besides it was accepted: $B = 10^{-23}$ m^2/neutron, that provides volume deformation in layers of the considered rod of 3–3.5%; for duralumin—$\mu = 1.6$ cm, for fluoroplastic—$\mu = 3.21$ cm.

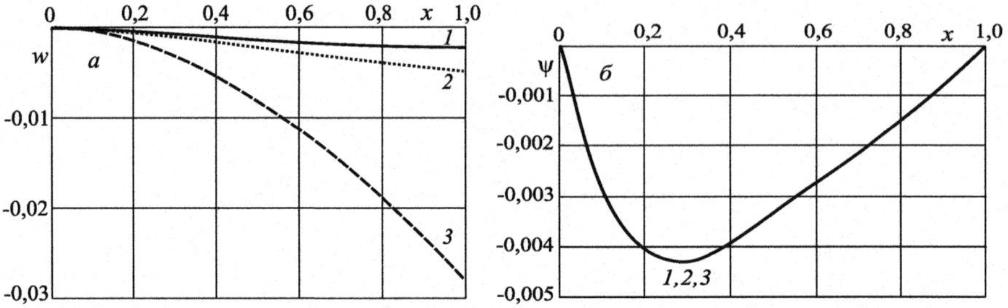

Fig. 6.2 Variation of deflection w (a) and relative shear ψ (b) along the rod axis

Figures 6.2, 6.3, 6.4 and 6.5 show the effect of neutron flux intensity on the stress–strain state of an elastic three-layer rod when cantilevered and when both ends are embedded. Load quantity $q' = -0.15$ MPa, $p' = 0$.

Figure 6.2 shows the variation of deflection w (a) and relative shear ψ (b) along the axis of a cantilevered rod when irradiated for one hour with neutron flux φ (neutron/(m^2s)) of different intensities: 1—$\varphi_1 = 0$, 2—$\varphi_2 = 10^{17}$, 3—$\varphi_3 = 10^{18}$.

Exposure to neutron flux leads to an increase in volumetric strain and an increase in elastic design deflection by a factor of about 1.5 and 5, respectively. The relative shift is not affected by irradiation.

The distribution of linear strains ε_x^k and normal stresses σ_x^k along the height of the cross-section at the embedment ($x = 0$) of the cantilever rod is shown in Figs. 6.3, 6.4, respectively. The numbering of the curves is the same as before. The strains on the layer glues are continuous, they go to the positive region with increasing neutron flux intensity and grow in modulus. The stresses in the load-bearing layers go to the positive region, while the stresses in the filler compressed by the outer layers become negative.

Fig. 6.3 Distribution over the height of the cross-section in the embedment ($x = 0$) of linear strains ε_x^k (cantilever)

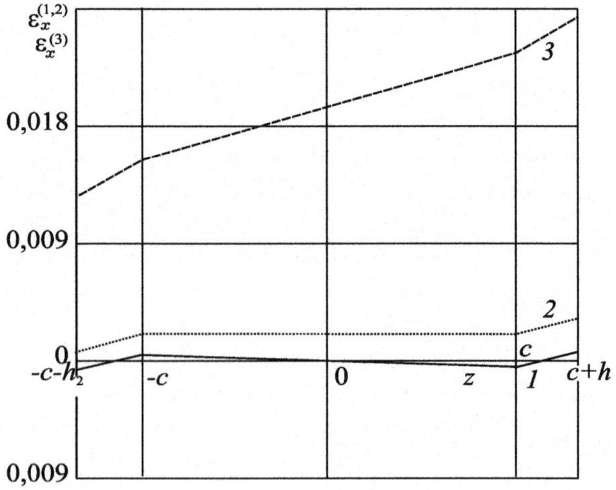

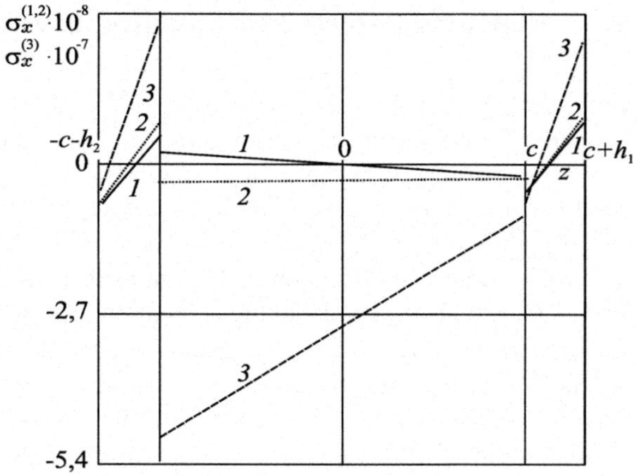

Fig. 6.4 Distribution over the height of the cross-section in the embedment ($x = 0$) of normal stresses σ_x^k (cantilever)

When both ends are embedded, displacements and strains do not depend on the neutron flux. Its influence on stresses is shown in Fig. 6.5.

An increase in the volumetric strain at constant rod length leads to its compression and a shift of stresses to the negative region, with the modulus of stresses growing faster in the upper load-bearing layer.

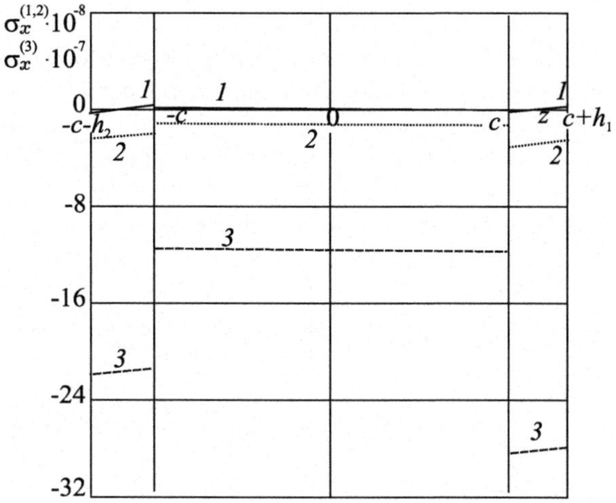

Fig. 6.5 Distribution over the height of the cross-section in the embedment (x = 0) of normal stresses σ_x^k (embedment–embedment)

6.3 Repeated Alternating Loading Elastoplastic Rod

Let, starting from the moment $t = t_1$, there is an instantaneous unloading and re-loading with forces of opposite sign p'', q'', changing according to the same law as at loading from the natural state. These forces will create a displacement field $u''(x)$, $w''(x)$, strains $\varepsilon_x^{''(k)}(x,z)$, $э_x^{''(k)}(x,z)$, $\varepsilon^{''(k)}(x,z)$, $\psi''(x,z)$ and stresses $\sigma_x^{''(k)}(x,z)$, $s_x^{''(k)}(x,z)$, $\sigma^{''(k)}(x,z)$, $\sigma_{xz}^{''(3)}(x,z)$ in the kth layer of the rod. In this case we will assume that during unloading and subsequent variable loading the neutron flux influence has stopped and the irradiation level in all points of the body remains unchanged. It coincides with the level at the start of unloading $I'(z) = I_1(z, t_1)$.

As before, we keep for all values of stress–strain state and load differences of the type (4.15), in which the values with one stroke are stresses, deformations and displacements in the rod before unloading, two strokes mark similar parameters during the second half-cycle. For example, deflection: $w^* = w' - w''$.

For the introduced stresses and strains with asterisks, we assume physical equations of state of the type (6.1):

$$\begin{aligned} s_x^{(k)*} &= 2G_k(1-\omega^{(k)*})\, э_x^{(k)*}, \\ s_{xz}^{(3)*} &= 2G_3(1-\omega)^{(3)*}\, э_{xz}^{(3)*}, \\ \sigma^{(k)*} &= 3K_k\varepsilon^{(k)*}\ (k = 1,\ 2,\ 3). \end{aligned} \tag{6.19}$$

According to Moskvitin's hypothesis, we assume the corresponding universal nonlinearity functions in the bearing layers to be expressed through the plasticity functions under loading from the natural state:

$$\omega^{(k)*} = \omega'^{(k)}(\varepsilon_u^{(k)*}, \varepsilon_y^{(k)*}, I_1). \tag{6.20}$$

Here nonlinearity functions of the type (3.39) are assumed:

$$\begin{aligned} \omega^{(k)*} &= \begin{cases} 0, & \varepsilon_u^{(k)*} \le \varepsilon_y^{(k)*}(I_1), \\ A_k^*\left(1 - \frac{\varepsilon_y^{(k)*}(I')}{\varepsilon_u^{(k)*}}\right)^{\alpha_k^*}, & \varepsilon_u^{(k)*} > \varepsilon_y^{(k)*}(I_1), \end{cases} \\ \varepsilon_y^{(k)*}(I_1) &= \varepsilon_{y0}\left[1 + A(1-\exp(-\xi I_1))^{1/2}\right], \end{aligned} \tag{6.21}$$

where for alloy D16-T: $A_k^* = 0.924$, $\alpha_k^* = 2.27$, $\varepsilon_y^{(k)*} = \alpha_2^{(k)}\varepsilon_y^{(k)}(I')$, $\alpha_2^{(k)} = 2.02$, $\varepsilon_y^{(k)*}(I_1)$, ε_{y0} are yield strengths of irradiated and non-irradiated material.

Singling out the nonlinear part in the stresses and internal forces with asterisks, carrying out a formal procedure of derivation of equilibrium equations for the values with asterisks, we finally obtain the following iteration system:

$$\begin{aligned} a_1 u,_{xx}^{(n)*} + a_6\psi,_{xx}^{(n)*} - a_7 w,_{xxx}^{(n)*} &= -p^* + p_\omega^{(n-1)*}, \\ a_6 u,_{xx}^{(n)*} + a_2\psi,_{xx}^{(n)*} - a_3 w,_{xxx}^{(n)*} - a_5\psi^{(n)*} &= h_\omega^{(n-1)*}, \end{aligned}$$

$$a_7 u,_{xxx}^{(n)*} + a_3 \psi,_{xxx}^{(n)*} - a_4 w,_{xxxx}^{(n)*} = -q^* + q_\omega^{(n-1)*}. \tag{6.22}$$

Additional "external" loads $p_\omega^{(n-1)*}, \ h_\omega^{(n-1)*}, \ q_\omega^{(n-1)*}$ in (6.22) are calculated by formulae (3.44), in which the nonlinearity function of the form (6.20,6.21). At the first step ($n = 1$) they are assumed to be equal to zero, further they are calculated according to the results of the previous approximation, they serve as corrections for plasticity and physical nonlinearity of layer materials under repeated loading and take into account the impact of neutron fluxes.

Thus, the system of Eq. (6.22) for the quantities with asterisks coincides with the system (6.9) when loading the rod from the natural state with precision to notation. Therefore, its analytical solution will be of the type (6.10):

$$\begin{aligned}
\psi^{(n)*}(x) &= C_2^{(n)*}\mathrm{sh}(\beta x) + C_3^{(n)*}\mathrm{ch}(\beta x) \\
&\quad + \frac{1}{\beta}\left[\mathrm{sh}(\beta x)\int g^{(n)*}\mathrm{ch}(\beta x)\mathrm{d}x - \mathrm{ch}(\beta x)\int g^{(n)*}\mathrm{sh}(\beta x)\mathrm{d}x\right]. \\
u^{(n)*}(x) &= \gamma_3\psi^{(n)*} + C_7^{(n)*}x + C_8^{(n)*} \\
&\quad + \frac{1}{\alpha_2}\left[-a_4 L_2^{-1}\left(p^* - p_\omega^{(n-1)*}\right) + a_7 L_3^{-1}\left(q - q_\omega^{(n-1)*}\right) + \frac{a_7}{2}C_1^{(n)*}x^2\right], \\
w^{(n)*}(x) &= \frac{1}{\alpha_2}\left[\alpha_1\int\psi^{(n)*}\mathrm{d}x - a_7 L_3^{-1}\left(p^* - p_\omega^{(n-1)*}\right)\right. \\
&\quad \left. + a_1 L_4^{-1}\left(q^* - q_\omega^{(n-1)*}\right) + \frac{1}{6}a_1 C_1^{(n)*}x^3\right] \\
&\quad + \frac{1}{2}C_4^{(n)*}x^2 + C_5^{(n)*}x + C_6^{(n)*}.
\end{aligned} \tag{6.23}$$

Here coefficients $\alpha_1, \ \alpha_2, \ \alpha_3, \ \beta^2, \ \gamma_1, \ \gamma_2, \ \gamma_3$ and linear integral operators $\mathrm{L}_2^{-1}, \ \mathrm{L}_3^{-1}, \ \mathrm{L}_4^{-1}$ are defined by formulas (3.19, 3.23); integration constants $C_1^{(n)*}, \ldots, C_8^{(n)*}$ at each step of the approximation follow from the conditions of rod fixation; the integrand function will depend on external and additional loads with asterisks:

$$g^{(n)*}(x) = \frac{\alpha_2}{\alpha_1}\gamma_1 h_\omega^{(n-1)*} + \gamma_2\left(p^* - p_\omega^{(n-1)*}\right) + \gamma_1\left(\int\left(q^* - q_\omega^{(n-1)*}\right)dx + C_1^{(n)*}\right)$$

Satisfying by the solution (6.23) the boundary conditions for the *cantilever* (4.25), *hinged rod* (4.26), *rigid embedding of both ends* (4.27), the integration constants $C_1^{(n)*}, \ldots, C_8^{(n)*}$ are obtained in the form (4.29–4.31), since the neutron flux impact has stopped before unloading. However, the nonlinearity functions included in the corresponding equations will be calculated here taking into account the magnitude of the neutron flux fixed before unloading.

Thus, the obtained formulas (6.23, 4.29–4.31) describe in iterations the analytical solution of the boundary value problem with asterisks under repeated alternating loading of a three-layer elastoplastic rod in neutron flow in case of different types of boundary conditions. The corresponding solution of the problem during the second half-cycle is obtained from relations (4.15) taking into account the solution (6.23).

6.4 Cyclic Loading in Neutron Flux

Quite similarly, we obtain the solution for any mth variable loading taking into account neutron irradiation. Let, after $(m - 1)$ alternating half-cycles of loading, starting from the moment t_{m-1}, an instantaneous unloading and a new loading by forces of the opposite sign $p^{(m)}$, $q^{(m)}$ (the upper index "m" is the number of loading) is carried out. These forces will create displacements $u^{(m)}$, $\psi^{(m)}$, $w^{(m)}$, strains $\varepsilon_{ij}^{(m,\,k)}$ and stresses $\sigma_{ij}^{(m,\,k)}$ in the kth layer of the rod.

Let for the time $(m - 1)$ of alternating half-cycles the integral neutron flux at all points of the body remained the same as at the beginning of the first unloading $I'(z) = I_1(z, t_1)$. For all parameters of the stress–strain state and loads we have differences, in which the values with index $(m-1)$ are stresses, strains, displacements and loads in the rod before unloading, index m denotes similar parameters in the process of the mth half-cycle:

$$
\begin{aligned}
\sigma_x^{(m,\,k)*} &= (-1)^n(\sigma_x^{(m-1,\,k)} - \sigma_x^{(m,\,k)}),\ \varepsilon_x^{(m,\,k)*} = (-1)^n(\varepsilon_x^{(m-1,\,k)} - \varepsilon_x^{(m,\,k)}),\\
u^{(m)*} &= (-1)^n(u^{(m-1)} - u^{(m)}),\ \psi^{(m)*} = (-1)^n(\psi^{(m-1)} - \psi^{(m)}),\\
w^{(m)*} &= (-1)^n(w^{(m-1)} - w^{(m)}),\\
q^{(m)*} &= (-1)^n(q^{(m-1)} - q^{(m)}),\ p^{(m)*} = (-1)^n(p^{(m-1)} - p^{(m)}).
\end{aligned}
\tag{6.24}
$$

The relations (4.16) relating strains and displacements with asterisks will be true for the values in (6.24) as well. For the stresses and strains marked with an asterisk, we assume physical equations of state of the type (6.19):

$$
\begin{aligned}
s_x^{(m,\,k)*} &= 2G_k\, э_x^{(m,\,k)*}\ (1 - \omega^{(m,\,k)*})(\varepsilon_u^{(m,\,k)*}, I_1),\\
s_{xz}^{(m,\,3)*} &= 2G_3\, э_{xz}^{(m,\,3)*}\ (1 - \omega^{(m,\,3)*})(\varepsilon_u^{(m,\,3)*}, I_1),\\
\sigma^{(m,\,k)*} &= 3K_k\varepsilon^{(m,\,k)*}\quad (k = 1,\ 2,\ 3)
\end{aligned}
\tag{6.25}
$$

The filler remains nonlinearly elastic. The corresponding universal nonlinearity functions in the bearing layers included in (6.25) are assumed to be expressed through the plasticity functions under loading from the natural state:

$$
\omega^{(m,\,k)*} = \omega'^{(k)}(\varepsilon_u^{(m,\,k)*},\ \varepsilon_y^{(m,\,k)*},\ I_1),
\tag{6.26}
$$

where, for example, for alloy D16T:

$$\omega^{(m,\,k)*} = \begin{cases} 0, & \varepsilon_u^{(m,\,k)*} \le \varepsilon_y^{(m,\,k)*}(I_1), \\ A_k^*\left(1 - \frac{\varepsilon_y^{(m,\,k)*}(I_1)}{\varepsilon_u^{(m,\,k)*}}\right)^{\alpha_k^*}, & \varepsilon_u^{(m,\,k)*} > \varepsilon_y^{(m,\,k)*}(I_1), \end{cases}$$

$$\varepsilon_y^{(m,k)*}(I_1) = \varepsilon_y^{(k)*}(I_1) = \varepsilon_{y0}\left[1 + A(1 - \exp(-\xi I_1))^{1/2}\right],$$

$A_k^{(m)*}$, $\alpha_k^{(m)*}$ are experimental parameters, ε_{y0}, $\varepsilon_y^{(m,\,k)*}(I_1)$ are yield strengths of unirradiated and irradiated material at the mth half-cycle.

The iterative system of equations of equilibrium of the rod at the nth step for the values with asterisks at the mth half-cycle will be of the type (6.22):

$$\begin{aligned} a_1 u,_{xx}^{(m,\,n)*} + a_6 \psi,_{xx}^{(m,\,n)*} - a_7 w,_{xxx}^{(m,\,n)*} &= -p^{(m)*} + p_\omega^{(m,\,n-1)*}, \\ a_6 u,_{xx}^{(m,\,n)*} + a_2 \psi,_{xx}^{(m,\,n)*} - a_3 w,_{xxx}^{(m,\,n)*} - a_5 \psi^{(m,\,n)*} &= h_\omega^{(m,\,n-1)*}, \\ a_7 u,_{xxx}^{(m,\,n)*} + a_3 \psi,_{xxx}^{(m,\,n)*} - a_4 w,_{xxxx}^{(m,\,n)*} &= -q^{(m)*} + q_\omega^{(m,\,n-1)*}. \end{aligned} \tag{6.27}$$

Additional "external" loads $p_\omega^{(m,\,n-1)*}$, $h_\omega^{(m,\,n-1)*}$, $q_\omega^{(m,\,n-1)*}$ at the first step ($n = 1$) are assumed to be equal to zero, and further are calculated according to the results of the previous approximation. They are corrections for plasticity and physical nonlinearity of layer materials and are formally calculated by formulas (4.40), in which the nonlinearity functions depend on the irradiation level fixed before the first unloading.

Adding to Eq. (6.27) the boundary conditions, for example, (4.41–4.43), we obtain the closed boundary value problem with asterisks introduced in (6.24). It coincides with the boundary value problem on loading of a rod from the natural state (6.9, 4.25–4.27) with accuracy up to notation. Therefore, its recurrence analytical solution will be as follows:

$$\begin{aligned} \psi^{(m,\,n)*}(x) &= C_2^{(m,\,n)*}\mathrm{sh}(\beta x) + C_3^{(m,\,n)*}\mathrm{ch}(\beta x) \\ &+ \frac{1}{\beta}\left[\mathrm{sh}(\beta x)\int g^{(m,\,n)*}\mathrm{ch}(\beta x)\mathrm{d}x - \mathrm{ch}(\beta x)\int g^{(m,\,n)*}\mathrm{sh}(\beta x)\mathrm{d}x\right], \\ u^{(m,\,n)*}(x) &= \frac{1}{\alpha_2}\left[-a_4 L_2^{-1}\left(p^{(m)*} - p_\omega^{(m,\,n)*}\right) + a_7 L_3^{-1}\left(q^{(m)*} - q_\omega^{(m,\,n)*}\right) + \frac{a_7}{2\alpha_2} C_1^{(m,\,n)*} x^2\right] \\ &+ C_7^{(m,\,n)*} x + C_8^{(m,\,n)*} + \gamma_3 \psi^{(m,\,n)*}, \\ w^{(m,\,n)*}(x) &= \frac{1}{\alpha_2}\left[\alpha_1 \int \psi^{(m,n)*}\mathrm{d}x - a_7 L_3^{-1}\left(p^{(m)*} - p_\omega^{(m,n)*}\right)\right. \\ &\left. + a_1 L_4^{-1}\left(q^{(m)*} - q_\omega^{(m,n)*}\right) + \frac{1}{6} a_1 C_1^{(m,\,n)*} x^3\right] \\ &+ \frac{1}{2} C_4^{(m,\,n)*} x^2 + C_5^{(m,\,n)*} x + C_6^{(m,\,n)*}. \end{aligned} \tag{6.28}$$

Here the coefficients α_1, α_2, α_3, β^2, γ_1, γ_2, γ_3 and linear integral operators L_2^{-1}, L_3^{-1}, L_4^{-1} are defined by formulas (3.29, 3.23) and are calculated at the value of neutron flux before the first unloading:

$$g^{(m,\,n)*}(x) = \frac{\alpha_2}{\alpha_1}\gamma_1 h_\omega^{(m,\,n-1)*} + \gamma_2 (p^{(m)*} - p_\omega^{(m,\,n-1)*})$$
$$+ \gamma_1\left(\int (q^{(m)*} - q_\omega^{(m,\,n-1)*})\mathrm{d}x + C_1^{(m,\,n)*}\right).$$

Satisfying by the solution (6.28) the boundary conditions for the *cantilever* (4.41), *hinged rod* (4.42), *rigid embedding of both ends* (4.43), the integration constants $C_1^{(n)*}$, ... , $C_8^{(n)*}$ are obtained in the form (4.46–4.48). However, the nonlinearity functions included in the corresponding formulae will be calculated here taking into account the magnitude of the neutron flux fixed before unloading.

Thus, the obtained formulas (6.28, 6.29–6.31) describe in iterations the analytical solution of the problem in displacements for the values with asterisks for an arbitrary *m*th half-cycle of alternating loading in the case of various types of boundary conditions. The required solution of the problem is obtained from relations (6.24) taking into account the solution (6.10) and (6.28).

6.5 Parametric Analysis of Stress–Strain State of Elastoplastic Rod

In this section, the dependence of the stress–strain state of an elastic–plastic three-layer rod on the neutron flux under single, repeated and cyclic loading is numerically analysed.

The mechanical and radiation characteristics of the materials of the rod layers (D16T–fluoroplastic–D16T) were taken as follows: $h_1 = h_2 = 0.03$, $c = 0{,}09$; $B = 10^{-23}$ m^2/neutron, which provides volumetric deformation in the layers of the considered rod 3–3.5%; for duralumin $\mu = 1.26$ cm, for fluoroplastic $\mu = 3.21$ cm; other necessary parameters of materials are given in Tables 2.1–2.3 and in Sect. 6.1. The plasticity function was described by relations (6.5.2, 6.21, 6.26). The dependence of strain yield strength on integral neutron flux I was taken into account by relations (2.40), of volumetric deformation—by relations (2.57).

Figures 6.6, 6.7, 6.8, 6.9, 6.10, 6.11 and 6.12 show the results of studies of the effect of irradiation with neutron flux intensity $\varphi_0 = 10^{18}$ neutrons/(m^{2}s) on the stress–strain state of an elastic–plastic three-layer rod under single loads. In case of cantilever fixing the intensity of the distributed load was taken $q' = -3.5$ MPa, in case of rigid embedding of both ends $q' = -10$ MPa; $p' = 0$. The time of neutron flux impact on the three-layer rod is taken $t = 3600$ s.

Figure 6.6 illustrates the variation of the integral neutron flux I along the thickness of the rod at different intensities φ_0: 1—10^{17}, 2—10^{18}. Passing through the rod, the magnitude of the flux decreases by a factor of about 2.

Figure 6.7 shows the increase in yield strength of D16T alloy with increasing integral neutron flux. In our case, at neutron flux intensity $\varphi_0 = 10^{18}$ neutrons/(m^{2}s), the integral flux will be $I = 3.6 \times 10^{21}$, which leads to an increase in strain yield

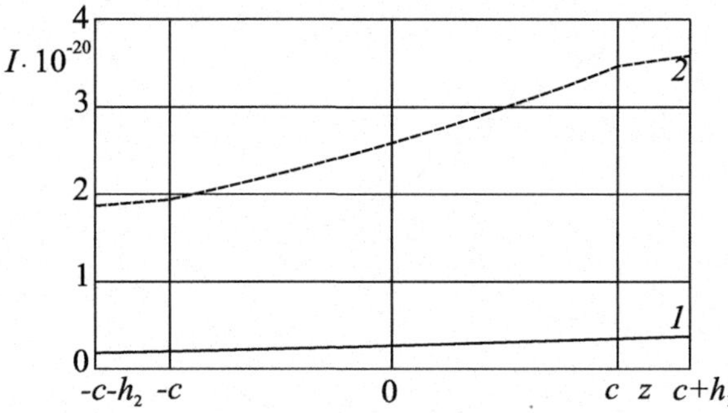

Fig. 6.6 Variation of integral neutron flux along the thickness of the rod

strength from a value of $\varepsilon_y = 0.73\%$ to $\varepsilon_y = 0.75\%$. When the neutron flux is further increased, the yield strength exceeds 1%.

Numerical analysis demonstrated the acceleration of convergence of the elastic solution method in the irradiated rod (Fig. 6.8): *a*—deflection, *b*—relative shear. This is explained by the growth of yield strength and weakening of plastic properties of metals under neutron exposure.

The 5th approximation is taken as the actual solution for the cantilever rod, and the difference from the elastic displacements is 13 and 16%, respectively. In the case of embedding-embedding, 4 approximations are sufficient, the difference from elastic displacements in this case is 7–9%.

Figure 6.9 shows a comparison of deflections (*a*) and relative shears (*b*) in three-layer unirradiated and irradiated rods at cantilever embedding of the left end: 1, 3—elastic and elastic–plastic displacements without irradiation; 2, 4—elastic and elastic–plastic displacements after irradiation. In both cases, curves 4 are located between graphs 2 and 3. Consequently, neutron irradiation increases displacement, due to volumetric deformation, but decreases plastic deformation by increasing the yield strength (by 3–4%).

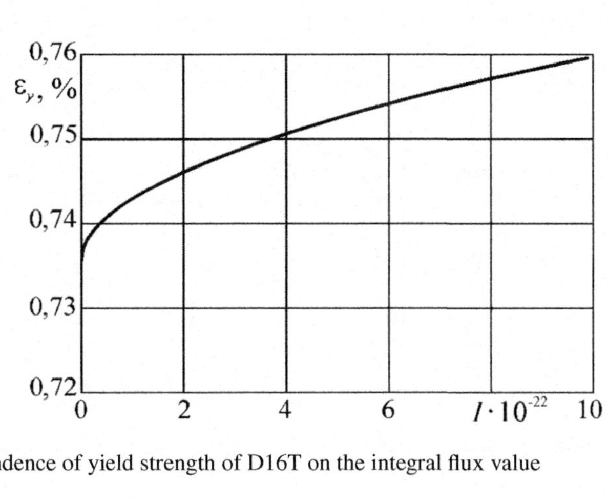

Fig. 6.7 Dependence of yield strength of D16T on the integral flux value

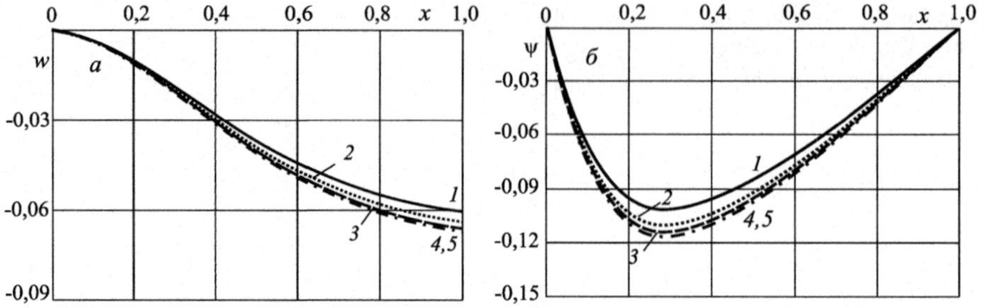

Fig. 6.8 Convergence of the elastic solution method for the irradiated rod: *a*—deflection, *b*—relative shear

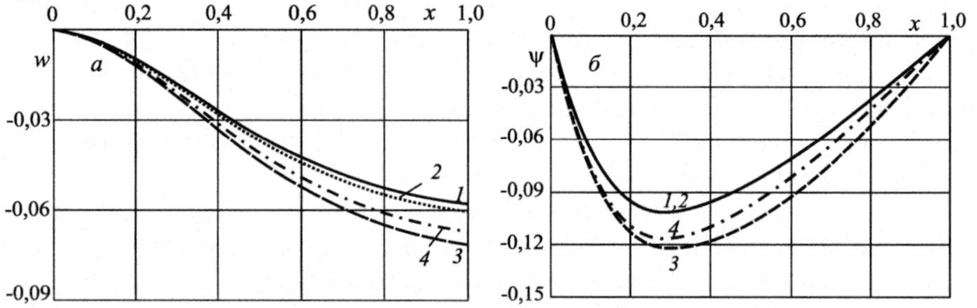

Fig. 6.9 Variation of deflection *w* (*a*) and relative shear ψ (*b*) along the axis (cantilever)

Figure 6.10 illustrates the variation along the rod axis of deflections (*a*) and relative shear (*b*) in three-layer unirradiated and irradiated rods when both ends are embedded: 1, 3—elastic and elastic–plastic displacements without irradiation; 2, 4—elastic and elastic–plastic displacements after irradiation. Here also in both cases, curves 4 are located between graphs 2 and 3. Consequently, in this case too, neutron irradiation increases displacements due to volumetric deformation, but decreases their plastic deformation by 2–3% due to the increase in yield strength.

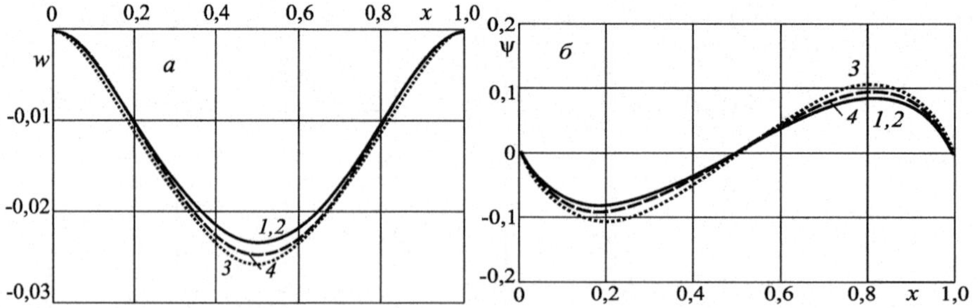

Fig. 6.10 Variation of deflection *w* (*a*) and relative shear ψ (*b*) along the axis ("embedding–embedding")

The distribution of the areas of physical nonlinearity of layer materials (grey fill) along the longitudinal section of a three-layer cantilever rod and under "embedding-embedding" boundary conditions are illustrated in Figs. 6.11 and 6.12, respectively. Here, the regions of nonlinearity occupy a slightly smaller volume than under isothermal loading (see Figs. 3.12, 3.15). It should be noted that volumetric deformation due to irradiation does not cause plastic deformation of the materials of the rod layers.

Figures 6.13, 6.14, 6.15 and 6.16 show the results of studies on the influence of irradiation with neutron flux intensity $\varphi = 10^{18}$ neutrons/(m^{2}s) on the stress–strain state of an elastoplastic three-layer rod under cyclic loads. The time of neutron flux impact on the rod under loading from the natural state is taken $t = 3600$ s. In the case of cantilever clamping, the load was assumed to be $q' = -3.5$ MPa, when both ends are embedded—$q' = -10$ MPa; $p' = 0$. Three-layer package: D16T–fluoroplastic–D16T.

Figures 6.13, 6.14 show the variation of deflection w (a) and shear ψ (b) along the axis of the considered three-layer rod after 8 half cycles of loading, calculated using different physical equations of state when both ends are embedded and when the left end is cantilevered, respectively. Curves with one stroke correspond to loading

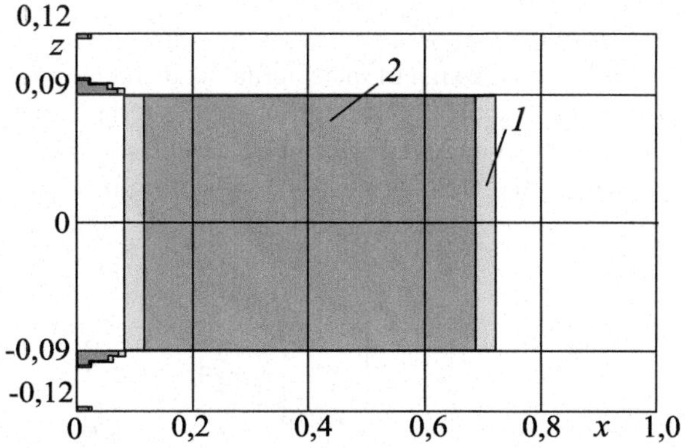

Fig. 6.11 Zones of physical nonlinearity (cantilever)

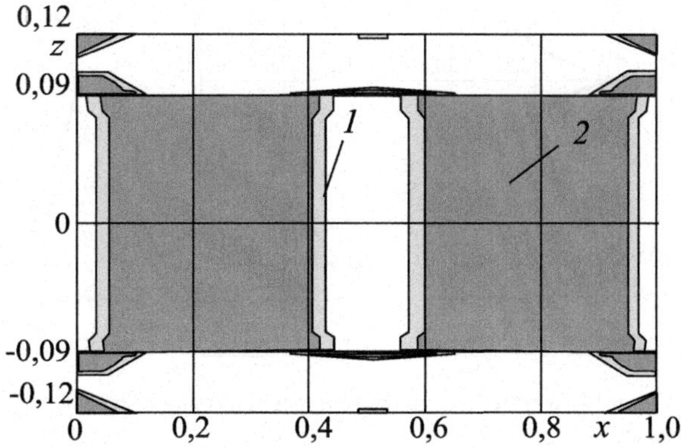

Fig. 6.12 Zones of physical nonlinearity ("embedding–embedding")

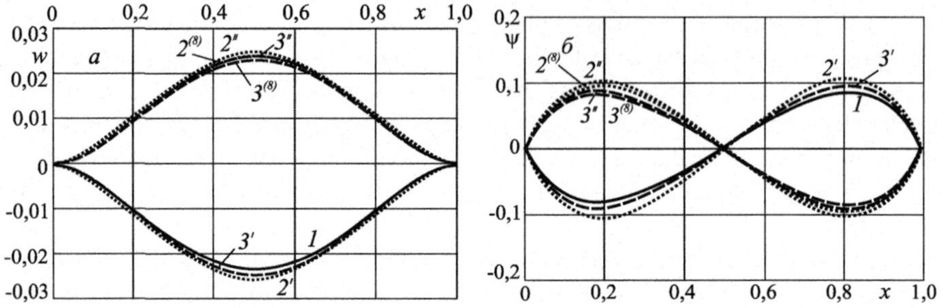

Fig. 6.13 Variation of deflection w *a* and relative shear ψ *b* along the axis under alternating loads ("embedding–embedding")

from the natural state, two strokes—repeated alternating bending: 1, 2—elastic and elastic–plastic rods without irradiation, 3—elastic–plastic rods at $\varphi = 10^{18}$ neutrons/(m^{2}s), 3^8—bending at the 8th half-cycle of the irradiated rod. At repeated loading the displacements decrease by 2–3%, at the 8th half-cycle the decrease is 4–5%, both for non-irradiated and irradiated rods, which is explained by cyclic hardening of the material at each change of the load sign.

Figure 6.15 shows the change in normal stresses σ_{xx} in the cantilever clamped of the rod when the sign of the external load changes. Graphs with one stroke—loading from natural state, two strokes—bending at the 2nd half-cycle: 1—stresses in elastic unirradiated rod, 2—stresses in elastic–plastic rod after irradiation.

Accounting for nonlinearity and neutron irradiation leads to an increase in stress extremes in the outer layers by 30%, in the filler they decrease by 50%.

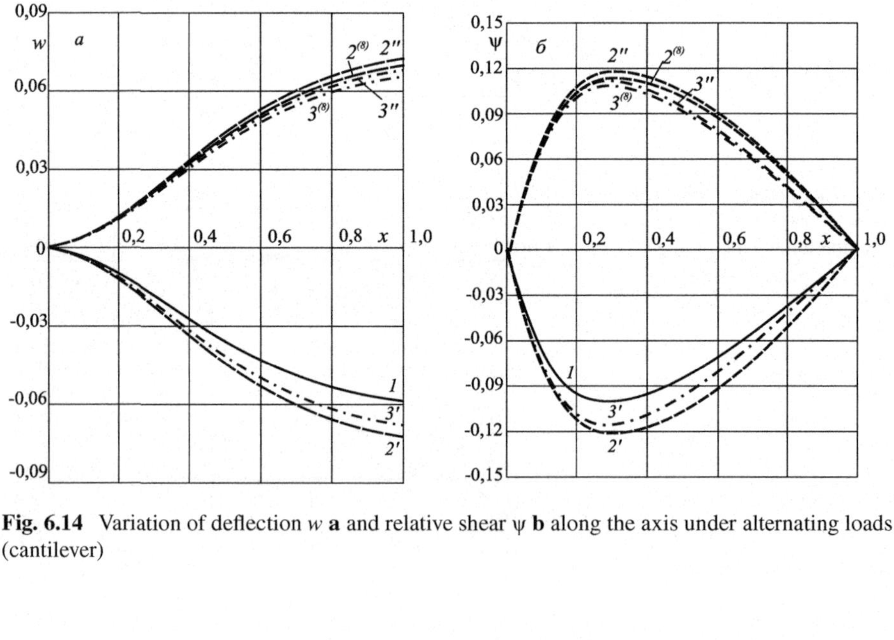

Fig. 6.14 Variation of deflection w **a** and relative shear ψ **b** along the axis under alternating loads (cantilever)

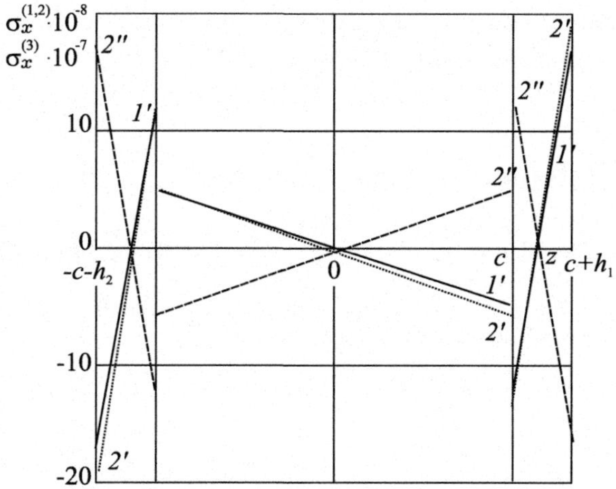

Fig. 6.15 Distribution of normal stresses σx at ends ($x = 0$) under alternating loading ("embedding-embedding")

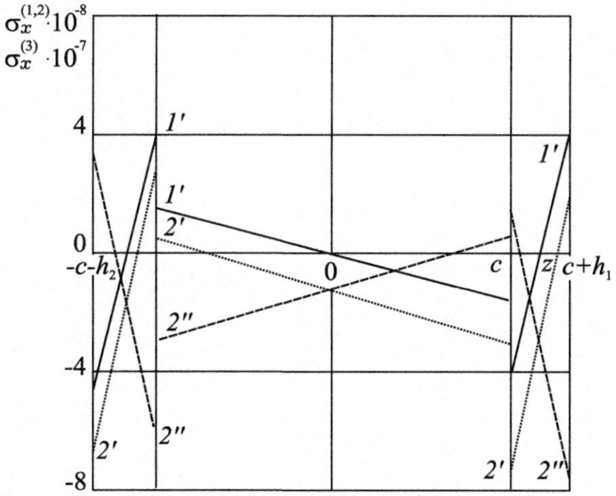

Fig. 6.16 Distribution of normal stresses σx at ends ($x = 0$) under alternating loading (cantilever)

Similar stress plots at the left end when both ends of the rod are embedded are shown in Fig. 6.16. In this case, the curves move to the negative region due to neutron impact, while the stresses in the bearing layers increase and in the filler decrease.

6.6 Conclusions

The methodology proposed in this section and the analytical solutions obtained allow us to carry out a parametric study of the stress–strain state of an elastic–plastic three-layer rod under single, repeated and cyclic radiation-force loading according to the known solution of the corresponding loading problem from the natural state. It should be emphasised that the above results are valid only in the range of small elastoplastic strains and under simple alternating loadings.

References

1. Ilyushin, A.A., and P.M. Ogibalov. 1960. *Elastic-plastic deformations of hollow cylinder*, 224. Moscow: State University (in Russian).
2. Platonov, P.A. 1971. *Effect of irradiation on the structure and properties of metals*, 40. Moscow: Mashinostroenie (in Russian).

Chapter 7
Three-Layer Rod with Compressible Filler

The bending of asymmetric in thickness elastic and elastoplastic three-layer rods with a rigid compressible filler in the temperature field is studied. Bernoulli hypothesis is accepted for the bearing layers, in the filler the exact relations of the theory of elasticity with linear approximation of displacements of its points along the height of the rod are true. The work of tangential and transverse stresses in the filler is taken into account. Numerical analysis of the obtained solutions was performed.

7.1 Statement of the Boundary Value Problem for a Thermoelastic Rod

A three-layer rod asymmetric in thickness is considered (Fig. 7.1). For isotropic bearing layers, Bernoulli hypotheses are adopted. Displacement continuity conditions are used at the contact boundaries. The materials of the load-bearing layers are incompressible in the transverse direction; in the filler its compression is taken into account; the strains are small. The x, y, z coordinate system is related to the median plane of the filler.

The surface force loads $p(x)$, $q(x)$ and the heat flow of intensity q_t, directed perpendicularly to the bearing layer *1*, act on the rod. At the ends of the rod in the first and second bearing layers the force loads are given: $N_p^{(1)}, Q_p^{(1)}, M_p^{(1)}, N_p^{(2)}, Q_p^{(2)}, M_p^{(2)}$; index p takes the values 0 and 1, indicating where the force loads is given. With $w_k(x)$ and $u_k(x)$ the deflections and longitudinal displacements of the middle surfaces of the bearing layers are indicated, h_k is thickness of the kth layer, $h_3 = 2c$ ($k = 1, 2, 3$—layer number), b_0 is rod width.

Consider that the surface $z = -c - h_2$ and the contour of the rod are heat insulated. This allows the inhomogeneous temperature field $T(z)$, counted from some initial temperature T_0, to be calculated with sufficient accuracy by formula (1.55).

S. Eduard et al., *Deformation of Three-layer Structural Elements in Thermal Radiation Fields*, https://doi.org/10.1007/978-981-97-7217-9_7

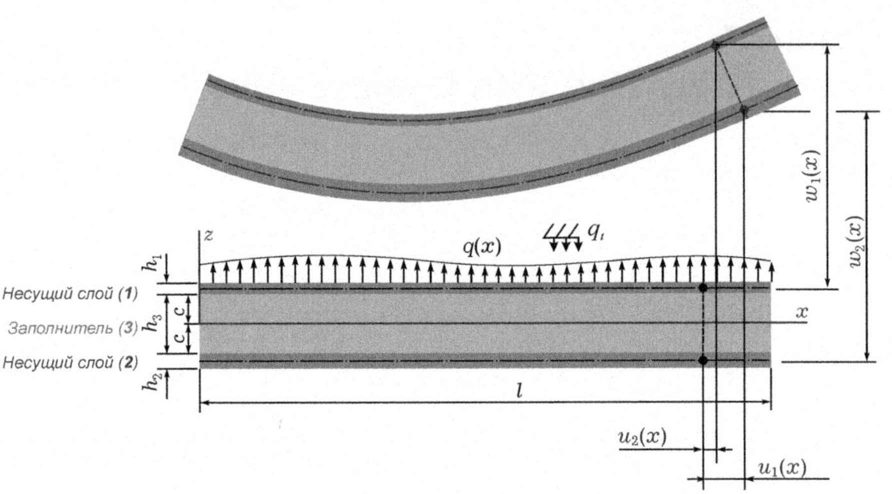

Fig. 7.1 Calculation diagram of the rod in the temperature field

The displacements in the layers $u^{(k)}(x, z)$ and $w^{(k)}(x, z)$ can be expressed through the four functions $w_1(x)$, $u_1(x)$, $w_2(x)$ и $u_2(x)$ (that we are looking for):

· in the bearing layers

$$u^{(1)} = u_1 - \left(z - c - \frac{h_1}{2}\right)w_{1,x}, \; w^{(1)} = w_1(c \le z \le c + h_1),$$
$$u^{(2)} = u_2 - \left(z + c + \frac{h_2}{2}\right)w_{2,x}, \; w^{(2)} = w_2(-c - h_2 \le z \le -c);$$

· in the filler

$$u^{(3)} = \left(1 + \frac{z}{c}\right)\left(\frac{1}{2}u_1 + \frac{h_1}{4}w_{1,x}\right) + \left(1 - \frac{z}{c}\right)\left(\frac{1}{2}u_2 - \frac{h_2}{4}w_{2,x}\right),$$
$$w^{(3)} = \frac{1}{2}\left(1 + \frac{z}{c}\right)w_1 + \frac{1}{2}\left(1 - \frac{z}{c}\right)w_2(-c \le z \le c), \tag{7.1}$$

where z is the coordinate of the fiber.

The components of the strain tensor follow from (7.1) and the Cauchy relations:

$$\varepsilon_x^{(1)} = u_{1,x} - \left(z - c - \frac{h_1}{2}\right)w_{1,xx}, \; \varepsilon_{xz}^{(1)} = 0(c \le z \le c + h_1),$$
$$\varepsilon_x^{(2)} = u_{2,x} - \left(z + c + \frac{h_2}{2}\right)w_{2,xx} \; \varepsilon_{xz}^{(2)} = 0(-c - h_2 \le z \le -c),$$
$$\varepsilon_x^{(3)} = \left(1 + \frac{z}{c}\right)\left(\frac{1}{2}u_{1,x} + \frac{h_1}{4}w_{1,xx}\right) + \left(1 - \frac{z}{c}\right)\left(\frac{1}{2}u_{2,x} - \frac{h_2}{4}w_{2,xx}\right),$$

$$\varepsilon_{xz}^{(3)} = \left(\frac{2z + h_1}{8c} + \frac{1}{4}\right) w_{1,x} + \left(\frac{-2z + h_2}{8c} + \frac{1}{4}\right) w_{2,x} + \frac{u_1 - u_2}{4c}$$

$$\varepsilon_z^{(3)} = \frac{1}{2c}(w_1 - w_2)(-c \le z \le c). \tag{7.2}$$

The spherical and deviator parts of the strain tensor in the case under consideration are ($\mathfrak{e}_{ij} = \varepsilon_{ij} - \varepsilon\delta_{ij} i, j = x,\ y,\ z$):

$$\begin{aligned} \varepsilon^{(k)} &= \frac{1}{3}\varepsilon_x^{(k)},\ \mathfrak{e}_x^{(k)} = \frac{2}{3}\varepsilon_x^{(k)} (k = 1, 2), \\ \varepsilon^{(3)} &= \frac{1}{3}(\varepsilon_x^{(3)} + \varepsilon_z^{(3)}),\ \mathfrak{e}_x^{(3)} = \frac{2}{3}\varepsilon_x^{(3)} - \frac{1}{3}\varepsilon_z^{(3)}. \end{aligned} \tag{7.3}$$

The internal forces and moments in the layers are introduced by the relations

$$\begin{aligned} N_x^{(k)} &= b_0 \int_{h_k} \sigma_x^{(k)} \mathrm{d}z,\ M_x^{(k)} = b_0 \int_{h_k} \sigma_x^{(k)} z\, \mathrm{d}z\ (k = 1,\ 2,\ 3), \\ N_z^{(3)} &= b_0 \int_{h_3} \sigma_z^{(3)} \mathrm{d}z,\ M_{xz}^{(3)} = b_0 \int_{h_3} \sigma_{xz}^{(3)} z\, \mathrm{d}z,\ Q^{(3)} = b_0 \int_{h_3} \sigma_{xz}^{(3)} \mathrm{d}z, \end{aligned} \tag{7.4}$$

where $\sigma_x^{(k)}, \sigma_{xz}^{(3)}, \sigma_z^{(3)}$ are stress tensor components.

We obtain equilibrium equations of the three-layer rod using the Lagrange's principle of variation:

$$\delta A = \delta W. \tag{7.5}$$

Here the variations of the work of external forces δA and the work of internal elastic forces δW are defined by the expressions

$$\begin{aligned} \delta A &= b_0 \int_0^l \left[p\left(\delta u_1 - \frac{h_1}{2}\delta w_{1,x}\right) + q\delta w_1 \right] dx, \\ \delta W &= b_0 \int_0^l \left[\sum_{k=1}^{3} \int_{h_k} \sigma_x^{(k)} \delta\varepsilon_x^{(k)} dz + 2 \int_{h_3} \sigma_{xz}^{(3)} \delta\varepsilon_{xz}^{(3)} dz + \int_{h_3} \sigma_z^{(3)} \delta\varepsilon_z^{(3)} dz \right] dx. \end{aligned} \tag{7.6}$$

Variations of displacements in the layers:

$$\delta u^{(1)} = \delta u_1 - \left(z - c - \frac{h_1}{2}\right)\delta w_{1,x},\ \delta w^{(1)} = \delta w_1 (c \le z \le c + h_1),$$

$$\delta u^{(2)} = \delta u_2 - \left(z + c + \frac{h_2}{2}\right)\delta w_{2,x},\ \delta w^{(2)} = \delta w_2 (-c - h_2 \le z \le -c),$$

$$\delta u^{(3)} = \left(1+\frac{z}{c}\right)\left(\frac{1}{2}\delta u_1 + \frac{h_1}{4}\delta w_{1,x}\right) + \left(1-\frac{z}{c}\right)\left(\frac{1}{2}\delta u_2 - \frac{h_2}{4}\delta w_{2,x}\right),$$
$$\delta w^{(3)} = \frac{1}{2}\left(1+\frac{z}{c}\right)\delta w_1 + \frac{1}{2}\left(1-\frac{z}{c}\right)\delta w_2 (-c \le z \le c).$$

The strain variations, according to (7.2), are as follows:

$$\delta\varepsilon_x^{(1)} = \delta u_{1,x} - \left(z - c - \frac{h_1}{2}\right)\delta w_{1,xx},\ \delta\varepsilon_{xz}^{(1)} = 0 (c \le z \le c + h_1),$$
$$\delta\varepsilon_x^{(3)} = \left(1+\frac{z}{c}\right)\left(\frac{1}{2}\delta u_{1,x} + \frac{h_1}{4}\delta w_{1,xx}\right)$$
$$+ \left(1-\frac{z}{c}\right)\left(\frac{1}{2}\delta u_{2,x} - \frac{h_2}{4}\delta w_{2,xx}\right),$$
$$\delta\varepsilon_z^{(3)} = \frac{1}{2c}(\delta w_1 - \delta w_2)(-c \le z \le c),$$
$$\delta\varepsilon_{xz}^{(3)} = \left(\frac{2z+h_1}{8c} + \frac{1}{4}\right)\delta w_{1,x} + \left(\frac{-2z+h_2}{8c} + \frac{1}{4}\right)\delta w_{2,x} + \frac{\delta u_1 - \delta u_2}{4c},$$
$$\delta\varepsilon_x^{(2)} = \delta u_{2,x} - \left(z + c + \frac{h_2}{2}\right)\delta w_{2,xx}\ \delta\varepsilon_{xz}^{(2)} = 0 (-c - h_2 \le z \le -c). \qquad (7.7)$$

Let us consider the integrals over the thickness of the layers included in expression (7.6). For the first layer, taking into account expressions (7.7), we obtain:

$$\int_{h_1} \sigma_x^{(1)}\delta\varepsilon_x^{(1)} dz = \int_{h_1} \sigma_x^{(1)}\left[\delta u_{1,x} - \left(z - c - \frac{h_1}{2}\right)\delta w_{1,xx}\right] dz$$
$$= \frac{1}{b_0}\left[\delta u_{1,x} N^{(1)} + \delta w_{1,xx}\left(c + \frac{h_1}{2}\right)N^{(1)} - \delta w_{1,xx} M^{(1)}\right].$$

Similarly for the other layers:

$$\int_{h_2} \sigma_x^{(2)}\delta\varepsilon_x^{(2)} dz = \frac{1}{b_0}\left[\delta u_{2,x} N^{(2)} - \delta w_{2,xx}\left(c + \frac{h_2}{2}\right)N^{(2)} - \delta w_{2,xx} M^{(2)}\right],$$
$$\int_{h_3} \sigma_x^{(3)}\delta\varepsilon_x^{(3)} dz = \frac{1}{b_0}\left[\frac{\delta u_{1,x}}{2}\left(N^{(3)} + \frac{M^{(3)}}{c}\right) + \frac{\delta u_{2,x}}{2}\left(N^{(3)} - \frac{M^{(3)}}{c}\right)\right.$$
$$\left. + \frac{\delta w_{1,xx} h_1}{4}\left(N^{(3)} + \frac{M^{(3)}}{c}\right) - \frac{\delta w_{2,xx} h_2}{4}\left(N^{(3)} - \frac{M^{(3)}}{c}\right)\right],$$
$$\int_{h_3} \sigma_{xz}^{(3)}\delta\varepsilon_{xz}^{(3)} dz = \frac{1}{8b_0}\left[\delta w_{1,x}\left(\left(2 + \frac{h_1}{c}\right)Q^{(3)} + \frac{2M_{xz}^{(3)}}{c}\right)\right.$$

$$+\delta w_{2,x}\left(\left(2+\frac{h_2}{c}\right)Q^{(3)}-\frac{2M_{xz}^{(3)}}{c}\right)+\frac{2\delta u_1}{c}Q^{(3)}-\frac{2\delta u_2}{c}Q^{(3)}\right]$$

$$\int_{h_3}\sigma_z^{(3)}\delta\varepsilon_z^{(3)}dz=\int_{h_3}\sigma_z^{(3)}\frac{\delta w_1-\delta w_2}{2c}dz=\frac{\delta w_1-\delta w_2}{2b_0c}N_z^{(3)}.$$

Taking into account these integrals, the variation of the work of the elastic forces will be as follows:

$$\delta W=\int_0^l\left[\frac{1}{b_0}\left[\delta u_{1,x}N^{(1)}+\delta w_{1,xx}\left(c+\frac{h_1}{2}\right)N^{(1)}-\delta w_{1,xx}M^{(1)}\right]\right.$$

$$+\frac{1}{b_0}\left[\delta u_{2,x}N^{(2)}-\delta w_{2,xx}\left(c+\frac{h_2}{2}\right)N^{(2)}-\delta w_{2,xx}M^{(2)}\right]$$

$$+\frac{1}{b_0}\left[\frac{\delta u_{1,x}}{2}\left(N^{(3)}+\frac{M^{(3)}}{c}\right)+\frac{\delta u_{2,x}}{2}\left(N^{(3)}-\frac{M^{(3)}}{c}\right)\right.$$

$$\left.+\frac{\delta w_{1,xx}h_1}{4}\left(N^{(3)}+\frac{M^{(3)}}{c}\right)-\frac{\delta w_{2,xx}h_2}{4}\left(N^{(3)}-\frac{M^{(3)}}{c}\right)\right]$$

$$+\frac{1}{4b_0}\left[\delta w_{1,x}\left(\left(2+\frac{h_1}{c}\right)Q^{(3)}+\frac{2M_{xz}^{(3)}}{c}\right)\right.$$

$$\left.+\delta w_{2,x}\left(\left(2+\frac{h_2}{c}\right)Q^{(3)}-\frac{2M_{xz}^{(3)}}{c}\right)+\frac{2\delta u_1}{c}Q^{(3)}-\frac{2\delta u_2}{c}Q^{(3)}\right]$$

$$\left.+\frac{\delta w_1-\delta w_2}{2bc}N_z^{(3)}\right]dx=\frac{1}{b_0}\int_0^l\left[\frac{Q^{(3)}}{2c}\delta u_1-\frac{Q^{(3)}}{2c}\delta u_2+\frac{N_z^{(3)}}{2c}\delta w_1-\frac{N_z^{(3)}}{2c}\delta w_2\right.$$

$$+\delta u_{1,x}\left(\frac{N^{(3)}}{2}+\frac{M^{(3)}}{2c}+N^{(1)}\right)+\delta u_{2,x}\left(\frac{N^{(3)}}{2}-\frac{M^{(3)}}{2c}+N^{(2)}\right)$$

$$+\delta w_{1,x}\left(\left(1+\frac{h_1}{2c}\right)\frac{Q^{(3)}}{2}+\frac{M_{xz}^{(3)}}{2c}\right)+\delta w_{2,x}\left(\left(1+\frac{h_2}{2c}\right)\frac{Q^{(3)}}{2}-\frac{M_{xz}^{(3)}}{2c}\right)$$

$$+\delta w_{1,xx}\left(\left(c+\frac{h_1}{2}\right)N^{(1)}-M^{(1)}+\frac{h_1}{4}N^{(3)}+\frac{h_1}{4c}M^{(3)}\right)$$

$$\left.+\delta w_{2,xx}\left(-\left(c+\frac{h_2}{2}\right)N^{(2)}-M^{(2)}-\frac{h_2}{4}N^{(3)}+\frac{h_2}{4c}M^{(3)}\right)\right]dx$$

$$=\frac{1}{b_0}\int_0^l(H_1\delta u_1-H_1\delta u_2+H_2\delta w_1-H_2\delta w_2+P_1\delta u_{1,x}+P_2\delta u_{2,x}$$

$$+T_1\delta w_{1,x}+T_2\delta w_{2,x}+S_1\delta w_{1,xx}+S_2\delta w_{2,xx})dx$$

$$=\frac{1}{b_0}\int_0^l[(H_1-P_{1,x})\delta u_1+(-H_1-P_{2,x})\delta u_2+(S_{1,xx}-T_{1,x}+H_2)\delta w_1$$

$$+ (S_{2},_{xx} - T_{2},_{x} - H_2)\delta w_2 + (P_1\delta u_1 + P_2\delta u_2 + (T_1 - S_{1},_{x})\delta w_1$$
$$+(T_2 - S_{2},_{x})\delta w_2 S_1\delta w_{1},_{x} + S_2\delta w_{2},_{x})dx,] \tag{7.8}$$

where the generalized internal forces are expressed through the internal forces (7.4):

$$H_1 = \frac{Q^{(3)}}{2c}, H_2 = \frac{N_z^{(3)}}{2c}, P_1 = \frac{N_x^{(3)}}{2} + \frac{M_x^{(3)}}{2c} + N_x^{(1)},$$
$$P_2 = \frac{N_x^{(3)}}{2} - \frac{M_x^{(3)}}{2c} + N_x^{(2)}, T_1 = \left(1 + \frac{h_1}{2c}\right)\frac{Q^{(3)}}{2} + \frac{M_{xz}^{(3)}}{2c},$$
$$S_1 = \left(c + \frac{h_1}{2}\right)N_x^{(1)} - M_x^{(1)} + \frac{h_1}{4}N_x^{(3)} + \frac{h_1}{4c}M_x^{(3)},$$
$$T_2 = \left(1 + \frac{h_2}{2c}\right)\frac{Q^{(3)}}{2} - \frac{M_{xz}^{(3)}}{2c},$$
$$S_2 = -\left(c + \frac{h_2}{2}\right)N_x^{(2)} - M_x^{(2)} - \frac{h_2}{4}N_x^{(3)} + \frac{h_2}{4c}M_x^{(3)}.$$

In obtaining formulas (7.8) we used transformations:

$$P_k\delta u_{k},_{x} = (P_k\delta u_k),_{x} - P_{k},_{x}\delta u_k, T_k\delta w_{k},_{x} = (T_k\delta w_k),_{x} - T_{k},_{x}\delta w_k$$
$$S_k\delta w_{k},_{xx} = (S_k\delta w_{k},_{x}),_{x} - (S_{k},_{x}\delta w_k),_{x} + S_{k},_{xx}\delta w_k (k = 1, 2).$$

Substituting (7.6), (7.8) into (7.5), we obtain the following system of equilibrium equations in forces:

$$\begin{cases} H_1 - P_{1},_{x} = b_0 p, \\ H_1 + P_{2},_{x} = 0, \\ S_{1},_{xx} + H_2 - T_{1},_{x} = b_0\left(q + \frac{p,_x h_1}{2}\right), \\ S_{2},_{xx} - H_2 - T_{2},_{x} = 0. \end{cases} \tag{7.9}$$

and force boundary conditions at $x = 0, l$:

$$P_1 = N_p^{(1)}, P_2 = N_p^{(2)}, T_1 - S_{1},_{x} = Q_p^{(1)}, T_2 - S_{2},_{x} = Q_p^{(2)},$$
$$S_1 = M_p^{(1)}, S_2 = M_p^{(2)}.$$

Here $N_p^{(1)}$, $Q_p^{(1)}$, $M_p^{(1)}$, $N_p^{(2)}$, $Q_p^{(2)}$, $M_p^{(2)}$ are specified forces at the ends of the rod in the bearing layers; index p takes the values 0 and l, indicating where the force loads is given.

The relations of Hooke's law in the temperature field are used as physical equations of state:

$$
\begin{aligned}
s_i^{(k)} &= 2G_k(T_k)\, \text{э}_i^{(k)},\ s_{xz}^{(3)} = 2G_3\, \text{э}_{xz}^{(3)},\\
\sigma^{(k)} &= 3K_k(T_k)(\varepsilon^{(k)} - \alpha_{0k}T_k)(i = x,\ z;\quad k = 1,\ 2\ ,\ 3),
\end{aligned}
\tag{7.10}
$$

where $s_i^{(k)}$, $\text{э}_i^{(k)}$, $\sigma^{(k)}$, $\varepsilon^{(k)}$ are deviator and spherical parts of the stress and strain tensors; $G_k(T_k)$, $K_k(T_k)$ are temperature-dependent elastic modules of the material of the kth layer, calculated by Bell's formula (1.37); α_{0k} is coefficient of linear thermal expansion; T_k is temperature increment from the initial value T_0.

Components of the stress tensor in the rod layers, taking into account the temperature:

$$
\begin{aligned}
\sigma_x^{(k)} &= K_k^+(T_k)\varepsilon_x^{(k)} - 3K_k\alpha_{0k}T_k(k = 1,\ 2),\\
\sigma_{xz}^{(3)} &= 2G_3(T_3)\varepsilon_{xz},\ \sigma_x^{(3)} = K_3^+(T_3)\varepsilon_x^{(3)} + K_3^-(T_3)\varepsilon_z^{(3)} - 3K_3\alpha_{03}T_3,\\
\sigma_z^{(3)} &= K_3^+(T_3)\varepsilon_z^{(3)} + K_3^-(T_3)\varepsilon_x^{(3)} - 3K_3\alpha_{03}T_3,\\
K_k^+(T_k) &= K_k(T_k) + \frac{4}{3}G_k(T_k),\ K_k^-(T_k) = K_k(T_k) - \frac{2}{3}G_k(T_k).
\end{aligned}
\tag{7.11}
$$

Using (7.3), (7.10), (7.11), let us distinguish the force (index "0") and temperature (index "t") components in the stress tensor components:

- in the bearing layers ($k = 1,2$)

$$
\begin{aligned}
\sigma_x^{(k)} &= \sigma_x^{(k)0} - \sigma_x^{(k)t},\ \sigma_x^{(k)0} = 2G_k\, \text{э}_x^{(k)} + 3K_k\varepsilon^{(k)}\\
&= K_k^+\varepsilon_x^{(k)},\ \sigma_x^{(k)t} = 3K_k\alpha_{0k}T_k;
\end{aligned}
$$

- in the filler

$$
\begin{aligned}
\sigma_x^{(3)} &= \sigma_x^{(3)0} - \sigma_x^{(3)t},\ \sigma_x^{(3)0} = 2G_3\, \text{э}_x^{(3)} + 3K_3\varepsilon^{(3)}\\
&= K_3^+\varepsilon_x^{(3)} + K_3^-\varepsilon_z^{(3)},\ \sigma_x^{(3)t} = 3K_3\alpha_{03}T_3,\\
\sigma_z^{(3)} &= \sigma_z^{(3)0} - \sigma_z^{(3)t},\ \sigma_z^{(3)0} = 2G_3\, \text{э}_z^{(3)}\\
&\quad + 3K_3\varepsilon^{(3)} = K_3^+\varepsilon_z^{(3)} + K_3^-\varepsilon_x^{(3)},\\
\sigma_z^{(3)t} &= 3K_3\alpha_{03}T_3,\ \sigma_{xz}^{(3)} = 2G_3\, \text{э}_{xz}^{(3)} = 2G_3\varepsilon_{xz}^{(3)}.
\end{aligned}
\tag{7.12}
$$

For the internal forces (7.4), similarly to (7.12), we obtain:

$$
\begin{aligned}
N_x^{(k)} &= N_x^{(k)0} - N_x^{(k)t},\ M_x^{(k)} = M_x^{(k)0} - M_x^{(k)t}(k = 1,\ 2,\ 3),\\
N_z^{(3)} &= N_z^{(3)0} - N_z^{(3)t},\ M_{xz}^{(3)} = M_{xz}^{(3)0},\ Q^{(3)} = Q^{(3)0}.
\end{aligned}
\tag{7.13}
$$

The linear components (index "0") and temperature terms (index "t") in (7.13) are calculated by the formulas:

- in the bearing layers ($k = 1, 2$)

$$
\begin{aligned}
N_x^{(k)0} &= b_0 \int_{h_k} \sigma_x^{(k)0} dz = b_0 \int_{h_k} K_k^+ \varepsilon_x^{(k)} dz, \\
N_x^{(k)t} &= b_0 \int_{h_k} \sigma_x^{(k)t} dz = 3b_0 \int_{h_k} K_k \alpha_{0k} T_k dz, \\
M_x^{(k)0} &= b_0 \int_{h_k} \sigma_x^{(k)0} z dz = b_0 \int_{h_k} K_k^+ \varepsilon_x^{(k)} z dz, \\
M_x^{(k)t} &= b_0 \int_{h_k} \sigma_x^{(k)t} z dz = 3b_0 \int_{h_k} K_k \alpha_{0k} T_k z dz,
\end{aligned}
$$

- in the filler

$$
\begin{aligned}
N_x^{(3)0} &= b_0 \int_{h_3} \sigma_x^{(3)0} dz = b_0 \int_{h_3} \left(K_3^+ \varepsilon_x^{(3)} + K_3^- \varepsilon_z^{(3)}\right) dz, \\
N_x^{(3)t} &= b_0 \int_{h_3} \sigma_x^{(3)t} dz = 3b_0 \int_{h_3} K_3 \alpha_{03} T_3 dz, \\
N_z^{(3)0} &= b_0 \int_{h_3} \sigma_z^{(3)0} dz = b_0 \int_{h_3} \left(K_3^+ \varepsilon_z^{(3)} + K_3^- \varepsilon_x^{(3)}\right) dz, \\
N_z^{(3)t} &= b_0 \int_{h_3} \sigma_z^{(3)t} dz = 3b_0 \int_{h_3} K_3 \alpha_{03} T_3 dz, \\
M_x^{(3)0} &= b_0 \int_{h_3} \sigma_x^{(3)0} z dz = b_0 \int_{h_3} \left(K_3^+ \varepsilon_x^{(3)} + K_3^- \varepsilon_z^{(3)}\right) z dz, \\
M_x^{(3)t} &= b_0 \int_{h_3} \sigma_x^{(3)t} z dz = 3b_0 \int_{h_3} K_3 \alpha_{03} T_3 z dz, \\
M_{xz}^{(3)0} &= b_0 \int_{h_3} \sigma_{xz}^{(3)0} z dz = 2b_0 \int_{h_3} G_3 \varepsilon_{xz}^{(3)} z dz, \\
Q^{(3)0} &= b_0 \int_{h_3} \sigma_{xz}^{(3)0} dz = 2b_0 \int_{h_3} G_3 \varepsilon_{xz}^{(3)} dz,
\end{aligned}
\tag{7.14}
$$

In the generalized forces introduced in (7.8), using relations (7.13), let us also distinguish the temperature components:

$$
H_1^0 = \frac{Q^{(3)0}}{2c}, H_2 = H_2^0 - H_2^t, H_2^0 = \frac{N_z^{(3)0}}{2c}, H_2^t = \frac{N_z^{(3)t}}{2c},
$$

$$
\begin{aligned}
T_1^0 =&\left(1+\frac{h_1}{2c}\right)\frac{Q^{(3)0}}{2}+\frac{M_{xz}^{(3)0}}{2c},\, T_2^0=\left(1+\frac{h_2}{2c}\right)\frac{Q^{(3)0}}{2}-\frac{M_{xz}^{(3)0}}{2c},\\
S_1 =&S_1^0-S_1^t,\, S_1^0=\left(c+\frac{h_1}{2}\right)N_x^{(1)0}\\
&-M_x^{(1)0}+\frac{h_1}{4}N_x^{(3)0}+\frac{h_1}{4c}M_x^{(3)0},\\
S_1^t =&\left(c+\frac{h_1}{2}\right)N_x^{(1)t}-M_x^{(1)t}+\frac{h_1}{4}N_x^{(3)t}+\frac{h_1}{4c}M_x^{(3)t},\\
S_2 =&S_2^0-S_2^t,\, S_2^0=-\left(c+\frac{h_2}{2}\right)N_x^{(2)0}\\
&-M_x^{(2)0}-\frac{h_2}{4}N_x^{(3)0}+\frac{h_2}{4c}M_x^{(3)0}\\
S_2^t =&-\left(c+\frac{h_2}{2}\right)N_x^{(2)t}-M_x^{(2)t}-\frac{h_2}{4}N_x^{(3)t}\\
&+\frac{h_2}{4c}M_x^{(3)t},\, P_1=P_1^0-P_1^t,\\
P_1^0 =&\frac{N_x^{(3)0}}{2}+\frac{M_x^{(3)0}}{2c}+N_x^{(1)0},\, P_1^t\\
=&\frac{N_x^{(3)t}}{2}+\frac{M_x^{(3)t}}{2c}+N_x^{(1)t},\, P_2=P_2^0-P_2^t,\\
P_2^0 =&\frac{N_x^{(3)0}}{2}-\frac{M_x^{(3)0}}{2c}+N_x^{(2)0},\\
P_2^t =&\frac{N_x^{(3)t}}{2}-\frac{M_x^{(3)t}}{2c}+N_x^{(2)t}.
\end{aligned}
\tag{7.15}
$$

The linear (index "0") and additional (index "t") components included in (7.15) are calculated through stresses and strains by the formulas (7.14).

By substituting the generalized internal forces (7.15) into the system of equations of equilibrium (7.9) and considering that the temperature changes only along the rod thickness $T_k(z)$, we reduce it to the form:

$$
\begin{cases}
H_1^0-P_1^0{}_{,x}=b_0p\,,\\
H_1^0+P_2^0{}_{,x}=0\,,\\
S_1^0{}_{,xx}+H_2^0-T_1^0{}_{,x}=b_0q+H_2^t+\frac{1}{2}b_0p_{,x}h_1\,,\\
S_2^0{}_{,xx}-H_2^0-T_2^0{}_{,x}=-H_2^t\,.
\end{cases}
\tag{7.16}
$$

The force boundary conditions will be treated in the same way. That is, at $x=0$ and l the following requirements must be satisfied:

$$
\begin{aligned}
&P_1^0=N_p^{(1)}+P_1^t,\, P_2^0=N_p^{(2)}+P_2^t,\, T_1^0-S_1^0{}_{,x}=Q_p^{(1)},\\
&T_2^0-S_2^0{}_{,x}=Q_p^{(2)},\, S_1^0=M_p^{(1)}+S_1^t,\, S_2^0=M_p^{(2)}+S_2^t.
\end{aligned}
$$

Here $N_p^{(1)}$, $Q_p^{(1)}$, $M_p^{(1)}$, $N_p^{(2)}$, $Q_p^{(2)}$, $M_p^{(2)}$ are the given forces at the ends of the rod in the load-bearing layers; index p takes the values 0 and 1, indicating where the force loads is given.

Using relations (7.14), (7.11) and (7.2), express linear internal forces and moments through the functions u_1, w_1, u_2, w_2:

$$
\begin{aligned}
N_x^{(1)0} &= b_0 \int_c^{c+h_1} \sigma_x^{(1)}\, dz = b_0 h_1 K_1^+ u_{1},_x\,,\ N_x^{(2)0} \\
&= b_0 \int_{-c-h_1}^{-c} \sigma_x^{(2)}\, dz = b_0 h_2 K_2^+ u_{2},_x\,, \\
N_x^{(3)0} &= b_0 \int_{-c}^{c} \sigma_x^{(3)}\, dz = b_0 K_3^- (w_1 - w_2) \\
&\quad + b_0 c K_3^+ \left(u_{1},_x + u_{2},_x + \frac{h_1}{2} w_{1},_{xx} - \frac{h_2}{2} w_{2},_{xx} \right), \\
N_z^{(3)0} &= b_0 \int_{-c}^{c} \sigma_x^{(3)}\, dz = b_0 K_3^+ (w_1 - w_2) \\
&\quad + b_0 c K_3^- \left(u_{1},_x + u_{2},_x + \frac{h_1}{2} w_{1},_{xx} - \frac{h_2}{2} w_{2},_{xx} \right), \\
M_x^{(1)0} &= b_0 \int_c^{c+h_1} \sigma_x^{(1)} z dz = b_0 K_1^+ \left[h_1 \left(c + \frac{h_1}{2} \right) u_{1},_x - \frac{h_1^3}{12} w_{1},_{xx} \right], \\
M_x^{(2)0} &= b_0 \int_{-c-h_2}^{-c} \sigma_x^{(2)}\, z\, dz = b_0 K_2^+ \left[-h_2 \left(c + \frac{h_2}{2} \right) u_{2},_x - \frac{h_2^3}{12} w_{2},_{xx} \right], \\
M_x^{(3)0} &= b_0 \int_{-c}^{c} \sigma_x^{(3)} z dz = b_0 c^2 K_3^+ \left[\frac{1}{3} u_{1},_x - \frac{1}{3} u_{2},_x - \frac{h_1}{6} w_{1},_{xx} + \frac{h_2}{6} w_{2},_{xx} \right], \\
M_{xz}^{(3)0} &= b_0 \int_{-c}^{c} \sigma_{xz}^{(3)}\, z\, dz = \frac{b_0 c G_3}{3} [w_{1},_x - w_{2},_x], \\
Q^{(3)0} &= b_0 \int_{-c}^{c} \sigma_{xz}^{(3)}\, dz = b_0 G_3 \left[u_1 - u_2 + \left(c + \frac{h_1}{2} \right) w_{1},_{xx} + \left(c + \frac{h_2}{2} \right) w_{2},_{xx} \right].
\end{aligned}
\tag{7.17}
$$

If in equilibrium Eq. (7.16) we substitute expressions of linear components of internal forces through functions $w_1(x)$, $u_1(x)$, $w_2(x)$ and $u_2(x)$ (7.17), then, at $p =$ const, we obtain a system of linear differential equations in displacements:

$$
\begin{aligned}
& a_1u_1 - a_1u_2 - a_4u_{1,xx} - a_5u_{2,xx} + a_2w_{1,x} \\
& + a_3w_{2,x} - 2a_6w_{1,xxx} + a_7w_{2,xxx} = p, \\
& - a_1u_1 + a_1u_2 - a_5u_{1,xx} - a_9u_{2,xx} \\
& - a_{10}w_{1,x} - a_{17}w_{2,x} - a_6w_{1,xxx} + 2a_7w_{2,xxx} = 0, \\
& - a_2u_{1,x} + a_{10}u_{2,x} + 2a_6u_{1,xxx} + a_6u_{2,xxx} + a_{11}w_{1,xx} - a_{12}w_{2,xx} \\
& + a_{15}w_{1,xxxx} - a_{16}w_{2,xxxx} + a_8w_1 - a_8w_2 = q - q_t, \\
& - a_3u_{1,x} + a_{17}u_{2,x} - a_7u_{1,xxx} - 2a_7u_{2,xxx} - a_{12}w_{1,xx} + a_{14}w_{2,xx} \\
& - a_{16}w_{1,xxxx} + a_{13}w_{2,xxxx} - a_8w_1 + a_8w_2 = q_t.
\end{aligned} \tag{7.18}
$$

Here the temperature terms q_t and coefficients $a_1, \ \ldots \ , a_{17}$ are determined taking into account the temperature dependence of the materials elasticity parameters over the thickness of the package:

$$
\begin{aligned}
q_t &= -\frac{H_2^t}{b_0} = -\frac{N_z^{(3)t}}{2cb_0}, \\
a_1 &= \frac{L_0(G_3)}{4c^2}, \; a_2 = \left(1 + \frac{h_1}{2c}\right)\frac{L_0(G_3)}{4c} \\
&+ \frac{L_1(G_3)}{4c^2} - \frac{L_0(K_3^-)}{4c} - \frac{L_1(K_3^-)}{4c^2}, \\
a_3 &= \left(1 + \frac{h_1}{2c}\right)\frac{L_0(G_3)}{4c} - \frac{L_1(G_3)}{4c^2} + \frac{L_0(K_3^-)}{4c} + \frac{L_1(K_3^-)}{4c^2}, \\
a_4 &= L_0(K_1^+) + \frac{L_1(K_3^+)}{2c} + \frac{L_0(K_3^+)}{4} + \frac{L_2(K_3^+)}{4c^2}, \\
a_5 &= \frac{L_0(K_3^+)}{4} - \frac{L_2(K_3^+)}{4c^2}, \\
a_6 &= \frac{L_1(K_1^+)}{2} - \frac{L_0(K_1^+)}{2}\left(c + \frac{h_1}{2}\right) \\
&+ \frac{h_1L_1(K_3^+)}{8c} + \frac{h_1L_0(K_3^+)}{16} + \frac{h_1L_2(K_3^+)}{16c^2}, \\
a_7 &= \frac{h_2}{8c^2}\left(c^2L_0(K_3^+) - L_2(K_3^+)\right), \; a_8 = \frac{L_0(K_3^+)}{4c^2}, \\
a_9 &= L_0(K_2^+) - \frac{L_1(K_3^+)}{2c} + \frac{L_0(K_3^+)}{4} + \frac{L_2(K_3^+)}{4c^2}, \\
a_{10} &= \left(1 + \frac{h_1}{2c}\right)\frac{L_0(G_3)}{4c} + \frac{L_1(G_3)}{4c^2} + \frac{L_0(K_3^-)}{4c} - \frac{L_1(K_3^-)}{4c^2},
\end{aligned}
$$

$$a_{11} = \frac{h_1 L_0(K_3^-)}{4c} + \frac{h_1 L_1(K_3^-)}{4c^2} - \left(1 + \frac{h_1}{2c}\right)^2 \frac{L_0(G_3)}{4}$$
$$- \left(1 + \frac{h_1}{2c}\right)\frac{L_1(G_3)}{2c} - \frac{L_2(G_3)}{4c^2},$$
$$a_{12} = \left(1 + \frac{h_1}{2c}\right)\left(1 + \frac{h_2}{2c}\right)\frac{L_0(G_3)}{4} - \frac{L_2(G_3)}{4c^2} + \frac{L_1(K_3^-)}{8c^2}(h_1 - h_2)$$
$$+ \frac{L_0(K_3^-)}{8c}(h_1 + h_2) - \frac{L_1(G_3)}{8c^2}(h_1 - h_2),$$
$$a_{13} = L_2(K_2^+) + L_1(K_2^+)\left(c + \frac{h_2}{2}\right) - \frac{h_2^2 L_1(K_3^+)}{8c}$$
$$+ \frac{h_2^2 L_0(K_3^+)}{16} + \frac{h_2^2 L_2(K_3^+)}{16c^2},$$
$$a_{14} = \frac{h_2 L_0(K_3^-)}{4c} - \frac{h_2 L_1(K_3^-)}{4c^2} - \left(1 + \frac{h_2}{2c}\right)^2 \frac{L_0(G_3)}{4}$$
$$+ \left(1 + \frac{h_2}{2c}\right)\frac{L_1(G_3)}{2c} - \frac{L_2(G_3)}{4c^2},$$
$$a_{15} = L_2(K_1^+) - L_1(K_1^+)\left(c + \frac{h_1}{2}\right) + \frac{h_1^2 L_1(K_3^+)}{8c}$$
$$+ \frac{h_1^2 L_0(K_3^+)}{16} + \frac{h_1^2 L_2(K_3^+)}{16c^2},$$
$$a_{16} = \frac{h_1 h_2}{16c^2}\left(c^2 L_0(K_3^+) - L_2(K_3^+)\right),$$
$$a_{17} = \left(1 + \frac{h_2}{2c}\right)\frac{L_0(G_3)}{4c} - \frac{L_1(G_3)}{4c^2} - \frac{L_0(K_3^-)}{4c} + \frac{L_1(K_3^-)}{4c^2} \tag{7.19}$$

Here integral operator $\mathrm{L}_n(f_k(z)) \equiv \int_{h_k} f_k(z) z^n \mathrm{d}z \, (n = 0, 1, 2)$.

Either force or kinematic conditions can be used as boundary conditions. In the case of kinematic conditions of the free support of the considered rod at the ends on rigid supports fixed in space in the end cross-sections x = 0, l (l is the length of the rod) the following requirements must be satisfied:

$$w_k = u_{k,x} = w_{k,xx} = 0 (k = 1, \ 2), \tag{7.20}$$

Thus, the system of Eq. (7.18) with coefficients (7.19) and boundary conditions (7.20) form a closed boundary problem for the deformation of an elastic three-layer rod with a compressible filler in the temperature field, which is calculated by formula (1.55).

7.2 General Solution of the Boundary Value Problem for a Thermoelastic Rod

The solution of the system of differential Eq. (7.18) is constructed in the form of expansion into trigonometric series, which automatically satisfy the boundary conditions for support on rigid supports (7.20):

$$\begin{gathered} u_1 = \sum_{m=1}^{\infty} U_{1m} \cos\left(\frac{\pi m x}{l}\right), u_2 = \sum_{m=1}^{\infty} U_{2m} \cos\left(\frac{\pi m x}{l}\right), \\ w_1 = \sum_{m=1}^{\infty} W_{1m} \sin\left(\frac{\pi m x}{l}\right), w_2 = \sum_{m=1}^{\infty} W_{2m} \sin\left(\frac{\pi m x}{l}\right), \end{gathered} \tag{7.21}$$

where $U_{1m}, U_{2m}, W_{1m}, W_{2m}$ are the amplitudes of displacements.

The external load and temperature components in the rod layers are represented as expansions in the following trigonometric series:

$$\begin{gathered} q = \sum_{m=1}^{\infty} q_m \sin\left(\frac{\pi m x}{l}\right), q_m = \frac{2}{l} \int_0^l q(x) \sin\left(\frac{\pi m x}{l}\right) dx, \\ p = \sum_{m=1}^{\infty} p_m \sin\left(\frac{\pi m x}{l}\right), p_m = \frac{2}{l} \int_0^l p(x) \sin\left(\frac{\pi m x}{l}\right) dx, \\ q_t = \sum_{m=1}^{\infty} q_{tm} \sin\left(\frac{\pi m x}{l}\right), q_{tm} = \frac{2}{l} \int_0^l q_t(x) \sin\left(\frac{\pi m x}{l}\right) dx. \end{gathered} \tag{7.22}$$

After substituting the displacements (7.21) and the loads (7.22) into the system of differential Eq. (7.18), we obtain the following system of linear algebraic equations for the required displacement amplitudes $U_{1m}, U_{2m}, W_{1m}, W_{2m}$ at the mth harmonic:

$$\begin{cases} b_1 U_{1m} + b_2 U_{2m} + b_3 W_{1m} + b_4 W_{2m} = p_m, \\ b_2 U_{1m} + b_5 U_{2m} + b_6 W_{1m} - b_7 W_{2m} = 0, \\ b_3 U_{1m} + b_6 U_{2m} + b_8 W_{1m} + b_9 W_{2m} = q_m - q_{tm}, \\ b_4 U_{1m} - b_7 U_{2m} + b_9 W_{1m} + b_{10} W_{2m} = q_{tm}. \end{cases} \tag{7.23}$$

where the coefficients b_i depend on the parameter m and on the temperature through the coefficients a_n defined in (7.19):

$$\begin{gathered} b_1 = a_1 + a_4\left(\frac{\pi m}{l}\right)^2; b_2 = -a_1 + a_5\left(\frac{\pi m}{l}\right)^2; \\ b_3 = a_2\frac{\pi m}{l} + 2a_6\left(\frac{\pi m}{l}\right)^3; b_4 = a_3\frac{\pi m}{l} - a_7\left(\frac{\pi m}{l}\right)^3; \end{gathered}$$

$$b_5 = -a_1 + a_9\left(\frac{\pi m}{l}\right)^2;\ b_6 = -a_{10}\frac{\pi m}{l} + a_6\left(\frac{\pi m}{l}\right)^3;$$
$$b_7 = a_{17}\frac{\pi m}{l} + 2a_7\left(\frac{\pi m}{l}\right)^3;\ b_8 = -a_{11}\left(\frac{\pi m}{l}\right)^2 + a_{15}\left(\frac{\pi m}{l}\right)^4 + a_8;$$
$$b_9 = a_{12}\left(\frac{\pi m}{l}\right)^2 - a_{16}\left(\frac{\pi m}{l}\right)^4 - a_8;\ b_{10} = a_{14}\left(\frac{\pi m}{l}\right)^2 - a_{13}\left(\frac{\pi m}{l}\right)^4 - a_8;$$

The solution of the system (7.23) can be written in the determinants:

$$U_{1m} = \frac{\Delta_{1m}}{\Delta},\ U_{2m} = \frac{\Delta_{2m}}{\Delta},\ W_{1m} = \frac{\Delta_{3m}}{\Delta},\ W_{3m} = \frac{\Delta_{4m}}{\Delta}, \tag{7.24}$$

where

$$\Delta = \begin{vmatrix} b_1 & b_2 & b_3 & b_4 \\ b_2 & b_5 & b_6 & -b_7 \\ b_8 & b_9 & b_{10} & b_{11} \\ b_{12} & -b_{13} & b_{11} & b_{14} \end{vmatrix},\ \Delta_{1m} = \begin{vmatrix} p_m & b_1 & b_3 & b_4 \\ 0 & b_2 & b_6 & -b_7 \\ q_m - q_{tm} & b_8 & b_{10} & b_{11} \\ q_{tm} & b_{12} & b_{11} & b_{14} \end{vmatrix},$$

$$\Delta_{2m} = \begin{vmatrix} b_1 & p_m & b_3 & b_4 \\ b_2 & 0 & b_6 & -b_7 \\ b_8 & q_m - q_{tm} & b_{10} & b_{11} \\ b_{12} & q_{tm} & b_{11} & b_{14} \end{vmatrix},\ \Delta_{3m} = \begin{vmatrix} b_1 & b_2 & p_m & b_4 \\ b_2 & b_5 & 0 & -b_7 \\ b_8 & b_9 & q_m - q_{tm} & b_{11} \\ b_{12} & -b_{13} & q_{tm} & b_{14} \end{vmatrix},$$

$$\Delta_{4m} = \begin{vmatrix} b_1 & b_2 & b_3 & p_m \\ b_2 & b_5 & b_6 & 0 \\ b_8 & b_9 & b_{10} & q_m - q_{tm} \\ b_{12} & -b_{13} & b_{11} & q_{tm} \end{vmatrix},$$

Calculating the determinants in (7.24), we obtain the coefficients U_{1m}, U_{2m}, W_{1m}, W_{2m} of the expansion in the series of displacements. The displacements in the load-bearing layers and the filler follow from relations (7.1), the strains—from (7.2), and the stresses—from (7.12). Thus, formulas (7.21)–(7.24) provide an analytical solution to the problem of deformation in temperature flow of an elastic three-layer rod with a compressible filler under a uniformly distributed load.

Constant temperature. Assume that after heating the rod, the temperature field in each of its layers is averaged, i.e. $T_k = \text{const}$. The elasticity parameters of the materials of the layers G_k, K_k are calculated by formula (1.37) at the temperature T_k. Then the temperature components of the internal forces (7.14) take the form ($k = 1, 2, 3$):

$$\begin{gathered} N_x^{(k)t} = 3b_0 h_k K_k \alpha_{0k} T_k,\ M_x^{(1)t} = 3b_0 K_1 \alpha_{01}\left(c + \frac{h_1}{2}\right)T_1, \\ M_x^{(2)t} = -3b_0 K_2 \alpha_{02}\left(c + \frac{h_2}{2}\right)T_2,\ M_x^{(3)t} = 0,\ N_z^{(3)t} = 3b_0 h_3 K_3 \alpha_{03} T_3. \end{gathered} \tag{7.25}$$

The generalized forces are still calculated by formulae (7.15), taking into account relations (7.25). The temperature terms q_t and coefficients $a_1, \ldots, a_{17}$ in the system of Eq. (7.18) at temperature T_k can be written out explicitly:

$$q_1 = -\frac{3b_0 h_3 K_3 \alpha_{03} T_3}{2cb_0},$$

$$a_1 = \frac{G_3}{2c}; \; a_2 = \frac{G_3}{2}\left(1 + \frac{h_1}{2c}\right) - \frac{k_3^-}{2},$$

$$a_3 = \frac{G_3}{2}\left(1 + \frac{h_2}{2c}\right) - \frac{k_3^-}{2}, \; a_4 = K_1^+ h_1 + \frac{2K_3^+ c}{3},$$

$$a_5 = \frac{K_3^+ c}{3}, \; a_6 = \frac{K_3^+ ch_1}{6}, \; a_7 = \frac{K_3^+ ch_2}{6}, \; a_8 = \frac{K_3^+}{2},$$

$$a_9 = K_2^+ h_2 + \frac{2K_3^+ c}{3}, \; a_{10} = \frac{G_3}{2}\left(1 + \frac{h_1}{2c}\right) + \frac{k_3^-}{2},$$

$$a_{11} = \frac{k_3^- h_1}{2} - \frac{G_3 c}{2}\left(1 + \frac{h_1}{2c}\right)^2 - \frac{G_3 c}{6},$$

$$a_{12} = \frac{k_3^- (h_1 + h_2)}{4} + \frac{G_3 c}{2}\left(1 + \frac{h_1}{2c}\right)\left(1 + \frac{h_1}{2c}\right) - \frac{G_3 c}{6},$$

$$a_{13} = \frac{K_2^+ h_2^3}{12} + \frac{K_3^+ h_2^2}{6}, \; a_{14} = \frac{K_3^- h_2}{2} - \frac{G_3 c}{2}\left(1 + \frac{h_2}{2c}\right)^2 - \frac{G_3 c}{6},$$

$$a_{15} = \frac{K_1^+ h_1^3}{12} + \frac{K_3^+ ch_1^2}{6}, \; a_{16} = \frac{K_3^+ ch_2 h_1}{12}, \; a_{17} = \frac{G_3}{2}\left(1 + \frac{h_1}{2c}\right) + \frac{k_3^-}{2},$$

$$K_k^+(T_k) = K_k(T_k) + \frac{4}{3}G_k(T_k), \; K_k^-(T_k) = K_k(T_k) - \frac{2}{3}G_k(T_k). \quad (7.26)$$

The form of the basic equations and solutions (7.21)–(7.24) remains the same.

7.3 The Case of a Uniformly Distributed Load

Consider the bending of a three-layer rod under the action of a uniformly distributed load $q = \text{const}$, $p = 0$. The decomposition coefficients of the power and heat components of the loads in series (7.22) will be:

$$q_m = \frac{2q_0}{\pi m}(1 - \cos(\pi m)),$$

$$q_m = \frac{2}{l}\int_0^l q_t(x)\sin\left(\frac{\pi m x}{l}\right)dx$$

$$= -\frac{3\alpha_{03}}{cl}\int_0^l\int_{h_3} K_3(z)T_3(z)\sin\left(\frac{\pi mx}{l}\right)dzdx, \tag{7.27}$$

where the temperature is calculated by the formula (4.1).

Substituting expressions (7.27) into the system of algebraic Eq. (7.23) and solving it numerically, we obtain for each value of parameter m the required displacement amplitudes U_{1m}, U_{2m}, W_{1m}, W_{2m}. The total displacements $w_1(x)$, $u_1(x)$, $w_2(x)$ and $u_2(x)$ are obtained by summing the series (7.21).

Numerical analysis in this section was carried out for a three-layered rod, the layers of which are made of D16T-fluoroplastic-D16T materials. The heat spent on heating the outside metal layer was not taken into account due to its thinness and low heat capacity. The temperatures of the load-bearing layers were assumed to be equal to the temperature of the filler at their bonding points: $T^{(1)} = T^{(3)}(c, t)$, $T^{(2)} = T^{(3)}(-c, t)$. The temperature field in the filler was calculated by formula (4.1), with $h_1 = 0$, $q_t = 5000$ Jw/(m^2 s).

Numerical results for an elastic rod in the case of uniformly distributed load are obtained at intensity $q_0 = -10$ MPa at time $t = 60$ min.

The corresponding calculated temperature change along the thickness of the filler and the first layer is shown in Fig. 7.2 at different points in time (min.) after the start of the heat flux: *1*–15; *2*–30; *3*–45; *4*–60.

Figure 7.3.*a* demonstrates the effect of the number of terms in series (7.21) on the calculated value of the maximum deflection of the layer 1: *1*—isothermal load, *2*—thermo-force load ($t = 60$ min.). The results of the calculations show that after 20 members of the series, the subsequent terms do not affect the final sum. Nevertheless, in further numerical calculations in the summation of trigonometric series was taken $m = 100$.

Figure 7.3*b* shows the isothermal elastic deflections: *1*, *2*—in the first and second layers with compressible filler; *3*—filler incompressible. Taking into account the

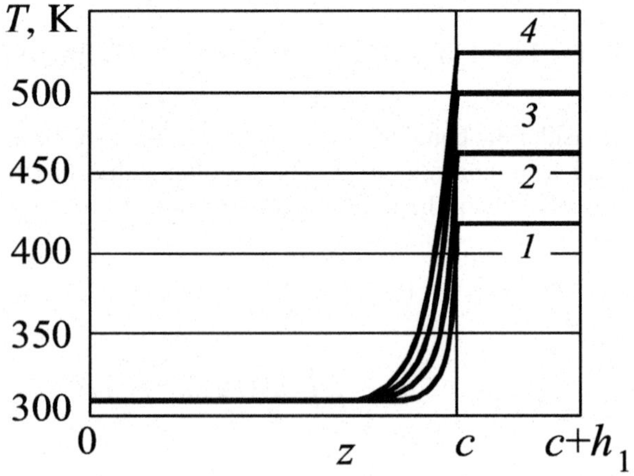

Fig. 7.2 Temperature variation across the thickness of the filler and the first layer

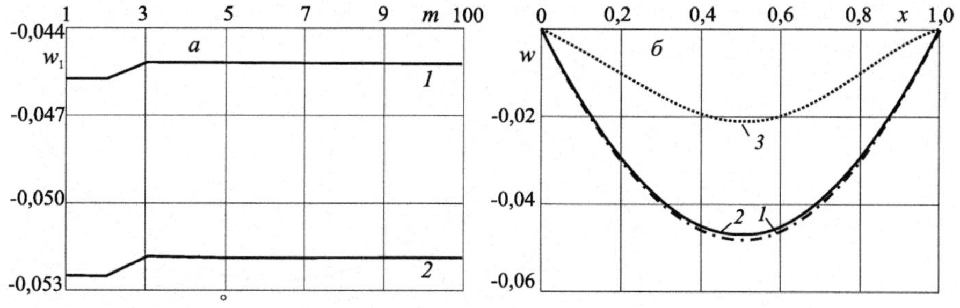

Fig. 7.3 Number of members of the summarized series

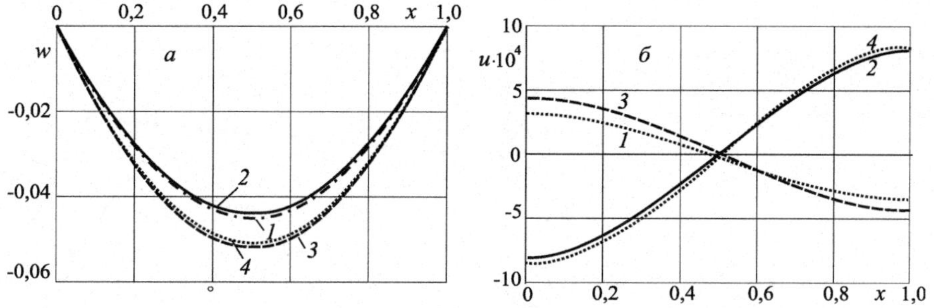

Fig. 7.4 Change of deflections w (*a*) and longitudinal displacements *u* (*b*) along the rod axis

compressibility of the filler makes the three-layer beam less rigid, which increases the calculated deflection by more than two times.

The effect of temperature on the calculated deflections w and longitudinal displacements u in the bearing layers is shown in Fig. 7.4*a*, *b*, respectively: *1, 2*—isothermal in the first and second layers; *3, 4*—thermoelastic in the first and second layers.

The effect of temperature leads to a decrease in the elastic modulus of the materials and, consequently, to an increase in the maximum deflections by about 15–16%. Longitudinal displacements in the first layer increase by 25.8%, in the second layer, the temperature of which remains constant, change weakly.

Figure 7.5 illustrates the change in compression strain ϵ_{zz} in the filler along the rod axis (in percent): *1*—isothermal loading, *2*—thermos-force loading. The temperature effect leads to an increase in the volumetric strain and, consequently, a decrease in the compression strain by 12%.

The change of normal stresses σ_{xx} in the median ($x = 0.5$) cross section is shown in Fig. 7.6: *1*—isothermal loading, *2*—thermal-force loading. In the second bearing layer the temperature does not change, so the stresses visually remain the same, their difference is on the order of 10%. In the filler, the stresses change by 19% in the area of gluing with the outer layer. In the first layer they shift to the negative area, their change reaches in absolute value 127%.

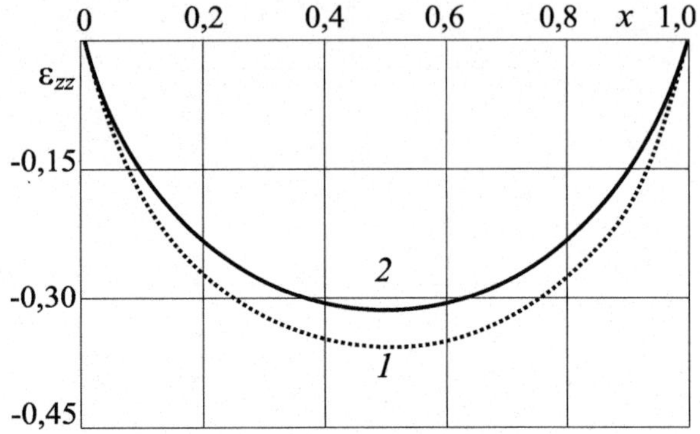

Fig. 7.5 Change of compressive strain ϵ_{zz} in the filler along the rod axis

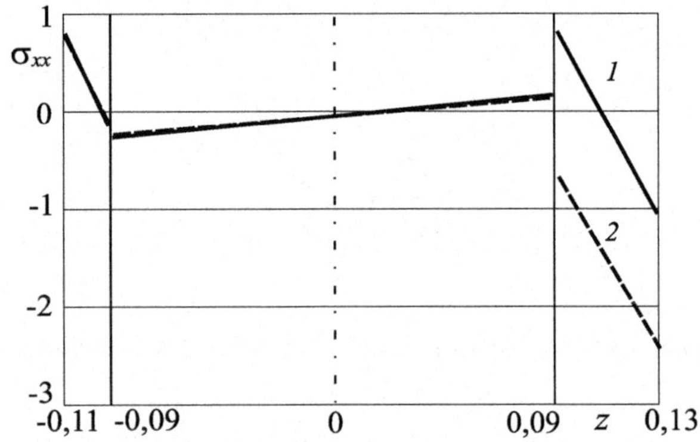

Fig. 7.6 Change of normal stresses σ_{xx} in the median ($x = 0.5$) cross section

7.4 Deformation of an Elastic Rod by Local Loads

The temperature condition of the three-layer rod package in this section was taken the same as in Sect. 7.3.

Local rectangular load. Let an elastic three-layer rod with a compressible filler be subjected to a transverse surface load $q(x)$ uniformly distributed in the interval $a \le X \le b$ in the temperature field (Fig. 7.7).

Its analytical form and the coefficients of expansion in series (7.22) are:

$$q(x) = q_0(H_0(b - x) - H_0(a - x)).$$

$$q_m = -\frac{2q_0}{\pi m}\left(\cos\frac{\pi mb}{l} - \cos\frac{\pi ma}{l}\right), \tag{7.28}$$

where $H_0(x)$ is Heaviside function, q_0 is load intensity.

After calculating the coefficients q_m, q_{tm} by formulas (7.27), (7.28) and introducing them into solution (7.24), we obtain the required displacement amplitudes

Fig. 7.7 Calculation scheme of loading by local load

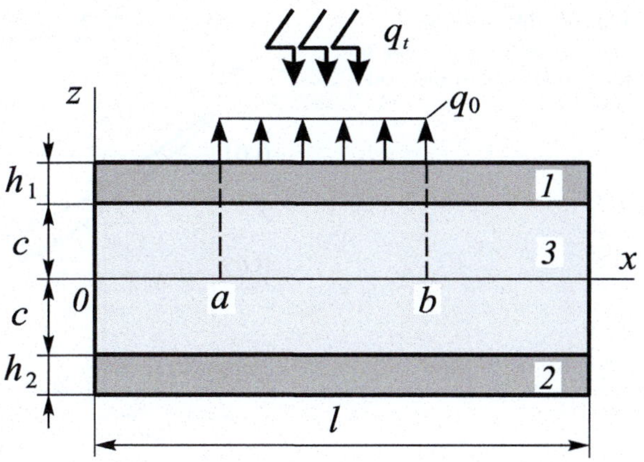

U_{1m}, U_{2m}, W_{1m}, W_{2m}. Total displacements are calculated by summing the series (7.21), strains are calculated by Eq. (7.2), stresses are calculated from Hooke's law (7.10).

Numerical results were obtained under the conditions outlined in Sect. 7.3, with temperature and load intensities, respectively $q_t = 5000$ J/(m^2·s), $q_0 = -10$ MPa.

Figures 7.8, 7.9 show the deflections of the bearing layers as a function of the edges of the load spot in the cross section in the middle of the rod ($x = 0.5$): *1*, *3*—deflection of the first layer isothermal and thermo-force; *2*, *4*—similar deflections of the second bearing layer.

Figure 7.8 illustrates the change in deflections depending on the length of the load interval *b*. The maximum, of course, is observed when the load is distributed over the entire rod. The difference of these deflections gives the value of compression of the filler, which in this case is not significant. The temperature component of the deflections reaches 16%.

Figure 7.9 shows the deflections of the load-bearing layers of the rod in the cross section with coordinate $x = 0.5$ as a function of the coordinate *a* of the left edge of

Fig. 7.8 Deflections of bearing layers depending on the boundary *b* of the load spot

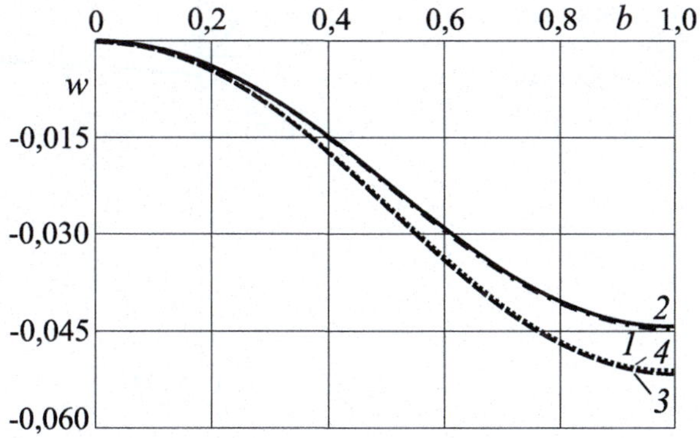

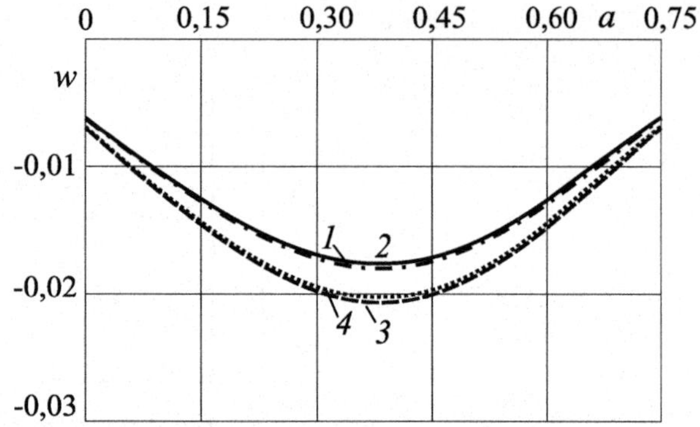

Fig. 7.9 Deflections of bearing layers depending on the boundary a of the load spot

the load interval at constant length $b–a = 0.25$. As the load moves from the left end of the rod to the right, the deflections increase, reach a maximum at $a = 0.37$, i.e. when the load is distributed symmetrically about the ends of the rod in the range {0.37; 0.62} and then decrease. The temperature component of the deflections is 23%. The compression of the filler here is also insignificant, which is explained by the small load.

The distribution of longitudinal stresses σ_{xx} in the cross section of the three-layer rod with coordinate $x = 0.5$ is shown in Fig. 7.10: *1*—isothermal loading, 2—thermal force loading. The load is distributed on a section of the rod $0.5 \le x \le 0.75$. Stresses in the first layer shift to the negative region due to thermal expansion of the material. In the second layer and filler, the temperature has little effect on stress distribution. Stress maximums are observed on the outer planes of the rod. In the first layer they are negative, in the second are positive, they are about the same in absolute value. Stress values in the bearing layers are related to the intensity of the surface load $q_1 = 10^9$ Pa, in the filler—to $q_3 = 10^8$ Pa.

Concentrated transverse force. Let us consider the case of bending of the considered three-layered rod by a concentrated force. For this purpose let's assume that the

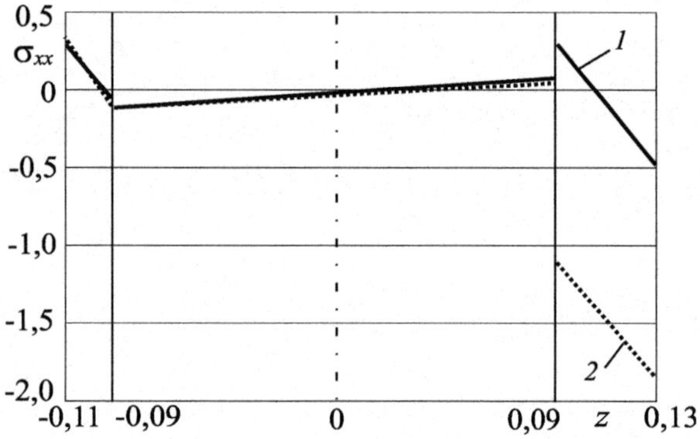

Fig. 7.10 Longitudinal stress distribution σ_{xx} in the cross section of the three-layer rod with coordinate $x = 0.5$

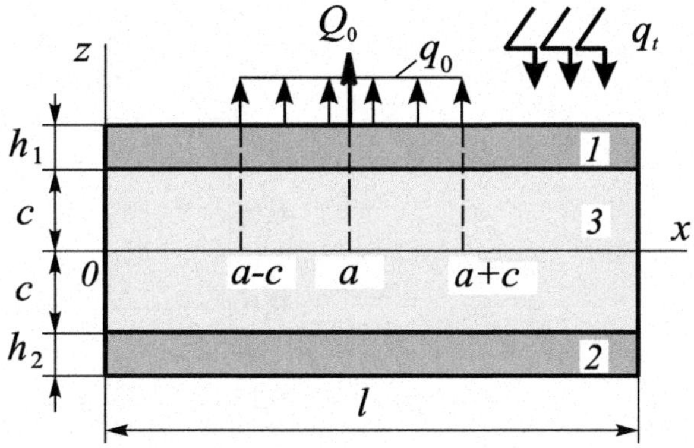

Fig. 7.11 Scheme of loading with a concentrated force

shear force (along the *y*-coordinate) of constant intensity Q_0 is applied in the section with coordinate $x = a$ (Fig. 7.11).

Let us solve the problem by applying the results obtained for the distributed surface load intensity q_0 on the interval $(a - c) \leq x \leq (a + c)$. Assume that *c* is small and introduce a replacement $q_0 = Q_0/(2c)$ in the coefficients of (7.28). After that we will "pull down" the load interval to the section x = *a*, for which we will decrease the value of *c* to zero, leaving Q_0 constant. As a result, we obtain the following coefficients of decomposition of the rectangular load in series:

$$q_m = \frac{2Q_0}{2c\pi m}\left(\cos\frac{\pi m(a+c)}{l} - \cos\frac{\pi m(a-c)}{l}\right) = \frac{2Q_0}{l}\sin\frac{\pi m a}{l}. \tag{7.29}$$

Calculating the coefficients q_m by the above formula (7.29), and adding them to the solution (7.24), we obtain the required displacement amplitudes U_{1m}, U_{2m}, W_{1m}, W_{2m}. The total displacements are then calculated by summing up the series (7.21), strains—according to the formulas (7.2), stresses—from Hooke's law (7.10).

Numerical results were obtained with the value of the external force $Q_0 = 10^6$ N.

Figure 7.12 shows the change along the rod axis of the maximum difference between the deflections of the bearing layers $\epsilon_{zz} = (w_1 - w_2)/(2c)$, if a concentrated force is applied in a cross section with coordinate $x = 0.5$: *1*—isothermal deformation, *2*—thermal force deformation. The maximum compression is reached in the middle of the rod. There is no compression on the supports. The jump in the middle of the rod is caused by a local downward force. With temperature effect, the compression in the range $0.4 < x < 0.6$ decreases, and the other part of the rod is tensile.

Figure 7.13 shows the change of deflections of the bearing layers w_1, w_2 in the middle of the rod ($x = 0.5$) depending on the coordinate *a* of the point of applying the concentrated force *Q*.

As the force moves away from the left edge of the rod, the deflections increase and reach a maximum at $a = 0.5$, then their value decreases. If the force is applied at the supports of the rod, the deflections are zero. Temperature increases the maximum deflections by 16%.

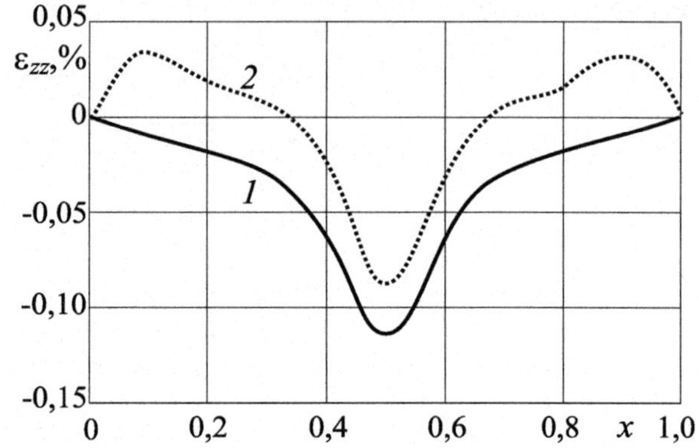

Fig. 7.12 Variation of filler compression along the rod axis

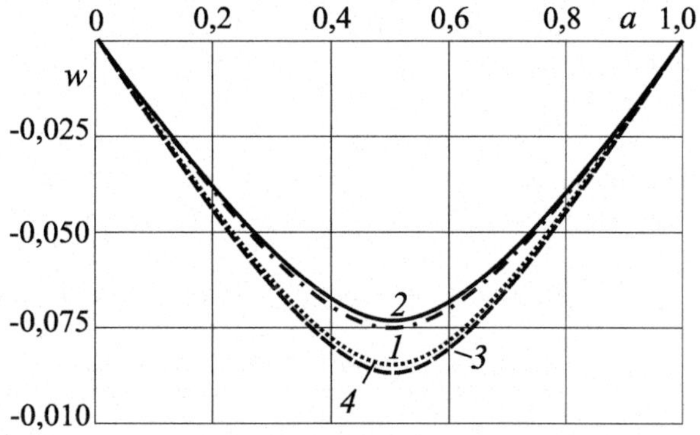

Fig. 7.13 Change of deflections of bearing layers in the middle of the rod ($x = 0.5$) depending on the coordinate a of the point of application of the concentrated force Q

The distribution of the corresponding normal stresses σ_{xx} along the rod axis on the first layer surfaces is shown in the Fig. 7.14: *1*, *3* – $z = c$ (internal surface), *2*, *4* – $z = c + h_1$ (external surface); graphs 3 and 4 taking into account the temperature. Stress values in the bearing layers are referred to the load intensity $q_1 = 10^9$ Pa, in the filler—to $q_3 = 10^8$ Pa. An increase in the temperature of the layer leads to its compression and an increase in stresses by five times in absolute value.

Peaks of strains and stresses are observed in the section of the rod with a concentrated force. It is interesting to note that the graph of strain δ in the section x = 0.5 on the coordinate of force application *b* will be the same in shape and magnitude as the graph in Fig. 7.12. This is because if we substitute the expression for the parameters (7.29) into the formula for the expansion of the load in series (7.22), the result will be the load as a function symmetric with respect to the arguments *x* and *a* ($f(x, a) = f(a, x)$):

$$q = \sum_{m=1}^{\infty} q_m \sin\left(\frac{\pi m x}{l}\right) = \frac{2Q_0}{l} \sum_{m=1}^{\infty} \sin\frac{\pi m a}{l} \sin\left(\frac{\pi m x}{l}\right).$$

Therefore, the corresponding graph is not given here.

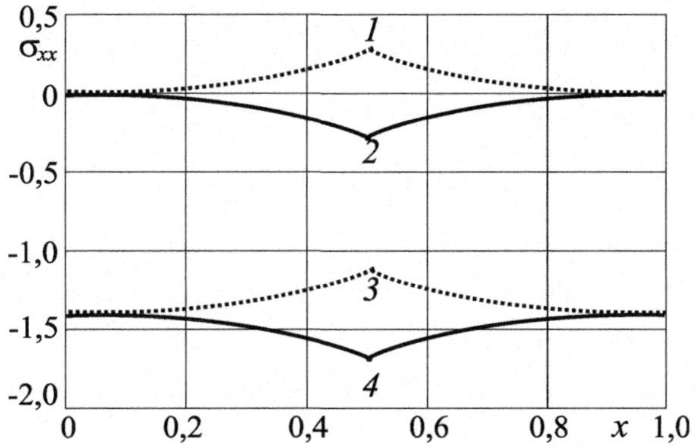

Fig. 7.14 The distribution of the normal stresses σ_{xx} along the rod axis on the first layer surfaces

Concentrated bending moment. Let the considered three-layered rod in the cross-section x $= a$ be subjected to the transverse moment of intensity $M_0 = \text{const}$ (Fig. 7.15). Solve the problem using the sum of solutions (7.29) for two concentrated forces, equal in magnitude, directed in opposite directions and acting in closely spaced sections with coordinates $x = a - c$ and $x = a + c$ (c is small). In this sum we substitute $Q_0 = M_0/(2c)$ and decrease the parameter c to zero, leaving M_0 constant. As a result, calculating the limit we obtain:

$$q = \frac{2M_0}{l} \lim_{c \to 0} \left[\frac{l}{2c\pi m} \left(\sin \frac{\pi m(a+c)}{l} - \sin \frac{\pi m(a-c)}{l} \right) \right]$$
$$= \frac{2M_0 \pi m}{l^2} \cos \frac{\pi m a}{l}.$$

Then, after calculating the parameters q_m using the obtained formula and the determinants in the solution (7.24), we obtain the required displacement amplitudes U_{1m}, U_{2m}, W_{1m}, W_{2m}. The displacements, strains and stresses in the layers are then calculated according to the previously discussed scheme.

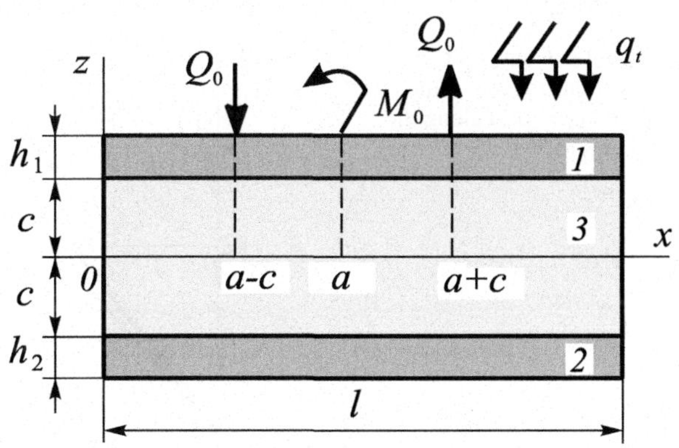

Fig. 7.15 Calculation scheme of concentrated moment loading

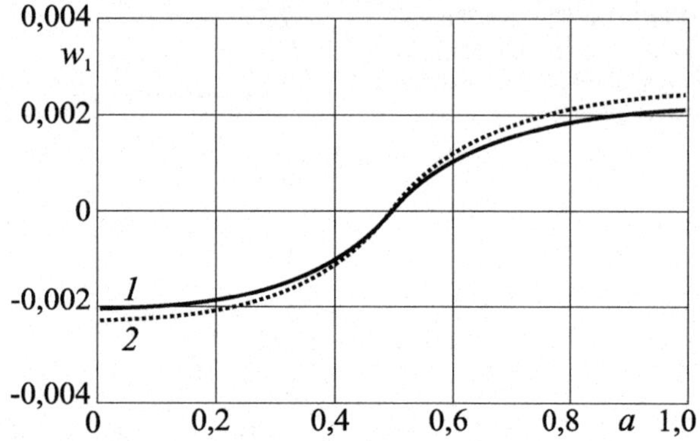

Fig. 7.16 Change of deflection w_1 in the middle of the rod ($x = 0.5$) depending on the coordinate a of the concentrated moment point of application

Numerical results were obtained at $M_0 = -10^5$ N·m.

Figure 7.16 illustrates the change in deflection of the first bearing layer in the middle of the rod ($x = 0.5$) depending on the coordinate a of the point of application of the concentrated moment: *1*—isothermal loading, *2*—thermal-force loading.

Figure 7.17 shows the change of deflections of the load-bearing layers along the axis of the rod if a bending moment is applied in the section: *1, 3*—first layer; *2, 4*—second layer (even curves—for thermal influence). In both cases, the heating of the rod leads to an increase in deflections by 20–23%.

Figure 7.18 shows the change along the rod axis of the maximum difference between the deflections of the load-bearing layers $\epsilon_{zz} = (w_1 - w_2)/(2c)$ if the bending moment in the cross-section x = 0.75: *1*—isothermal deformation, *2*—thermal force deformation. The maximums of compression and tension of the filler are reached in the vicinity of the moment application ($x = 0.68$; 0.82). When heated, the compression decreases and the stretching of the filler increases by 70%.

The distribution of the corresponding normal stresses σ_{xx} along the rod axis on the surfaces of the first layer is shown in Fig. 7.19: *1, 3* – $z = c$—internal layer surface, *2, 4* – $z = c + h_1$—external surface (*3, 4*—with thermal influence). Stress values in the bearing layers are referred to the load intensity $q_1 = 10^9$ Pa, in the filler to $q_3 =$

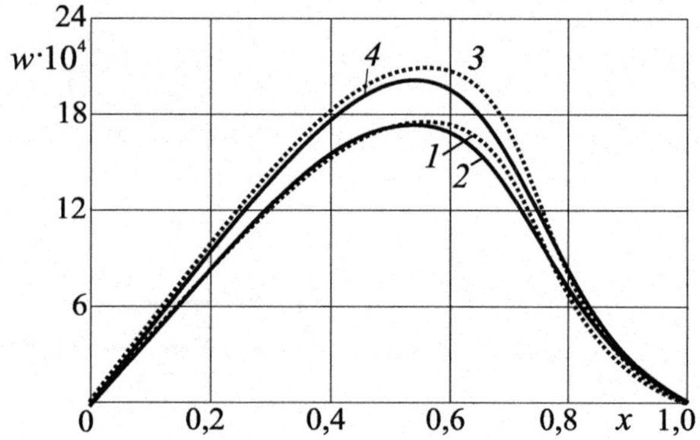

Fig. 7.17 Change of deflections in the middle of the rod ($x = 0.5$) along the rod axis ($a = 0.75$)

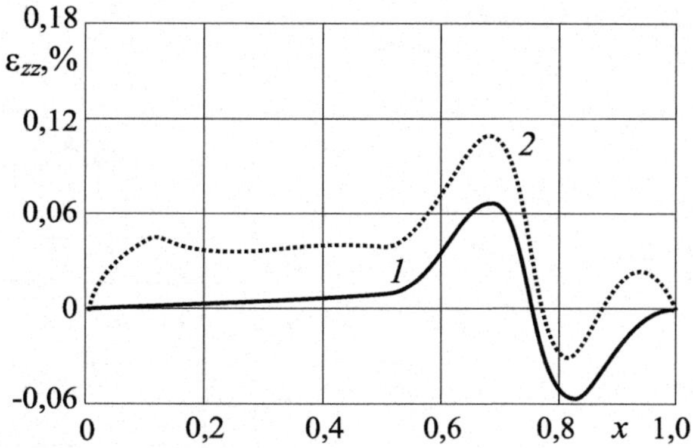

Fig. 7.18 Change of filler compression along the rod axis under the action of the moment

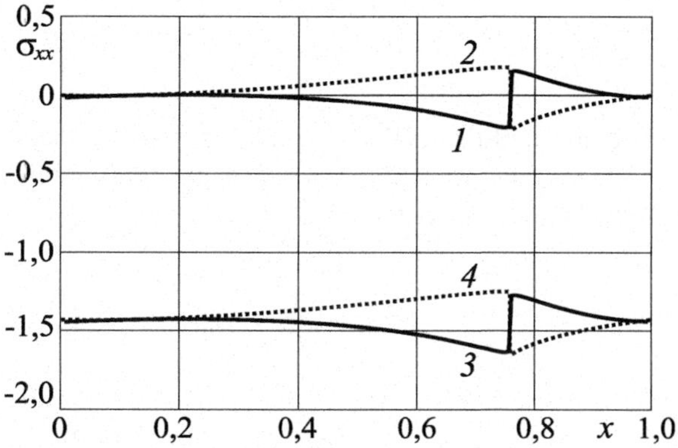

Fig. 7.19 Distribution of normal stresses σ_{xx} along the rod axis on the surfaces of the first layer

10^8 Pa. In this case, the acting of a concentrated moment leads to a stress jump in the moment cross-section. Increasing the temperature of the rod shifts the stresses to the negative region and increases them by six times in absolute value.

Local distributed sinusoidal load. Suppose that load $q(x)$, distributed over a *half-wave of a sinusoid* (Fig. 7.20), acts on the surface of a three-layer rod in the region $0 \leq x \leq b$:

$$q(x) = q_0 H_0(b - x) \sin\left(\frac{\pi x}{b}\right), \tag{7.30}$$

where $H_0(x)$ is Heaviside function; q_0 is load amplitude.

By substituting the load (7.30) into the corresponding formula from (7.22), we obtain expressions for the parameters q_m:

$$q_m = \frac{bq_0}{\pi}\left[\frac{1}{bm - l}\sin\frac{\pi(bm - l)}{l} - \frac{1}{bm + l}\sin\frac{\pi(bm + l)}{l}\right.\quad (m = 1, 2, 3, \cdots). \tag{7.31}$$

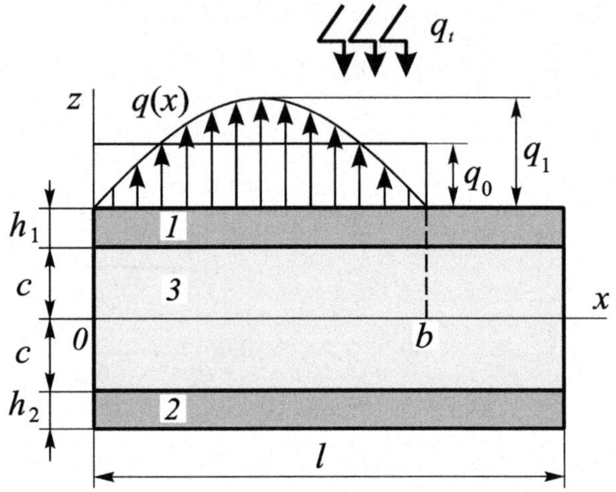

Fig. 7.20 Calculation scheme of sinusoidal loading

Here the condition $bm \neq l$ must be satisfied so that division by zero does not occur. In rod sections where $b = l\,/\,p$ (p is a natural number), if $m \neq p$ the formula (7.31) is fulfilled, if $m = p$, then this formula:

$$q_p = \frac{2q_0}{l}\int_0^l H_0(b-x)\left(\sin\frac{\pi px}{l}\right)^2 dx$$

$$= \frac{q_0}{l}\int_0^b \left(1-\cos\frac{2\pi px}{l}\right) dx = \frac{q_0 b}{l}.$$

The magnitude of the amplitude of the sinusoidal load, the resultant force is statically equivalent to a rectangular load acting on the same surface of the rod, is obtained from the condition of equality of the areas occupied by the loads in question:

$$q_0' = q_0 l \bigg/ \int_0^l \sin\left(\frac{\pi x}{l}\right) dx = \frac{1}{2}\pi q_0.$$

Numerical results were obtained at load amplitudes q_0–$5{\cdot}10^6$ Pa and $q_0' = 0.5\,\pi q_0$.

Figure 7.21 shows the change of deflections (a) and longitudinal displacements (b) in the first layer along the longitudinal axis of the three-layer rod under the action of statically equivalent distributed loads ($b = 1$): *1*—rectangular; *2*—sinusoidal (both isothermal); *3*—sinusoidal thermo-force. With the same equal force, the maximum displacement is greater from a sinusoidal load. The maximum difference between

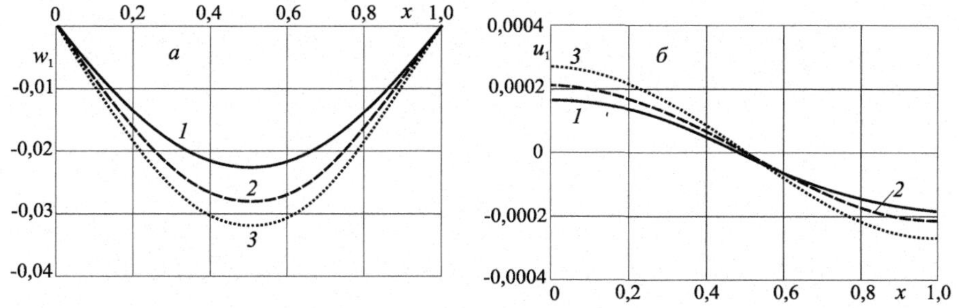

Fig. 7.21 Change of deflections w (a) and longitudinal displacements u (b) along the rod axis under local sinusoidal load

deflections under isothermal loading is 22%. Temperature effect adds another 15% to deflection under sinusoidal load.

A similar situation is observed for the maximum deflections, which are achieved in the section x $= 0.5$ of the three-layer rod, when the load interval length b increases (Fig. 7.22). Numbering of curves *1, 2* corresponds to the number of the layer; the dash at the top corresponds to displacements from sinusoidal load, without the dash—displacements from rectangular load; $3'$, $4'$—displacements of bearing layers under thermal-force action. Here the difference of deflections of the bearing layers is not great, because the load is reduced by 2 times compared to the previous one.

Consequently, with the same static equivalent of the considered distributed loads, the sinusoidal load is more dangerous, from the point of view of strength, because it causes deflections of layers larger in magnitude than the uniformly distributed one.

Let a local sinusoidal load acts on the rod surface inside the section a $G_k(T_k),\ K_k(T_k)\ x \le b$:

$$q(x) = q_0[H_0(b - x) - H_0(a - x)] \sin\left[\frac{\pi(x - a)}{b - a}\right]. \tag{7.32}$$

The load decomposition coefficients (7.32) into a series will be

Fig. 7.22 Change of deflections w in the middle of the rod depending on the size of the load interval

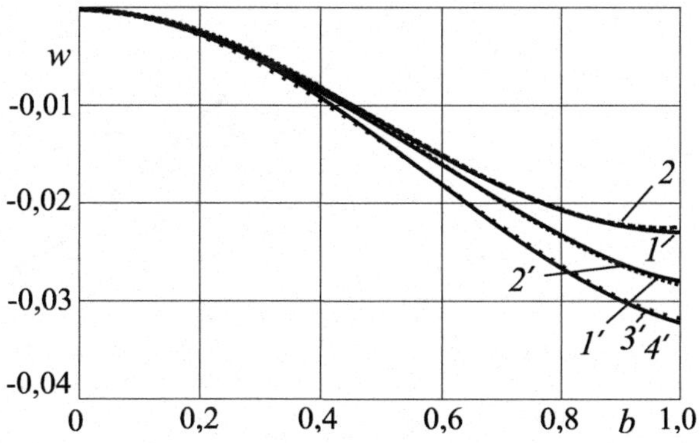

$$q_m = \frac{2}{l}\int_0^l q_0[H_0(b-x) - H_0(a-x)]\sin\left[\frac{\pi(x-a)}{b-a}\right]\sin\frac{\pi m x}{l}dx = q'_m q_0.$$

Here

$$\begin{aligned} q'_m &= \frac{1}{l}\int_a^b \left[\cos\left(\frac{\pi(x-a)}{b-a} + \frac{\pi m x}{l}\right) - \cos\left(\frac{\pi(x-a)}{b-a} + \frac{\pi m x}{l}\right)\right]dx \\ &= \frac{(b-a)}{\pi}\left\{\frac{1}{m(b-a)-l}\left[\sin\frac{\pi(bm-l)}{l} - \sin\frac{\pi a m}{l}\right]\right. \\ &\quad \left. - \frac{1}{m(b-a)+l}\left[\sin\frac{\pi(bm+l)}{l} - \sin\frac{\pi a m}{l}\right]\right\}(m = 1, 2, 3, \ldots). \end{aligned} \quad (7.33)$$

For the coefficients q'_m, the condition $m(b - a) \neq l$ must be satisfied so that division by zero does not occur. In the sections of the rod where $(b - a) = l / p$ (p is a natural number), if the condition $m \neq p$ is met, formula (7.33) should be used, if $m = p$, we use the following parameters

$$\begin{aligned} q'_p &= \frac{2}{l}\int_0^l [H_0(b-x) - H_0(a-x)]\sin\left[\frac{\pi p(x-a)}{l}\right]\sin\frac{\pi p x}{l}dx \\ &= \left\{\frac{1}{p}\cos\frac{\pi p a}{l} - \frac{1}{2\pi p}\left[\sin\frac{\pi p(2b-a)}{l} - \sin\frac{\pi p a}{l}\right]\right\}. \end{aligned}$$

When $a = 0$, the solution (7.31) follows.

Figure 7.23 shows the variation of the deflection of the first load-bearing layer of the considered rod in section x = 0.5 depending on the coordinate of the left edge of the load spot, length $a - b = 0.25$: *1*—rectangular load q_0, 2—sinusoidal load q'_0, *3*—sinusoidal thermal-force load.

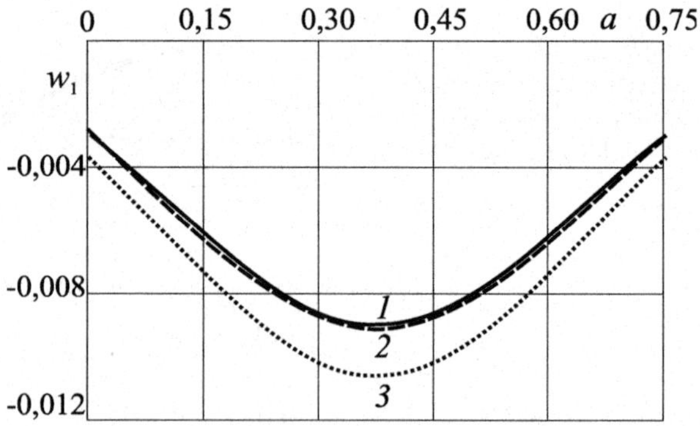

Fig. 7.23 Variation of deflection of the first load-bearing layer in cross-section x = 0.5

The Fig. 7.23 shows that the heat adds 17% to the deflection. The maximum is observed at $a = 0.37$, i.e., when the load is distributed symmetrically about the ends of the rod in the interval $\{0.37; 0.62\}$. The minimum is observed at the load spot adjacent to the rod ends. The small differences of the curves with the same resultant forces are explained by the small area of action of the loads, which levels out their differences.

A local distributed parabolic load $q(x)$ acts on the surface of the three-layer rod at Sect. $0 \le x \le b$, where $b \le l$ (Fig. 7.24). Its analytical form and coefficients of series expansion have the following representation

$$
\begin{aligned}
q(x) &= \frac{4q_0 x}{b}\left(1 - \frac{x}{b}\right)H_0(b - x),\\
q_m &= \frac{8lq_0}{b(\pi m)^2}\left[\frac{2l}{b\pi m}\left(1 - \cos\frac{\pi mb}{l}\right) - \sin\frac{\pi mb}{l}\right],
\end{aligned} \tag{7.34}
$$

where $H_0(x)$ is Heaviside function, q_0 is load amplitude.

The amplitude of the parabolic load q_0', whose equal force is statically equivalent to the load q_0 uniformly distributed on the same surface, will be as follows

$$
q_0' = q_0 b \Bigg/ \int_0^b \frac{4x}{b}\left(1 - \frac{x}{b}\right)dx = 1.5q_0.
$$

Numerical results. Variation of the magnitude of the filler compression strain along the axis of the rod at statically equivalent load values of different shapes is shown in Fig. 7.25: *1, 3*—parabolic load; *2, 4*—rectangular load (curves *3, 4* taking into account temperature). With increasing temperature, the compression decreases in both cases by 14–15%.

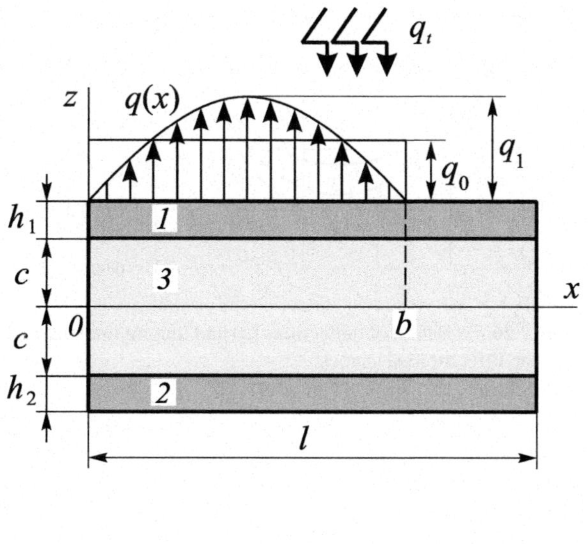

Fig. 7.24 Calculation scheme of parabolic loading

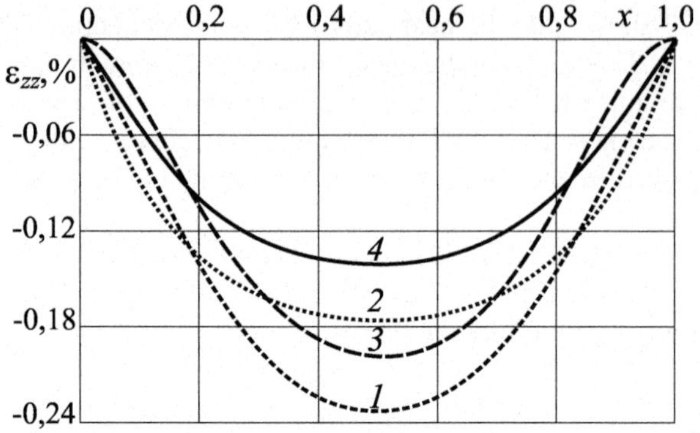

Fig. 7.25 Variation of filler compression along the rod axis

Figure 7.26 shows the variation of deflections (*a*) and longitudinal displacements (*b*) in the first layer along the axis of a three-layer rod under differently shaped loads distributed over the entire surface of the first layer ($b = 1$): *1*—rectangular load, *2*—parabolic load, *3*—parabolic load with temperature.

With the same equal force, the maximum displacements from a parabolic load are larger in magnitude by about 17% of the corresponding displacements from a uniformly distributed load. Temperature adds another 18% to the displacements.

If the local parabolic load is applied on the surface inside the rod section $a \leq x \leq b$, then

$$q(x) = \frac{4q_0}{(b-a)^2}(b-x)(x-a)[H_0(b-x) - H_0(a-x)]. \tag{7.35}$$

The load decomposition coefficients (7.35) into a series will be

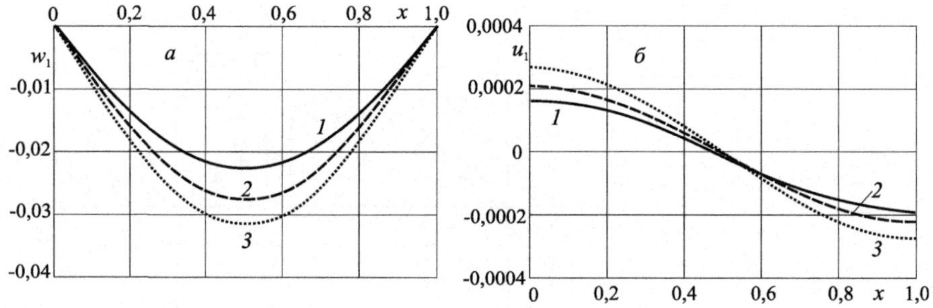

Fig. 7.26 Variation of deflections (*a*) and longitudinal displacements (*b*) in the first layer along the axis at different load shapes

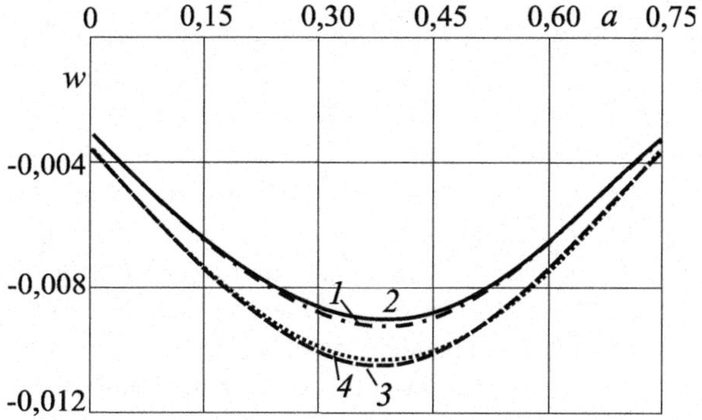

Fig. 7.27 Deflections of bearing layers of the rod in cross section x = 0.5 depending on the coordinate *a*

$$q_m = \frac{8q_0}{l(b-a)^2}\int_0^l (b-x)(x-a)[H_0(b-x) - H_0(a-x)]\sin\frac{\pi m x}{l}dx = q'_m q_0,$$

where

$$q'_m = \frac{8l}{(b-a)^2(\pi m)^3}\left\{2l\left(\cos\frac{\pi ma}{l} - \cos\frac{\pi mb}{l}\right)\right.$$
$$\left. + \pi m(a-b)\left(\sin\frac{\pi mb}{l} + \sin\frac{\pi ma}{l}\right)\right\}.$$

At $a = 0$, the coefficients (7.34) follow from these expressions.

Numerical results. Figure 7.27 shows the deflections of the bearing layers of the rod in the cross section x = 0.5 as a function of the coordinate a of the left edge of the parabolic load spot q'_0 ($a - b = 0.25$): *1*, *3* is first layer; *2*, *4* is second layer (curves *3*, *4*—taking into account temperature). The maxima are observed when the load is centrally located ($a = 0.37$).

If we compare displacements from statically equivalent sinusoidal and convex parabolic loads, they are almost identical. This indicates that they are equally dangerous from the point of view of the strength of structural elements.

7.5 Elastoplastic Rod Under Loading from the Natural State

Physical equations of state corresponding to the theory of small elastoplastic strains, taking into account temperature, are used to describe the strain in the layers of the rod:

$$s_i^{(k)} = 2G_k(1 - \omega^{(k)}(\varepsilon_u^{(k)}, T_k))\, э_i^{(k)}, s_{xz}^{(3)} = 2G_3(1 - \omega^{(k)}(\varepsilon_u^{(k)}, T_k))\, э_{xz}^{(3)},$$
$$\sigma^{(k)} = 3K_k(\varepsilon^{(k)} - \alpha_{0k}T_k)\ (i = x,\ z;\quad k = 1,\ 2\ ,\ 3), \tag{7.36}$$

where $s\prime_x^{(k)}$, $\sigma\prime^{(k)}$ and $э_{xx}^{\prime(k)}$, $\varepsilon^{\prime(k)}$ are deviator and spherical parts of stress and strain tensors of the kth layer; $s_{xz}^{(3)}$, $э_{xz}^{(3)}$ are tangential stress and shear strain in the filler; $G_k(T)$, $K_k(T)$ are temperature-dependent shear and volume deformation modules; α_{0k} is coefficient of linear thermal expansion of the material; $\varepsilon_u^{(k)}$ is strain intensity; $\varepsilon_y^{(k)}$ is deformation yield strength of the bearing layer material; $\omega^{(k)}\left(\varepsilon_u^{(k)}, T_k\right)$ is temperature-dependent nonlinearity function (1.38), (1.39) under loading from a natural state; we do not put a dash at the top to make expressions less cumbersome.

Based on the relations (7.36), let us distinguish elastic (index "e") and nonlinear (index "ω") components in the stress tensor, which will include temperature additives as well. For the rod under consideration we obtain:

- in the bearing layers ($k = 1{,}2$)

$$\sigma_x^{(k)} = \sigma_x^{(k)e} - \sigma_x^{(k)\omega}, \sigma_x^{(k)e} = 2G_k\, э_x^{(k)} + 3K_k\varepsilon^{(k)} = K_k^{+}\varepsilon_x^{(k)},$$
$$\sigma_x^{(k)\omega} = 2G_k\, э_x^{(k)}\, \omega^{(k)} + 3K_k\alpha_{0k}T_k = \frac{4}{3}G_k\varepsilon_x^{(k)}\omega^{(k)} + 3K_k\alpha_{0k}T_k;$$

- in filler

$$\sigma_x^{(3)} = \sigma_x^{(3)e} - \sigma_x^{(3)\omega}, \sigma_x^{(3)e} = 2G_3\, э_x^{(3)} + 3K_3\varepsilon^{(3)} = K_3^{+}\varepsilon_x^{(3)} + K_3^{-}\varepsilon_z^{(3)},$$
$$\sigma_x^{(3)\omega} = 2G_3\, э_x^{(3)}\, \omega^{(3)} + 3K_3\alpha_{03}T_3$$
$$= \frac{2}{3}G_3(2\varepsilon_x^{(3)} - \varepsilon_z^{(3)})\omega^{(3)} + 3K_3\alpha_{03}T_3,$$
$$\sigma_z^{(3)} = \sigma_z^{(3)e} - \sigma_z^{(3)\omega}, \sigma_z^{(3)e} = 2G_3\, э_z^{(3)} + 3K_3\varepsilon^{(3)} = K_3^{+}\varepsilon_z^{(3)} + K_3^{-}\varepsilon_x^{(3)},$$
$$\sigma_z^{(3)\omega} = 2G_3\, э_z^{(3)}\, \omega^{(3)} + 3K_3\alpha_{03}T_3$$
$$= \frac{2}{3}G_3(2\varepsilon_z^{(3)} - \varepsilon_x^{(3)})\omega^{(3)} + 3K_3\alpha_{03}T_3,$$
$$\sigma_{xz}^{(3)} = \sigma_{xz}^{(3)e} - \sigma_{xz}^{(3)\omega}, \sigma_{xz}^{(3)e} = 2G_3\, э_{xz}^{(3)} = 2G_3\varepsilon_{xz}^{(3)},$$
$$\sigma_{xz}^{(3)\omega} = 2G_3\, э_{xz}^{(3)}\, \omega^{(3)} = 2G_3\varepsilon_{xz}^{(3)}\omega^{(3)},$$
$$K_k^{+} = K_k + \frac{4}{3}G_k, K_k^{-} = K_k - \frac{2}{3}G_k. \tag{7.37}$$

Carrying out a similar operation (7.37) with the internal forces (7.4), we obtain:

$$N_x^{(k)} = N_x^{(k)e} - N_x^{(k)\omega}, M_x^{(k)} = M_x^{(k)e} - M_x^{(k)\omega}, (k = 1,\ 2,\ 3)$$
$$N_z^{(3)} = N_z^{(3)e} - N_z^{(3)\omega}, M_{xz}^{(3)} = M_{xz}^{(3)e} - M_{xz}^{(3)\omega}, Q^{(3)} = Q^{(3)e} - Q^{(3)\omega}. \tag{7.38}$$

Linear (index "e") and additional (index "ω") components in the internal forces (7.38) are calculated by the following formulas:

- in the bearing layers

$$N_x^{(k)e} = b_0 \int_{h_k} \sigma_x^{(k)e} \mathrm{d}z = b_0 \int_{h_k} K_k^+ \varepsilon_x^{(k)} \mathrm{d}z,$$

$$M_x^{(k)e} = b_0 \int_{h_k} \sigma_x^{(k)e} z \mathrm{d}z = b_0 \int_{h_k} K_k^+ \varepsilon_x^{(k)} z \mathrm{d}z,$$

$$N_x^{(k)\omega} = b_0 \int_{h_k} \sigma_x^{(k)\omega} \mathrm{d}z = \frac{4}{3} b_0 \int_{h_k} G_k \varepsilon_x^{(k)} \omega^{(k)} \mathrm{d}z + 3b_0 \int_{h_k} K_k \alpha_{0k} T_k \mathrm{d}z,$$

$$M_x^{(k)\omega} = b_0 \int_{h_k} \sigma_x^{(k)\omega} z \mathrm{d}z = \frac{4}{3} b_0 \int_{h_k} G_k \varepsilon_x^{(k)} \omega^{(k)} z \mathrm{d}z + 3b_0 \int_{h_k} K_k \alpha_{0k} T_k z \mathrm{d}z;$$

- in filler

$$N_x^{(3)e} = b_0 \int_{h_3} \sigma_x^{(3)e} \mathrm{d}z = b_0 \int_{h_3} \left(K_3^+ \varepsilon_x^{(3)} + K_3^- \varepsilon_z^{(3)}\right) \mathrm{d}z,$$

$$N_x^{(3)\omega} = b_0 \int_{h_3} \sigma_x^{(3)\omega} \mathrm{d}z = \frac{2}{3} b_0 \int_{h_3} G_3 (2\varepsilon_x^{(3)} - \varepsilon_z^{(3)}) \omega^{(3)} \mathrm{d}z$$

$$+ 3b_0 \int_{h_3} K_3 \alpha_{03} T_3 \mathrm{d}z,$$

$$N_z^{(3)e} = b_0 \int_{h_3} \sigma_z^{(3)e} \mathrm{d}z = b_0 \int_{h_3} \left(K_3^+ \varepsilon_z^{(3)} + K_3^- \varepsilon_x^{(3)}\right) \mathrm{d}z,$$

$$N_z^{(3)\omega} = b_0 \int_{h_3} \sigma_z^{(3)\omega} \mathrm{d}z = \frac{2}{3} b_0 \int_{h_3} G_3 (2\varepsilon_z^{(3)} - \varepsilon_x^{(3)}) \omega^{(3)} \mathrm{d}z$$

$$+ 3b_0 \int_{h_3} K_3 \alpha_{03} T_3 \mathrm{d}z,$$

$$M_x^{(3)e} = b_0 \int_{h_3} \sigma_x^{(3)0} z \mathrm{d}z = b_0 \int_{h_3} \left(K_3^+ \varepsilon_x^{(3)} + K_3^- \varepsilon_z^{(3)}\right) z \mathrm{d}z,$$

$$M_x^{(3)\omega} = b_0 \int_{h_3} \sigma_x^{(3)\omega} z \mathrm{d}z = \frac{2}{3} b_0 \int_{h_3} G_3 (2\varepsilon_x^{(3)} - \varepsilon_z^{(3)}) \omega^{(3)} z \mathrm{d}z$$

$$+ 3b_0 \int_{h_3} K_3 \alpha_{03} T_3 z \mathrm{d}z,$$

$$Q^{(3)e} = b_0 \int\limits_{h_3} \sigma_{xz}^{(3)e} \mathrm{d}z = 2b_0 \int\limits_{h_3} G_3 \varepsilon_{xz}^{(3)} \mathrm{d}z,$$

$$Q^{(3)\omega} = b_0 \int\limits_{h_3} \sigma_{xz}^{(3)\omega} \mathrm{d}z = 2b_0 \int\limits_{h_3} G_3 \varepsilon_{xz}^{(3)} \omega^{(3)} \mathrm{d}z.$$

$$M_{xz}^{(3)e} = b_0 \int\limits_{h_3} \sigma_{xz}^{(3)e} z\mathrm{d}z = 2b_0 \int\limits_{h_3} G_3 \varepsilon_{xz}^{(3)} z\mathrm{d}z,$$

$$M_{xz}^{(3)\omega} = b_0 \int\limits_{h_3} \sigma_{xz}^{(3)\omega} z\mathrm{d}z = 2b_0 \int\limits_{h_3} G_3 \varepsilon_{xz}^{(3)} \omega^{(3)} z\mathrm{d}z. \tag{7.39}$$

In efforts $Q^{(3)\omega}$, $M_{xz}^{(3)\omega}$, temperature components are absent.

The generalized forces introduced in (7.8) are also decomposed into linear and nonlinear components:

$$H_1 = H_1^e - H_1^\omega, H_1^e = \frac{2Q^{(3)e}}{c}, H_1^\omega = \frac{2Q^{(3)\omega}}{c},$$

$$H_2 = H_2^e - H_2^\omega, H_2^e = \frac{N_z^{(3)e}}{2c}, H_2^\omega = \frac{N_z^{(3)\omega}}{2c},$$

$$T_1 = T_1^e - T_1^\omega, T_1^e = \left(1 + \frac{h_1}{2c}\right)\frac{Q^{(3)e}}{2} + \frac{M_{xz}^{(3)e}}{2c},$$

$$T_1^\omega = \left(1 + \frac{h_1}{2c}\right)\frac{Q^{(3)\omega}}{2} + \frac{M_{xz}^{(3)\omega}}{2c},$$

$$T_2 = T_2^e - T_2^\omega, T_2^e = \left(1 + \frac{h_2}{2c}\right)\frac{Q^{(3)e}}{2} - \frac{M_{xz}^{(3)e}}{2c},$$

$$T_2^\omega = \left(1 + \frac{h_2}{2c}\right)\frac{Q^{(3)\omega}}{2} - \frac{M_{xz}^{(3)\omega}}{2c},$$

$$S_1 = S_1^e - S_1^\omega, S_1^e = \left(c + \frac{h_1}{2}\right)N_x^{(1)0}$$

$$- M_x^{(1)0} + \frac{h_1}{4}N_x^{(3)0} + \frac{h_1}{4c}M_x^{(3)0},$$

$$S_1^\omega = \left(c + \frac{h_1}{2}\right)N_x^{(1)\omega} - M_x^{(1)\omega} + \frac{h_1}{4}N_x^{(3)\omega} + \frac{h_1}{4c}M_x^{(3)\omega},$$

$$S_2 = S_2^e - S_2^\omega, S_2^e = -\left(c + \frac{h_2}{2}\right)N_x^{(2)e}$$

$$- M_x^{(2)e} - \frac{h_2}{4}N_x^{(3)e} + \frac{h_2}{4c}M_x^{(3)e},$$

$$S_2^\omega = \left(c + \frac{h_2}{2}\right)N_x^{(2)\omega} - M_x^{(2)\omega} + \frac{h_2}{4}N_x^{(2)\omega} + \frac{h_2}{4c}M_x^{(3)\omega},$$

$$
\begin{aligned}
& P_1 = P_1^e - P_1^\omega,\, P_1^e = \frac{N_x^{(3)e}}{2} + \frac{M_x^{(3)e}}{2c} + N_x^{(1)e},\\
& P_1^\omega = \frac{N_x^{(3)\omega}}{2} + \frac{M_x^{(3)\omega}}{2c} + N_x^{(1)\omega},\, P_2 = P_2^e - P_2^\omega,\\
& P_2^e = \frac{N_x^{(3)e}}{2} - \frac{M_x^{(3)e}}{2c} + N_x^{(2)e},\\
& P_2^\omega = \frac{N_x^{(3)\omega}}{2} - \frac{M_x^{(3)\omega}}{2c} + N_x^{(2)\omega}.
\end{aligned}
\tag{7.40}
$$

The linear (index "e") and additional (index "ω") components of internal forces included in (7.39), (7.40) are calculated through the corresponding stress components introduced by the formulas (7.37).

The system of equations of equilibrium in the efforts is reduced to the form coinciding with (7.9). Therefore, the corresponding system of nonlinear differential equations in displacements describing the deformation of an elastic–plastic rod in the temperature field will be similar to (7.18) with nonlinear additions in the right-hand side. For its solution the previously discussed method of elastic solutions is applied. Then in iterative form the system of equations in displacements will be:

$$
\begin{aligned}
& a_1 u_1^n - a_1 u_2^n - a_4 u_1^n{}_{,xx} - a_5 u_2^n{}_{,xx} + a_2 w_1^n{}_{,x}\\
& + a_3 w_2^n{}_{,x} - 2a_6 w_2^n{}_{,xxx} = p + p_\omega^{(n-1)},\\
& - a_1 u_1^n + a_1 u_2^n - a_5 u_1^n{}_{,xx} - a_9 u_2^n{}_{,xx} - a_{10} w_1^n{}_{,x}\\
& - a_{17} w_2^n{}_{,x} - a_6 w_1^n{}_{,xxx} + 2a_7 w_2^n{}_{,xxx} = h_\omega^{(n-1)},\\
& - a_2 u_1^n{}_{,x} + a_{10} u_2^n{}_{,x} + 2a_6 u_1^n{}_{,xxx} + a_6 u_2^n{}_{,xxx} + a_{11} w_1^n{}_{,xx} - a_{12} w_2^n{}_{,xx}\\
& + a_{15} w_1^n{}_{,xxxx} - a_{16} w_2^n{}_{,xxxx} + a_8 w_1^n - a_8 w_2^n = q + q_\omega^{(n-1)},\\
& - a_3 u_1^n{}_{,x} + a_{17} u_2^n{}_{,x} - a_7 u_2^n{}_{,xxx} - 2a_7 u_2^n{}_{,xxx} - a_{12} w_1^n{}_{,xx} + a_{14} w_2^n{}_{,xx}\\
& - a_{16} w_1^n{}_{,xxxx} + a_{13} w_2^n{}_{,xxxx} - a_8 w_1^n + a_8 w_2^n = g_\omega^{(n-1)}.
\end{aligned}
\tag{7.41}
$$

Here n is linear approximation number; coefficients $a_1, \ldots, a_{17}$ are determined by formulas (7.19); in additional loads $p_\omega^{(n-1)}$, $h_\omega^{(n-1)}$, $q_\omega^{(n-1)}$, $g_\omega^{(n-1)}$ at the first approximation step the nonlinear terms are assumed to be equal to zero ($\omega^{(k)} = 0$), and further they are calculated according to the results of the previous iteration:

$$
\begin{aligned}
& p_\omega^{(n-1)} = \frac{1}{b_0}\left(H_1^{\omega(n-1)} - P_1^{\omega(n-1)}{}_{,x}\right),\\
& h_\omega^{n-1} = \frac{1}{b_0}\left(H_1^{\omega(n-1)} + P_1^{\omega(n-1)}{}_{,x}\right),\\
& q_\omega^{(n-1)} = \frac{1}{b_0}\left(S_1^{\omega(n-1)}{}_{,xx} + H_2^{\omega(n-1)} - T_1^{\omega(n-1)}{}_{,x}\right),\\
& g_\omega^{(n-1)} = \frac{1}{b_0}\left(S_2^{\omega(n-1)}{}_{,xx} - H_2^{\omega(n-1)} - T_2^{\omega(n-1)}{}_{,x}\right),
\end{aligned}
\tag{7.42}
$$

where

$$
\begin{aligned}
H_1^{\omega(n-1)} &= \frac{2Q^{(3)\omega(n-1)}}{c}, H_2^{\omega(n-1)} = \frac{N_z^{(3)\omega(n-1)}}{2c},\\
T_1^{\omega(n-1)} &= \left(1+\frac{h_1}{2c}\right)\frac{Q^{(3)\omega(n-1)}}{c} + \frac{M_{xz}^{(3)\omega(n-1)}}{2c},\\
T_2^{\omega(n-1)} &= \left(1+\frac{h_2}{2c}\right)\frac{Q^{(3)\omega(n-1)}}{c} - \frac{M_{xz}^{(3)\omega(n-1)}}{2c},\\
S_1^{\omega(n-1)} &= \left(c+\frac{h_1}{2}\right)N_x^{(1)\omega(n-1)} - M_x^{(3)\omega(n-1)}\\
&+ \frac{h_1}{4}N_x^{(3)\omega(n-1)} + \frac{h_1}{4c}M_x^{(3)\omega(n-1)},\\
S_2^{\omega(n-1)} &= -\left(c+\frac{h_2}{2}\right)N_x^{(2)\omega(n-1)} - M_x^{(2)\omega(n-1)}\\
&- \frac{h_2}{4}N_x^{(3)\omega(n-1)} + \frac{h_2}{4c}M_x^{(3)\omega(n-1)},\\
P_1^{\omega(n-1)} &= \frac{N_x^{(3)\omega(n-1)}}{2} + \frac{M_x^{(3)\omega(n-1)}}{2c} + N_x^{(1)\omega(n-1)},\\
P_2^{\omega(n-1)} &= \frac{N_x^{(3)\omega(n-1)}}{2} - \frac{M_x^{(3)\omega(n-1)}}{2c} + N_x^{(2)\omega(n-1)},\\
N_x^{(k)\omega(n-1)} &= \frac{4}{3}b_0\int_{h_k} G_k\varepsilon_x^{(k)(n-1)}\omega^{(k)}(\varepsilon_u^{(k)(n-1)},\ T_k)\mathrm{d}z\\
&+ 3b_0\int_{h_k} K_k\alpha_{0k}T_k\mathrm{d}z,\\
M_x^{(k)\omega(n-1)} &= \frac{4}{3}b_0\int_{h_k} G_k\varepsilon_x^{(k)(n-1)}\omega^{(k)}(\varepsilon_u^{(k)(n-1)},\ T_k)z\mathrm{d}z\\
&+ 3b_0\int_{h_k} K_k\alpha_{0k}T_kz\mathrm{d}z,\\
N_x^{(3)\omega(n-1)} &= \frac{2}{3}b_0\int_{h_3} G_3(2\varepsilon_x^{(3)(n-1)} - \varepsilon_z^{(3)(n-1)})\omega^{(3)}(\varepsilon_u^{(3)(n-1)},\ T_k)\mathrm{d}z\\
&+ 3b_0\int_{h_3} K_3\alpha_{03}T_3\mathrm{d}z,\\
N_z^{(3)\omega(n-1)} &= \frac{2}{3}b_0\int_{h_3} G_3(2\varepsilon_z^{(3)(n-1)} - \varepsilon_x^{(3)(n-1)})\omega^{(3)}(\varepsilon_u^{(3)(n-1)},\ T_k)\mathrm{d}z\\
&+ 3b_0\int_{h_3} K_3\alpha_{03}T_3\mathrm{d}z,
\end{aligned}
$$

$$M_x^{(3)\omega(n-1)} = \frac{2}{3} b_0 \int\limits_{h_3} G_3 (2\varepsilon_x^{(3)(n-1)} - \varepsilon_z^{(3)(n-1)}) \omega^{(3)} (\varepsilon_u^{(3)(n-1)}, T_k) z \mathrm{d}z$$

$$+ 3b_0 \int\limits_{h_3} K_3 \alpha_{03} T_3 z \mathrm{d}z,$$

$$Q^{(3)\omega(n-1)} = 2b_0 \int\limits_{h_3} G_3 \varepsilon_{xz}^{(3)(n-1)} \omega^{(3)} (\varepsilon_u^{(3)(n-1)}, T_3) \mathrm{d}z \; (k = 1, 2).$$

The application of the formulated method of thermoelastic solutions makes it possible at each approximation step to reduce the nonlinear boundary value problem of thermoelastic-plastic bending of the rod under consideration to the corresponding linear thermoelastic problem (7.41) with additional loads (7.42). The first approximation will be an analytical or numerical solution of the thermoelasticity problem (7.21), (7.24) for the rod under consideration.

As boundary conditions we take kinematic conditions of free supporting (7.20) at the ends on rigid supports fixed in space. Then in the end sections x = 0, l (l is the rod length) at each iteration step n the following requirements with respect to the displacements in the bearing layers must be fulfilled ($k = 1,\ 2$):

$$w_k^n = u_{k},_x^n = w_{k},_{xx}^n = 0. \tag{7.43}$$

The solution of the system of differential Eq. (7.41) is taken as a decomposition into trigonometric series, which automatically satisfy the boundary conditions (7.43):

$$\begin{gathered} u_1^n = \sum_{m=1}^{\infty} U_{1m}^n \cos\left(\frac{\pi m x}{l}\right),\ u_2^n = \sum_{m=1}^{\infty} U_{2m}^n \cos\left(\frac{\pi m x}{l}\right), \\ w_1^n = \sum_{m=1}^{\infty} W_{1m}^n \sin\left(\frac{\pi m x}{l}\right),\ w_2^n = \sum_{m=1}^{\infty} W_{2m}^n \sin\left(\frac{\pi m x}{l}\right), \end{gathered} \tag{7.44}$$

where U_{1m}^n, U_{2m}^n, W_{1m}^n, W_{2m}^n are unknown displacement amplitudes.

The transverse load and additional forces in the rod layers are also represented as decompositions into trigonometric series:

$$q = \sum_{m=1}^{\infty} q_m \sin\left(\frac{\pi m x}{l}\right),\ q_m = \frac{2}{l} \int\limits_0^l q(x) \sin\left(\frac{\pi m x}{l}\right) \mathrm{d}x,$$

$$p = \sum_{m=1}^{\infty} p_m \sin\left(\frac{\pi m x}{l}\right),\ p_m = \frac{2}{l} \int\limits_0^l p(x) \sin\left(\frac{\pi m x}{l}\right) \mathrm{d}x,$$

$$q_\omega^{n-1} = \sum_{m=1}^{\infty} q_{\omega m}^{n-1} \sin\left(\frac{\pi m x}{l}\right), \; q_{\omega m}^{n-1} = \frac{2}{l}\int_0^l q_\omega^{n-1}(x)\sin\left(\frac{\pi m x}{l}\right)\mathrm{d}x,$$

$$g_\omega^{n-1} = \sum_{m=1}^{\infty} g_{\omega m}^{n-1} \sin\left(\frac{\pi m x}{l}\right), \; g_{\omega m}^{n-1} = \frac{2}{l}\int_0^l g_\omega^{n-1}(x)\sin\left(\frac{\pi m x}{l}\right)\mathrm{d}x,$$

$$h_\omega^{n-1} = \sum_{m=1}^{\infty} h_{\omega m}^{n-1} \cos\left(\frac{\pi m x}{l}\right), \; h_{\omega m}^{n-1} = \frac{2}{l}\int_0^l h_\omega^{n-1}(x)\cos\left(\frac{\pi m x}{l}\right)\mathrm{d}x,$$

$$p_\omega^{n-1} = \sum_{m=1}^{\infty} p_{\omega m}^{n-1} \cos\left(\frac{\pi m x}{l}\right), \; p_{\omega m}^{n-1} = \frac{2}{l}\int_0^l p_\omega^{n-1}(x)\cos\left(\frac{\pi m x}{l}\right)\mathrm{d}x. \quad (7.45)$$

After substituting the displacements (7.44) and additional forces (7.45) into the equations of equilibrium (7.32), we obtain a system of linear algebraic equations to determine the required amplitudes of displacements U_{1m}^n, U_{2m}^n, W_{1m}^n, W_{2m}^n:

$$\begin{cases} b_1 U_{1m}^n + b_2 U_{2m}^n + b_3 W_{1m}^n + b_4 W_{2m}^n = p_m + p_{\omega m}^{n-1}, \\ b_2 U_{1m}^n + b_5 U_{2m}^n + b_6 W_{1m}^n - b_7 W_{2m}^n = h_{\omega m}^{n-1}, \\ b_3 U_{1m}^n + b_6 U_{2m}^n + b_8 W_{1m}^n + b_9 W_{2m}^n = q_m + q_{\omega m}^{n-1}, \\ b_4 U_{1m}^n - b_7 U_{2m}^n + b_9 W_{1m}^n + b_{10} W_{2m}^n = g_{\omega m}^{n-1}. \end{cases} \quad (7.46)$$

Here the b_i coefficients are calculated by the formulas introduced in (7.23).

The solution of the system (7.46) can be written out in determinants

$$U_{1m}^n = \frac{\Delta_{1m}^{n-1}}{\Delta}, \; U_{2m}^n = \frac{\Delta_{2m}^{n-1}}{\Delta}, \; W_{1m}^n = \frac{\Delta_{3m}^{n-1}}{\Delta}, \; W_{2m}^n = \frac{\Delta_{4m}^{n-1}}{\Delta}, \quad (7.47)$$

where

$$\Delta = \begin{vmatrix} b_1 & b_2 & b_3 & b_4 \\ b_2 & b_5 & b_6 & -b_7 \\ b_3 & b_6 & b_8 & b_9 \\ b_4 & -b_7 & b_9 & b_{10} \end{vmatrix}, \; \Delta_{1m}^{n-1} = \begin{vmatrix} p_m + p_{\omega m}^{n-1} & b_2 & b_3 & b_4 \\ h_{\omega m}^{n-1} & b_5 & b_6 & -b_7 \\ q_m + q_{\omega m}^{n-1} & b_6 & b_8 & b_9 \\ g_{\omega m}^{n-1} & -b_7 & b_9 & b_{10} \end{vmatrix},$$

$$\Delta_{2m}^{n-1} = \begin{vmatrix} b_1 & p_m + p_{\omega m}^{n-1} & b_3 & b_4 \\ b_2 & h_{\omega m}^{n-1} & b_6 & -b_7 \\ b_3 & q_m + q_{\omega m}^{n-1} & b_{10} & b_{11} \\ b_4 & g_{\omega m}^{n-1} & b_{11} & b_{14} \end{vmatrix},$$

$$\Delta_{3m}^{n-1} = \begin{vmatrix} b_1 & b_2 & p_m + p_{\omega m}^{n-1} & b_4 \\ b_2 & b_5 & h_{\omega m}^{n-1} & -b_7 \\ b_3 & b_6 & q_m + q_{\omega m}^{n-1} & b_{11} \\ b_4 & -b_7 & g_{\omega m}^{n-1} & b_{14} \end{vmatrix}, \; \Delta_{4m}^{n-1} = \begin{vmatrix} b_1 & b_2 & b_3 & p_m + p_{\omega m}^{n-1} \\ b_2 & b_5 & b_6 & h_{\omega m}^{n-1} \\ b_3 & b_6 & b_8 & q_m + q_{\omega m}^{n-1} \\ b_4 & -b_7 & b_9 & g_{\omega m}^{n-1} \end{vmatrix},$$

Calculating the determinants in (7.47) by any of the standard methods, we obtain the amplitudes U_{1m}^n, U_{2m}^n, W_{1m}^n, W_{2m}^n. Then the required functions are calculated using the formulas (7.44). Displacements in bearing layers and filler follow from relations (7.1), strains—from (7.2), and stresses—from (7.37).

Thus, formulas (7.44) taking into account (7.47) and (7.42) give in iterations the analytical solution of the problem on deformation of a three-layer elastic–plastic rod with a compressible filler in the temperature field.

Uniformly distributed load. Let a uniformly distributed transverse surface load act on an elastoplastic rod in the temperature field $q(x) = \text{const}$. In this case, the deformation of the rod is described by a system of equilibrium equations in iterative form (7.41). Its solution (7.44) will take into account the load constancy, temperature field and physical nonlinearity of the layer materials by introducing into the system of equations for the required displacement amplitudes U_{1m}, U_{2m}, W_{1m}, W_{2m} (7.46) the load and temperature decomposition coefficients and nonlinear additions (7.45).

Numerical results for an elastic–plastic rod are obtained at the intensity of uniformly distributed load $q_0 = -5{\cdot}\text{MPa}$ ($p = 0$). Relative thicknesses of layers were taken as: $h_1 = 0.04$, $h_2 = 0.02$, $h_3 = 0.18$. The thermal state of the rod is assumed to be the same as in Sect. 7.3. The temperature distribution over the thickness of the filler and the first layer is shown in Fig. 7.2.

The convergence of the elastic solution process is shown in Fig. 7.28: *a*—deflection of the first layer, *b*—relative shear in the filler material. The curve number corresponds to the iteration number. Here the 5th approximation is taken as the solution, it differs from the previous one by less than 1%.

Figure 7.29 shows the variation along the axis of the rod: *a*—deflection in the first layer, *b*—shear in the filler; 1—elastic behavior of the rod, 2—elastic–plastic behavior of the rod, 3—thermoelastic behavior of the, 4—thermoelastic-plastic behavior of the rod. Temperature influence increases elastic displacements by 12–15%. Taking into account the non-linearity, this difference reaches 26–30%, which is due to the influence of temperature on the mechanical properties of materials.

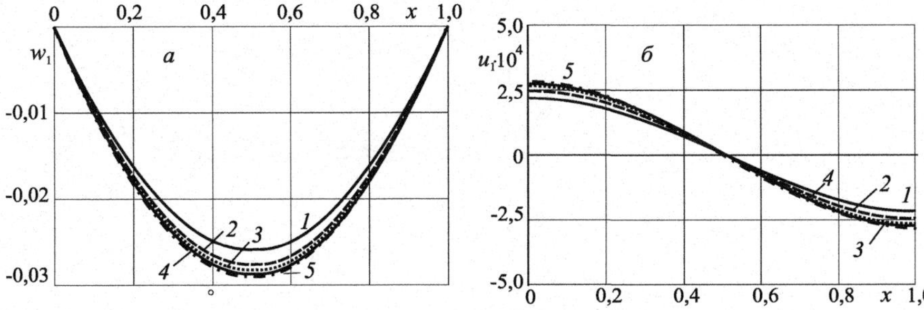

Fig. 7.28 Convergence of the elastic solution process: *a*—deflection of the first layer, *b*—relative shear in the filler material

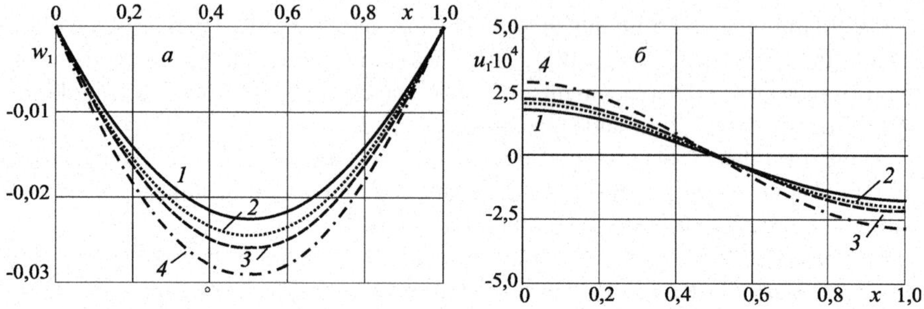

Fig. 7.29 Variation of deflection w_1 (a) and longitudinal displacement u_1 (b) along the rod axis

The distribution of normal stresses σ_{xx} over the median cross-section ($x = 0.5$) is illustrated in Fig. 7.30: 1—elastic rod, 2—thermo-elastic rod, 3—thermo-elastic–plastic rod. Due to the volumetric thermal expansion of the material constrained by the supports, the stresses in the first layer of the rod are shifted to the negative region. Taking into account the plasticity of the material increases the thermoelastic stresses in the bond with the filler, but reduces such stresses on the external surface. The nonlinearity of the materials in the filler and the second layer has little effect on the stresses. Stress values in the outer layers are reduced by a factor of 10^8 and in the filler—by a factor of 10^7.

Figure 7.31 shows the distribution of normal stresses $\sigma_{xx}^{(1)}$ along the boundary planes of the first layer: *1, 3, 5* – $z = c$; *2, 4, 6* – $z = c + h_1$ (*1, 4*—elastic rod; *2, 5*—thermoelastic rod; *3, 6*—thermoelastic-plastic rod). On the external plane the stresses are negative, the temperature shifts them further into the compression region, taking plasticity into account changes their magnitude by 10%. The effect of plasticity is much weaker in filler bonding.

The stress variation along the rod axis in the second bearing layer is shown in Fig. 7.32: *1, 3, 5* – $z = -c-h_2$; *2, 4, 6* – $z = -c$ (*1, 4*—elastic rod; *2, 5*—thermoelastic rod; *3, 6*—thermoelastic-plastic rod). In this case the influence of temperature on

Fig. 7.30 Distribution of normal stresses σ_{xx} across the medial cross section ($x = 0.5$)

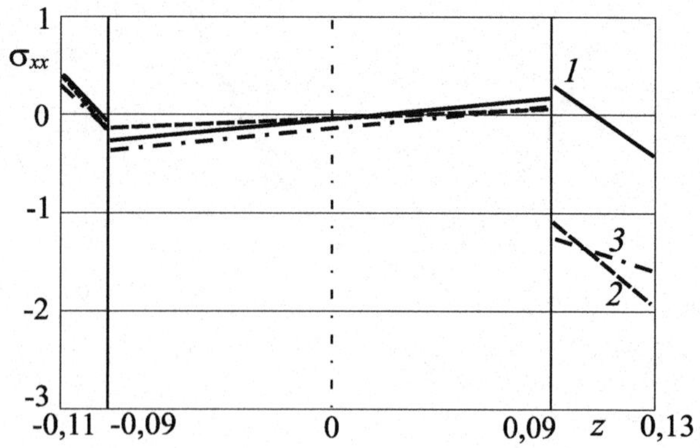

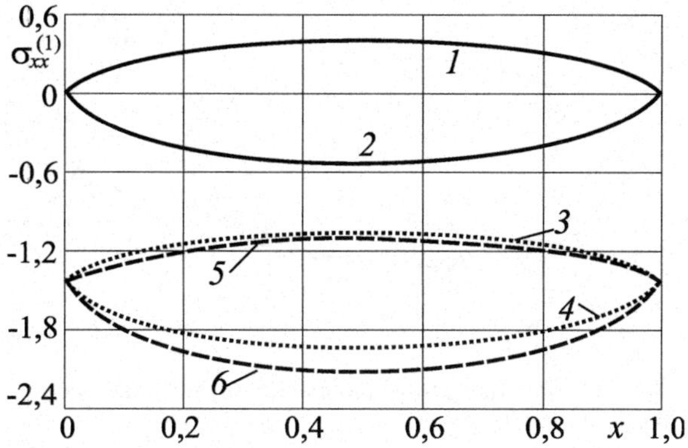

Fig. 7.31 Distribution of normal stresses $\sigma_{xx}^{(1)}$ along the boundary planes of the first layer

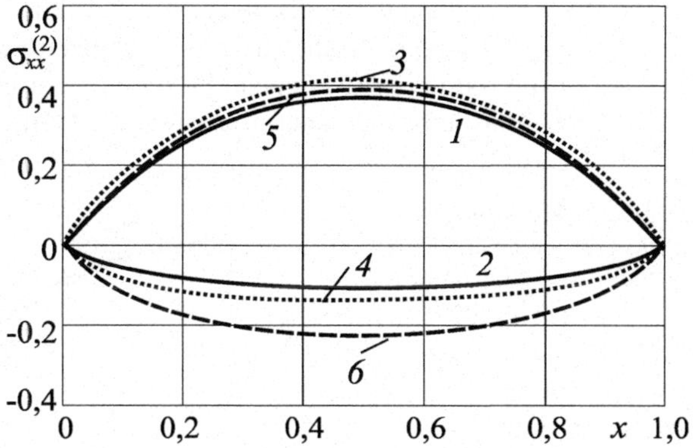

Fig. 7.32 Distribution of normal stresses $\sigma_{xx}^{(2)}$ along the boundary planes of the second layer

stresses is practically not felt, taking into account plasticity gives results similar to the previous ones.

The variation of longitudinal $\sigma_{xx}^{(3)}$ (*a*) and transverse $\sigma_{zz}^{(3)}$ (*b*) normal stresses at the boundary planes of the filler is illustrated in Fig. 7.33: *1, 3, 5* – $z = -c$; *2, 4, 6* – $z = c$ (*1, 4*—elastic rod; *2, 5*—thermoelastic rod; *3, 6*—thermoelastic-plastic rod). The corresponding graphs of both stresses are quite similar. Temperature and physical nonlinearity have a particularly strong influence on the stress variation in the bonding with the first layer.

Local rectangular load. Let a transverse surface load $q(x)$ uniformly distributed in the interval $a \leq x \leq b$ act on an elastoplastic rod in the temperature field (see Fig. 7.7). The analytical form of the load and the series expansion coefficients are given by expressions (7.28). The deformation of the rod is described by a system of equations of equilibrium in iterative form (7.41). The solution (7.44) of this system will take into account the load locality, temperature field and physical nonlinearity of the layer materials by introducing into the system of equations for the required displacement amplitudes U_{1m}, U_{2m}, W_{1m}, W_{2m} (7.46) coefficients (7.37) and nonlinear additions (7.45).

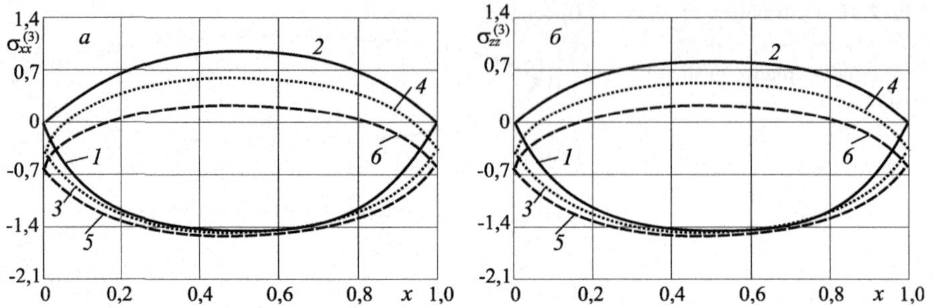

Fig. 7.33 Variation of longitudinal $\sigma_{xx}^{(3)}$ (*a*) and transverse $\sigma_{zz}^{(3)}$ (*b*) normal stresses at the boundary planes of the filler material

Numerical results are obtained at $q_0 = -5 \cdot \text{MPa}$.

Figures 7.34, 7.35 show the variation of deflections of the first layer in the cross-section of the rod ($x = 0.5$) depending on the coordinates of the edges of the load interval: *1*—elastic rod, *2*—thermoelastic rod, *3*—thermoelastic-plastic rod.

Fig. 7.34 Variation of deflections of the first layer in the cross section of the rod ($x = 0.5$) as a function of coordinate *b*

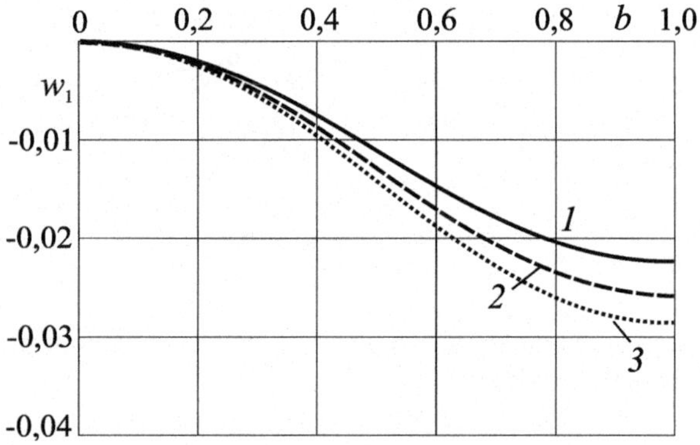

Fig. 7.35 Variation of deflections of the first layer in the cross section of the rod ($x = 0.5$) as a function of coordinate *a*

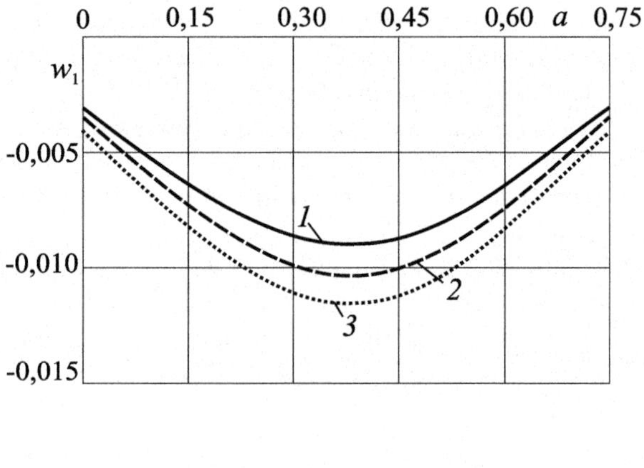

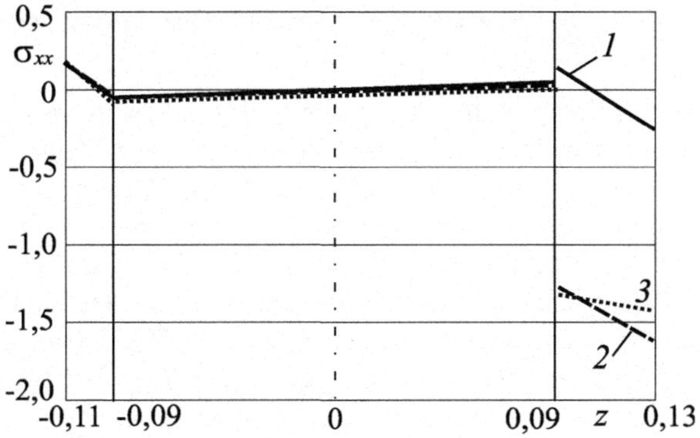

Fig. 7.36 Variation of longitudinal stresses σ_{xx} along the height of the cross section ($x = 0.5$)

Figure 7.34 shows the variation of deflections along the axis of the rod as a function of the length of the load interval b. The maximum, as before, is observed at the load distributed over the whole rod ($b = 1$). The temperature component of the deflection is 16% and the nonlinearity adds 12% to the maximum thermoelastic deflection.

Figure 7.35 shows the deflections of the load-bearing layers of the rod in the cross-section ($x = 0.5$) as a function of the coordinate a of the left edge of the load interval at its constant length ($b - a = 0.25$). As the load moves from left to right, the deflections increase, reach a maximum at $a = 0.37$, i.e., when the load is distributed symmetrically about the ends of the rod in the interval $\{0.37; 0.62\}$, and then decrease. The temperature component of the deflections is 23%, non-linearity adds 8% to the thermoelastic deflection.

The graphs of longitudinal stresses σ_{xx} in the cross section ($x = 0.5$) of the considered rod are shown in Fig. 7.36: *1*—elastic rod, *2*—thermoelastic rod, *3*—thermoelastic-plastic rod.

The load is distributed over a section $7.0,7.5 \leq x \leq 0.75$. The stresses in the first layer are shifted to the negative region due to the thermal expansion of the material. In the second layer and the filler, the temperature has little effect on the magnitude and sign of the stresses. The maxima are observed on the external planes. Stress values in bearing layers are related to the surface load intensity $q_1 = 10^9$ Pa, in the filler—to $q_3 = 10^8$ Pa.

In case of other local loads (concentrated forces and moments, sinusoidal, parabolic) the solution will take them into account by entering into the system of equations for the required displacement amplitudes (7.46) the corresponding coefficients of decomposition of these loads into series.

7.6 Alternating Loading in the Temperature Field

Let, starting from the moment $t = t_1$, there is an instantaneous unloading and reloading with forces of opposite sign p", q", changing according to the same law as at loading from the natural state. These forces will create in the kth layer of the rod a field of displacements $u_1''(x)$, $w_1''(x)$, $u_2''(x)$, $w_2''(x)$, strains $\varepsilon_x''^{(k)}(x,z)$, $\varepsilon_z''^{(3)}(x,z)$, $\varepsilon_{xz}''^{(3)}(x,z)$ and stresses $\sigma_x''^{(k)}(x,z)$, $\sigma_z''^{(3)}(x,z)$, $\sigma_{xz}''^{(3)}(x,z)$. In this case, we will assume that during unloading and subsequent alternating loading the temperature in all points of the body remains unchanged. It coincides with the temperature field at the moment of the beginning of unloading, i.e. $T'(z) = T(z,t_1)$ and the elastic modules of the materials of the layers have taken the values $G_k(z) \equiv G_k(T'(z)), K_k(z) = K_k(T'(z))$, and the corresponding loading trajectories still belong to the class of simple ones.

As before, for all values of the stress–strain state and load we keep the differences, in which the values with one stroke are stresses, strains and displacements in the rod before unloading, two strokes mark similar parameters during the second half cycle:

$$
\begin{aligned}
&\sigma_x^{(k)*} = \sigma_x'^{(k)} - \sigma_x''^{(k)}, \sigma_z^{(3)*} = \sigma_z'^{(3)} - \sigma_z''^{(3)}, \sigma_{xz}^{(3)*} = \sigma_{xz}'^{(3)} - \sigma_{zx}''^{(3)},\\
&\varepsilon_x^{(3)*} = \varepsilon_x'^{(3)} - \varepsilon\sigma_x''^{(k)}, \varepsilon_z^{(3)*} = \varepsilon_z'^{(3)} - \varepsilon\sigma_z''^{(3)}, \varepsilon_{xz}^{(3)*} = \varepsilon_{xz}'^{(3)} - \varepsilon\sigma_{xz}''^{(3)},\\
&q^* = q' - q'', p^* = p' - p''(k = 1,2,3),\\
&u_1^* = u_1' - u_1'', w_1^* = w_1' - w_1'', u_2^* = u_2' - u_2'', w_2^* = w_2' - w_2''.
\end{aligned}
\tag{7.48}
$$

Due to the linearity of the relationship between strains and displacements in the layers of the rod (7.2), similar relations will be valid for the corresponding values with asterisks:

$$
\begin{aligned}
\varepsilon_x^{(1)*} &= u_{1,x}^* \left(z - c - \frac{h_1}{4}\right) w_{1,xx}^*, \varepsilon_{xz}^{(1)*} = 0 (c \le z \le c + h_1),\\
\varepsilon_x^{(3)*} &= \left(1 + \frac{z}{c}\right)\left(\frac{1}{2}u_{1,x}^* + \frac{h_1}{4}w_{1,xx}^*,\right) + \left(1 - \frac{z}{c}\right)\left(\frac{1}{2}u_{2,x}^* - \frac{h_2}{4}w_{2,xx}^*,\right),\\
\varepsilon_{xz}^{(3)*} &= \left(\frac{2z + h_1}{8c} + \frac{1}{4}\right)w_{1,x}^* + \left(\frac{-2z + h_2}{8c} + \frac{1}{4}\right)w_{2,x}^* + \frac{u_1^* - u_2^*}{4c},\\
\varepsilon_z^{(3)*} &= \frac{1}{2c}(w_1 - w_2)(-c \le z \le c),\\
\varepsilon_x^{(2)*} &= u_{2,x}^* \left(z + c + \frac{h_2}{4}\right) w_{2,xx}^*,\\
\varepsilon_{xz}^{(2)*} &= 0(-c - h_2 \le z \le -c).
\end{aligned}
\tag{7.49}
$$

For stresses and strains (7.48), (7.49), marked with an asterisk, let us assume physical equations of state of type (7.36):

$$
\begin{aligned}
s_i^{(k)*} &= 2G_k(1-\omega^{(k)*}(\varepsilon_u^{(k)*},\ T_k))\ э_i^{(k)*},\\
s_{xz}^{(3)*} &= 2G_3(1-\omega^{(k)*}(\varepsilon_u^{(k)*},\ T_k))\ э_i^{(3)*},\\
\sigma^{(k)*} &= 3K_k(\omega^{(k)}-\alpha_{0k}T_k)\ (i=x,z;\ k=1,2,3).
\end{aligned}
\tag{7.50}
$$

According to Moskvitin's theory, we assume that universal nonlinearity functions $\omega^{(k)*}$ are expressed through plasticity functions under loading from the natural state by relations (4.34), (4.35). Let us represent the components of stresses and strains with asterisks, using relations (7.50), in the following form:

- in bearing layers ($k = 1, 2$)

$$
\begin{aligned}
&\sigma_x^{(k)*} = \sigma_x^{(k)e*} - \sigma_x^{(k)\omega*},\ \sigma_x^{(k)e*} = 2G_k\ э_x^{(k)*} + 3K_k\varepsilon^{(k)*} = K_k^+\varepsilon_x^{(k)*},\\
&\sigma_x^{(k)\omega*} = 2G_k\ э_x^{(k)*}\ \omega^{(k)*} = \frac{4}{3}G_k\varepsilon_x^{(k)*}\omega^{(k)*};
\end{aligned}
$$

- in filler

$$
\begin{aligned}
&\sigma_x^{(3)*} = \sigma_x^{(3)e*} - \sigma_x^{(3)\omega*},\ \sigma_x^{(3)e*} = 2G_3\ э_x^{(3)*} + 3K_3\varepsilon^{(3)*} = K_3^+\varepsilon_x^{(3)*} + K_3^-\varepsilon_z^{(3)*},\\
&\sigma_x^{(3)\omega*} = 2G_3\ э_x^{(3)*}\ \omega^{(3)*} = \frac{2}{3}G_3(2\varepsilon_x^{(3)*} - \varepsilon_z^{(3)*})\omega^{(3)*},\\
&\sigma_z^{(3)*} = \sigma_z^{(3)e*} - \sigma_z^{(3)\omega*},\ \sigma_z^{(3)e*}\\
&\quad = 2G_3\ э_z^{(3)*} + 3K_3\varepsilon^{(3)*} = K_3^+\varepsilon_z^{(3)*} + K_3^-\varepsilon_x^{(3)*},\\
&\sigma_z^{(3)\omega*} = 2G_3\ э_z^{(3)*}\ \omega^{(3)*} = \frac{2}{3}G_3(2\varepsilon_z^{(3)*} - \varepsilon_x^{(3)*})\omega^{(3)*},\\
&\sigma_{xz}^{(3)*} = \sigma_{xz}^{(3)e*} - \sigma_{xz}^{(3)\omega*},\ \sigma_{xz}^{(3)e*} = 2G_3\ э_{xz}^{(3)*} = 2G_3\varepsilon_{xz}^{(3)*},\\
&\sigma_{xz}^{(3)\omega*} = 2G_3\ э_{xz}^{(3)*}\ \omega^{(3)*} = 2G_3\varepsilon_{xz}^{(3)*}\omega^{(3)*}.
\end{aligned}
\tag{7.51}
$$

Performing the similar operation (7.51) with the internal forces, we obtain:

$$
\begin{aligned}
&N_x^{(k)*} = N_x^{(k)e*} - N_x^{(k)\omega*},\ M_x^{(k)*} = M_x^{(k)e*} - M_x^{(k)\omega*},\ (k=1,2,3)\\
&N_z^{(3)*} = N_z^{(3)e*} - N_z^{(3)\omega*},\ M_{xz}^{(3)*} = M_{xz}^{(3)e*} - M_{xz}^{(3)\omega*},\ Q^{(3)*} = Q^{(3)e*} - Q^{(3)\omega*}.
\end{aligned}
\tag{7.52}
$$

Linear (index "e") and additional (index "ω") components in the internal forces (7.52) are calculated by the following formulas:

- in bearing layers ($k = 1, 2$)

$$
\begin{aligned}
N_x^{(k)e^*} &= b_0\int_{b_k}\sigma_x^{(k)e^*}\mathrm{d}z = b_0\int_{b_k}K_k^+\varepsilon_x^{(k)^*}\mathrm{d}z,\\
M_x^{(k)e^*} &= b_0\int_{h_k}\sigma_x^{(k)e^*}z\mathrm{d}z = b_0\int_{h_k}K_k^+\varepsilon_x^{(k)^*}z\mathrm{d}z,
\end{aligned}
$$

$$N_x^{(k)\omega^*} = b_0 \int_{h_k} \sigma_x^{(k)\omega^*} \mathrm{d}z = \frac{4}{3} b_0 \int_{h_k} G_k \varepsilon_x^{(k)^*} \omega^{(k)^*} \mathrm{d}z,$$

$$M_x^{(k)\omega^*} = b_0 \int_{h_k} \sigma_x^{(k)\omega^*} z\mathrm{d}z = \frac{4}{3} b_0 \int_{h_k} G_k \varepsilon_x^{(k)^*} \omega^{(k)^*} z\mathrm{d}z;$$

· in filler

$$\begin{aligned}
N_x^{(3)e*} &= b_0 \int\limits_{h_3} \sigma_x^{(3)e*} \mathrm{d}z = b_0 \int\limits_{h_3} \left(K_3^+ \varepsilon_x^{(3)*} + K_3^- \varepsilon_x^{(3)*}\right) \mathrm{d}z, \\
N_x^{(3)\omega*} &= b_0 \int\limits_{h_3} \sigma_x^{(3)\omega*} \mathrm{d}z = \frac{2}{3} b_0 \int\limits_{h_3} G_3 \left(2\varepsilon_x^{(3)*} - \varepsilon_z^{(3)*}\right) \omega^{(3)*} \mathrm{d}z, \\
N_z^{(3)e*} &= b_0 \int\limits_{h_3} \sigma_z^{(3)e*} \mathrm{d}z = b_0 \int\limits_{h_3} \left(K_3^+ \varepsilon_z^{(3)*} + K_3^- \varepsilon_z^{(3)*}\right) \mathrm{d}z, \\
M_z^{(3)\omega*} &= b_0 \int\limits_{h_3} \sigma_z^{(3)\omega*} \mathrm{d}z = \frac{2}{3} b_0 \int\limits_{h_3} G_3 \left(2\varepsilon_z^{(3)*} - \varepsilon_x^{(3)*}\right) \omega^{(3)*} \mathrm{d}z, \\
M_x^{(3)e*} &= b_0 \int\limits_{h_3} \sigma_x^{(3)e*} z\mathrm{d}z = b_0 \int\limits_{h_3} \left(K_3^+ \varepsilon_x^{(3)*} + K_3^- \varepsilon_z^{(3)*}\right) z\mathrm{d}z, \\
M_x^{(3)\omega*} &= b_0 \int\limits_{h_3} \sigma_x^{(3)\omega*} z\mathrm{d}z = \frac{2}{3} b_0 \int\limits_{h_3} G_3 \left(2\varepsilon_x^{(3)*} - \varepsilon_z^{(3)*}\right) \omega^{(3)*} z\mathrm{d}z, \\
Q^{(3)\omega*} &= b_0 \int\limits_{h_3} \sigma_{xz}^{(3)\omega*} \mathrm{d}z = 2b_0 \int\limits_{h_3} G_3 \varepsilon_{xz}^{(3)*} \omega^{(3)*} \mathrm{d}z, \\
M_{xz}^{(3)e*} &= b_0 \int\limits_{h_3} \sigma_{xz}^{(3)e*} z\mathrm{d}z = 2b_0 \int\limits_{h_3} G_3 \varepsilon_{xz}^{(3)*} z\mathrm{d}z, \\
M_{xz}^{(3)\omega*} &= b_0 \int\limits_{h_3} \sigma_{xz}^{(3)\omega*} z\mathrm{d}z = 2b_0 \int\limits_{h_3} G_3 \varepsilon_{xz}^{(3)*} \omega^{(3)*} z\mathrm{d}z.
\end{aligned} \tag{7.53}$$

Using the internal forces (7.53), we obtain a system of nonlinear differential equations of equilibrium for displacements with asterisks, which will be similar to the system (7.41). Applying for the solution the elastic solution method considered earlier, let us write it out in the following iterative form:

$$\begin{aligned}
& a_1 u_1^{*n} - a_1 u_2^{*n} - a_4 u_1^{*n},_{xx} - a_5 u_2^{*n},_{xx} + a_2 w_1^{*n},_x \\
& + a_3 w_2^{*n},_x - 2a_6 w_1^{*n},_{xxx} + a_7 w_2^{*n}, x_{xx} = p_\omega^{*(n-1)}, \\
& - a_1 u_1^{*n} + a_1 u_2^{*n} - a_5 u_1^{*n},_{xx} - a_9 u_2^{*n},_{xx} - a_{10} w_1^{*n},_x \\
& - a_{17} w_2^{*n},_x - a_6 w_1^{*n},_{xxx} + 2a_7 w_2^{*n},_{xxx} = h_\omega^{*(n-1)},
\end{aligned}$$

$$
\begin{aligned}
&- a_2 u_1^{*n},_x + a_{10} u_2^{*n},_x + 2a_6 u_1^{*n},_{xx} + a_6 u_2^{*n},_{xxx} + a_{11} w_1^{*n},_{xx} \\
&- a_{12} w_2^{*n},_{xx} + a_{15} w_1^{*n},_{xxxx} - a_{16} w_2^{*n},_{xxxx} + a_8 w_1^{*n} - a_8 w_2^{*n} = q + q_\omega^{*(n-1)}, \\
&- a_3 u_1^{*n},_x + a_{17} u_2^{*n},_x - a_7 u_1,_{xxx} - 2a_7 u_2^{*n},_{xxx} - a_{12} w_1^{*n},_{xx} \\
&+ a_{14} w_2^{*n},_{xx} - a_{16} w_1^{*n},_{xxxx} + a_{13} w_2^{*n},_{xxxx} - a_8 w_1^{*n} + a_8 w_2^{*n} = g_\omega^{*(n-1)}.
\end{aligned}
\tag{7.54}
$$

Here n is the approximation number; the coefficients $a_1, \ldots, a_{19}$ are determined by the formulas (7.19); in the additional loads $p_\omega^{*(n-1)}$, $h_\omega^{*(n-1)}$, $q_\omega^{*(n-1)}$, $g_\omega^{*(n-1)}$ at the first approximation step the nonlinear terms are assumed to be equal to zero, and further they are calculated according to the results of the previous iteration:

$$
\begin{aligned}
p_\omega^{*(n-1)} &= \frac{1}{b_0}\left(H_1^{\omega^*(n-1)} - P_1^{\omega^*(n-1)},_x\right),\ h_\omega^{*(n-1)} = \frac{1}{b_0}\left(H_1^{\omega^*(n-1)} - P_2^{\omega*(n-1)},_x\right), \\
q_\omega^{*(n-1)} &= \frac{1}{b_0}\left(S_1^{\omega*(n-1)},_{xx} + H_2^{\omega*(n-1)} - T_1^{\omega*(n-1)},_x\right), \\
g_\omega^{*(n-1)} &= \frac{1}{b_0}\left(S_2^{\omega^*(n-1)},_{xx} + H_2^{\omega^*(n-1)} - T_2^{\omega^*(n-1)},_x\right),
\end{aligned}
\tag{7.55}
$$

where

$$
\begin{aligned}
H_1^{\omega^*(n-1)} &= \frac{2Q^{(3)\omega^*(n-1)}}{c},\ H_2^{\omega^*(n-1)} = \frac{N_z^{(3)\omega*(n-1)}}{2c}, \\
T_1^{\omega^*(n-1)} &= \left(1 + \frac{h_1}{2c}\right)\frac{Q^{(3)\omega^*(n-1)}}{2} + \frac{M_{xz}^{(3)\omega^*(n-1)}}{2c}, \\
T_2^{\omega^*(n-1)} &= \left(1 + \frac{h_2}{2c}\right)\frac{Q^{(3)\omega^*(n-1)}}{2} - \frac{M_{xz}^{(3)\omega^*(n-1)}}{2c}, \\
S_1^{\omega^*(n-1)} &= \left(c + \frac{h_1}{2}\right)N_x^{(1)\omega^*(n-1)} - M_x^{(1)\omega^*(n-1)} \\
&\quad + \frac{h_1}{4}N_x^{(3)\omega^*(n-1)} + \frac{h_1}{4c}M_x^{(3)\omega^*(n-1)}, \\
S_2^{\omega^*(n-1)} &= -\left(c + \frac{h_2}{2}\right)N_x^{(2)\omega^*(n-1)} - M_x^{(2)\omega^*(n-1)} \\
&\quad - \frac{h_2}{4}N_x^{(3)\omega^*(n-1)} + \frac{h_2}{4c}M_x^{(3)\omega^*(n-1)}, \\
S_2^{\omega^*(n-1)} &= -\left(c + \frac{h_2}{2}\right)N_x^{(2)\omega^*(n-1)} - M_x^{(2)\omega^*(n-1)} \\
&\quad - \frac{h_2}{4}N_x^{(3)\omega^*(n-1)} + \frac{h_2}{4c}M_x^{(3)\omega^*(n-1)}, \\
P_2^{\omega^*(n-1)} &= \frac{N_x^{(3)\omega^*(n-1)}}{2} - \frac{M_x^{(3)\omega^*(n-1)}}{2c} + N_x^{(2)\omega^*(n-1)}, \\
N_x^{(k)\omega^*(n-1)} &= \frac{4}{3}b_0 \int_{h_k} G_k \varepsilon_x^{(k)*(n-1)} \omega^{(k)*}\left(\varepsilon_u^{(k)*(n-1)}, T_k\right)\mathrm{d}z,
\end{aligned}
$$

$$
\begin{aligned}
M_x^{(k)\omega^*(n-1)} &= \frac{4}{3} b_0 \int_{h_k} G_k \varepsilon_x^{(k)^*(n-1)} \omega^{(k)^*}\left(\varepsilon_u^{(k)^*(n-1)}, T_k\right) z dz, \\
N_x^{(3)^*(n-1)} &= \frac{2}{3} b_0 \int_{b_3} G_3\left(2\varepsilon_x^{(3)^*(n-1)} - \varepsilon_z^{(3)^*(n-1)}\right)\omega^{(3)^*}\left(\varepsilon_u^{(3)^*(n-1)}, T_k\right) dz, \\
N_z^{(3)\omega^*(n-1)} &= \frac{2}{3} b_0 \int_{h_3} G_3\left(2\varepsilon_z^{(3)^*(n-1)} - \varepsilon_x^{(3)*(n-1)}\right)\omega^{(3)*}\left(\varepsilon_u^{(3)*(n-1)}, T_k\right) dz, \\
M_x^{(3)^*(n-1)} &= \frac{2}{3} b_0 \int_{h_3} G_3\left(2\varepsilon_x^{(3)*(n-1)} - \varepsilon_z^{(3)*(n-1)}\right)\omega^{(3)^*}\left(\varepsilon_l^{(3)*(n-1)}, T_k\right) z dz, \\
M_{xz}^{(3)\omega^*(n-1)} &= 2b_0 \int_{h_3} G_3 \varepsilon_{xz}^{(3)^*(n-1)} \omega^{(3)^*(n-1)}\left(\varepsilon_u^{(3)^*(n-1)}, T_3\right) z dz, \\
Q^{(3)\omega^*(n-1)} &= 2b_0 \int_{h_3} G_3 \varepsilon_{xz}^{(3)^*(n-1)} \omega^{(3)^*}\left(\varepsilon_u^{(3)^*(n-1)}, T_3\right) dz \; (k = 1, \cdot 2).
\end{aligned}
$$

Application of the method of elastic solutions allows at each approximation step to reduce the nonlinear boundary value problem with asterisks to the corresponding linear system (7.54) with additional loads (7.55). The first approximation will be the analytical or numerical solution of the thermoelasticity problem (7.21)–(7.24) for the rod under consideration.

Under kinematic conditions of free resting of the rod at the ends on rigid supports fixed in space, in the end sections x $= 0, l$ at each step the following requirements with respect to the displacements with asterisks must be fulfilled:

$$
w_k^{*n} = u_k^{*n},_x = w_k^{*n},_{xx} = 0 \, (k = 1, 2). \tag{7.56}
$$

The solution of the system of differential Eq. (7.54) is taken as a decomposition into trigonometric series, which automatically satisfy the boundary conditions (7.56):

$$
\begin{aligned}
u_1^{n^*} &= \sum_{m=1}^{\infty} U_{1m}^{n^*} \cos\left(\frac{\pi m x}{l}\right), \; u_2^{n^*} = \sum_{m=1}^{\infty} U_{2m}^{n^*} \cos\left(\frac{\pi m x}{l}\right), \\
w_1^{n^*} &= \sum_{m=1}^{\infty} W_{1m}^{n^*} \sin\left(\frac{\pi m x}{l}\right), \; w_2^{n^*} = \sum_{m=1}^{\infty} W_{2m}^{n^*} \sin\left(\frac{\pi m x}{l}\right),
\end{aligned} \tag{7.57}
$$

where U_{1m}^{n*}, U_{2m}^{n*}, W_{1m}^{n*}, W_{2m}^{n*} are required amplitudes of displacements with asterisks.

External loads and additional loads with asterisks are also represented as trigonometric series decompositions:

$$
\begin{aligned}
q^* &= \sum_{m=1}^{\infty} q_m^* \sin\left(\frac{\pi m x}{l}\right), \; q_m^* = \frac{2}{l} \int_0^l q^*(x) \sin\left(\frac{\pi m x}{l}\right) dx, \\
p^* &= \sum_{m=1}^{\infty} p_m^* \sin\left(\frac{\pi m x}{l}\right), \; p_m^* = \frac{2}{l} \int_0^l p^*(x) \sin\left(\frac{\pi m x}{l}\right) dx,
\end{aligned}
$$

$$q_{\omega}^{(n-1)^*} = \sum_{m=1}^{\infty} q_{\omega m}^{(n-1)^*} \sin\left(\frac{\pi m x}{l}\right), q_{\omega m}^{(n-1)^*} = \frac{2}{l}\int_0^l q_{\omega}^{(n-1)^*}(x)\sin\left(\frac{\pi m x}{l}\right)dx,$$

$$g_{\omega}^{(n-1)^*} = \sum_{m=1}^{\infty} g_{\omega m}^{(n-1)^*} \sin\left(\frac{\pi m x}{l}\right), g_{\omega m}^{(n-1)^*} = \frac{2}{l}\int_0^l g_{\omega}^{(n-1)^*}(x)\sin\left(\frac{\pi m x}{l}\right)dx,$$

$$h_{\omega}^{(n-1)^*} = \sum_{m=1}^{\infty} h_{\omega m}^{(n-1)^*} \cos\left(\frac{\pi m x}{l}\right), h_{\omega m}^{(n-1)^*} = \frac{2}{l}\int_0^l h_{\omega}^{(n-1)^*}(x)\cos\left(\frac{\pi m x}{l}\right)dx,$$

$$p_{\omega}^{(n-1)^*} = \sum_{m=1}^{\infty} p_{\omega m}^{(n-1)^*} \cos\left(\frac{\pi m x}{l}\right), p_{\omega m}^{(n-1)^*} = \frac{2}{l}\int_0^l p_{\omega}^{(n-1)^*}(x)\cos\left(\frac{\pi m x}{l}\right)dx. \quad (7.58)$$

After substituting displacements (7.57) and loads (7.58) into (7.54), we obtain a system of linear algebraic equations to determine the amplitudes of displacements U_{1m}^{n}, U_{2m}^{*n}, W_{1m}^{*n}, W_{2m}^{*n}:

$$\begin{cases} b_1 U_{1m}^{*n} + b_2 U_{2m}^{*n} + b_3 W_{1m}^{*n} + b_4 W_{2m}^{*n} = p_m^* + p_{\omega m}^{*(n-1)}, \\ b_2 U_{1m}^{*n} + b_5 U_{2m}^{*n} + b_6 W_{1m}^{*n} - b_7 W_{2m}^{*n} = h_{\omega m}^{*(n-1)}, \\ b_3 U_{1m}^{*n} + b_6 U_{2m}^{*n} + b_8 W_{1m}^{*n} + b_9 W_{2m}^{*n} = q_m^* + q_{\omega m}^{*(n-1)}, \\ b_4 U_{1m}^{*n} - b_7 U_{2m}^{*n} + b_9 W_{1m}^{*n} + b_{10} W_{2m}^{*n} = g_{\omega m}^{*(n-1)}, \end{cases} \quad (7.59)$$

where coefficients b_i are calculated by the formulas introduced in (7.23).

The view of the solution of the system (7.59) in determinants is:

$$U_{1m}^{*n} = \frac{\Delta_{1m}^{*(n-1)}}{\Delta}, U_{2m}^{*n} = \frac{\Delta_{2m}^{*(n-1)}}{\Delta}, W_{1m}^{*n} = \frac{\Delta_{3m}^{*(n-1)}}{\Delta}, W_{2m}^{*n} = \frac{\Delta_{4m}^{*(n-1)}}{\Delta}, \quad (7.60)$$

where

$$\Delta = \begin{vmatrix} b_1 & b_2 & b_3 & b_4 \\ b_2 & b_5 & b_6 & -b_7 \\ b_3 & b_6 & b_8 & b_9 \\ b_4 & -b_7 & b_9 & b_{10} \end{vmatrix}, \Delta_{1m}^{*(n-1)} = \begin{vmatrix} p_m^* + p_{\omega m}^{*(n-1)} & b_2 & b_3 & b_4 \\ h_{\omega m}^{*(n-1)} & b_5 & b_6 & -b_7 \\ q_m^* + q_{\omega m}^{n-1} & b_6 & b_8 & b_9 \\ g_{\omega m}^{*(n-1)} & -b_7 & b_9 & b_{10} \end{vmatrix},$$

$$\Delta_{2m}^{n-1} = \begin{vmatrix} b_1 & p_m^* + p_{\omega m}^{*(n-1)} & b_3 & b_4 \\ b_2 & h_{\omega m}^{*(n-1)} & b_6 & -b_7 \\ b_3 & q_m^* + q_{\omega m}^{*(n-1)} & b_8 & b_9 \\ b_4 & g_{\omega m}^{*(n-1)} & b_9 & b_{10} \end{vmatrix},$$

$$\Delta_{3m}^{n-1} = \begin{vmatrix} b_1 & b_2 & p_m^* + p_{\omega m}^{*(n-1)} & b_4 \\ b_2 & b_5 & h_{\omega m}^{*(n-1)} & -b_7 \\ b_3 & b_6 & q_m^* + q_{\omega m}^{*(n-1)} & b_9 \\ b_4 & -b_7 & g_{\omega m}^{*(n-1)} & b_{10} \end{vmatrix}, \Delta_{4m}^{n-1} = \begin{vmatrix} b_1 & b_2 & b_3 & p_m^* + p_{\omega m}^{*(n-1)} \\ b_2 & b_5 & b_6 & h_{\omega m}^{*(n-1)} \\ b_3 & b_6 & b_8 & q_m^* + q_{\omega m}^{*(n-1)} \\ b_4 & -b_7 & b_9 & g_{\omega m}^{*(n-1)} \end{vmatrix}.$$

Calculating the determinants in (7.60) by any of the standard methods, we obtain the amplitudes U_{1m}^{n}, U_{2m}^{*n}, W_{1m}^{*n}, W_{2m}^{*n}. Then by formulas (7.55) the required functions with asterisks are calculated. The corresponding solution of the problem of rod deformation in the process of the second half-cycle is obtained from relations (7.48), taking into account the solution (7.57)–(7.60):

$$\sigma_x^{\prime\prime(k)}(x,z) = \sigma_x^{\prime(k)}(x,z) - \sigma_x^{(k)*}(x,z), \sigma_z^{\prime\prime(3)}(x,z) = \sigma_z^{\prime(3)}(x,z) - \sigma_z^{(3)*}(x,z),$$
$$\sigma_{xz}^{\prime\prime(3)}(x,z) = \sigma_{xz}^{\prime(3)}(x,z) - \sigma_{xz}^{(3)*}(x,z), \varepsilon_x^{\prime\prime(k)}(x,z) = \varepsilon_x^{\prime(k)}(x,z) - \varepsilon_x^{(k)*}(x,z),$$
$$\varepsilon_z^{\prime\prime(3)}(x,z) = \varepsilon_z^{\prime(3)}(x,z) - \varepsilon_z^{(3)*}(x,z), \varepsilon_{xz}^{\prime\prime(3)}(x,z) = \varepsilon_{xz}^{\prime(3)}(x,z) - \varepsilon_{xz}^{(3)*}(x,z),$$
$$u_k^{\prime\prime}(x,z) = u_k^{\prime}(x,z) - u_k^{*}(x,z), w_k^{\prime}(x,z) = w_k^{\prime}(x,z) - w_k^{*}(x,z), (k = 1,\ 2). \tag{7.61}$$

Here the quantities with one stroke are the stresses, strains and displacements in the rod before unloading, calculated by formulas (7.44)–(7.47), in which the stroke at the top must be formally added.

For numerical realisation of the solution (7.61) the value of load intensity $q = -5$ MPa, $p = 0$ was taken.

Figure 7.37 shows graphs of deflection (*a*) and longitudinal displacement (*b*) along the length of the rod in the first layer under a variable uniformly distributed thermoforce load: *1*—elastic rod, *2*—thermoelastic rod, *3*—thermoelastic-plastic rod; one stroke—solution of the problem when loading the rod from the natural state, two strokes—repeated alternating loading.

Figures 7.38, 7.39 show the change of deflections of the first layer in the cross-section of the rod ($x = 0.5$) depending on the coordinates of the edges of the rectangular loading interval: *1*—elastic rod, *2′*—thermoelastic-plastic rod, under loading from the natural state, *2″*—thermoelastic-plastic rod, under repeated alternating loading.

It should be noted that under repeated alternating loading and uniformly distributed load, as well as local rectangular load, the yield strength of the material of bearing layers changes only due to force hardening of the material, because the temperature after changing the sign of the load does not change. Therefore, the

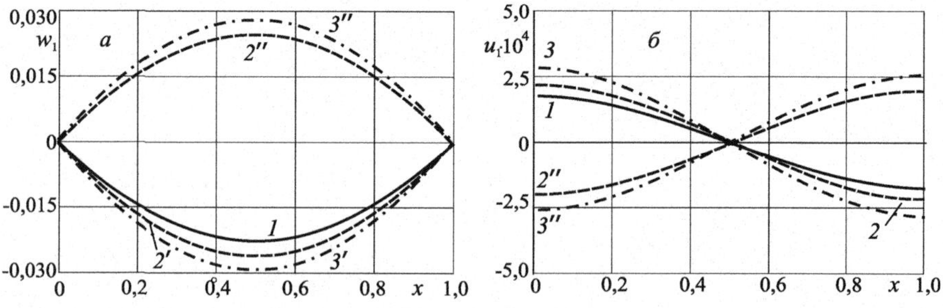

Fig. 7.37 Variation of deflection w_1 (*a*) and longitudinal displacement u_1 (*b*) along the rod axis

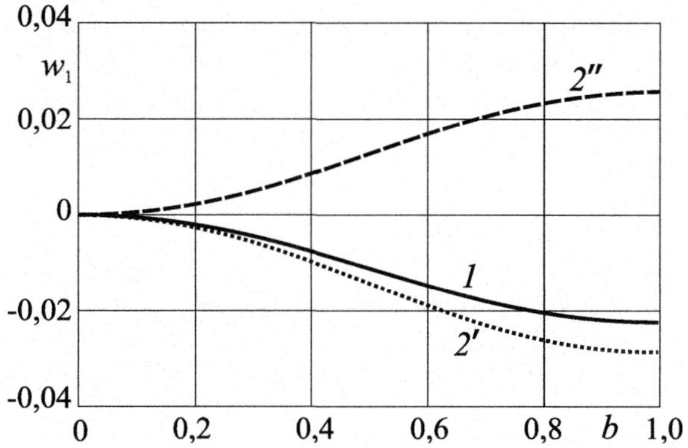

Fig. 7.38 Variation of the first layer deflection in the cross-section of the rod ($x = 0.5$) as a function of b

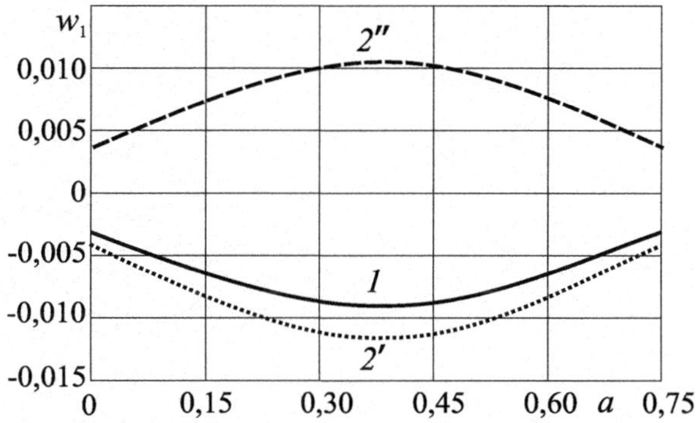

Fig. 7.39 Variation of the first layer deflection in the cross-section of the rod ($x = 0.5$) as a function of a

maximum value of deflections at repeated loadings in the temperature field decreases by 1–2%, however, in case of subsequent cyclic load changes, this divergence will accumulate.

7.7 Elastoplastic Rod in a Thermoradiation Field

Let an elastoplastic three-layer rod be in the temperature field $T_k(z)$ and irradiated with neutron flux $I = \phi t$ (ϕ—intensity in neutrons/s, t—time). The layers use the corresponding physical equations of state of the theory of small elastoplastic deformations:

$$\begin{aligned}
s_x^{(k)} &= 2G_k\left(1 - \omega^{(k)}\left(\varepsilon_u^{(k)}, T_k, I\right)\right) \text{э}_x^{(k)}, \\
\sigma^{(k)} &= K_k(T_k)\left(3\varepsilon^{(k)} - 3\alpha_{0k}T_k - B_k I\right) (k = 1, 2, 3), \\
s_z^{(3)} &= 2G_3\left(1 - \omega^{(3)}\left(\varepsilon_u^{(3)}, T_3, I\right)\right) \text{э}_z^{(3)}, \\
s_{xz}^{(3)} &= 2G_3\left(1 - \omega^{(k)}\left(\varepsilon_u^{(k)}, T_k\right)\right) \text{э}_{xz}^{(3)} .
\end{aligned} \tag{7.62}$$

The difference of these relations from relations (7.36) is an additional dependence of the plasticity function $\omega^{(k)}\left(\varepsilon_u^{(k)}, T_k, I\right)$ on the neutron flux value.

Based on relations (7.62), let us distinguish elastic (index "e") and nonlinear (index "ω") components in the stress tensor, which will include thermoradiation additives as well:

· in bearing layers ($k = 1{,}2$)

$$\sigma_x^{(k)} = \sigma_x^{(k)e} - \sigma_x^{(k)\omega}, \sigma_x^{(k)e} = 2G_k \text{э}_x^{(k)} + 3K_k\varepsilon^{(k)} = K_k^+\varepsilon_x^{(k)},$$

$$\sigma_x^{(k)\omega} = 2G_k \text{э}_x^{(k)} \omega^{(k)} + K_k(3\alpha_{0k}T_k + B_kI) = \frac{4}{3}G_k\varepsilon_x^{(k)}\omega^{(k)} + K_k(3\alpha_{0k}T_k + B_kI);$$

in filler

$$\begin{aligned}
\sigma_x^{(3)} &= \sigma_x^{(3)e} - \sigma_x^{(3)\omega}, \sigma_x^{(3)e} = 2G_3 \text{э}_x^{(3)} + 3K_3\varepsilon^{(3)} = K_3^+\varepsilon_x^{(3)} + K_3^-\varepsilon_z^{(3)},\\
\sigma_x^{(3)\omega} &= 2G_3 \text{э}_x^{(3)} \omega^{(3)} + K_3(3\alpha_{03}T_3 + B_3I)\\
&= \frac{2}{3}G_3\left(2\varepsilon_x^{(3)} - \varepsilon_z^{(3)}\right)\omega^{(3)} + K_3(3\alpha_{03}T_3 + B_3I),\\
\sigma_z^{(3)} &= \sigma_z^{(3)e} - \sigma_z^{(3)\omega}, \sigma_z^{(3)e} = 2G_3 \text{э}_z^{(3)} + 3K_3\varepsilon^{(3)} = K_3^+\varepsilon_z^{(3)} + K_3^-\varepsilon_x^{(3)},\\
\sigma_z^{(3)\omega} &= 2G_3 \text{э}_z^{(3)} \omega^{(3)} + K_3(3\alpha_{03}T_3 + B_3I)\\
&= \frac{2}{3}G_3\left(2\varepsilon_z^{(3)} - \varepsilon_x^{(3)}\right)\omega^{(3)} + K_3(3\alpha_{03}T_3 + B_3I),\\
\sigma_{xz}^{(3)} &= \sigma_{xz}^{(3)e} - \sigma_{xz}^{(3)\omega}, \sigma_{xz}^{(3)e} = 2G_3 \text{э}_{xz}^{(3)} = 2G_3\varepsilon_{xz}^{(3)},\\
\sigma_{xz}^{(3)\omega} &= 2G_3 \text{э}_{xz}^{(3)} \omega^{(3)} = 2G_3\varepsilon_{xz}^{(3)}\omega^{(3)}.
\end{aligned} \tag{7.63}$$

Carrying out an operation similar to (7.63) with the internal forces (7.4), we obtain relations coinciding with (7.38). The linear components in them (index "e") are calculated by formulae of the type (7.39). Of the nonlinear terms (index "ω"), only those that account for the effects of neutron flux will change:

· in bearing layers ($k = 1{,}2$)

$$N_x^{(k)\omega} = b_0\int_{h_k}\sigma_x^{(k)\omega}\mathrm{d}z = \frac{4}{3}b_0\int_{h_k}G_k\varepsilon_x^{(k)}\omega^{(k)}\mathrm{d}z + b_0\int_{h_k}K_k(3\alpha_{0k}T_k + B_kI)\mathrm{d}z,$$

$$M_x^{(k)\omega} = b_0\int_{h_k}\sigma_x^{(k)\omega}z\mathrm{d}z = \frac{4}{3}b_0\int_{h_k}G_k\varepsilon_x^{(k)}\omega^{(k)}z\mathrm{d}z + b_0\int_{h_k}K_k(3\alpha_{0k}T_k + B_kI)z\mathrm{d}z;$$

· in filler

$$N_x^{(3)\omega} = b_0\int_{h_3}\sigma_x^{(3)\omega}\mathrm{d}z = \frac{2}{3}b_0\int_{h_3}G_3\left(2\varepsilon_x^{(3)} - \varepsilon_z^{(3)}\right)\omega^{(3)}\mathrm{d}z + b_0\int_{h_3}K_3(3\alpha_{03}T_3 + B_3I)\mathrm{d}z,$$

$$
\begin{aligned}
N_z^{(3)\omega} &= b_0 \int_{h_3} \sigma_z^{(3)\omega} \mathrm{d}z = \frac{2}{3} b_0 \int_{h_3} G_3 \left(2\varepsilon_z^{(3)} - \varepsilon_x^{(3)} \right) \omega^{(3)} \mathrm{d}z + b_0 \int_{h_3} K_3 (3\alpha_{03} T_3 + B_3 I) \mathrm{d}z, \\
M_x^{(3)\omega} &= b_0 \int_{h_3} \sigma_x^{(3)\omega} z \mathrm{d}z = \frac{2}{3} b_0 \int_{h_3} G_3 \left(2\varepsilon_x^{(3)} - \varepsilon_z^{(3)} \right) \omega^{(3)} z \mathrm{d}z + b_0 \int_{h_3} K_3 (3\alpha_{03} T_3 + B_3 I) z \mathrm{d}z,
\end{aligned}
\tag{7.64}
$$

Let us perform calculations similar to Sect. 7.2. As a result, we obtain a system of nonlinear differential equations in displacements, which is similar to the system describing the deformation of an elastic–plastic rod in a thermoradiation field. Its iterative form will coincide with (7.41):

$$
\begin{aligned}
& a_1 u_1^n - a_1 u_2^n - a_4 u_1^n{}_{,xx} - a_5 u_2^n{}_{,xx} + a_2 w_1^n{}_{,x} \\
& + a_3 w_2^n{}_{,x} - 2a_6 w_1^n{}_{,xxx} + a_7 w_2^n{}_{,xxx} = p + p_\omega^{(n-1)}, \\
& - a_1 u_1^n + a_1 u_2^n - a_5 u_1^n{}_{,xx} - a_9 u_2^n{}_{,xx} - a_{10} w_1^n{}_{,x} \\
& - a_{17} w_2^n{}_{,x} - a_6 w_1^n{}_{,xxx} + 2a_7 w_2^n{}_{,xxx} = h_\omega^{(n-1)}, \\
& - a_2 u_1^n{}_{,x} + a_{10} u_2^n{}_{,x} + 2a_6 u_1^n{}_{,xx} + a_6 u_2^n{}_{,xxx} + a_{11} w_1^n{}_{,xx} \\
& - a_{12} w_2^n{}_{,xx} + a_{15} w_1^n{}_{,xxxx} - a_{16} w_2^n{}_{,xxxx} + a_8 w_1^n - a_8 w_2^n = q + q_\omega^{(n-1)}, \\
& - a_3 u_1^n{}_{,x} + a_{17} u_2^n{}_{,x} - a_7 u_1{}_{,xxx} - 2a_7 u_2^n{}_{,xxx} - a_{12} w_1^n{}_{,xx} \\
& + a_{14} w_2^n{}_{,xx} - a_{16} w_1^n{}_{,xxxx} + a_{13} w_2^n{}_{,xxxx} - a_8 w_1^n + a_8 w_2^n = g_\omega^{(n-1)}.
\end{aligned}
\tag{7.65}
$$

The additional loads $p_\omega^{(n-1)}$, $h_\omega^{(n-1)}$, $q_\omega^{(n-1)}$, $g_\omega^{(n-1)}$ are calculated by formulae (7.42), where, by virtue of relations (7.64), only the following nonlinear additives have changed:

$$
\begin{aligned}
N_x^{(k)\omega(n-1)} &= \frac{4}{3} b_0 \int_{h_k} G_k \varepsilon_x^{(k)(n-1)} \omega^{(k)} (\varepsilon_u^{(k)(n-1)}, T_k, I) \mathrm{d}z \\
&+ b_0 \int_{h_k} K_k (3\alpha_{0k} T_k + B_k I) \mathrm{d}z, \\
M_x^{(k)\omega(n-1)} &= \frac{4}{3} b_0 \int_{h_k} G_k \varepsilon_x^{(k)(n-1)} \omega^{(k)} (\varepsilon_u^{(k)(n-1)}, T_k, I) z \mathrm{d}z \\
&+ b_0 \int_{h_k} K_k (3\alpha_{0k} T_k + B_k I) z \mathrm{d}z, \\
N_x^{(3)\omega(n-1)} &= \frac{2}{3} b_0 \int_{h_3} G_3 (2\varepsilon_x^{(3)(n-1)} - \varepsilon_z^{(3)(n-1)}) \omega^{(3)} (\varepsilon_u^{(3)(n-1)}, T_k, I) \mathrm{d}z \\
&+ b_0 \int_{h_3} K_3 (3\alpha_{03} T_3 + B_3 I) \mathrm{d}z,
\end{aligned}
$$

$$
\begin{aligned}
N_z^{(3)\omega(n-1)} =& \frac{2}{3} b_0 \int\limits_{h_3} G_3 (2\varepsilon_z^{(3)(n-1)} - \varepsilon_x^{(3)(n-1)}) \omega^{(3)} (\varepsilon_u^{(3)(n-1)}, T_k, I) \mathrm{d}z \\
&+ b_0 \int\limits_{h_3} K_3 (3\alpha_{03} T_3 + B_3 I) \mathrm{d}z, \\
M_x^{(3)\omega(n-1)} =& \frac{2}{3} b_0 \int\limits_{h_3} G_3 (2\varepsilon_x^{(3)(n-1)} - \varepsilon_z^{(3)(n-1)}) \omega^{(3)} (\varepsilon_u^{(3)(n-1)}, T_k, I) z \mathrm{d}z \\
&+ b_0 \int\limits_{h_3} K_3 (3\alpha_{03} T_3 + B_3 I) z \mathrm{d}z, \\
M_x^{(3)\omega(n-1)} =& 2b_0 \int\limits_{h_3} G_3 (2\varepsilon_{xz}^{(3)(n-1)} \omega^{(3)(n-1)} (\varepsilon_u^{(3)(n-1)}, T_3, I) z \mathrm{d}z, \\
Q^{(3)\omega(n-1)} =& 2b_0 \int\limits_{h_3} G_3 \varepsilon_{xz}^{(3)(n-1)} \omega^{(3)} (\varepsilon_u^{(3)(n-1)}, T_3, I) \mathrm{d}z (k = 1, 2).
\end{aligned}
\tag{7.66}
$$

In additional loadings (7.65), (7.66) at the first approximation step the nonlinear terms are taken equal to zero ($\omega^{(k)} = 0$), and further they are calculated according to the results of the previous iteration. The first approximation will be the analytical solution of the elasticity problem (7.21), (7.24).

Application of the method of elastic solutions makes it possible to reduce the problem of elastic–plastic bending of a three-layered rod with a compressible filler in a thermoradiation field to the corresponding linear problem of elasticity theory with additional loads at each approximation step.

As boundary conditions we take the kinematic conditions (7.43) of free support of the rod at the ends on rigid supports fixed in space. The solution can be represented in trigonometric series (7.44). The transverse load $q(x)$ and the additional forces are also represented as series. (7.45). Substituting them into the equations of equilibrium, we obtain a system of linear algebraic equations coinciding in form with (7.46) for calculating the required amplitudes of displacements U_{1m}^n, U_{2m}^n, W_{1m}^n, W_{2m}^n. The difference here is in the nonlinear additions, which also take into account the influence of neutron irradiation. Its solution can be taken in the form (7.47).

Numerical realisation of the solution was carried out in the case of isothermal loading under neutron irradiation of the considered three-layer rod. In this case the coefficients of the system of Eq. (7.65) are calculated by the formulas (7.26).

The impact of neutron flux was modelled by increasing the yield strength of the outer layer material by 20% and the physical nonlinearity limit of the filler by 10%.

A qualitative picture of deflection (a) and longitudinal displacement (b) changes in the first layer in case of uniformly distributed load is shown in Fig. 7.40: *1*—elastic rod, *2*—elastic–plastic rod, *3*—elastic–plastic rod in neutron flow.

Neutron exposure reduces elastic–plastic deflection by 4% due to weakening of plastic properties of the material and increase of its stiffness. Longitudinal elastic–plastic displacements change insignificantly.

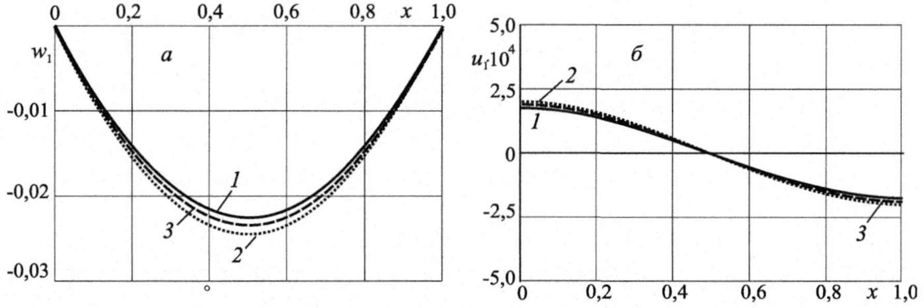

Fig. 7.40 Variation of deflection (*a*) and longitudinal displacement (*b*) in the first layer along the axis of the rod

7.8 Repeated Alternating Loading in Neutron Flux

Consider an elastoplastic three-layer rod with compressible filler, which is irradiated by neutron flux $I = \phi t$ (ϕ—intensity, neutron/s, t—time) in the absence of thermal influence. The solution of this boundary value problem follows from the solution obtained in Sect. 7.7, if in it the influence of temperature is not taken into account.

Let, starting from the moment $t = t_1$, there is an instantaneous unloading and re-loading with forces of opposite sign p'', q'', changing according to the same law as at loading from the natural state. These forces will create in the kth layer of the rod a field of displacements $u''(x)$, $w''(x)$, strains $\varepsilon_x''^{(k)}(x, z)$, $э_x''^{(k)}(x, z)$, $\varepsilon_x''^{(k)}(x, z)$, $\psi''(x, z)$, and stresses $\sigma_x''^{(k)}(x, z)$, $s_x''^{(k)}(x, z)$, $\sigma''^{(k)}(x, z)$, $\sigma_{xz}''^{(3)}(x, z)$. We assume that during the time of unloading and subsequent variable loading the neutron flux influence has ceased and the integral neutron flux remains unchanged at all points of the body. It (neutron flux) coincides with the flux level at the time of the start of unloading $I'(z) = I_1(z, t_1)$.

As before, for all parameters of the stress–strain state and loading we introduce parameters with asterisks of the type (7.48), in which the values with one Schrich are stresses, deformations and displacements in the rod before unloading, two schriches mark similar parameters during the second half-cycle. For example, the deflection of the first layer will be $w^* = w\prime - w\prime\prime$. For the introduced stresses and strains with asterisks, let us assume physical equations of state (5.19).

After appropriate transformations, we obtain a system of nonlinear differential equations of equilibrium in iterations for displacements with asterisks, which is similar to the system (7.54):

$$
\begin{aligned}
& a_1 u_1^{*n} - a_1 u_2^{*n} - a_4 u_1^{*n},_{xx} - a_5 u_2^{*n},_{xx} + a_2 w_1^{*n},_x \\
& \quad + a_3 w_2^{*n},_x - 2a_6 w_1^{*n},_{xxx} + a_7 w_2^{*n},_{xxx} = p_\omega^{*(n-1)} \\
& - a_1 u *_1^n + a_1 u_2^{*n} - a_5 u_1^{*n},_{xx} - a_9 u_2^{*n},_{xx} - a_{10} w_1^{*n},_x \\
& - a_{17} w *_{2,x}^n - a_6 w_1^{*n},_{xxx} + 2a_7 w_2^{*n},_{xxx} = h_\omega^{*(n-1)}, \\
& - a_2 u_1^{*n},_x + a_{10} u_2^{*n},_x + 2a_6 u_1^{*n},_{xxx} + a_6 u_2^{*n},_{xxx} + a_{11} w_1^{*n},_{xx}
\end{aligned}
$$

$$- a_{12}w_2^{*n},_{xx} + a_{15}w_1^{*n},_{xxxx} - a_{16}w_2^{*n},_{xxxx} + a_8 w_1^{*n} - a_8 w_2^{*n} = q + q_\omega^{*(n-1)},$$
$$- a_3 u_1^{*n},_x + a_{17}u_2^{*n},_x - a_7 u_1^{*n},_{xxx} - 2a_7 u_2^{*n},_{xxx} - a_{12}w_1^{*n},_{xx} + a_{14}w_2^{*n},_{xx}$$
$$- a_{16}w_1^{*n},_{xxxx} + a_{13}w_2^{*n},_{xxxx} - a_8 w_1^{*n} + a_8 w_2^{*n} = g_\omega^{*(n-1)}, \tag{7.67}$$

where n is approach number; coefficients $a_1, \ldots, a_{19}$ are determined by formulas (7.26).

Additional "external" loads in (7.67) $p_\omega^{(n-1)*}$, $h_\omega^{(n-1)*}$, $q_\omega^{(n-1)*}$ are calculated by formulas (7.55), in which nonlinear components of internal forces will depend not on temperature, but on the value of the integral neutron flux:

$$N_x^{(k)\omega*(n-1)} = \frac{4}{3}b_0 \int_{h_k} G_k \varepsilon_x^{(k)*(n-1)} \omega^{(k)*}(\varepsilon_u^{(k)*(n-1)}, I_1)\mathrm{d}z,$$

$$M_x^{(k)\omega*(n-1)} = \frac{4}{3}b_0 \int_{h_k} G_k \varepsilon_x^{(k)*(n-1)} \omega^{(k)*}(\varepsilon_u^{(k)*(n-1)}, I_1)z\mathrm{d}z,$$

$$N_x^{(3)\omega*(n-1)} = \frac{2}{3}b_0 \int_{h_3} G_3(2\varepsilon_x^{(3)*(n-1)} - \varepsilon_z^{(3)*(n-1)})\omega^{(3)*}(\varepsilon_u^{(3)*(n-1)}, I_1)\mathrm{d}z,$$

$$N_z^{(3)\omega*(n-1)} = \frac{2}{3}b_0 \int_{h_3} G_3(2\varepsilon_z^{(3)*(n-1)} - \varepsilon_x^{(3)*(n-1)})\omega^{(3)*}(\varepsilon_u^{(3)*(n-1)}, I_1)\mathrm{d}z,$$

$$M_x^{(3)\omega*(n-1)} = \frac{2}{3}b_0 \int_{h_3} G_3(2\varepsilon_x^{(3)*(n-1)} - \varepsilon_z^{(3)*(n-1)})\omega^{(3)*}(\varepsilon_u^{(3)*(n-1)}, I_1)z\mathrm{d}z,$$

$$M_{xz}^{(3)\omega*(n-1)} = 2b_0 \int_{h_3} G_3 \varepsilon_{xz}^{(3)*(n-1)} \omega^{(3)*(n-1)}(\varepsilon_u^{(3)*(n-1)}, I_1)z\mathrm{d}z,$$

$$Q^{(3)\omega*(n-1)} = 2b_0 \int_{h_3} G_3 \varepsilon_{xz}^{(3)*(n-1)} \omega^{(3)*}(\varepsilon_u^{(3)*(n-1)}, I_1)\mathrm{d}z(k = 1, 2).$$

Additional "external" loads at the first step ($n = 1$) are assumed to be zero, further they are calculated from the results of the previous approximation. They are corrections for plasticity and physical nonlinearity of layer materials under repeated loading and take into account the impact of the neutron flux. Here, the nonlinearity function $\omega^{(k)*}(\varepsilon_u^{(k)*(n-1)}, I_1)$ depends only on the magnitude of the strain intensity at (n–1) step and the integral neutron flux received by the rod before unloading. It is determined by relations (5.20), (5.21).

Thus, the system of Eq. (7.67) for the values with asterisks coincides with the system (7.65) when loading the rod from the natural state, with the accuracy of notation. Therefore, in the case of the boundary conditions of free support of the rod at the ends on rigid supports fixed in space (7.43), the analytical solution of the boundary value problem can be taken in trigonometric series (7.44). The transverse load $q(x)$ and additional loads are also represented as series (7.45). Substituting them into the equations of equilibrium, we obtain a system of linear algebraic equations coinciding in form with (7.46) for calculating the required amplitudes of displacements U^n_{1m}, U^n_{2m}, W^n_{1m}, $\omega^{(k)*}$. The difference here is in the nonlinear additives, which take into account the effect of neutron irradiation, but do not take into account temperature. The solution of this system can be taken in the form (7.47). Numerical investigation of this solution was carried out under the same conditions as in Sect. 7.7.

Figures 7.41, 7.42 show the change of deflections of the first layer in the cross-section of the rod ($x = 0.5$) depending on the coordinates of the edges of the rectangular loading interval: *1*—elastic, $2'$, $2''$—elastoplastic rod, under direct and reverse loading in the neutron flux. Figure 7.40 illustrates the variation of deflection along the axis of the rod as a function of the length of the load interval *b*. Figure 7.41 shows graphs of deflection as a function of the coordinate *a* of the left edge of the load interval at its constant length ($b - a = 0.25$).

It should be noted that due to the impact of neutron flux on the first half-cycle of loading there is a weakening of plastic properties of the material and an increase in its stiffness. This leads to a 3–4% reduction in elastic–plastic deflection. Under repeated alternating loading and uniformly distributed and local rectangular loads, the yield strength of the material of bearing layers changes only due to force hardening of the

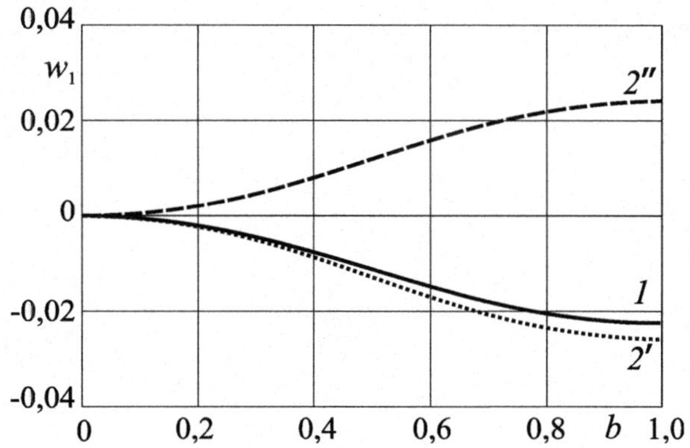

Fig. 7.41 Variation of the deflection of the first layer in the cross section of the rod ($x = 0.5$) as a function of coordinate *b*

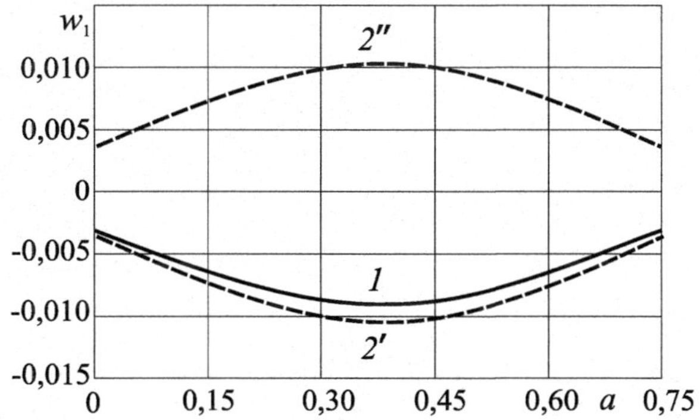

Fig. 7.42 Variation of the deflection of the first layer in the cross section of the rod ($x = 0.5$) as a function of coordinate *a*

material, since the effect of neutron flux stops. Therefore, the maximum value of deflections during repeated loading in neutron flux decreases by 1–2%, however, in case of subsequent cyclic load changes this discrepancy will accumulate.

Thus, the formulations and solutions of boundary value problems given in this section allow us to take into account the compressibility of the filler during deformation of physically nonlinear three-layered rods in thermoradiation fields under both single and variable loads.

Let us note some problems of quasi-statics for three-layer rods that are waiting to be solved. Here, first of all, we should point out the possibility of taking into account viscoelastic properties of polymeric materials and creep of metals, especially at elevated temperatures. It would also be interesting to study the deformation of such structural elements in electromagnetic fields.

Chapter 8
Elastic Circular Three-Layer Plates in a Temperature Field

Axisymmetric bending of asymmetric in thickness circular elastic three-layer plates with rigid incompressible filler in the temperature field is investigated. The kinematic assumptions are based on the “broken” line hypothesis. Loads are continuous and localised. Strains are small. The work of tangential stresses in the filler is taken into account. Numerical analysis of the dependence of the obtained solutions on temperature and boundary conditions was carried out.

8.1 Formulation of the Boundary Value Problem of Thermoelasticity

The problem formulation and its solution are carried out in the cylindrical coordinate system r, φ, z. The median plane of the filler is taken as the coordinate plane, with the z-axis pointing perpendicularly upwards to it. For thin outer bearing layers of thickness $h_1 \neq h_2$ Kirchhoff’s hypotheses are accepted; for a thick rigid filler ($h_3 = 2c$), taking the load in the tangential direction, the hypothesis of rectilinearity and incompressibility of the deformed normal, which is rotated by some additional angle $\psi(r)$, is valid.

It is considered that arbitrary distributed loads $q(r), p(r)$ and heat flux q_t are applied to the outer surface of the first load-bearing layer (Fig. 8.1), and the linear forces and moments $T_r^0, H_r^0, M_r^0, Q_r^0$ are applied to the contour of the plate. We consider the surface $z = -c - h_2$ and the contour of the plate to be thermally insulated. This allows the inhomogeneous temperature field $T(z)$, counted from some initial temperature T_0, to be calculated with sufficient accuracy by formula (2.55). Due to the symmetry of the load, there are no tangential displacements in the layers: $u_\varphi{}^{(k)} = 0$ (k is layer number), and the plate deflection w, the relative shear in the filler ψ and the radial displacement of the coordinate surface u do not depend on the coordinate φ, that is $w(r)$, $u(r)$, $\psi(r)$. In the following we consider these functions to be sought functions.

S. Eduard et al., *Deformation of Three-layer Structural Elements in Thermal Radiation Fields*, https://doi.org/10.1007/978-981-97-7217-9_8

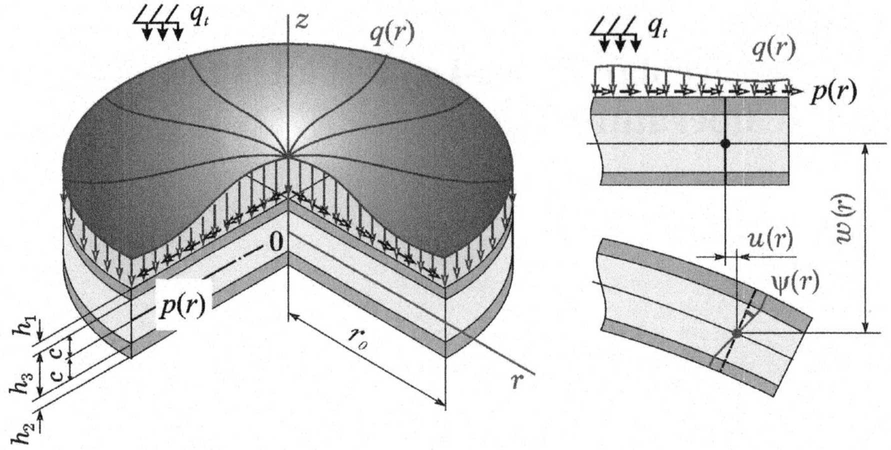

Fig. 8.1 Scheme of thermal force loading of a three-layer plate

A rigid diaphragm is assumed on the plate contour to prevent relative shear of the layers ($\psi = 0$ at $r = 1$). All displacements and linear dimensions of the plate are referred to its radius r_0.

Based on the hypothesis of straightness of the filler normal, we can write:

$$2\varepsilon_{rz}^{(3)} = u_r^{(3)},_z + w,_r = \psi.$$

After integration, using the accepted hypotheses, we obtain the following expressions for radial displacements in the layers $u_r^{(k)}$:

$$\begin{aligned} u_r^{(1)} &= u + c\psi - zw,_r, & c \le z \le c + h_1, \\ u_r^{(3)} &= u + z\psi - zw,_r, & -c \le z \le c, \\ u_r^{(2)} &= u - c\psi - zw,_r, & -c - h_2 \le z \le -c, \end{aligned} \tag{8.1}$$

where z is distance from the fibre in consideration to the median plane of the filler; $u + c\psi$ is displacement value of the outer bearing layer due to the deformation of the filler, for the second bearing layer this displacement will be $u - c\psi$; comma in the lower index denotes the differentiation operation on the coordinate following it.

The strains in the layers follow from (8.1) and Cauchy relations

$$\begin{aligned} \varepsilon_r^{(1)} &= u,_r + c\psi,_r - zw,_{rr}, & \varepsilon_\varphi^{(1)} &= \frac{1}{r}(u + c\psi - zw,_r); & \varepsilon_{rz}^{(1)} &= 0, \\ \varepsilon_r^{(3)} &= u,_r + z\psi,_r - zw,_{rr}, & \varepsilon_\varphi^{(3)} &= \frac{1}{r}(u + z\psi - zw,_r); & \varepsilon_{rz}^{(3)} &= \frac{1}{2}\psi, \\ \varepsilon_r^{(2)} &= u,_r - c\psi,_r - zw,_{rr}, & \varepsilon_\varphi^{(2)} &= \frac{1}{r}(u - c\psi - zw,_r); & \varepsilon_{rz}^{(2)} &= 0. \end{aligned} \tag{8.2}$$

Using the components of the stress tensor $\sigma_\alpha^{(k)}$ ($\alpha = r, \varphi$), we introduce generalised internal loads (forces) and moments in the plate:

$$T_\alpha \equiv \sum_{k=1}^{3} T_\alpha^{(k)} = \sum_{k=1}^{3} \int_{h_k} \sigma_\alpha^{(k)} \mathrm{d}z, \; M_\alpha \equiv \sum_{k=1}^{3} M_\alpha^{(k)} = \sum_{k=1}^{3} \int_{h_k} \sigma_\alpha^{(k)} z \mathrm{d}z,$$

$$H_\alpha = M_\alpha^{(3)} + c\left(T_\alpha^{(1)} - T_\alpha^{(2)}\right), \; Q = \int_{-c}^{c} \sigma_{rz}^{(3)} \mathrm{d}z. \tag{8.3}$$

The equations of equilibrium of the three-layer plate are obtained using the Lagrange principle of possible displacements (2.45):

$$\delta A = \delta W, \tag{8.4}$$

where δA, δW are variations of external forces and internal stresses.

When determining the work of external forces, we consider that arbitrary distributed loads are applied to the medial surface of the filler, and linear forces and moments are applied to the contour of the plate. The variation of the external surface load work is represented as ($dS = r\mathrm{d}r\mathrm{d}\varphi$):

$$\begin{aligned} \delta A_1 &= \iint_S (q\delta w + p\delta u) r\mathrm{d}r\mathrm{d}\varphi. \\ \delta A_2 &= \int_0^{2\pi} (T_r^0 \delta u + H_r^0 \delta\psi + M_r^0 \delta w,_r + Q^0 \delta w)\mathrm{d}\varphi. \end{aligned} \tag{8.5}$$

Variation of the work of elastic forces

$$\delta W = \iint_S \left[\sum_{k=1}^{3} \int_{h_k} (\sigma_r^{(k)} \delta\varepsilon_r^{(k)} + \sigma_\varphi^{(k)} \delta\varepsilon_\varphi^{(k)})\mathrm{d}z + \int_{-c}^{c} \sigma_{rz}^{(3)} \delta\psi \, \mathrm{d}z \right] r\mathrm{d}r\mathrm{d}\varphi. \tag{8.6}$$

Here the double integral is spread over the entire medial surface of the filler S.

Variations of displacements in layers follow from relations (8.1):

$$\begin{aligned} \delta u^{(1)} &= \delta u + c\delta\psi - z\delta w,_r \; ; & c \le z \le c + h_1, \\ \delta u^{(3)} &= \delta u + z\delta\psi - z\delta w,_r ; & -c \le z \le c, \\ \delta u^{(2)} &= \delta u - c\delta\psi - z\delta w,_r ; & -c - h_2 \le z \le -c, \end{aligned}$$

and the strain variations follow from (8.2):

$$\delta\varepsilon_r^{(1)} = \delta u,_r + c\delta\psi,_r - z\delta w,_{rr}, \quad \delta\varepsilon_\varphi^{(1)} = \frac{1}{r}(\delta u + c\delta\psi - z\delta w,_r),$$
$$\delta\varepsilon_r^{(3)} = \delta u,_r + z\delta\psi,_r - z\delta w,_{rr}, \quad \delta\varepsilon_\varphi^{(3)} = \frac{1}{r}(\delta u + z\psi - z\delta w,_r),$$
$$\delta\varepsilon_r^{(2)} = \delta u,_r - c\delta\psi,_r - z\delta w,_{rr}, \quad \delta\varepsilon_\varphi^{(2)} = \frac{1}{r}(\delta u - c\delta\psi - z\delta w,_r), \quad \delta\varepsilon_{rz}^{(3)} = \frac{1}{2}\delta\psi. \tag{8.7}$$

Let us consider the total integral over the thickness of layers, included in the virtual work of elasticity forces (8.6) Using variations of strains (8.7) for radial components in the first layer, we obtain the following expression

$$\int_{h_1} \sigma_r^{(1)}\delta\varepsilon_r^{(1)}dz = \int_{h_1} \sigma_r^{(1)}(\delta u,_r + c\delta\psi,_r - z\delta w,_{rr})dz$$
$$= T_r^{(1)}\delta u,_r + cT_r^{(1)}\delta\psi,_r - M_r^{(1)}\delta w,_{rr}.$$

Likewise.

$$\int_{h_2} \sigma_r^{(2)}\delta\varepsilon_r^{(2)}dz = T_r^{(2)}\delta u,_r - cT_r^{(2)}\delta\psi,_r - M_r^{(2)}\delta w,_{rr},$$
$$\int_{h_3} \sigma_r^{(3)}\delta\varepsilon_r^{(3)}dz = T_r^{(3)}\delta u,_r + M_r^{(3)}\delta\psi,_r - M_r^{(3)}\delta w,_{rr},$$
$$\int_{h_1} \sigma_\varphi^{(1)}\delta\varepsilon_\varphi^{(1)}dz = \int_{h_1} \sigma_\varphi^{(1)}\frac{1}{r}(\delta u + c\delta\psi - z\delta w,_r)dz$$
$$= \frac{1}{r}\left(T_\varphi^{(1)}\delta u + cT_\varphi^{(1)}\delta\psi - M_\varphi^{(1)}\delta w,_r\right),$$
$$\int_{h_2} \sigma_\varphi^{(2)}\delta\varepsilon_\varphi^{(2)}dz = \frac{1}{r}\left(T_\varphi^{(2)}\delta u - cT_\varphi^{(2)}\delta\psi - M_\varphi^{(2)}\delta w,_r\right).$$
$$\int_{h_3} \sigma_\varphi^{(3)}\delta\varepsilon_\varphi^{(3)}dz = \frac{1}{r}\left(T_\varphi^{(3)}\delta u + M_\varphi^{(3)}\delta\psi - M_\varphi^{(3)}\delta w,_r\right)$$

Let's sum up the obtained expressions:

$$\sum_{k=1}^{3}\int_{h_k}(\sigma_r^{(k)}\delta\varepsilon_r^{(k)} + \sigma_\varphi^{(k)}\delta\varepsilon_\varphi^{(k)})dz + \int_{-c}^{c}\sigma_{rz}^{(3)}\delta\varepsilon dz$$
$$= T_r\delta u,_r + H_r\delta\psi,_r - M_r\delta w,_{rr} + Q\delta\psi + \frac{1}{r}\left(T_\varphi\delta u + H_\varphi\delta\psi - M_\varphi\delta w,_r\right),$$

where the internal forces T_α, M_α, H_α and Q are introduced by relations (8.3).

The variation of the potential energy of deformation is obtained by substituting the last expression into (8.6) and introducing integration in the polar coordinate system. Then we have:

$$\delta W = \int_r \int_\varphi [r(T_r \delta u,_r + H_r \delta \psi,_r - M_r \delta w,_{rr} + Q\delta\psi) + T_\varphi \delta u + H_\varphi \delta \psi - M_\varphi \delta w,_r] \mathrm{d}\varphi \mathrm{d}r.$$

Let us represent the terms in the last integrand in the following form:

$$\begin{aligned}
&rT_r\delta u,_r = (rT_r\delta u),_r - (rT_r),_r \delta u, \quad rH_r\delta\psi,_r = (rH_r\delta\psi),_r - (rH_r),_r \delta\psi, \\
&rM_r\delta w,_{rr} = (rM_r\delta w,_r),_r - [(rM_r),_r \delta w],_r + (rM_r),_{rr} \delta w, \\
&M_\varphi\delta w,_r = (M_\varphi\delta w),_r - M_\varphi,_r \delta w.
\end{aligned}$$

The variation of the stress work δW can then be decomposed into two integrals, taking the differentiation operation in the first of them out of the common bracket, and grouping the terms at the same virtual displacements in the second one:

$$\delta W = \int_r \int_\varphi \left\{ rT_r\delta u + rH_r\delta\psi - rM_r\delta w,_r + [(rM_r),_r - M_\varphi]\delta w \right\},_r \mathrm{d}\varphi \mathrm{d}r -$$

$$\int_r \int_\varphi \left\{ [(rT_r),_r - T_\varphi]\delta u + [(rH_r),_r - H_\varphi]\delta\psi - [(rM_r),_{rr} - M_\varphi,_r]\delta w \right\} \mathrm{d}\varphi \mathrm{d}r.$$

From here:

$$\delta W = \int_0^{2\pi} \left\{ rT_r\delta u + rH_r\delta\psi - rM_r\delta w,_r + [(rM_r),_r - M_\varphi]\delta w \right\} \mathrm{d}\varphi -$$

$$\int_r \int_\varphi \left\{ [(rT_r),_r - T_\varphi]\delta u + [(rH_r),_r - H_\varphi - rQ]\delta\psi + [(rM_r),_{rr} - M_\varphi,_r]\delta w \right\} \mathrm{d}\varphi \mathrm{d}r.$$

Let's substitute the obtained expression and virtual work of external load (8.5) into (8.4) and require the fulfilment of this equality at any values of varying displacements. It is possible at equality to zero of coefficients at independent variations of required functions. From here follows a system of differential equations of equilibrium in forces describing the deformation of a circular three-layer plate:

$$\begin{aligned}
&T_r,_r + \frac{1}{r}(T_r - T_\varphi) = -p, \\
&H_r,_r + \frac{1}{r}(H_r - H_\varphi) - Q = 0, \\
&M_r,_{rr} + \frac{1}{r}(2M_r,_r - M_\varphi,_r) = -q.
\end{aligned} \tag{8.8}$$

At the boundary $r = 1$ the force conditions must be fulfilled:

$$T_r = T_r^0, \ \ H_r = H_r^0, \ \ M_r = M_r^0, \ \ M_{r},_r + \frac{1}{r}(M_r - M_\varphi) = Q^0, \qquad (8.9)$$

where T_r^0, M_r^0, Q^0 are boundary forces.

Thermoelastic relations of Hooke's law (2.16) in deviator-spherical form are used to obtain equations of equilibrium in displacements:

$$\begin{aligned} s_\alpha^{(k)} &= 2G_k(T_k)\ э_\alpha^{(k)}, \quad \sigma^{(k)} = 3K_k(T_k)(\varepsilon^{(k)} - \alpha_{0k}\Delta T_k) \quad (k = 1, 2, 3), \\ s_{rz}^{(3)} &= 2G_k(T_k)\ э_{rz}^{(3)} \quad (\alpha = r, \varphi), \end{aligned} \qquad (8.10)$$

where $s_\alpha^{(k)}$, $э_\alpha^{(k)}$ are deviator parts and $\sigma^{(k)}$, $\varepsilon^{(k)}$ are spherical parts of stress and strain tensors, $G_k(T_k)$, $K_k(T_k)$ are temperature-dependent shear and volume deformation modules; T_k is temperature, ΔT_k is temperature increment, α_{0k} is coefficient of linear thermal expansion of the kth layer material.

The expression of the generalised forces through the three required functions $u(r)$, $\psi(r)$, $w(r)$ follows after substituting (8.10) into (8.3), taking into account the strains (8.2). A detailed derivation is given for T_r:

$$\begin{aligned} T_r &\equiv \sum_{k=1}^{3} T_r^{(k)} = \sum_{k=1}^{3} \int_{h_k} \sigma_r^{(k)} \mathrm{d}z = \sum_{k=1}^{3} \int_{h_k} (2G_k\ э_r^{(k)} + 3K_k(\varepsilon^{(k)} - \alpha_{0k}T_k))\mathrm{d}z \\ &= \sum_{k=1}^{3} \int_{h_k} [2G_k(\varepsilon_r^{(k)} - \varepsilon^{(k)}) + 3K_k(\varepsilon^{(k)} - \alpha_{0k}T_k)]\mathrm{d}z \\ &= \sum_{k=1}^{3} \int_{h_k} [2G_k\varepsilon_r^{(k)} + (K_k - \frac{2}{3}G_k)(\varepsilon_r^{(k)} + \varepsilon_\varphi^{(k)}) - 3K_k\alpha_{0k}T_k]\mathrm{d}z \\ &= \sum_{k=1}^{3} \int_{h_k} [(K_k + \frac{4}{3}G_k)\varepsilon_r^{(k)} + (K_k - \frac{2}{3}G_k)\varepsilon_\varphi^{(k)} - 3K_k\alpha_{0k}T_k]\mathrm{d}z \\ &= \sum_{k=1}^{3} \int_{h_k} (K_k^+\varepsilon_r^{(k)} + K_k^-\varepsilon_\varphi^{(k)}T_\varphi - 3K_k\alpha_{0k}T_k)\mathrm{d}z \end{aligned}$$

where the following notations are introduced:

$$K_k + \frac{4}{3}G_k \equiv K_k^+, \ \ K_k - \frac{2}{3}G_k \equiv K_k^-.$$

Then, decomposing the last sum by layers, we have:

$$T_r = \int_{c}^{c+h_1} \left[K_1^+ (u,_r + c\psi,_r - zw,_{rr}) + \frac{1}{r} K_1^- (u + c\psi - zw,_r) \right] \mathrm{d}z$$

$$- 3K_1\alpha_{01} \int_{c}^{c+h_1} T_1 \mathrm{d}z$$

$$+ \int_{-c-h_2}^{-c} \left[K_2^+ (u,_r - c\psi,_r - zw,_{rr}) + \frac{1}{r} K_2^- (u - c\psi - zw,_r) \right] \mathrm{d}z$$

$$- 3K_2\alpha_{02} \int_{-c-h_2}^{-c} T_2 \mathrm{d}z$$

$$+ \int_{-c}^{c} \left[K_3^+ (u,_r + z\psi,_r - zw,_{rr}) + \frac{1}{r} K_3^- (u + z\psi - zw,_r) \right] \mathrm{d}z$$

$$- 3K_3\alpha_{03} \int_{-c}^{c} T_3 \mathrm{d}z \tag{8.11}$$

Let's write out the simplest integrals that we need later on:

$$\int_{h_k} dz = h_k, \quad \int_{c}^{c+h_1} zdz = h_1\left(c + \frac{h_1}{2}\right), \quad \int_{-c}^{c} zdz = 0, \quad \int_{-c-h_2}^{-c} zdz = -h_2\left(c + \frac{h_2}{2}\right),$$

$$\int_{c}^{c+h_1} z^2dz = h_1\left(c^2 + ch_1 + \frac{h_1^2}{3}\right), \quad \int_{-c}^{c} z^2dz = \frac{2}{3}c^3, \quad \int_{-c-h_2}^{-c} z^2dz = h_2\left(c^2 + ch_2 + \frac{h_2^2}{3}\right).$$

With the help of these relations, after integration of expression (8.11) and adding similar ones, we obtain:

$$T_r = \sum_{k=1}^{3} \left[h_k(Ku,_r + \frac{u}{r} K_k^-) - 3K_k\alpha_{0k} \int_{h_k} T_k \mathrm{d}z \right]$$
$$+ c(K_1^+ h_1 - K_2^+ h_2)\psi,_r + c(K_1^- h_1 - K_2^- h_2)\frac{\psi}{r}$$
$$- \left[K_1^+ h_1 \left(c + \frac{h_1}{2}\right) - K_2^+ h_2 \left(c + \frac{h_2}{2}\right) \right] w,_{rr}$$
$$- \left[K_1^- h_1 \left(c + \frac{h_1}{2}\right) - K_2^- h_2 \left(c + \frac{h_2}{2}\right) \right] \frac{w,_r}{r}. \tag{8.12}$$

The expression for T_φ follows from (8.12), if the coefficients K_k^+ and K_k^- are swapped:

$$T_\varphi = \sum_{k=1}^{3} \int_{h_k} (K_k^- \varepsilon_r^{(k)} + K_k^+ \varepsilon_\varphi^{(k)} - 3K_k \alpha_{0k} T_k)\mathrm{d}z +$$
$$\sum_{k=1}^{3} \left[h_k (K_k^- u,_r + \frac{u}{r} K_k^+) - 3K_k \alpha_{0k} \int_{h_k} T_k \mathrm{d}z \right] +$$
$$c(K_1^- h_1 - K_2^- h_2)\psi,_r + c(K_1^+ h_1 - K_2^+ h_2)\frac{\psi}{r} - \left[K_1^- h_1 \left(c + \frac{h_1}{2}\right) - K_2^- h_2 \left(c + \frac{h_2}{2}\right) \right] w,_{rr}$$
$$- \left[K_1^+ h_1 \left(c + \frac{h_1}{2}\right) - K_2^+ h_2 \left(c + \frac{h_2}{2}\right) \right] \frac{w,_r}{r} \qquad (8.13)$$

Similarly we obtain the generalised moments M_r, H_r:

$$\begin{aligned}
M_r = &\left[K_1^+ h_1 \left(c + \frac{h_1}{2}\right) - K_2^+ h_2 \left(c + \frac{h_2}{2}\right) \right] u,_r + \left[K_1^- h_1 \left(c + \frac{h_1}{2}\right) - K_2^- h_2 \left(c + \frac{h_2}{2}\right) \right] \frac{u}{r} \\
&+ \left[cK_1^+ h_1 \left(c + \frac{h_1}{2}\right) + cK_2^+ h_2 \left(c + \frac{h_2}{2}\right) + \frac{2}{3} c^3 K_3^+ \right] \psi,_r \\
&+ \left[cK_1^- h_1 \left(c + \frac{h_1}{2}\right) + cK_2^- h_2 \left(c + \frac{h_2}{2}\right) + \frac{2}{3} c^3 K_3^- \right] \frac{\psi}{r} \\
&- \left[K_1^+ h_1 \left(c^2 + ch_1 + \frac{h_1^2}{3}\right) + K_2^+ h_2 \left(c^2 + ch_2 + \frac{h_2^2}{3}\right) + \frac{2}{3} c^3 K_3^+ \right] w,_{rr} \\
&- \left[K_1^- h_1 \left(c^2 + ch_1 + \frac{h_1^2}{3}\right) + K_2^- h_2 \left(c^2 + ch_2 + \frac{h_2^2}{3}\right) + \frac{2}{3} c^3 K_3^- \right] \frac{w,_r}{r} \\
&- \sum_{k=1}^{3} 3K_k \alpha_{0k} \int_{h_k} T_k z \mathrm{d}z \\
H_r = &\, c(K_1^+ h_1 - K_2^+ h_2) u,_r + c(K_1^- h_1 - K_2^- h_2) \frac{u}{r} \\
&+ \left[c^2 (K_1^+ h_1 + K_2^+ h_2) + \frac{2}{3} c^3 K_3^+ \right] \psi,_r + \left[c^2 (K_1^- h_1 + K_2^- h_2) + \frac{2}{3} c^3 K_3^- \right] \frac{\psi}{r} \\
&- \left[c \left(K_1^+ h_1 (c + \frac{h_1}{2}) + K_2^+ h_2 (c + \frac{h_2}{2}) \right) + \frac{2}{3} c^3 K_3^+ \right] w,_{rr} \\
&- \left[c \left(K_1^- h_1 (c + \frac{h_1}{2}) + K_2^- h_2 (c + \frac{h_2}{2}) \right) + \frac{2}{3} c^3 K_3^- \right] \frac{w,_r}{r} \\
&- 3K_3 \alpha_{03} \int_{h_3} T_3 z \mathrm{d}z - 3c \left(K_1 \alpha_{01} \int_{h_1} T_1 \mathrm{d}z - K_2 \alpha_{02} \int_{h_2} T_2 \mathrm{d}z \right), \\
Q = &\, 2cG_3 \psi.
\end{aligned} \qquad (8.14)$$

The relations for M_φ, H_φ follow from (8.14), if in the same-named efforts to swap the values K_k^+ and K_k^-.

After substituting the obtained expressions for the generalised internal forces and moments (8.12–8.14) into Eq. (8.8) and carrying out the necessary transformations, we obtain in displacements the following system of differential equations of equilibrium for a circular three-layer plate located in the temperature field:

$$\begin{aligned} &\mathrm{L}_2(a_1 u + a_2 \psi - a_3 w,_r) = -p, \\ &\mathrm{L}_2(a_2 u + a_4 \psi - a_5 w,_r) - 2cG_3 \psi = 0\,, \\ &\mathrm{L}_3(a_3 u + a_5 \psi - a_6 w,_r) = -q. \end{aligned} \tag{8.15}$$

Here linear differential operators L_2 (Bessel operator), L_3 are as follows:

$$\begin{aligned} &\mathrm{L}_2(g) \equiv \left(\frac{1}{r}(rg),_r\right),_r \equiv g,_{rr} + \frac{g,_r}{r} - \frac{g}{r^2}, \\ &\mathrm{L}_3(g) \equiv \frac{1}{r}(r\mathrm{L}_2(g)),_r \equiv g,_{rrr} + \frac{2g,_{rr}}{r} - \frac{g,_r}{r^2} + \frac{g}{r^3}. \end{aligned} \tag{8.16}$$

The coefficients a_i are determined by the integral relations following from (8.11–8.14), since the elastic modules of materials in the layers change along their thickness and in time together with the temperature:

$$a_1 = \sum_{k=1}^{3} K_{k0}, \quad a_2 = c(K_{10} - K_{20}), \quad a_3 = \sum_{k=1}^{3} K_{k1},$$

$$a_4 = K_{32} + c^2(K_{10} + K_{20}), \quad a_5 = K_{32} + c(K_{11} + K_{21}), \quad a_6 = \sum_{k=1}^{3} K_{k2}, \tag{8.17}$$

$$K_{km} = \int_{h_k} \left[K_k(T_k) + \frac{4}{3}G_k(T_k)\right] z^m \mathrm{d}z, \quad (m = 0,\ 1,\ 2).$$

If the temperature field is averaged over the thickness of the kth layer of the plate, the values of elastic parameters are the same inside the layer:

$$G_k(T_k) \equiv G_k(T_{0k}), \; K_k(T_k) \equiv K_k(T_{0k}), \; T_{0k}(t) = \frac{1}{h_k}\int_{h_k} T_k(z,\ t)\,\mathrm{d}z.$$

In this case, after integration in (8.17), we obtain the coefficients:

$$a_1 = \sum_{k=1}^{3} h_k K_k^+, \quad a_2 = c(h_1 K_1^+ - h_2 K_2^+),$$

$$a_3 = h_1\left(c + \frac{1}{2}h_1\right)K_1^+ - h_2\left(c + \frac{1}{2}h_2\right)K_2^+,\ a_4 = c^2\left(h_1K_1^+ + h_2K_2^+ + \frac{2}{3}cK_3^+\right),$$
$$a_5 = c\left[h_1\left(c + \frac{1}{2}h_1\right)K_1^+ + h_2\left(c + \frac{1}{2}h_2\right)K_2^+ + \frac{2}{3}c^2K_3^+\right],$$
$$a_6 = h_1\left(c^2 + ch_1 + \frac{1}{3}h_1^2\right)K_1^+ + h_2\left(c^2 + ch_2 + \frac{1}{3}h_2^2\right)K_2^+ + \frac{2}{3}c^3K_3^+. \tag{8.18}$$

The boundary value problem of determining the functions $u(r)$, $\psi(r)$, $w(r)$ is closed by joining to the system of Eq. (8.15) the force (8.9) or kinematic boundary conditions:

- when the plate contour is rigidly embedded

$$u = \psi = w = w,_r = 0,\ at\, r = 1;$$

- when the plate is hinged

$$u = \psi = w = M_r = 0,\ at\, r = 1. \tag{8.19}$$

It should be noted that the temperature is not explicitly included in the equations of equilibrium (8.15), as in the case of a three-layer rod with incompressible filler. The corresponding temperature terms in the internal generalised forces are reduced after substituting (8.12–8.14) into Eq. (8.8).

8.2 General Solution of the Boundary Value Problem of the Theory of Elasticity

Let us give the method of analytical solution of the system of differential Eq. (8.15), which will be used in the future as well. After one-time integration from the third Eq. (8.15) follows:

$$L_2(a_3u + a_5\psi - a_6w,_r) = -\frac{1}{r}\left(\int qr\,\mathrm{d}r - C_1\right).$$

The integration constant C_1 will be defined below.

Using the obtained equation and the first one from (8.15), we can exclude the radial displacement $u(r)$ and deflection $w(r)$ from the second equation of the same system. As a result, we obtain the inhomogeneous modified Bessel equation for finding the function $\psi(r)$:

$$L_2(\psi) - \beta^2\psi = f. \tag{8.20}$$

Here

$$\beta^2 = \frac{2cb_3G_3}{b_1b_3 - b_2^2},\ b_1 = \frac{a_1a_4 - a_2^2}{a_1},\ b_2 = \frac{a_1a_5 - a_2a_3}{a_1},\ b_3 = \frac{a_1a_6 - a_3^2}{a_1},$$
$$f(r) = \frac{b_3}{b_1b_3 - b_2^2}\left[\frac{p(a_2b_3 - a_3b_2)}{a_1b_3} + \frac{b_2}{b_3r}\left(\int q r \mathrm{d}r - C_1\right)\right], \tag{8.21}$$

where the integration constant C_1 includes numerical multipliers.

The solution of the obtained equation can be represented as the sum of the general solution of the corresponding homogeneous equation ψ_0 and the partial solution ψ_r of the inhomogeneous Eq. (8.20):

$$\psi(r) = \psi_0(r) + \psi_r(r), \tag{8.22}$$

Herewith

$$\mathrm{L}_2(\psi_0) - \beta^2\psi_0 = 0.$$

The solution to this equation is known [1]:

$$\psi_0 = C_2I_1(\beta r) + C_3K_1(\beta r). \tag{8.23}$$

Here $I_1(\beta r)$ is the modified first order Bessel function, $K_1(\beta r)$ is the Macdonald function.

As it is known from the theory of linear differential equations of the second order, the partial solution of the Eq. (8.20) can be obtained using two linearly independent fundamental solutions of the corresponding homogeneous equation. It is written as:

$$\psi_r(r) = K_1(\beta r)\int \frac{I_1(\beta r) f(r)}{W}\mathrm{d}r - I_1(\beta r)\int \frac{K_1(\beta r) f(r)}{W}\mathrm{d}r, \tag{8.24}$$

where W is Vronsky 's determinant:

$$W\{I_1,\ K_1\} = K_{1,r}I_1 - I_{1,r}\ K_1 = -\frac{1}{r}.$$

Thus, the partial solution (8.24) in our problem takes the form:

$$\psi_r = -K_1(\beta r)\int I_1(\beta r) f(r) r\mathrm{d}r + I_1(\beta r)\int K_1(\beta r) f(r) r\mathrm{d}r. \tag{8.25}$$

Substituting (8.23, 8.25) into (8.22) we obtain the required solution for the shear. Then deflection and radial displacement follow from the remaining equations of the system (8.15). The bottom line is:

$$\begin{aligned}
\psi &= C_2 I_1(\beta r) + C_3 K_1(\beta r) - K_1(\beta r)\int I_1(\beta r) f(r) r\mathrm{d}r + I_1(\beta r)\int K_1(\beta r) f(r) r\mathrm{d}r,\\
w &= \frac{b_2}{b_3}\int \psi\,\mathrm{d}r - \frac{a_3}{b_3 a_1}\int L_2^{-1}(p)\mathrm{d}r + \frac{1}{b_3}\int L_3^{-1}(q)\mathrm{d}r + \frac{1}{4}C_1 r^2(\ln r - 1)\\
&\quad + \frac{C_5 r^2}{4b_3} + C_6 \ln r + C_4,\\
u &= \frac{a_3}{a_1}w_{,r} - \frac{a_2}{a_1}\psi - \frac{1}{a_1}L_2^{-1}(p) + \frac{C_7 r}{2} + \frac{C_8}{r},
\end{aligned} \tag{8.26}$$

where C_1, C_3, ..., C_8 are integration constants, L_2^{-1}, L_3^{-1} are linear integral operators inverse to differential operators (8.16):

$$\begin{aligned}
L_2^{-1}(f) &\equiv \frac{1}{r}\int r\int f\,\mathrm{d}r\mathrm{d}r,\\
\mathrm{L}_3^{-1}(f) &\equiv \frac{1}{r}\int r\int \frac{1}{r}\int rf\,\mathrm{d}r\mathrm{d}r\mathrm{d}r.
\end{aligned}$$

All integrals occurring in the obtained solution and further on should be understood as defined with variable upper limit, i.e.:

$$\int f(r)dr \equiv \int_0^r f(z)\mathrm{d}z.$$

Proceeding from the condition of boundedness of displacements in the center of the plate, it is necessary in expressions (8.21, 8.25, 8.26) for solid plates to put the constants of integration $C_1 = C_3 = C_6 = C_8 = 0$, since the logarithm, the Macdonald function and the function r^{-1} at the origin of coordinates ($r = 0$) are unbounded. The other four integration constants are determined from the boundary conditions of the type (8.19)

As a result, the general analytical solution of the problem of thermoelasticity theory on deformation of a solid circular three-layered plate in the temperature field takes the following form:

$$\begin{aligned}
\psi &= C_2 I_1(\beta r) - K_1(\beta r)\int I_1(\beta r) f(r) r\mathrm{d}r + I_1(\beta r)\int K_1(\beta r) f(r) r\mathrm{d}r,\\
w &= \frac{1}{b_3}\left[b_2\left(\frac{C_2}{\beta}I_0(\beta r) + \int \psi_r \mathrm{d}r\right) - \int\left(\frac{a_3}{a_1}L_2^{-1}(p) - L_3^{-1}(q)\right)\mathrm{d}r + \frac{C_5 r^2}{4} + C_4\right],\\
u &= \frac{a_3}{a_1 a_6 - a_3^2}\left[L_3^{-1}(q) - \frac{a_6}{a_3}L_2^{-1}(p) + \left(a_5 - \frac{a_2 a_6}{a_3}\right)\psi + C_7 r\right],
\end{aligned} \tag{8.27}$$

If the plate contour is embedded, then, satisfying the first condition (8.19), we obtain the following integration constants:

$$C_2 = -\frac{\psi_r}{I_1(\beta)}\bigg|_{r=1},$$

$$C_4 = -b_2\left(\frac{C_2}{\beta}I_0(\beta) + \int \psi_r \mathrm{d}r\bigg|_{r=1}\right) + \int \left(\frac{a_3}{a_1}\mathrm{L}_2^{-1}(p) - \mathrm{L}_3^{-1}(q)\right)\mathrm{d}r\bigg|_{r=1} - \frac{C_5}{4},$$

$$C_5 = 2\left(\frac{a_3}{a_1}\mathrm{L}_2^{-1}(p) - \mathrm{L}_3^{-1}(q)\right)\bigg|_{r=1}, C_7 = \left(\frac{a_6}{a_3}\mathrm{L}_2^{-1}(p) - \mathrm{L}_3^{-1}(q)\right)\bigg|_{r=1}. \quad (8.28)$$

In this case, the solution (8.27) does not depend explicitly on temperature. It will affect only the value of the elastic modules of the materials of the layers of the considered three-layer plate.

8.3 Case of Uniformly Distributed Surface Load

At bending of a plate by continuous uniformly distributed transverse load ($p = 0, q = const$) from (8.21) and (8.25) follows:

$$f(r) = \frac{b_2 q}{2(b_1 b_3 - b_2^2)}r, \quad \psi_r(r) = -\frac{b_2 q}{4cb_3 G_3}r.$$

Substituting these expressions into formulas (8.27, 8.28), we obtain the solution of the bending problem of a circular thermoelastic three-layer plate clamped along its contour:

$$\psi = \frac{b_2 q}{4cb_3 G_3}\left(\frac{I_1(\beta r)}{I_1(\beta)} - r\right),$$

$$u = \frac{a_3}{a_1 a_6 - a_3^2}\left[\left(a_5 - \frac{a_2 a_6}{a_3}\right)\psi + \frac{qr}{16}(r^2 - 1)\right], \quad (8.29)$$

$$w = \frac{b_2^2 q}{4cb_3^2 G_3}\left(\frac{I_0(\beta r) - I_0(\beta)}{\beta I_1(\beta)} - \frac{1}{2}(r^2 - 1)\right) + \frac{q}{64b_3}(r^2 - 1)^2.$$

If the contour of the plate is hinge supported, then in bending by a uniformly distributed transverse load, the integration constants follow from the second of conditions (8.19):

$$C_2 = \frac{b_2 q}{4cb_3 G_3 I_1(\beta)}, \quad C_4 = -\frac{b_2^2 q}{4cb_3 G_3}\left(\frac{I_0(\beta)}{\beta I_1(\beta)} - \frac{1}{2}\right) - \frac{q}{64} - \frac{C_5}{4},$$

$$C_5 = \frac{2b_3}{a_6 + a_7}\left[a_3 u,_r(1) + \psi,_r(1)\left(a_5 - \frac{a_6 b_2}{b_3}\right)\right.$$

$$\left. - \frac{q}{16b_3}(3a_6 + a_7) - 3\sum_{k=1}^{3}\alpha_{0k}\int_{h_k} K_k T_k z \mathrm{d}z\right], C_7 = -\frac{q}{16}, \quad (8.30)$$

where

$$\psi,_r(1) = \frac{b_2 q}{4cb_3G_3}\left(\frac{\beta I_0(\beta)}{I_1(\beta)} - 2\right),\; u,_r(1) = \frac{a_3}{a_1a_6 - a_3^2}\left[\left(a_5 - \frac{a_2a_6}{a_3}\right)\psi,_r(1) + \frac{q}{8}\right],$$

$$a_7 = h_1\left(c^2 + ch_1 + \frac{h_1^2}{3}\right)K_1^- + h_2\left(c^2 + ch_2 + \frac{h_2^2}{3}\right)K_2^- + \frac{2}{3}c^3K_3^-.$$

In this case, the solution will depend explicitly on the temperature through the integration constants C_4, C_5 (8.30). At the free edge of the plate the integration constants are determined from the conditions $\psi = 0$, $T_r = M_r = M_r,_r = 0$, which will also entail their direct dependence on temperature.

Numerical results in this section are given for a circular three-layer asymmetric in thickness plate, the layers of which are made of materials D16T–fluoroplastic–D16T. Mechanical characteristics of materials are given in Tables 2.1 and 2.2 Load intensity value $q_0 = 1$ MPa. Geometrical parameters of the plate are referred to its radius r_0.

The dependence of shear (*a*) and deflection (*b*) on the thickness asymmetry of a hinged along the contour plate at $T_1 = 293$ K is shown in Fig. 8.2: 1—$h_1 = 0.04$, $h_2 = 0.04$; 2—$h_1 = 0.03$, $h_2 = 0.05$; 3—$h_1 = 0.02$, $h_2 = 0.06$; 4—$h_1 = 0.01$, $h_2 = 0.07$ ($h_3 = 0.4$) For the same total thickness of the bearing layers, the symmetrical plate has the lowest stiffness.

Similar results are obtained for a three-layer plate at temperature $T_2 = 323$ K (Fig. 8.3). The numbering of the curves is the same, i.e. *a*—shear, *b*—deflection. Increasing temperature makes the asymmetric plate less rigid, extreme deflections are reached at $h_1 = 0.03$, $h_2 = 0.05$ (2).

Further, the following relative layer thicknesses were used: $h_1 = 0.02$; $h_2 = 0.06$; $h_3 = 0.4$. Figure 8.4 *a, b* shows the variation of shear in the filler and deflection of a rigidly embedding plate along its radius at different temperatures, obtained according to the solution (8.29): 1—$T_1 = 293$ K, 2—$T_2 = 323$ K, 3—$T_3 = 373$ K. Increasing the temperature by 30° above room temperature leads to an increase in the absolute values of maximum displacements by about 30%. When the temperature

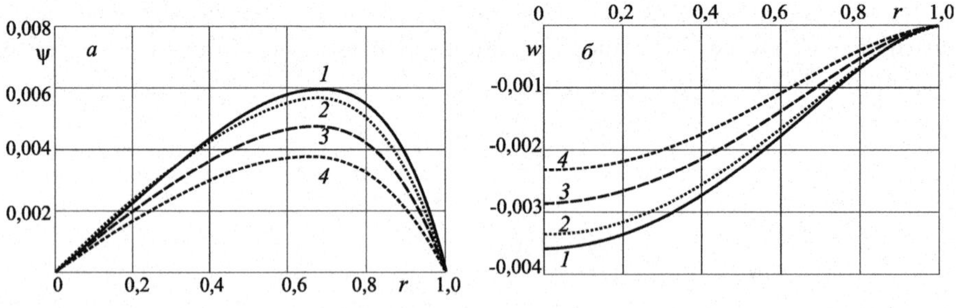

Fig. 8.2 Dependence of displacements on thickness asymmetry in a three-layer plate hinged along the contour ($T_1 = 293$ K)

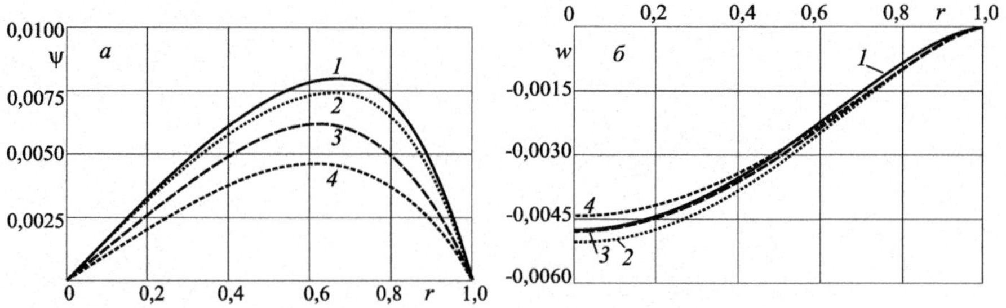

Fig. 8.3 Dependence of displacements on thickness asymmetry in a three-layer plate hinged along the contour ($T_2 = 323$ K)

is increased by 80°, an increase in absolute values of maximum displacements is observed in the order of 70%.

Similar curves of shear and deflection variation in a hinged plate are shown in Fig. 8.5 *a*, *b*. The numbering of the curves is the same as in Fig. 8.4. The shear plots at all temperatures practically coincide with the analogous ones when the plate contour is embedded (see Fig. 8.4), since the temperature affects mainly the spherical parts of the stress and strain tensors (8.10). The deflections increase with increasing temperature to a much greater extent than with contour clamping: by 66 and 175%, respectively, compared to isothermal loading.

Figure 8.6 graphs illustrate the variation of relative shear in the filler (*a*) and deflection in the plate (*b*) at temperature $T_1 = 293$ K in case of different boundary conditions: 1—embedded plate contour, 2—hinged plate contour. Shear is almost identical, deflections differ by about 3%.

The distribution of linear strains ε_r^k along the height of the cross-section of a plate with a rigidly embedded contour in the embedment ($r = 1$) is shown in Fig. 8.7: 1—$T_1 = 293$ K; 2—$T_2 = 323$ K, 3—$T_3 = 373$ K. Here, the strains at the layer glues are continuous. They are extreme on the outer surface of the bottom layer. When the plate is heated, the maximum strains in the upper layer shift to the negative region,

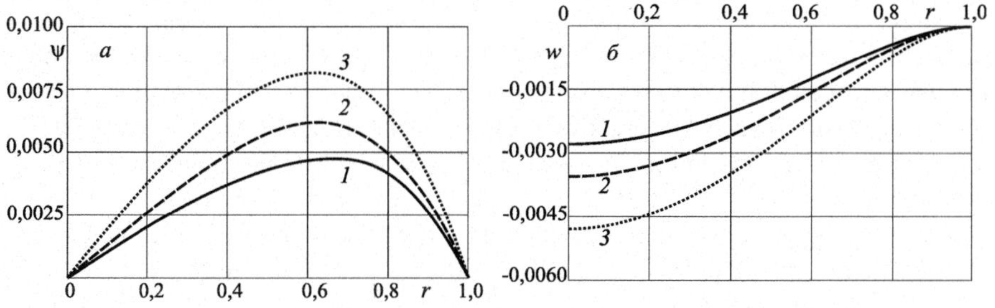

Fig. 8.4 Variation of displacements in a three-layer plate rigidly clamped along the contour at different temperatures ($T_1 = 293$ K, $T_2 = 323$ K, $T_3 = 373$ K)

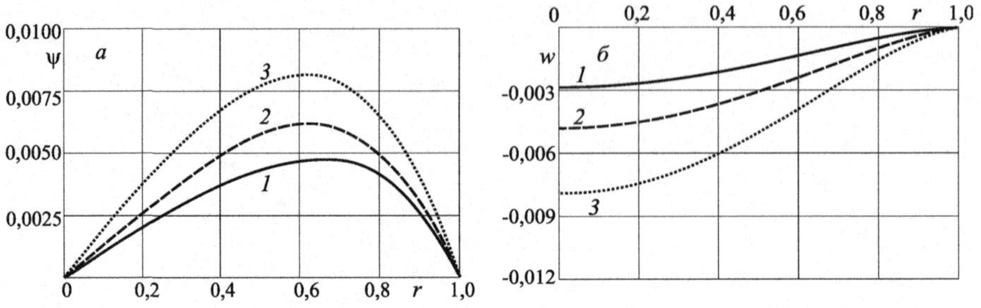

Fig. 8.5 Variation of displacements in a three-layer plate hinged along the contour at different temperatures ($T_1 = 293$ K, $T_2 = 323$ K, $T_3 = 373$ K)

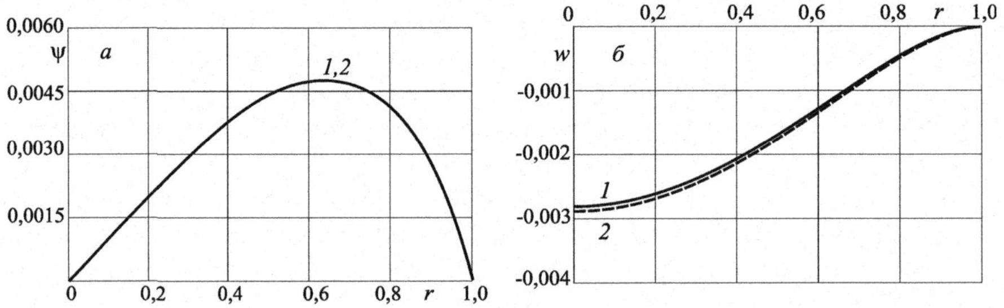

Fig. 8.6 Dependence of displacements in a three-layer plate on the type of boundary conditions ($T_2 = 323$ K)

decreasing in magnitude by 63%. Maximum modulus strains on the outer surface of the bottom layer increase by 112%.

Similar curves of radial strain variation at the center of the plate $r = 0$ at hinged support of the contour are shown in Fig. 8.8. The numbering of the graphics remains the same. The effect of temperature on strains in the plate is stronger when the contour is hinged, an increase of 5 times is observed.

The graphs in Fig. 8.9 show the distribution of normal stresses σ_r along the thickness of the plate: *a*—in the plate center, *b*—on a rigidly clamped contour at different temperatures: 1—$T_1 = 293$ K, 2—$T_2 = 323$ K, 3—$T_3 = 373$ K. When the plate is heated, the stresses at the center shift in parallel to the negative region, increasing in modulus by several times. On the contour, they change to a much lesser extent. The stress extremums in both cases are reached on the outer surface of the bottom layer.

Similar graphics when the plate contour is hinged are shown in Fig. 8.10: *a*—in the center, *b*—on the plate contour. The difference from the previous case is that when the plate is heated, the stresses in the center reach a maximum in modulus on the outer surface of the upper bearing layer. On the contour, as before, they change to a lesser extent, reaching a maximum on the outer surface of the second layer.

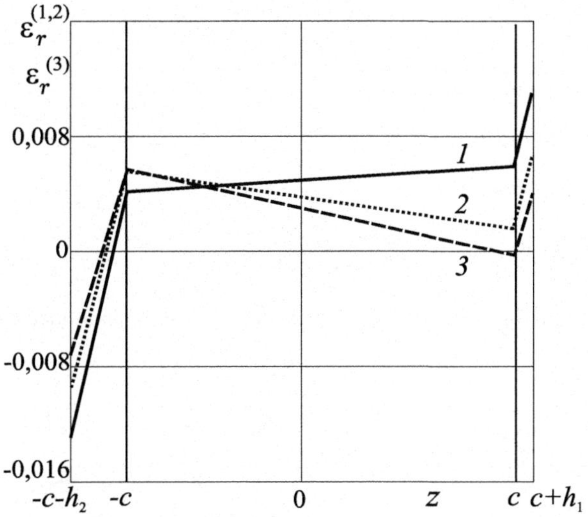

Fig. 8.7 Distribution of radial strains along the plate height on a rigidly embedded contour

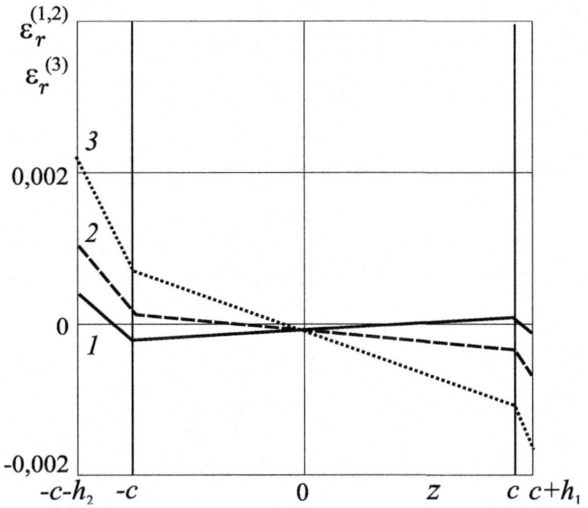

Fig. 8.8 Distribution of radial strains in the center of the plate with a hinged contour

Stresses and strains in the plate at hinge fixation of the contour are reduced by an order of magnitude. At the layer boundaries, the stress curves are discontinuous, which is explained by the different modules of elasticity of the layer materials.

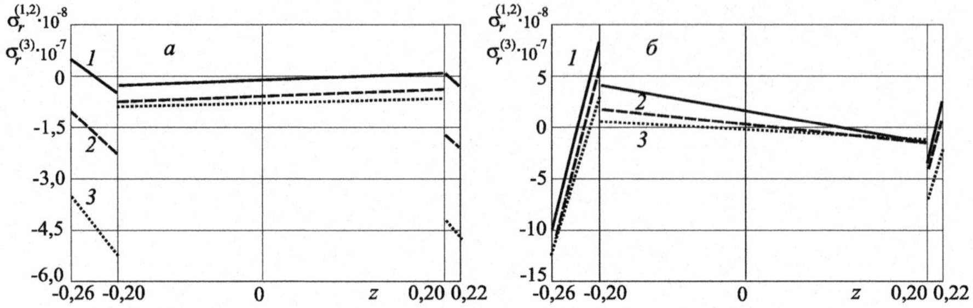

Fig. 8.9 Radial stress distribution along the plate height at rigidly embedded contour

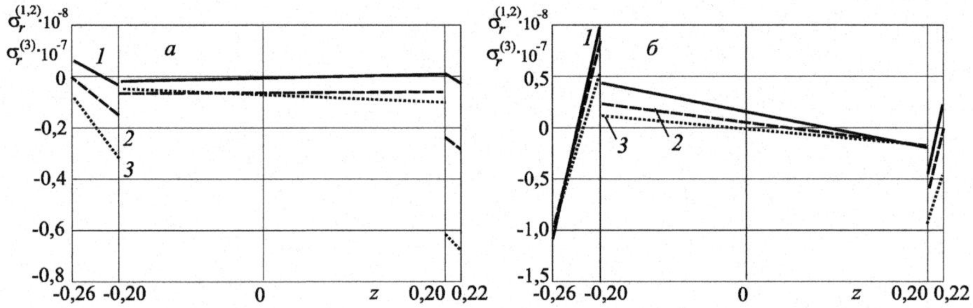

Fig. 8.10 Radial stress distribution along the plate height at hinged fixation of the contour

8.4 Plate Bending by Local Loads

Let us consider some examples of axisymmetric deformation of a circular three-layer plate by local loads. For convenience of analytical writing we will use the zero-order Heaviside function:

$$H_0(x) = \begin{cases} 1, & x \geq 0, \\ 0, & x < 0. \end{cases} \tag{8.31}$$

Load uniformly distributed around the circle [0, *b*]. Let a local transverse surface load uniformly distributed over a circle of relative radius $r \leq b$ act on the considered three-layer plate (Fig. 8.11). Using function (8.31), it can be written in the following form.

$$q(r) = q_0 H_0(b - r), \quad p = 0. \tag{8.32}$$

To transform the solution (8.27) to the load case (8.32), it is necessary to perform some integral transformations beforehand. Thus, the integral operator L_3^{-1} from the Heaviside function will be:

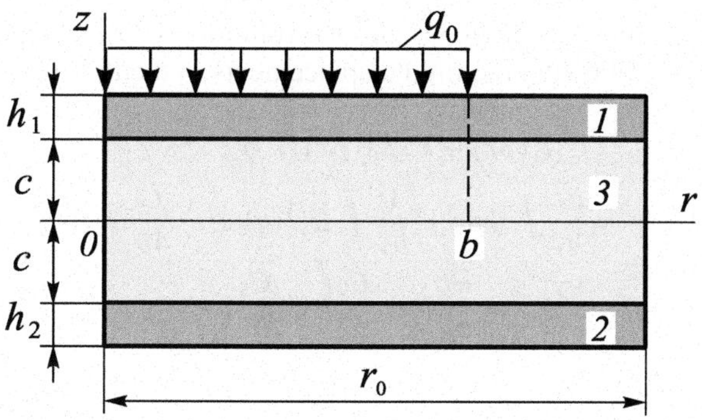

Fig. 8.11 Circular loading scheme

$$L_3^{-1}(H_0(b-r)) = \left(\frac{r^3}{16} - \frac{b^4}{16r} - \frac{b^2 r}{4}\ln\left(\frac{r}{b}\right)\right)H_0(b-r). \tag{8.33}$$

The function (8.21), which is included in the equation for determining the relative shear (8.20), under the load (8.32) takes the form:

$$f(r) = \frac{b_2}{(b_1 b_3 - b_2^2)r}\left(\int q_0 H_0(b-r) r\mathrm{d}r - C_1\right)$$
$$= \frac{\gamma_1}{r}\left(\frac{q_0\left(r^2 - b^2\right)}{2}H_0(b-r) - C_1\right),\ \gamma_1 = \frac{b_2}{(b_1 b_3 - b_2^2)}.$$

Introducing it into the integrals included in the partial solution (8.25), we obtain:

$$\int I_1(\beta r) f(r) r\mathrm{d}r = \frac{\gamma_1 q_0}{2\beta}\left(r^2 I_2(\beta r) - b^2 I_2(\beta b)\right)H_0(b-r) -$$
$$-\frac{\gamma_1 q_0 b^2}{2\beta}(I_0(\beta r) - I_0(\beta b))H_0(b-r) - \frac{C_1\gamma_1}{\beta}I_0(\beta r) =$$
$$= \frac{\gamma_1 q_0}{2\beta}H_0(b-r)\left[\left(r^2 - b^2\right)I_0(\beta r) + \frac{2}{\beta}(bI_1(\beta b) - rI_1(\beta r))\right] - \frac{C_1\gamma_1}{\beta}I_0(\beta r),$$
$$\int K_1(\beta r) f(r) r\mathrm{d}r = \frac{\gamma_1 q_0}{2\beta}H_0(b-r)\left[\left(b^2 - r^2\right)K_0(\beta r) + \frac{2}{\beta}(bK_1(\beta b) - rK_1(\beta r))\right] +$$
$$+\frac{C_1\gamma_1}{\beta}K_0(\beta r).$$

As a result, the partial solution (8.25) takes the form:

$$\psi_r = \frac{\gamma_1 q_0}{2\beta^2}H_0(b-r)\left[\frac{b^2}{r} - r + 2b(K_1(\beta b)I_1(\beta r) - I_1(\beta b)K_1(\beta r))\right] + \frac{C_1\gamma_1}{\beta^2 r}. \tag{8.34}$$

Substituting (8.33) into the solution (8.26), taking into account (8.34), we obtain the following required displacements in the three-layer circular plate:

$$\psi = C_2 I_1(\beta r) + C_3 K_1(\beta r) + \psi_r,$$
$$w = \frac{b_2}{b_3}\int \psi \mathrm{d}r + \frac{1}{b_3}\int L_3^{-1}(q)\mathrm{d}r - \frac{C_1}{4b_3}r^2(\ln r - 1) + \frac{C_5 r^2}{4b_3} + C_6 \ln r + C_4,$$
$$u = \frac{a_3}{a_1}w,_r - \frac{a_2}{a_1}\psi + \frac{C_7 r}{2} + \frac{C_8}{r}, \tag{8.35}$$

where

$$\int L_3^{-1}(q)\mathrm{d}r = q_0\left(\frac{r^4 - 5b^4}{64} - \frac{b^4}{16}\ln\left(\frac{r}{b}\right) - \frac{b^2 r^2}{8}\ln\left(\frac{r}{b}\right) + \frac{b^2 r^2}{16}\right)H_0(b - r),$$
$$\int \psi \mathrm{d}r = \frac{C_2 I_0(\beta r)}{\beta} - \frac{C_3 K_0(\beta r)}{\beta} + \frac{C_1 \gamma_1}{\beta^2}\ln(r) + \frac{\gamma_1 q_0}{2\beta^2}H_0(b - r)\times$$
$$\left[\frac{b^2 - r^2}{2} + b^2\ln\left(\frac{r}{b}\right) + \frac{2b}{\beta}(K_1(\beta b)I_0(\beta r) + I_1(\beta b)K_0(\beta r)) - \frac{2}{\beta^2}\right].$$

The integration constants are determined from the conditions of plate fixation and boundedness of the solution at its centre. From the condition at the centre of the plate it follows:

$$C_1 = -\frac{q_0 b^2}{2},\ C_3 = \frac{q_0 \gamma_1 b I_1(\beta b)}{\beta^2},\ C_6 = \frac{q_0 b^4}{16 b_3},\ C_8 = 0.$$

Satisfying by the solution (8.35) of the plate contour termination condition (8.19), we obtain:

$$C_2 = \frac{\gamma_1 q_0}{\beta^2 I_1(\beta)}\left(\frac{b^2}{2} - bK_1(\beta)I_1(\beta b)\right),$$
$$C_4 = -\frac{b_2 \gamma_1 q_0}{b_3 \beta^4 I_1(\beta)}\left(\frac{b^2 I_0(\beta)\beta}{2} - bI_1(\beta b)\right) + \frac{q_0 b^4}{32 b_3} + \frac{q_0 b^2}{16 b_3},$$
$$C_5 = \frac{q_0 b^2}{8}(2 - b^2),\ C_7 = 0.$$

At hinged support of the contour from conditions (8.19) we have:

$$C_2 = \frac{\gamma_1 q_0}{\beta^2 I_1(\beta)}\left(\frac{b^2}{2} - bK_1(\beta)I_1(\beta b)\right),$$
$$C_4 = \frac{b_2}{b_3}\left(-\frac{C_2 I_0(\beta)}{\beta} + \frac{C_3 K_0(\beta)}{\beta}\right) - \frac{1}{4b_3}(C_1 + C_5),$$

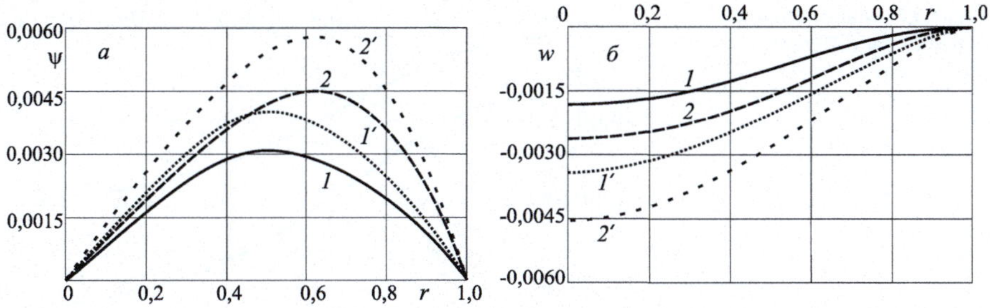

Fig. 8.12 Displacements in the plate at different spot radiuses

$$C_5 = \frac{q_0 b^2}{4}\left(1 - \frac{b^2}{2}\right)\frac{a_3^2 - a_1 b_3 + a_7 a_1}{a_1(a_6 + a_7)} - \frac{6b_3}{a_6 + a_7}\sum_{k=1}^{3}\alpha_{0k}\int_{h_k} K_k T_k z\mathrm{d}z,$$

$$C_7 = \frac{2a_3}{a_1(a_6 + a_7)}\left(3\sum_{k=1}^{3}\alpha_{0k}\int_{h_k} K_k T_k z\mathrm{d}z + \frac{q_0 b^2}{4}\left(1 - \frac{b^2}{2}\right)\right). \qquad (8.36)$$

Here the solution of the boundary value problem clearly depends on temperature through the integration constants C_4, C_5, C_7 in (8.36).

Numerical results are obtained for an elastic circular three-layer plate hinged along the contour, the layers of which are made of D16T–fluoroplastic–D16T materials. Load $q_0 = 1$ MPa, relative thicknesses of layers:$h_1 = 0, 02$; $h_2 = 0, 06$; $h_3 = 0, 4$.

Figures 8.12 *a, b* show the change in relative shear and deflection of the plate along its radius. The curves are plotted at different radiuses b of the spot of local uniformly distributed surface load: 1—$b = 0.5$; 2—$b = 0.17$; without a stroke—$T_1 = 293$ K, with a stroke—$T_2 = 323$ K. As the load area increases, the extremum in Fig. 8.12 *a* shifts to the contour of the plate. When heated, displacements increase in proportion to each other.

Figure 8.13 illustrates the dependence of the maximum deflection of the plate under consideration on the radius b of the surface load spot at different temperatures: 1—$T_1 = 293$ K, 2—$T^2 = 323$ K.

The deflection increases non-linearly with increasing load spot, reaching a maximum at the load acting on the entire plate surface.

Load uniformly distributed over the ring [*a*, *b*]. Let the plate be subjected to a local load uniformly distributed over an ring of relative radius $a \le r \le b$ (Fig. 8.14):

$$q = q_0(H_0(b - r) - H_0(a - r)) \qquad (8.37)$$

Due to the linearity of the boundary value problem, the required solution can be obtained as the difference of two solutions (8.35) corresponding to the load components (8.37). As a result, after making appropriate additions to the partial solution (8.34), the shear in the filler will be:

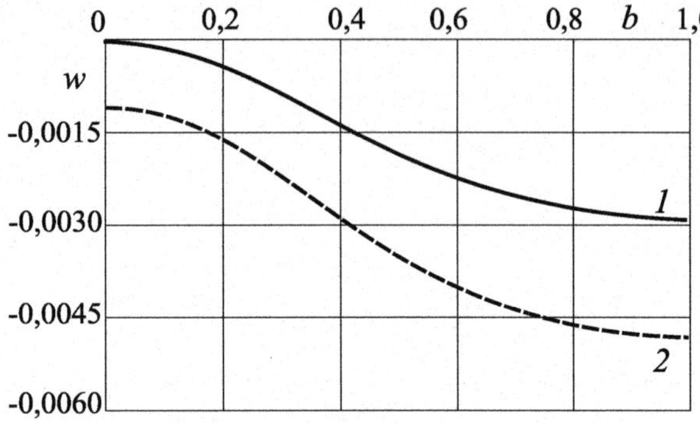

Fig. 8.13 Maximum plate deflection at different temperatures

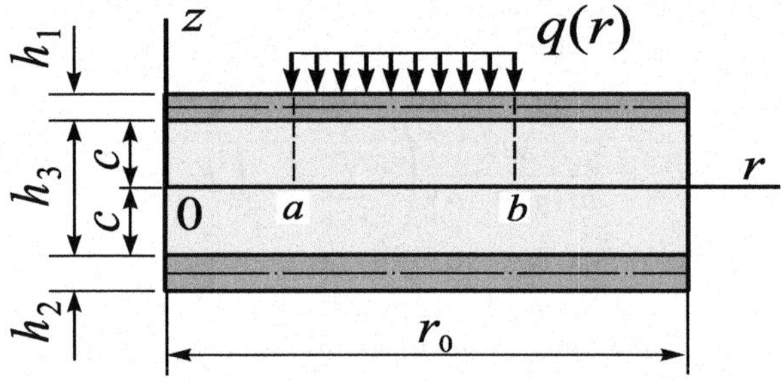

Fig. 8.14 Scheme of loading by ring load

$$\psi = C_2 I_1(\beta r) + C_3 K_1(\beta r) + \frac{\gamma_1 q_0}{2\beta^2} H_0(b-r)\left[\frac{b^2}{r} - r + 2b(K_1(\beta b)I_1(\beta r) - I_1(\beta b)K_1(\beta r))\right] - \frac{\gamma_1 q_0}{2\beta^2} H_0(a-r)\left[\frac{a^2}{r} - r + 2a(K_1(\beta a)I_1(\beta r) - I_1(\beta a)K_1(\beta r))\right] + \frac{C_1 \gamma_1}{\beta^2 r} \quad (8.38)$$

The deflection $w(r)$ and radial displacement $u(r)$ of the considered three-layer plate formally retain the form (8.35), but the integrals included in them will change, i.e.

$$w = \frac{b_2}{b_3}\int \psi \mathrm{d}r + \frac{1}{b_3}\int L_3^{-1}(q)\mathrm{d}r - \frac{C_1}{4b_3} r^2(\ln r - 1) + \frac{C_5 r^2}{4b_3} + C_6 \ln r + C_4,$$
$$u = \frac{a_3}{a_1} w,_r - \frac{a_2}{a_1}\psi + \frac{C_7 r}{2} + \frac{C_8}{r},$$

where

$$\int L_3^{-1}(q)\mathrm{d}r = q_0\left[\frac{r^4-5b^4}{64}-\frac{b^4}{16}\ln\left(\frac{r}{b}\right)-\frac{b^2r^2}{8}\ln\left(\frac{r}{b}\right)+\frac{b^2r^2}{16}\right]H(b-r)-$$
$$q_0\left[\frac{r^4-5a^4}{64}-\frac{a^4}{16}\ln\left(\frac{r}{a}\right)-\frac{a^2r^2}{8}\ln\left(\frac{r}{a}\right)+\frac{a^2r^2}{16}\right]H(a-r),$$
$$\int \psi\mathrm{d}r = \frac{C_2I_0(\beta r)}{\beta}-\frac{C_3K_0(\beta r)}{\beta}+\frac{C_1\gamma_1}{\beta^2}\ln(r)+\frac{\gamma_1 q_0}{2\beta^2}H_0(b-r)$$
$$\times\left[\frac{b^2-r^2}{2}+b^2\ln\left(\frac{r}{b}\right)+\frac{2b}{\beta}(K_1(\beta b)I_0(\beta r)+I_1(\beta b)K_0(\beta r))-\frac{2}{\beta^2}\right]-\frac{\gamma_1 q_0}{2\beta^2}\times$$
$$H_0(a-r)\left[\frac{a^2-r^2}{2}+a^2\ln\left(\frac{r}{a}\right)+\frac{2a}{\beta}(K_1(\beta a)I_0(\beta r)+I_1(\beta a)K_0(\beta r))-\frac{2}{\beta^2}\right].$$

Integration constants at rigid embedding of the plate contour will be:

$$C_1=-\frac{q_0(b^2-a^2)}{2},\ C_2=\frac{\gamma_1 q_0}{\beta^2 I_1(\beta)}\left(\frac{b^2-a^2}{2}-bK_1(\beta)I_1(\beta b)+aK_1(\beta)I_1(\beta a)\right),$$
$$C_3=\frac{q_0\gamma_1}{\beta^2}(bI_1(\beta b)-aI_1(\beta a)),$$
$$C_4=-\frac{b_2\gamma_1 q_0}{b_3\beta^4 I_1(\beta)}\left(\frac{I_0(\beta)\beta\left(b^2-a^2\right)}{2}-bI_1(\beta b)+aI_1(\beta a)\right)+$$
$$\frac{q_0\left(b^4-a^4\right)}{32b_3}+\frac{q_0\left(b^2-a^2\right)}{16b_3},\ C_5=\frac{q_0}{8}\left(2b^2-b^4-2a^2+a^4\right),$$
$$C_6=\frac{q_0}{16b_3}\left(b^4-a^4\right),\ C_7=\frac{a_3q_0}{4b_3a_1}\left(b^2-a^2\right),\ C_8=0. \tag{8.39}$$

Integration constants for hinged support of the plate contour:

$$C_4=\frac{b_2}{b_3}\left(-\frac{C_2I_0(\beta)}{\beta}+\frac{C_3K_0(\beta)}{\beta}\right)-\frac{1}{4b_3}(C_1+C_5),$$
$$C_5=q_0\frac{a_3^2-a_1b_3+a_7a_1}{4a_1(a_6+a_7)}\left(b^2\left(1-\frac{b^2}{2}\right)-a^2\left(1-\frac{a^2}{2}\right)\right)$$
$$-\frac{6b_3}{a_6+a_7}\sum_{k=1}^{3}\alpha_{0k}\int\limits_{h_k}K_kT_kz\mathrm{d}z,$$
$$C_7=\frac{2a_3}{a_1(a_6+a_7)}\left(3\sum_{k=1}^{3}\alpha_{0k}\int\limits_{h_k}K_kT_kz\mathrm{d}z+\frac{q_0b^2}{4}\left(1-\frac{b^2}{2}\right)-\frac{q_0a^2}{4}\left(1-\frac{a^2}{2}\right)\right).$$

In this case the integration constants C_1, C_2, C_3, C_6, C_8 keep the form (8.39).

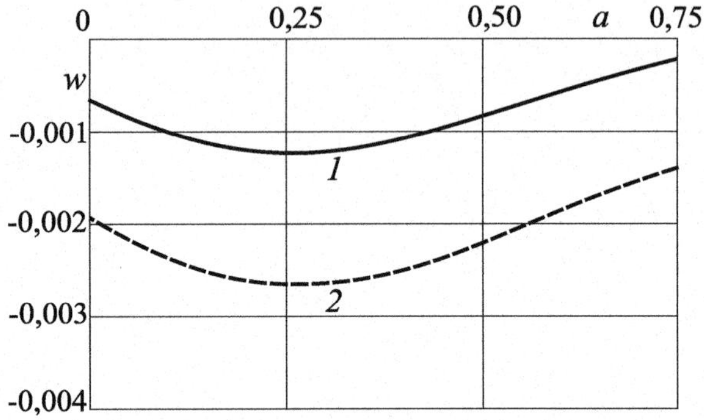

Fig. 8.15 Variation of plate deflection depending on the position of the ring load spot ($T_1 = 293$ K, $T_2 = 233$ K)

Figure 8.15 shows the variation in plate deflection as a function of the position of an ring load spot of constant thickness (b—a) $=$ 0.25, as it moves towards the contour: 1—$T_1 =$ 293 K, 2—$T_2 = 233$ K the load is distributed on a circle of radius b, at $a =$ 0.75 the load ring is adjacent to the contour of the plate. The maximum deflection is observed for ring load with hole radius $a = 0.25$

Linear transverse load. Let a linear transverse load $a = 0.25$ $Q(r)$ of constant intensity Q_0, applied along a circle of radius $r = a$, act on the considered three-layer plate (Fig. 8.16).

Let us solve the problem by applying the results obtained for the ring load $a - \xi \leq r \leq a + \xi$ with intensity q_0. Suppose that the ring is thin, i.e. x << 1. c Then we substitute $q_0 = Q_0/(2\xi)$ in function (8.38) and set the parameter ξ to zero, leaving Q_0 constant. As a result, for the relative shear we obtain the following expression:

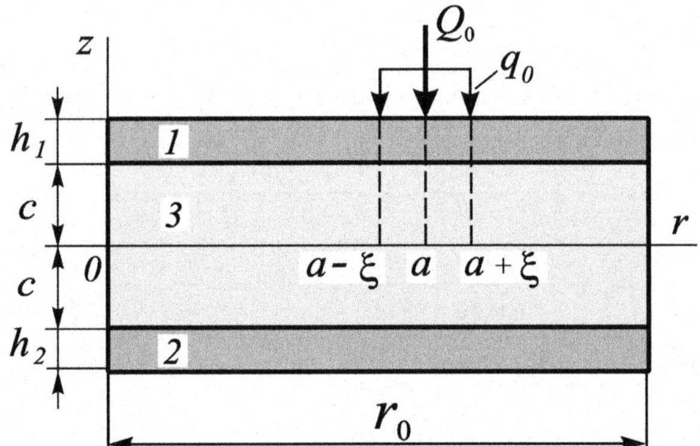

Fig. 8.16 Scheme of loading by linear force load

$$\psi = \lim_{\xi \to 0} \left[C_2 I_1(\beta r) + C_3 K_1(\beta r) + \frac{\gamma_1 Q_0}{4\xi\beta^2} H_0(a + \xi - r) \times \right.$$
$$\times \left[\frac{(a+\xi)^2}{r} - r + 2(a+\xi)(K_1(\beta(a+\xi)) I_1(\beta r) - I_1(\beta(a+\xi)) K_1(\beta r)) \right] -$$
$$- \frac{\gamma_1 Q_0}{4\xi\beta^2} H_0(a - \xi - r) \left[\frac{(a-\xi)^2}{r} - r + 2(a-\xi)(K_1(\beta(a-\xi)) I_1(\beta r) - \right.$$
$$\left. \left. - I_1(\beta(a-\xi)) K_1(\beta r)) \right] + \frac{C_1 \gamma_1}{\beta^2 r} \right].$$

After taking the limit, the solution of the problem on bending of a circular three-layer plate under the action of a linear load will be:

$$\psi = C_2 I_1(\beta r) + C_3 K_1(\beta r) +$$
$$+ \frac{\gamma_1 Q_0 a}{\beta} H_0(a - r) \left(\frac{1}{\beta r} - K_1(\beta r) I_0(\beta a) - I_1(\beta r) K_0(\beta a) \right) + \frac{C_1 \gamma_1}{\beta^2 r},$$
$$w = \frac{b_2}{b_3} \int \psi \mathrm{d}r + \frac{1}{b_3} \int L_3^{-1}(Q_0) \mathrm{d}r - \frac{C_1}{4b_3} r^2 (\ln r - 1) + \frac{C_5 r^2}{4b_3} + C_6 \ln r + C_4, \quad (8.40)$$
$$u = \frac{a_3}{a_1} w,_r - \frac{a_2}{a_1} \psi + \frac{C_7 r}{2} + \frac{C_8}{r},$$

where

$$\int L_3^{-1}(Q_0) \mathrm{d}r = \lim_{\xi \to 0} \left[\frac{Q_0}{2\xi} \left(\frac{r^4 - 5(a+\xi)^4}{64} - \frac{(a+\xi)^4}{16} \ln\left(\frac{r}{a+\xi} \right) - \right. \right.$$
$$\left. - \frac{(a+\xi)^2 r^2}{8} \ln\left(\frac{r}{a+\xi} \right) + \frac{(a+\xi)^2 r^2}{16} \right) H(a + \xi - r) -$$
$$- \left(\frac{r^4 - 5(a-\xi)^4}{64} - \frac{(a-\xi)^4}{16} \ln\left(\frac{r}{a-\xi} \right) - \frac{(a-\xi)^2 r^2}{8} \ln\left(\frac{r}{a-\xi} \right) + \frac{(a-\xi)^2 r^2}{16} \right) \times$$
$$\left. \times H_0(a - \xi - r) \right) \Big] = \frac{Q_0 a}{4} \left(r^2 - a^2 - a^2 \ln\left(\frac{r}{a} \right) - r^2 \ln\left(\frac{r}{a} \right) \right) H_0(a - r)$$
$$\int \psi \mathrm{d}r = \lim_{\xi \to 0} \left(\frac{C_2 I_0(\beta r)}{\beta} - \frac{C_3 K_0(\beta r)}{\beta} + \frac{C_1 \gamma_1}{\beta^2} \ln(r) + \frac{\gamma_1 Q_0}{4\xi\beta^2} H_0(a + \xi - r) \times \right.$$
$$\times \left(\frac{(a+\xi)^2 - r^2}{2} + (a+\xi)^2 \ln\left(\frac{r}{a+\xi} \right) + \frac{2(a+\xi)}{\beta} (K_1(\beta(a+\xi)) I_0(\beta r) + \right.$$
$$\left. + I_1(\beta(a+\xi)) K_0(\beta r)) - \frac{2}{\beta^2} \right) - \frac{\gamma_1 Q_0}{4\xi\beta^2} H_0(a - \xi - r) \left(\frac{(a-\xi)^2 - r^2}{2} + (a-\xi)^2 \times \right.$$
$$\left. \left. \times \ln\left(\frac{r}{a-\xi} \right) + \frac{2(a-\xi)}{\beta} (K_1(\beta(a-\xi)) I_0(\beta r) + I_1(\beta(a-\xi)) K_0(\beta r)) - \frac{2}{\beta^2} \right) \right)$$
$$= \frac{C_2 I_0(\beta r)}{\beta} - \frac{C_3 K_0(\beta r)}{\beta} + \frac{C_1 \gamma_1}{\beta^2} \ln(r) + \frac{\gamma_1 Q_0 a}{\beta^2} H_0(a - r) \times$$
$$s'^{(k)}_{\alpha} = 2G_k(T)(1 - \omega'^{(k)}(\varepsilon_u^{(k)}, T)) \ni'^{(k)}_{\alpha}.$$

The integration constants for the clamped contour are as follows:

$$C_1 = -Q_0 a,\ C_2 = \frac{\gamma_1 Q_0 a}{\beta^2 I_1(\beta)}(1 - \beta K_1(\beta) I_0(\beta a)),\ C_3 = \frac{\gamma_1 Q_0 a I_0(\beta a)}{\beta},$$
$$C_4 = -\frac{b_2 \gamma_1 Q_0 a}{b_3 \beta^3 I_1(\beta)}(I_0(\beta) - I_0(\beta a)) + \frac{Q_0 a(a^2 + 1)}{8 b_3},\ C_5 = \frac{Q_0 a}{2}(1 - a^2),$$
$$C_6 = \frac{Q_0 a^3}{4 b_3},\ C_7 = 0,\ C_8 = 0. \tag{8.41}$$

When the plate contour is hinged, the integration constants are equal to:

$$C_4 = \frac{b_2}{b_3}\left(-\frac{C_2 I_0(\beta)}{\beta} + \frac{C_3 K_0(\beta)}{\beta}\right) - \frac{1}{4 b_3}(C_1 + C_5),$$
$$C_5 = \frac{Q_0 a}{2}(1 - a^2)\frac{a_3^2 - a_1 b_3 + a_7 a_1}{a_1(a_6 + a_7)} - \frac{6 b_3}{a_6 + a_7}\sum_{k=1}^{3} \alpha_{0k} \int_{h_k} K_k T_k z \mathrm{d}z,$$
$$C_7 = \frac{2 a_3}{a_1(a_6 + a_7)}\left(3\sum_{k=1}^{3} \alpha_{0k} \int_{h_k} K_k T_k z \mathrm{d}z + \frac{Q_0 a}{2}(1 - a^2)\right). \tag{8.42}$$

The constants C_1, C_2, C_3, C_6, C_8 keep the form (8.41).

Figure 8.17 *a*, *b* show the variation of shear in the fill and deflection of the plate along its radius. The curves are plotted according to formula (8.42) at different radiuses of the force circle: 1—$a = 0.5$; 2—$a = 0.75$; without a stroke—$T = 293$ K; with a stroke—$T = 323$ K. The intensity of the linear force $Q_0 = 3$ MP/m. As the radius of the force circle increases, the relative shear extrema shift to the plate contour.

The deflections reach their maximum values when the radius of the force circle is approximately $a = 0.38$ This follows from the plots of the variation of deflection at the centre of the plate as a function of the radius of the force circle a at different temperatures: 1—$T_1 = 293$ K, 2—$T_2 = 233$ K (Fig. 8.18). As the ring load further advances towards the contour, the deflection decreases to zero.

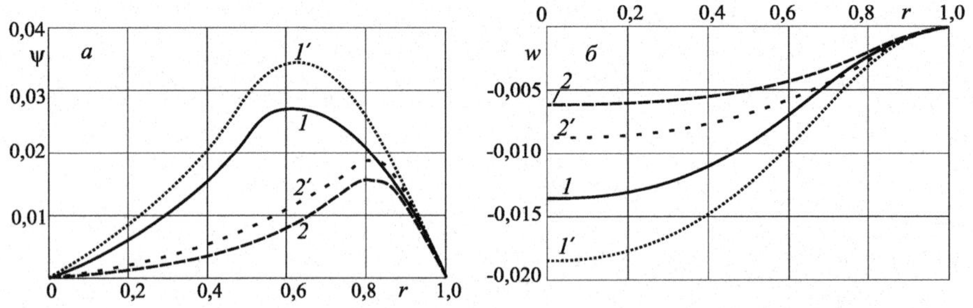

Fig. 8.17 Displacements at different radiuses of the force circle (Without a Stroke—$T = 293$ K, with a Stroke—$T = 323$ K)

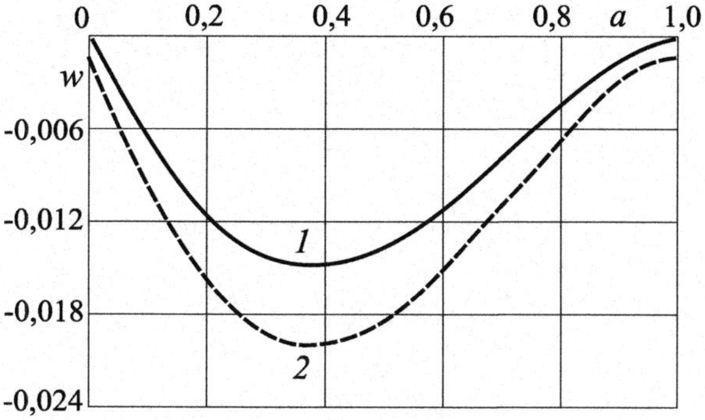

Fig. 8.18 Variations of deflection in the centre of the plate as a function of on the radius of the force circle (T_1 = 293 K, T_2 = 233 K)

When the plate is heated the deflections and shear in both cases (see Figs. 8.17 and 8.18) increase proportionally.

Formula (8.40) do not describe the deformation of the plate under the action of a concentrated force in its centre, since at b = 0 the circle degenerates and the resultant of linear force is zero. To avoid this, suppose that the mentioned resultant force $Q = 2\pi a Q_0$ remains constant when the radius of the circle along which it is applied changes. This is possible if the intensity Q_0 is variable, compensating for changes in the radius a. To obtain the corresponding solution, let us substitute in (8.40) as follows: $Q_0 = 2\pi a/Q$. Then we obtain the displacements:

$$\begin{aligned}
&\psi = C_2 I_1(\beta r) + C_3 K_1(\beta r) + \\
&\frac{\gamma_1 Q}{2\pi\beta} H_0(a-r)\left(\frac{1}{\beta r} - K_1(\beta r) I_0(\beta a) - I_1(\beta r) K_0(\beta a)\right) + \frac{C_1\gamma_1}{\beta^2 r}, \\
&w = \frac{b_2}{b_3}\int \psi dr + \frac{1}{b_3}\int L_3^{-1}(Q) dr - \frac{C_1}{4b_3} r^2(\ln r - 1) + \frac{C_5 r^2}{4b_3} + C_6 \ln r + C_4, \\
&u = \frac{a_3}{a_1} w_{,r} - \frac{a_2}{a_1}\psi + \frac{C_7 r}{2} + \frac{C_8}{r},
\end{aligned} \tag{8.43}$$

where

$$\begin{aligned}
&\int L_3^{-1}(Q) \mathrm{d}r = \frac{Q}{8\pi}\left(r^2 - a^2 - (r^2 + a^2)\ln\left(\frac{r}{a}\right)\right) H_0(a-r) \\
&\int \psi \mathrm{d}r = \frac{C_2 I_0(\beta r)}{\beta} - \frac{C_3 K_0(\beta r)}{\beta} + \frac{C_1\gamma_1}{\beta^2}\ln(r) + \frac{\gamma_1 Q}{2\pi\beta^2} H_0(a-r) \\
&\times\left(In\left(\frac{r}{a}\right) + I_0(\beta a) K_0(\beta r) - K_0(\beta a) I_0(\beta r)\right).
\end{aligned}$$

Integration constants at the clamped contour of the plate will be:

$$C_1 = -\frac{Q}{2\pi},\ C_2 = \frac{\gamma_1 Q}{2\pi\beta^2 I_1(\beta)}(1 - \beta K_1(\beta) I_0(\beta a)),\ C_3 = \frac{\gamma_1 Q I_0(\beta a)}{2\pi\beta},$$

$$C_4 = -\frac{b_2\gamma_1 Q}{2\pi b_3\beta^3 I_1(\beta)}(I_0(\beta) - I_0(\beta a)) + \frac{Q(a^2+1)}{16\pi b_3}, \quad C_5 = \frac{Q}{4}(1-a^2),$$
$$C_6 = \frac{Qa^2}{8\pi b_3}, \; C_7 = 0, \; C_8 = 0. \tag{8.44}$$

When the plate contour is hinged, the integration constants are equal to:

$$C_4 = \frac{b_2}{b_3}\left(-\frac{C_2 I_0(\beta)}{\beta} + \frac{C_3 K_0(\beta)}{\beta}\right) - \frac{1}{4b_3}(C_1 + C_5),$$
$$C_5 = \frac{Q}{4\pi}(1-a^2)\frac{a_3^2 - a_1 b_3 + a_7 a_1}{a_1(a_6+a_7)} - \frac{6b_3}{a_6+a_7}\sum_{k=1}^{3}\alpha_{0k}\int_{h_k} K_k T_k z\mathrm{d}z,$$
$$C_7 = \frac{2a_3}{a_1(a_6+a_7)}\left(3\sum_{k=1}^{3}\alpha_{0k}\int_{h_k} K_k T_k z\mathrm{d}z + \frac{Q}{4\pi}(1-a^2)\right). \tag{8.45}$$

The constants C_1, C_2, C_3, C_6, C_8 keep the form (8.44).

Variation of the maximum deflection of the plate (8.43, 8.45) as a function of the radius of the circle along which a transverse linear force load with constant equal force is applied is shown in Fig. 8.19: 1—$T_1 = 293$ K, 2—$T_2 = 323$ K. The maximum here is observed at $b = 0$, which corresponds to a concentrated force applied at the centre of the plate. As the load advances towards the contour, the deflection decreases to zero. The deflection increases proportionally with heating.

Linear moment. Let the plate under study be subjected to a linear moment distributed on a circle of radius $r = a$ (Fig. 8.20). The moment intensity is constant: $M_0 = \text{const}$.

Let us solve the problem using the difference of solutions (8.40) for two identical in value linear transverse forces acting on closely spaced circles of the radius $r = a - \xi$ and $r = a + \xi$, but directed in opposite directions. In this difference we substitute $Q_0 = M_0/(2\xi)$ and set ξ to zero. Then:

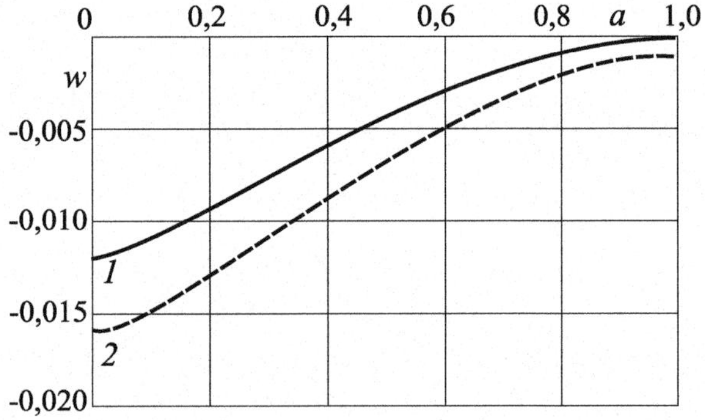

Fig. 8.19 Deflection of the plate as a function of the radius of the force circle of a linear load with constant resultant force ($T_1 = 293$ K, $T_2 = 233$ K)

Fig. 8.20 Loaded by linear moment scheme

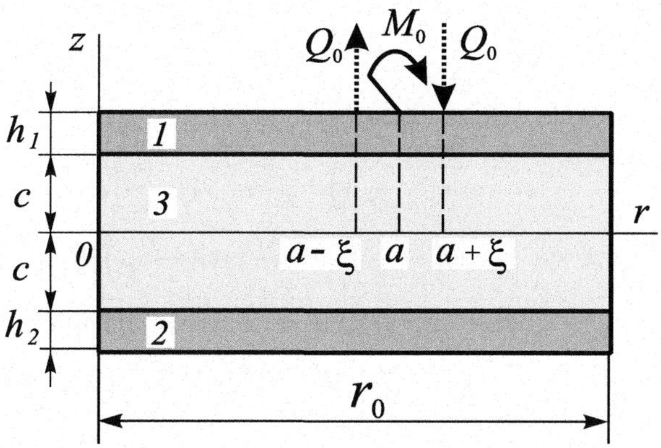

$$\psi = \lim_{\xi\to 0}\Bigg[C_2I_1(\beta r) + C_3K_1(\beta r) + \frac{\gamma_1 M_0 a}{2\xi\beta}H_0(a+\xi-r)\Bigg[\frac{1}{\beta r} - (K_0(\beta(a+\xi))\times$$
$$I_1(\beta r) + I_0(\beta(a+\xi))K_1(\beta r))] - \frac{\gamma_1 M_0 a}{2\xi\beta}H_0(a-\xi-r)\times$$
$$\Bigg[\frac{1}{\beta r} - \Big(K_0(\beta(a-\xi))I_1(\beta r) + I_0(\beta(a-\xi))K_1(\beta r))\Big] + \frac{C_1\gamma_1}{\beta^2 r}\Bigg].$$

After calculating the limit we obtain:

$$\psi = C_2I_1(\beta r) + C_3K_1(\beta r) + \frac{\gamma_1 M_0}{\beta}H_0(a-r)\Big(\frac{1}{\beta r} - K_1(\beta r)(I_0(\beta a)+$$
$$+I_1(\beta a)\beta a) - I_1(\beta r)(K_0(\beta a) - K_1(\beta a)\beta a)) + \frac{C_1\gamma_1}{\beta^2 r},$$
$$w = \frac{b_2}{b_3}\int \psi \mathrm{d}r + \frac{1}{b_3}\int L_3^{-1}(M_0)\mathrm{d}r - \frac{C_1}{4b_3}r^2(\ln r - 1) + \frac{C_5 r^2}{4b_3} + C_6\ln r + C_4,$$
$$u = \frac{a_3}{a_1}w_{,r} - \frac{a_2}{a_1}\psi + \frac{C_7 r}{2} + \frac{C_8}{r},$$
$$(8.46)$$

where

$$
\begin{aligned}
&\int L_3^{-1}(M_0)\mathrm{d}r = \lim_{\xi\to 0}\left[\frac{M_0}{8\xi}\left(r^2(a+\xi)-(a+\xi)^3-(a+\xi)^3\ln\left(\frac{r}{a+\xi}\right)-\right.\right.\\
&\left.-r^2(a+\xi)\ln\left(\frac{r}{a+\xi}\right)\right)H(a+\xi-r)-\frac{M_0}{8\xi}\left[r^2(a-\xi)-(a-\xi)^3-\right.\\
&\left.\left.-(a-\xi)^3\ln\left(\frac{r}{a-\xi}\right)-r^2(a-\xi)\ln\left(\frac{r}{a-\xi}\right)\right]H_0(a-\xi-r)\right]\\
&=\frac{M_0}{4}\left(2r^2-2a^2-3a^2\ln\left(\frac{r}{a}\right)-r^2\ln\left(\frac{r}{a}\right)\right)H_0(a-r),\\
&\int\psi\mathrm{d}r=\lim_{\xi\to 0}\left(\frac{C_2I_0(\beta r)}{\beta}-\frac{C_3K_0(\beta r)}{\beta}+\frac{C_1\gamma_1}{\beta^2}\ln(r)+\frac{\gamma_1M_0(a+\xi)}{2\xi\beta^2}H_0(a+\xi-r)\times\right.\\
&\times\left(\ln\left(\frac{r}{a+\xi}\right)+(I_0(\beta(a+\xi))K_0(\beta r)\ -K_0(\beta(a+\xi))I_0(\beta r))\right)-\frac{\gamma_1M_0(a-\xi)}{2\xi\beta^2}\times\\
&\left.\times H_0(a-\xi-r)\left(\ln\left(\frac{r}{a+\xi}\right)+(I_0(\beta(a+\xi))K_0(\beta r)\ -\ K_0(\beta(a+\xi))I_0(\beta r))\right)\right)=\\
&=\frac{C_2I_0(\beta r)}{\beta}-\frac{C_3K_0(\beta r)}{\beta}+\frac{C_1\gamma_1}{\beta^2}\ln(r)+\frac{\gamma_1M_0}{\beta^2}H_0(a-r)\times\\
&\times\left(\ln\left(\frac{r}{a}\right)-1+K_0(\beta r)(I_0(\beta a)+I_1(\beta a)\beta a)-I_0(\beta r)(K_0(\beta a)-K_1(\beta a)\beta a)\right).
\end{aligned}
$$

The integration constants at the clamped contour of the plate are as follows:

$$
\begin{aligned}
&C_1=-M_0,\ C_2=\frac{\gamma_1M_0}{\beta^2I_1(\beta)}(1-\beta K_1(\beta)(I_0(\beta a)+I_1(\beta a)\beta a)),\\
&C_3=\frac{\gamma_1M_0}{\beta}(I_0(\beta a)+I_1(\beta a)\beta a),\\
&C_4=-\frac{b_2\gamma_1M_0}{b_3\beta^3I_1(\beta)}(I_0(\beta)-(I_0(\beta a)-I_1(\beta a)\beta a))+\frac{3M_0a^2}{8b_3}+\frac{M_0}{8b_3},\\
&C_5=\frac{M_0}{2}(1-3a^2),\ C_6=\frac{3M_0a^2}{4b_3},\ C_7=0,\ C_8=0.
\end{aligned}
\tag{8.47}
$$

At a hinged contour, the integration constants C_1, C_2, C_3, C_6, C_8 keep the form (8.47), the others take the following form:

$$
\begin{aligned}
&C_4=\frac{b_2}{b_3}\left(-\frac{C_2I_0(\beta)}{\beta}+\frac{C_3K_0(\beta)}{\beta}\right)-\frac{1}{4b_3}(C_1+C_5),\\
&C_5=\frac{M_0}{2}(1-3a^2)\frac{a_3^2-a_1b_3+a_7a_1}{a_1(a_6+a_7)}-\frac{6b_3}{a_6+a_7}\sum_{k=1}^{3}\alpha_{0k}\int_{h_k}K_kT_kz\mathrm{d}z,\\
&C_7=\frac{2a_3}{a_1(a_6+a_7)}\left(3\sum_{k=1}^{3}\alpha_{0k}\int_{h_k}K_kT_kz\mathrm{d}z+\frac{M_0}{2}(1-3a^2)\right).
\end{aligned}
\tag{8.48}
$$

The functions (8.46) formally describe the case of a concentrated moment in the centre of the plate, since at $a = 0$ the solution does not degenerate.

Figure 8.21, *a* and *b* show the variation of shear in the filler and deflection of a circular three-layer plate along its radius. The curves are plotted according to formulas (8.46, 8.48) at different radiuses of the linear moment circle: 1—$a = 0.5$; 2—$a = 0.75$; without a stroke—$T = 293$ K, with a stroke—$T = 323$ K. Intensity of linear moments $M_0 = 0.3$ MH. As the moment circle moves towards the plate contour, the deflection decreases in magnitude and becomes of the same sign. In the case of shear, the maximum peaks are observed in the cross-sections of the application of linear moments. An increase in temperature leads to an increase in shear but a decrease in deflection.

The variation of the deflection at the centre of the plate as a function of the radius of the moment circle is illustrated in Fig. 8.22: 1—$T_1 = 293$ K, 2—$T_2 = 323$ K. The maximum here is observed at the load applied at the centre of the plate. As the moment circle moves towards the plate contour, the deflection decreases in magnitude, reaches zero in the vicinity of $a = 0.4$, changes sign and goes to zero at the contour load (at $T = 293$ K).

Upon heating, deflections and shear in both cases (see Figs. 8.21 and 8.22) increase proportionally.

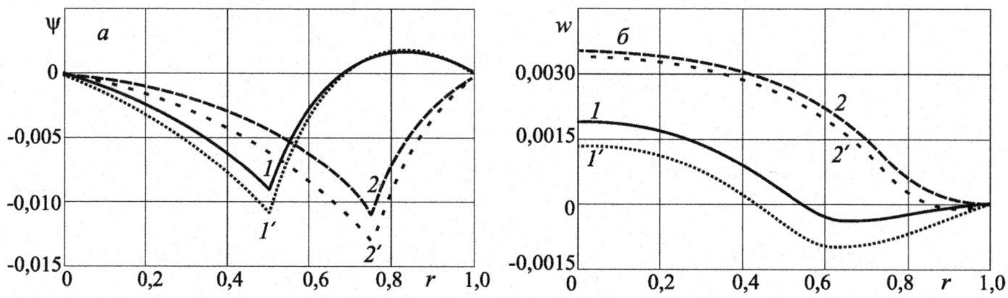

Fig. 8.21 Displacements at different radiuses of the linear moment circle (Without Stroke—$T = 293$ K, with Stroke—$T = 323$ K)

Fig. 8.22 Variations of deflection in the centre of the plate as a function of the radius of the moment circle with a constant resultant ($T_1 = 293$ K, $T_2 = 233$ K)

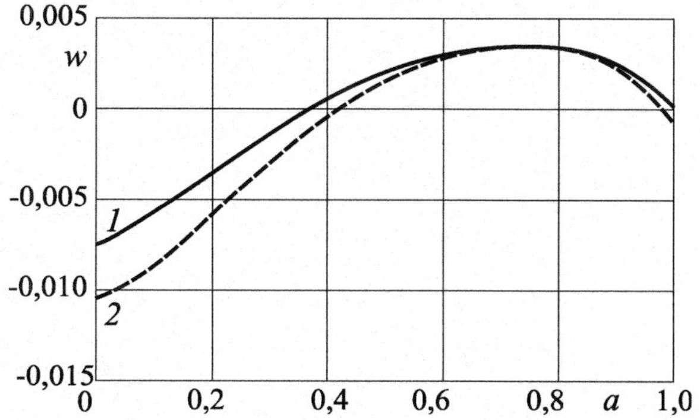

Now let us assume the constancy of the resultant of linear moments $M = 2\pi a M_0$ at any radius of the circle along which they act. The intensity M_0 will be variable, compensating for changes in radius a. Then from (8.46), by substituting $M_0 = M/(2\pi a)$, we obtain the expression for the required functions:

$$\psi = C_2 I_1(\beta r) + C_3 K_1(\beta r) + \frac{C_1\gamma_1}{\beta^2 r} + \frac{\gamma_1 M}{2\pi a\beta} H_0(a-r)\times$$
$$\times\left(\frac{1}{\beta r} - K_1(\beta r)(I_0(\beta a) + I_1(\beta a)\beta a) - I_1(\beta r)(K_0(\beta a) - K_1(\beta a)\beta a)\right),$$
$$w = \frac{b_2}{b_3}\int \psi \mathrm{d}r + \frac{1}{b_3}\int L_3^{-1}(M)\mathrm{d}r - \frac{C_1}{4b_3}r^2(\ln r - 1) + \frac{C_5 r^2}{4b_3} + C_6\ln r + C_4,$$
$$u = \frac{a_3}{a_1}w_{,r} - \frac{a_2}{a_1}\psi + \frac{C_7 r}{2} + \frac{C_8}{r}, \qquad (8.49)$$

where

$$\int L_3^{-1}(M_0)dr = \frac{M}{8\pi a}\left(2r^2 - 2a^2 - 3a^2\ln\left(\frac{r}{a}\right) - r^2\ln\left(\frac{r}{a}\right)\right)H_0(a-r),$$
$$\int \psi dr = \frac{C_2 I_0(\beta r)}{\beta} - \frac{C_3 K_0(\beta r)}{\beta} + \frac{C_1\gamma_1}{\beta^2}\ln(r) + \frac{\gamma_1 M}{2\pi a\beta^2}H_0(a-r)\times$$
$$\times\left(\ln\left(\frac{r}{a}\right) - 1 + K_0(\beta r)(I_0(\beta a) + I_1(\beta a)\beta a) - I_0(\beta r)(K_0(\beta a) - K_1(\beta a)\beta a)\right).$$

Integration constants at the clamped contour of the plate:

$$C_1 = -\frac{M}{2\pi a},\quad C_2 = \frac{\gamma_1 M}{2\pi a\beta^2 I_1(\beta)}(1 - \beta K_1(\beta)(I_0(\beta a) + I_1(\beta a)\beta a)),$$
$$C_3 = \frac{\gamma_1 M}{2\pi a\beta}(I_0(\beta a) + I_1(\beta a)\beta a),$$
$$C_4 = -\frac{b_2\gamma_1 M}{2\pi a b_3\beta^3 I_1(\beta)}(I_0(\beta) - (I_0(\beta a) - I_1(\beta a)\beta a)) + \frac{3M_0 a}{16\pi b_3} + \frac{M_0}{16\pi a b_3},$$
$$C_5 = \frac{M}{4\pi}\left(\frac{1}{a} - 3a\right),\quad C_6 = \frac{3Ma}{8\pi b_3},\quad C_7 = 0,\ C_8 = 0. \qquad (8.50)$$

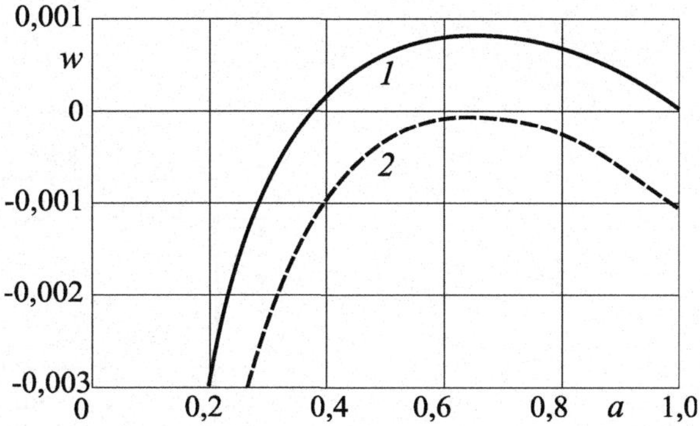

Fig. 8.23 Variations of deflection in the centre of the plate as a function of the radius of the moment circle ($T_1 = 293$ K, $T_2 = 233$ K)

When the contour is hinged:

$$C_4 = \frac{b_2}{b_3}\left(-\frac{C_2 I_0(\beta)}{\beta} + \frac{C_3 K_0(\beta)}{\beta}\right) - \frac{1}{4b_3}(C_1 + C_5),$$

$$C_5 = \frac{M}{4\pi a}(1 - 3a^2)\frac{a_3^2 - a_1 b_3 + a_7 a_1}{a_1(a_6 + a_7)} - \frac{6b_3}{a_6 + a_7}\sum_{k=1}^{3}\alpha_{0k}\int_{h_k} K_k T_k z \mathrm{d}z,$$

$$C_7 = \frac{2a_3}{a_1(a_6 + a_7)}\left(3\sum_{k=1}^{3}\alpha_{0k}\int_{h_k} K_k T_k z \mathrm{d}z + \frac{M}{4\pi a}(1 - 3a^2)\right). \quad (8.51)$$

The constants C_1, C_2, C_3, C_6, C_8 retain the form (8.50).

The variation of deflection (8.49, 8.51) at the centre of the plate as a function of the radius of the circle along which a moment load with constant resultant is applied is shown in Fig. 8.23. The variation of deflection (8.49, 8.51) in the centre of the plate as a function of the radius of the circle along which a moment load with constant resultant is applied is shown in Fig. 8.23: 1—$T_1 = 293$ K, 2—$T_2 = 323$ K. The extremum here is observed at radius = 0.65. As the temperature rises, the deflection becomes of the same sign.

Parabolic load distributed over a circle [0, ***a*****].** Let a local vertical surface load of parabolic shape (convex shape) distributed over a circle of relative radius $r \le a$ (Fig. 8.24) act on the circular three-layer plate. An analytical form can be taken for the load:

$$q(r) = q_0 H_0(a - r)\left(1 - \left(\frac{r}{a}\right)^2\right). \quad (8.52)$$

Fig. 8.24 Scheme of loading by local parabolic distributed load

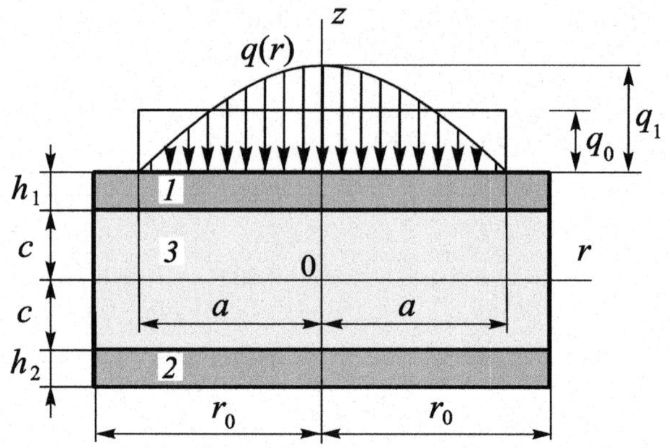

The third-order integral operator from the load (8.52), the function $g(r)$ and integrals from it included in the solution of (8.35) will be as follows:

$$L_3^{-1}(q)\mathrm{d}r = q_0 H_0(a-r)\left(\frac{3r^4-a^4}{48r} - \frac{r^5}{96a^2} - \frac{a^2r}{8}\ln\left(\frac{r}{a}\right) - \frac{a^2r^2}{32}\right),$$

$$f(r) = \frac{\gamma_1}{r}\left(\int q_0\left(1-\left(\frac{r}{a}\right)^2\right)H_0(a-r)r\mathrm{d}r - C_1\right)$$

$$= \frac{\gamma_1}{r}\left(\frac{q_0}{4}\left(2r^2 - a^2 - \frac{r^4}{a^2}\right)H_0(a-r) - C_1\right)$$

$$\int I_1(\beta r)f(r)r\mathrm{d}r = \frac{\gamma_1 q_0}{4\beta^2}H_0(a-r)\left[I_0(\beta r)\left(2r^2\beta - \frac{r^4\beta}{a^2} - \frac{8r^2}{a^2\beta}\right)+\right.$$

$$\left.+\frac{8I_0(\beta a)}{\beta} - \frac{16I_1(\beta a)}{\beta^2 a} - 4I_1(\beta r)\left(r - \frac{r^3}{a^2} - \frac{4}{a^2\beta^2}\right)\right] - \frac{C_1\gamma_1}{\beta}I_0(\beta r)$$

$$\int K_1(\beta r)f(r)r\mathrm{d}r = \frac{\gamma_1 q_0}{4\beta^2}H_0(a-r)\left[-K_0(\beta r)\left(2r^2\beta - \frac{r^4\beta}{a^2} - \frac{8r^2}{a^2\beta}\right)-\right.$$

$$\left.-\frac{8K_0(\beta a)}{\beta} - \frac{16K_1(\beta a)}{\beta^2 a} - 4K_1(\beta r)\left(r - \frac{r^3}{a^2} - \frac{4}{a^2\beta^2}\right)\right] - \frac{C_1\gamma_1}{\beta}I_0(\beta r).$$

Then the partial solution of ψ_r and the necessary integrals can be written in the following form:

$$
\begin{aligned}
&\psi_r = \frac{\gamma_1 q_0}{4\beta^4 a^2} H_0(a-r)\left[\beta^2 r^3 + \frac{\beta^2 a^4}{r} + r(8 - 2\beta^2 a^2) - 8\beta a^2 (K_1(\beta r) I_0(\beta a) + \right.\\
&\left. + I_1(\beta r) K_0(\beta a)) + 16a(K_1(\beta r) I_1(\beta a) - I_1(\beta r) K_1(\beta a))\right] + \frac{C_1\gamma_1}{\beta^2 r}\\
&\int L_3^{-1}(q(r))\mathrm{d}r = H_0(a-r)\left(\frac{r^4}{64} - \frac{r^6}{576a^2} + \frac{a^2 r^2}{64} - \frac{a^2 r^2}{16}\ln\left(\frac{r}{a}\right) - \frac{a^4}{48}\ln\left(\frac{r}{a}\right) - \frac{17a^4}{576}\right)\\
&\int \psi \mathrm{d}r = \frac{C_2 I_0(\beta r)}{\beta} - \frac{C_3 K_0(\beta r)}{\beta} + \frac{C_1\gamma_1}{\beta^2}\ln(r) + \frac{\gamma_1 q_0}{4\beta^4 a^2} H_0(a-r)\times\\
&\times\left[\beta^2 a^4 \ln\left(\frac{r}{a}\right) - \frac{16a}{\beta}(K_1(\beta a) I_0(\beta r) + I_1(\beta a) K_0(\beta r)) + \frac{16}{\beta^2} + 4(r^2 - a^2) + \right.\\
&\left. + \frac{\beta^2 r^4}{4} + \frac{3\beta^2 a^4}{4} - \beta^2 a^2 r^2 - 8a^2 (K_0(\beta a) I_0(\beta r) - I_0(\beta a) K_0(\beta r))\right].
\end{aligned}
\tag{8.53}
$$

The general form of the solution (8.35) formally remains unchanged in this case if we take into account the relations (8.53):

$$
\begin{aligned}
&\psi = C_2 I_1(\beta r) + C_3 K_1(\beta r) + \psi_r,\\
&w = \frac{b_2}{b_3}\int \psi \mathrm{d}r + \frac{1}{b_3}\int L_3^{-1}(q)\mathrm{d}r - \frac{C_1}{4b_3} r^2(\ln r - 1) + \frac{C_5 r^2}{4b_3} + C_6 \ln r + C_4,\\
&u = \frac{a_3}{a_1} w_{,r} - \frac{a_2}{a_1}\psi + \frac{C_7 r}{2} + \frac{C_8}{r}.
\end{aligned}
\tag{8.54}
$$

However, the integration constants obtained from the conditions of embedding of the plate contour ($u = \psi = w = w_{,r} = 0$ at $r = 1$) and the condition of boundedness of the solution at its centre will be:

$$
\begin{aligned}
&C_1 = -\frac{q_0 a^2}{4},\ C_2 = \frac{2\gamma_1 q_0}{\beta^3 I_1(\beta)}\left(\frac{\beta a^2}{8} - K_1(\beta)\left(I_0(\beta a) - \frac{2}{\beta a} I_1(\beta a)\right)\right),\\
&C_3 = \frac{2q_0\gamma_1}{\beta^3}\left(I_0(\beta a) - \frac{2}{\beta a} I_1(\beta a)\right),\\
&C_4 = -\frac{\gamma_1 b_2 q_0}{\beta^3 b_3 I_1(\beta)}\left(\frac{a^2 I_0(\beta)}{4} + \frac{2}{\beta^2}\left(I_0(\beta a) - \frac{2}{\beta a} I_1(\beta a)\right)\right) + \frac{q_0 a^4}{96b_3} + \frac{q_0 a^2}{32b_3},\\
&C_5 = \frac{q_0 a^2}{24}(3 + a^2)\ C_6 = \frac{q_0 a^4}{48b_3},\ C_7 = 0,\ C_8 = 0.
\end{aligned}
\tag{8.55}
$$

At a hinged contour, it is necessary to add constants to the solution (8.53):

$$
C_4 = \frac{b_2}{b_3}\left(1 - \frac{a^2}{3}\right)\frac{a_3^2 - a_1 b_3 + a_7 a_1}{a_1(a_6 + a_7)} - \frac{6b_3}{a_6 + a_7}\sum_{k=1}^{3}\alpha_{0k}\int_{hk} K_k T_k z \mathrm{d}z,
$$

$$
C_5 = \frac{q_0 a^2}{8}\left(1 - \frac{a^2}{3}\right)\frac{a_3^2 - a_1 b_3 + a_7 a_1}{a_1(a_6 + a_7)} - \frac{6b_3}{a_6 + a_7}\sum_{k=1}^{3}\alpha_{0k}\int_{h_k} K_k T_k z \mathrm{d}z,
$$

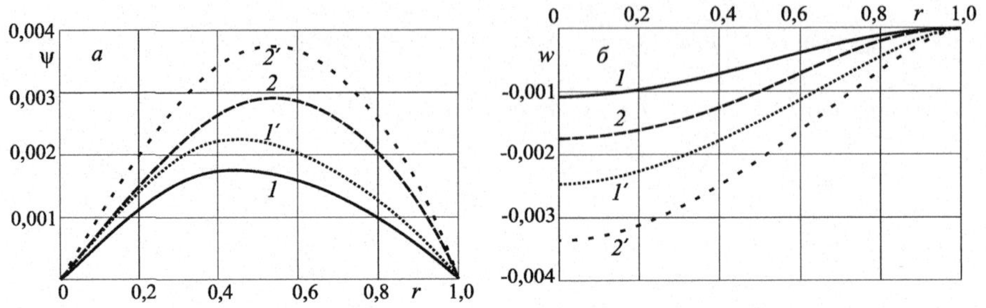

Fig. 8.25 Displacements in the plate at different radiuses of the parabolic load spot and temperatures

$$C_7 = \frac{2a_3}{a_1(a_6 + a_7)}\left(3\sum_{k=1}^{3}\alpha_{0k}\int_{h_k} K_k T_k z\mathrm{d}z + \frac{q_0 a^2}{8}\left(1 - \frac{a^2}{3}\right)\right).$$

The constants C_1, C_2, C_3, C_6, C_8 retain the form (8.55).

Figure 8.25 *a*, *b* shows the variation of relative shear in the fill and deflection of an hinged plate along its radius. The curves are plotted at different values of the radius b of the local parabolic surface load spot: 1—$b = 0.5$; 2—$b = 0.75$; without a stroke—$T_1 = 293$ K, with a stroke—$T_2 = 323$ K.

Figure 8.26 shows the dependence of the deflection of a circular three-layer plate in the centre on the radius of the force spot of different shapes and magnitudes of loads: 1—parabolic with amplitude q_0; 2—rectangular with intensity q_0; 3—parabolic with amplitude $q_1 = 2q_0$. Without a stroke—room temperature $T_1 = 293$ K, with a stroke—$T_2 = 323$ K. A parabolic load with amplitude q_1 is statically equivalent to a rectangular load $2q_0$, because when integrated over the area A occupied by the load, we have:

$$q_1\int_A\left(1 - \left(\frac{r^2}{a}\right)^2\right)H_0(a - r)r\mathrm{d}r\mathrm{d}\varphi = q_0\int_A H_0(a - r)r\mathrm{d}r\mathrm{d}\varphi.$$

From here we have $q_1 = 2q_0$.

As the radius of the load spot increases, the deflection increases non-linearly, reaching a maximum at the load distributed over the entire plate surface. For the same amplitude (1, 2), the rectangular load causes more deflection. In the case of statically equivalent loads (2, 3) the deflection from the parabolic load is larger by 47%, when the plate is heated the difference decreases to 40%.

Thus, the results presented in this section have shown that the stress–strain state of asymmetric in thickness elastic circular three-layer plates significantly depends on the effect of the temperature field, type and shape of the load and boundary conditions.

Fig. 8.26 Deflection in the centre of the plate as a function of the radius of the force spot of different shapes and magnitudes of loads

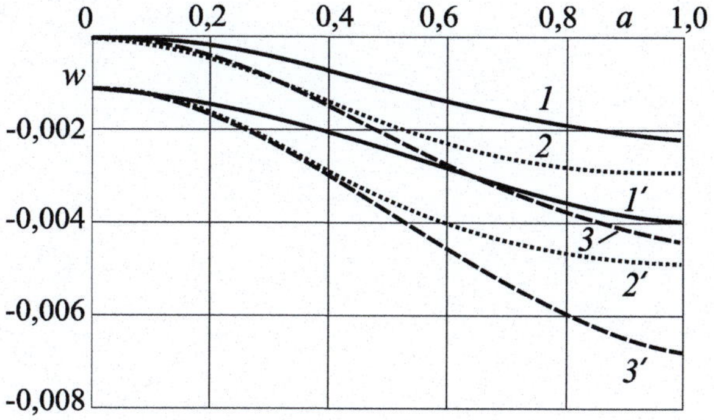

Reference

1. Korn, G., and T. Korn. 1978. *Mathematical handbook for scientists and engineers*, 832. Moscow, Nauka (in Russian).

Chapter 9
Cyclic Loading of Three-Layer Elastoplastic Circular Plates

Single and cyclic bending of asymmetric in thickness three-layer circular plates with rigid incompressible filler in the temperature field is investigated. The bearing layers are assumed to be elastoplastic, the filler is nonlinearly elastic. The kinematic assumptions are based on the "broken" line hypothesis. Loads are continuous and localised. Strains are small. The work of tangential stresses in the filler is taken into account. Parametric analysis of the obtained solutions was carried out.

9.1 Formulation of the Boundary Value Problem of Thermoplasticity

It is assumed that materials in a three-layer circular plate can exhibit elastic–plastic (bearing layers) and nonlinear-elastic (filler) properties during the deformation process. When describing the kinematics of the package, the broken line hypotheses are adopted (see Sect. 8.1). The plate is in a temperature field and a distributed load $p'(x)$, $q'(x)$ acts on it, and the corresponding loading trajectories belong to the class of simple trajectories (according to Ilyushin). Deformation of layer materials is described by nonlinear physical equations of state of the theory of small elastoplastic deformations of the type (2.32):

$$
\begin{aligned}
s'^{(k)}_{\alpha} &= 2G_k(T)(1-\omega'^{(k)}(\varepsilon^{(k)}_u,\ T))\,э'^{(k)}_{\alpha},\\
s'^{(3)}_{rz} &= 2G_3(T)(1-\omega'^{(3)}(\varepsilon^{(3)}_u,\ T))\,э'^{(3)}_{rz}\\
\sigma'^{(k)} &= 3K_k(T)(\varepsilon'^{(k)}-\alpha_{0k}T)\quad (k=1,\ 2,\ 3;\ \alpha=r,\varphi),
\end{aligned}
\tag{9.1}
$$

S. Eduard et al., *Deformation of Three-layer Structural Elements in Thermal Radiation Fields*, https://doi.org/10.1007/978-981-97-7217-9_9

where $\omega'_k(\varepsilon'^k_u, T_k)$ is function of physical nonlinearity of the material of the kth layer, $0 \le \omega'_k \le 1$, herewith $(\omega'_k = 0)$, if $\varepsilon'^{(k)}_u \le \varepsilon'^{(k)}_y(T_k)$; $\varepsilon'^{(k)}_u$ is strain intensity, $\varepsilon'^{(k)}_y$ is deformation yield strength (physical nonlinearity) of the material, the stroke denotes loading from the natural state.

We express the components of the stress tensor in the layers of the plate through the deviator and spherical parts of the strain tensor:

$$\sigma'^{(k)}_\alpha = 2G_k \ni'^{(k)}_\alpha(1 - \omega'^{(k)}) + 3K_k(\varepsilon'^{(k)} - \alpha_k T_k)(k = 1, 2, 3),$$
$$\sigma'^{(3)}_{rz} = s'^{(3)}_{rz} = 2G_3 \ni'^{(3)}_{rz}(1 - \omega'^{(3)})(\alpha = r, \varphi)\cdot$$

Let us distinguish linear and non-linear components in these stresses:

$$\sigma'^{(k)}_\alpha = \sigma'^{(k)}_{\alpha e} - \sigma'^{(k)}_{\alpha\omega}, \quad \sigma'^{(3)}_{rz} = \sigma'^{(3)}_{rze} - \sigma'^{(3)}_{rz\omega},$$
$$\sigma'^{(k)}_{\alpha e} = 2G_k \ni'^{(k)}_\alpha + 3K_k(\varepsilon'^{(k)} - \alpha_k T_k), \quad \sigma'^{(k)}_{\alpha\omega} = 2G_k\omega_k(\varepsilon'^{(k)}_u, T_k) \ni'^{(k)}_\alpha,$$
$$\sigma'^{(3)}_{rze} = 2G_3 \ni'^{(3)}_{rz}, \quad \sigma'^{(3)}_{rz\omega} = 2G_3\omega_3(\varepsilon'^{(3)}_u, T_k) \ni'^{(3)}_{rz} \quad (\alpha = r, \varphi). \tag{9.2}$$

Let us also represent the internal forces and moments in the layers of the plate as the difference of the linear and non-linear parts:

$$T'^{(k)}_\alpha = T'^{(k)}_{\alpha e} - T'^{(k)}_{\alpha\omega}, \quad M'^{(k)}_\alpha = M'^{(k)}_{\alpha e} - M'^{(k)}_{\alpha\omega},$$
$$H'^{(k)}_\alpha = H'^{(k)}_{\alpha e} - H'^{(k)}_{\alpha\omega}, \quad Q' = Q'_e - Q'_\omega, \quad (\alpha = r, \varphi).$$

The values $T'^{(k)}_{\alpha e}$, $T'^{(k)}_{\alpha\omega}$, $M'^{(k)}_{\alpha e}$, $M'^{(k)}_{\alpha\omega}$, $H'^{(k)}_{\alpha e}$, $H'^{(k)}_{\alpha\omega}$, Q'_e, Q'_ω are calculated by formulae (8.3), in which the stresses $\sigma'^{(k)}_\alpha$ should be replaced respectively by the components $\sigma'^{(k)}_{\alpha e}$, $\sigma'^{(k)}_{\alpha\omega}$, $\sigma'^{(3)}_{rze}$, $\sigma'^{(3)}_{rz\omega}$ from relations (9.2).

With the generalised internal loads (8.3) we proceed in the same way:

$$T'_\alpha = T'_{\alpha e} - T'_{\alpha\omega} = \sum_{k=1}^{3} T'^{(k)}_{\alpha e} - \sum_{k=1}^{3} T'^{(k)}_{\alpha\omega},$$
$$M'_\alpha = M'_{\alpha e} - M'_{\alpha\omega} = \sum_{k=1}^{3} M'^{(k)}_{\alpha e} - \sum_{k=1}^{3} M'^{(k)}_{\alpha\omega},$$
$$H'_{\alpha e} = M'^{(3)}_{\alpha e} + c\left(T'^{(1)}_{\alpha e} - T'^{(2)}_{\alpha e}\right),$$
$$H'_{\alpha\omega} = M'^{(3)}_{\alpha\omega} + c\left(T'^{(1)}_{\alpha\omega} - T'^{(2)}_{\alpha\omega}\right). \tag{9.3}$$

The system of equations of equilibrium of a circular three-layer plate in the forces (8.8) was obtained without using the physical equations of state, so it remains valid in the considered case as well. Substituting into it the expressions for internal loads (9.3), we reduce it to the form:

$$T'_{r,r} + \frac{1}{r}(T'_r - T'_\varphi) = -p' + p'_\omega,$$

$$H\prime_{r},_{r} + \frac{1}{r}(H\prime - H\prime_{\varphi}) - Q\prime = h\prime_{\omega},$$
$$M\prime,_{rr} + \frac{1}{r}(2M\prime_{r},_{r} - M\prime_{\varphi},_{r}) = -q\prime + q\prime_{\omega} \tag{9.4}$$

The left-hand side of these equations gathers the linear components of the generalised forces, with the index "*e*" omitted for convenience. The nonlinear additives are concentrated on the right-hand side and are included in the terms with the lower index "ω":

$$p\prime_{\omega} = T\prime_{r\omega},_{r} + \frac{1}{r}(T\prime_{r\omega} - T\prime_{\varphi\omega}),$$
$$q\prime_{\omega} = M\prime_{r\omega},_{rr} + \frac{1}{r}(2M\prime_{r\omega},_{r} - M\prime_{\varphi\omega},_{r})$$
$$q\prime_{\omega} = M\prime_{r\omega},_{rr} + \frac{1}{r}(2M\prime_{r\omega},_{r} - M\prime_{\varphi\omega},_{r}) \tag{9.5}$$

Linear components of generalised internal forces are still expressed through displacements by formulas of the type (8.12)–(8.14), therefore the system of equilibrium Eq. (9.4) in displacements can be reduced to the following form:

$$L_2(a_1u\prime + a_2\psi\prime - a_3w\prime,_{r}) = -p\prime + p\prime_{\omega},$$
$$L_2(a_2u\prime + a_4\psi\prime - a_5w\prime,_{r}) - 2cG_3\psi\prime = h\prime_{\omega},$$
$$L_3(a_3u\prime + a_5\psi\prime - a_6w\prime,_{r}) = -q' + q\prime_{\omega}. \tag{9.6}$$

Coefficients a_i and differential operators L_2, L_3 are defined by relations (8.16), (8.17). All displacements and linear dimensions of the plate are referred to its radius r_0.

To close the boundary value problem of thermoplasticity in the considered case it is necessary to add force (8.9) or kinematic (8.19) boundary conditions to Eqs. (9.6).

9.2 Physically Nonlinear Plate Under Continuous and Locally Applied Loads

The constructed system of differential Eqs. (9.6) is essentially nonlinear. Numerical or approximate methods must be used to solve it. In the further, when solving specific problems, the method of elastic solutions discussed in Sect. 2.2 will be used. Then in the iterative form the system (9.6) is rewritten in the form

$$L_2(a_1u\prime^{(n)} + a_2\psi\prime^{(n)} - a_3w\prime,_{r}^{(n)}) = -p\prime + p\prime_{\omega}^{(n-1)},$$
$$L_2(a_2u\prime^{(n)} + a_4\psi\prime^{(n)} - a_5w\prime,_{r}^{(n)}) - 2cG_3\psi\prime^{(n)} = h\prime_{\omega}^{(n-1)}$$
$$L_3(a_3u\prime^{(n)} + a_5\psi\prime^{(n)} - a_6w\prime,_{r}^{(n)}) = -q\prime + q\prime_{\omega}^{(n-1)}. \tag{9.7}$$

Here n is the number of approximation, the values $p\prime_{\omega}^{(n-1)}$, $h\prime_{\omega}^{(n-1)}$, $q\prime_{\omega}^{(n-1)}$, are called additional "external" loads and are assumed to be equal to zero at the first step, and further calculated by the results of the previous approximation. In this case, formulas like (9.5) are used, in which the index "n–1" must be added at the top of all terms:

$$
\begin{aligned}
p\prime_{\omega}^{(n-1)} &= T\prime_{r\omega}^{(n-1)},_r + \frac{1}{r}(T\prime_{r\omega}^{(n-1)} - T\prime_{\varphi\omega}^{(n-1)}),\\
h\prime_{\omega}^{(n-1)} &= H\prime_{r\omega}^{(n-1)},_r + \frac{1}{r}(H\prime_{r\omega}^{(n-1)} - H\prime_{\varphi\omega}^{(n-1)}) - Q\prime_{\omega}^{(n-1)},\\
q\prime_{\omega}^{(n-1)} &= M\prime_{r\omega}^{(n-1)},_{rr} + \frac{1}{r}(2M\prime_{r\omega}^{(n-1)},_r - M\prime_{\varphi\omega}^{(n-1)},_r).
\end{aligned}
\tag{9.8}
$$

Nonlinear additives in (9.8) are calculated by the following formulas:

$$
\begin{aligned}
T\prime_{\alpha\omega}^{(n-1)} &\equiv \sum_{k=1}^{3}\int_{h_k} \sigma\prime_{\alpha\omega}^{(k,n-1)}dz = \sum_{k=1}^{3}\int_{h_k} 2G_k\omega_k(\varepsilon\prime_u^{(k,n-1)}, T_k)\,\ni\prime_{\alpha}^{(k,n-1)}dz,\\
M\prime_{\alpha\omega}^{(n-1)} &\equiv \sum_{k=1}^{3}\int_{h_k} \sigma\prime_{\alpha\omega}^{(k,n-1)}zdz = \sum_{k=1}^{3}\int_{h_k} 2G_k\omega_k(\varepsilon\prime_u^{(k,n-1)}, T_k)\,\ni\prime_{\alpha}^{(k,n-1)}zdz,\\
H\prime_{\alpha\omega}^{(n-1)} &= M\prime_{\alpha\omega}^{(3,n-1)} + c\big(T\prime_{\alpha\omega}^{(1,n-1)} - T\prime_{\alpha\omega}^{(2,n-1)}\big),\\
Q\prime_{\omega}^{(n-1)} &= \int_{-c}^{c} 2G_k\omega_k(\varepsilon\prime_u^{(k,n-1)}, T_k)\,\ni\prime_{rz}^{(k,n-1)}dz \quad (\alpha = r, \phi).
\end{aligned}
\tag{9.9}
$$

The boundary value problem for determining the functions $u\prime^{(n)}(r)$, $\psi\prime^{(n)}(r), w\prime,_r^{(n)}(r)$ is closed by joining to the system of Eqs. (9.7) the boundary conditions that follow from (8.19):

when embedding the plate contour

$$
u\prime = \psi\prime^{(n)} = w\prime^{(n)} = w\prime,_r^{(n)} = 0 \; at \quad r = 1. \tag{9.10}
$$

when the plate is hinged

$$
u\prime^{(n)} = \psi\prime^{(n)} = w\prime^{(n)} = 0, \quad M\prime_r^{(n)} = M\prime_r^{(n-1)} \; at \quad r = 1. \tag{9.11}
$$

Thus, at each step of approximation we have a linear problem of elasticity theory for the considered plate with known additional "external" loads, which are calculated from the results of the previous approximation using the formulas (9.8), (9.9).

Let us consider some examples of the solution of the boundary value problem of thermoelastic-plasticity under different types of external loading.

Continuously distributed load. The procedure for finding the solution of the system (9.7), taking into account its restriction at the centre of the plate, does not differ from that applied earlier for the corresponding problem in the theory of elasticity (see Sect. 8.2). As a result, we have the following general solution of the problem of the theory of small elastic–plastic deformations in recurrence form:

$$\begin{aligned}
\psi\prime^{(n)} &= C_2^{(n)} I_1(\beta r) + \psi\prime_r^{(n)},\\
u\prime^{(n)} &= \frac{a_3}{a_1 a_6 - a_3^2}\left[L_3^{-1}(q\prime - q\prime_\omega^{(n-1)}) - \frac{a_6}{a_3} L_2^{-1}(p - p\prime_\omega^{(n-1)}) + \left(a_5 - \frac{a_2 a_6}{a_3}\right)\psi\prime^{(n)} + C_7^{(n)} r\right],\\
w\prime^{(n)} &= \frac{1}{b_3}\left[b_2\left(\frac{C_2^{(n)}}{\beta} I_0(\beta r) + \int \psi\prime_r^{(n)} dr\right) -\right.\\
&\left. - \int\left(\frac{a_3}{a_1} L_2^{-1}(p\prime - p\prime_\omega^{(n-1)}) - L_3^{-1}(q\prime - q\prime_\omega^{(n-1)})\right) dr + \frac{C_5^{(n)} r^2}{4} + C_4^{(n)}\right]
\end{aligned} \tag{9.12}$$

The partial solution of the inhomogeneous modified Bessel equation included in the solution (9.12) has the form:

$$\psi\prime_r^{(n)} = -K_1(\beta r)\int I_1(\beta r) f\prime^{(n)} r dr + I_1(\beta r)\int K_1(\beta r) f\prime^{(n)} r dr, \tag{9.13}$$

where the integrand function will be

$$f\prime^{(n)}(r) = \frac{b_3}{b_1 b_3 - b_2^2}\left[h\prime_\omega^{(n-1)} + \frac{a_2 b_3 - a_3 b_2}{a_1 b_3}(p\prime - p\prime_\omega^{(n-1)}) + \frac{b_2}{b_3 r}\int (q\prime - q\prime_\omega^{(n-1)}) r dr\right].$$

In the case of the plate contour embedding, from the boundary conditions (9.10) the recurrence formulas for the integration constants follow:

$$\begin{aligned}
C_2^{(n)} &= -\frac{\psi\prime_r^{(n)}\big|_{r=1}}{I_1(\beta)},\quad C_4^{(n)} = -b_2\left(\frac{C_2^{(n)} I_0(\beta)}{\beta} + \int \psi\prime_r^{(n)} dr\big|_{r=1}\right) +\\
&+ \int\left(\frac{a_3}{a_1} L_2^{-1}(p\prime - p\prime_\omega^{(n-1)}) - L_3^{-1}(q' - q\prime_\omega^{(n-1)})\right) dr\bigg|_{r=1} - \frac{C_5^{(n)}}{4},\\
C_5^{(n)} &= 2\left(\frac{a_3}{a_1} L_2^{-1}(p' - p\prime_\omega^{(n-1)}) - L_3^{-1}(q' - q\prime_\omega^{(n-1)})\right)\bigg|_{r=1},\\
C_7^{(n)} &= \left(\frac{a_6}{a_3} L_2^{-1}(p' - p\prime_\omega^{(n-1)}) - L_3^{-1}(q' - q\prime_\omega^{(n-1)})\right)\bigg|_{r=1}.
\end{aligned} \tag{9.14}$$

When the contour is hinged, the integration constants $C_2^{(n)}$, $C_4^{(n)}$, $C_7^{(n)}$ will be the same as in (9.14), and $C_5^{(n)}$ are determined from (9.11):

$$C_5^{(n)} = \frac{2b_3}{a_6 + a_7}\left[a_3 u'^{(n)},_r(1) + \psi'^{(n)},_r(1)\left(a_5 - \frac{a_6 b_2}{b_3}\right)\right.$$
$$+ \left(\frac{a_3}{a_1} L_2^{-1}(p' - p'^{(n-1)}_{\omega}) - L_3^{-1}(q' - q'^{(n-1)}_{\omega})\right),_r\Bigg|_{r=1}$$
$$\left. + \left(\frac{a_3}{a_1} L_2^{-1}(p\prime - p\prime^{(n-1)}_{\omega}) - L_3^{-1}(q\prime - q\prime^{(n-1)}_{\omega})\right)\Bigg|_{r=1} - 3\sum_{k=1}^{3} \alpha_{0k} \int_{h_k} K_k T_k z dz - M_r^{(n-1)}\right]. \tag{9.15}$$

Consequently, the recurrence solution (9.12)–(9.15) describes the deformation of a three-layer elastoplastic circular plate in the temperature field under a continuously distributed surface load.

Numerical results further in the section are given for a circular three-layer plate asymmetric in thickness, the layers of which are made of materials D16T-fluoroplastic-D16T. Nonlinearity functions of these materials are taken in the form (2.39), (2.43). The corresponding nonlinearity parameters are given in Tables 2.1, 2. 2. Load magnitude $q_0 = 5$ MPa, unless otherwise specifie. The boundary conditions correspond to the hinged support of the plate contour. Geometrical parameters are referred to its radius r_0, relative thicknesses of layers are: $h_1 = 0{,}02$; $h_2 = 0{,}06$; $h_3 = 0{,}4$.

Graphs in Fig. 9.1: relative shear in the filler (*a*), deflection (*b*). The graphs confirm the convergence of the elastic solution method at temperature T2 = 323 K. The maximum difference of the results in the 5th approximation, which is taken as the required solution, from the previous one is about 0.5% and from the original one is 19%.

The distribution of areas of physical nonlinearity of layer materials (grey fill) along the cross-section of the plate under isothermal (a) and thermo-force (b) loading is illustrated in Fig. 9.2. Here we should note the growth of nonlinearity areas in the temperature field.

Figure 9.3a, b shows the variation of shear in the filler and deflection of a hinged plate along its radius at different temperatures: *1*, *2*—elastic and physically nonlinear at $T_1 = 293$ K; *3*, *4*—elastic and physically nonlinear at $T_2 = 323$ K.

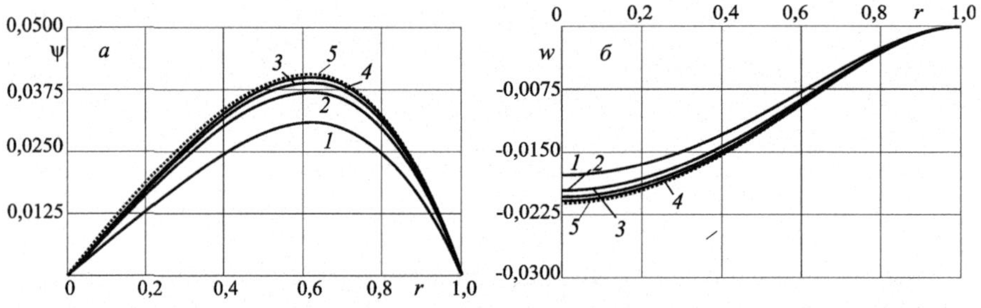

Fig. 9.1 Convergence of the elastic solution method ($T_2 = 323$ K)

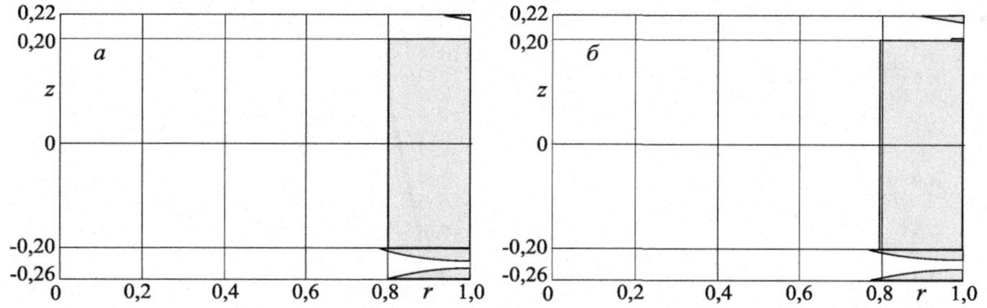

Fig. 9.2 Areas of physical nonlinearity of layer materials

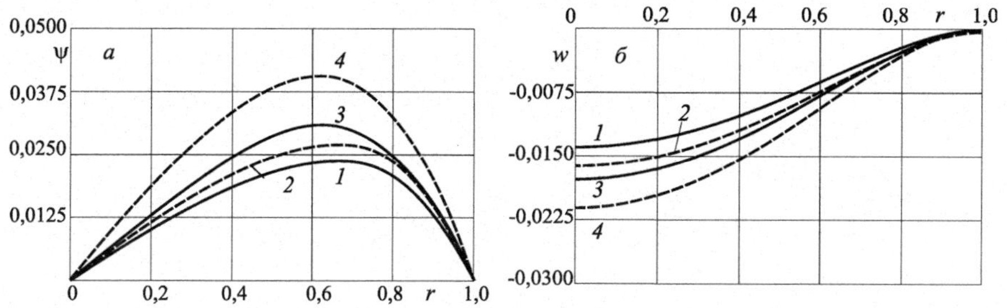

Fig. 9.3 Variation of displacements along the plate radius at different temperatures ($T_1 = 293$ K, $T_2 = 323$ K)

Under isothermal loading, taking into account the physical nonlinearity of the layer materials increases the calculated elastic displacements by 14%, at elevated temperature ($T_2 = 323$ K)—by 19%.

Graphs showing the distribution of normal stresses σ_r in a plate with a hinged contour at different temperatures are shown in Fig. 9.4: *a*—in the centre, *b*—on the contour of the plate. Curves *1*, *2*—elastic and physically nonlinear at $T_1 = 293$ K; *3*, *4*—elastic and physically nonlinear at $T_2 = 323$ K. When the physical nonlinearity of the layer materials is taken into account, the stresses in the centre of the plate change to a lesser extent than on the contour.

Local load uniformly distributed over the circle [0, *b*]. Let the three-layer plate under consideration be subjected to a local vertical surface load uniformly distributed over a circle of relative radius $0 \le b \le r$ (see Fig. 8.11). Using the Heaviside function (8.31), it can be written as follows:

$$q^{\prime}(r) = q_0 H_0(b - r), \quad p^{\prime} = 0. \tag{9.16}$$

The integrand function, introduced in (9.13), at load (9.16) takes the form:

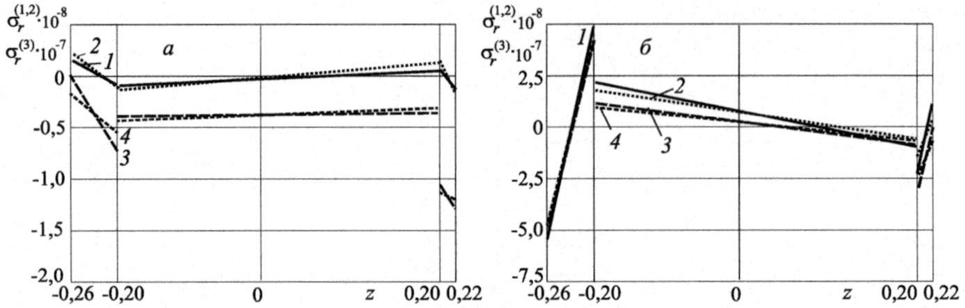

Fig. 9.4 Variation of normal stresses σ_r across the plate thickness in the centre and on the contour at temperatures $T_1 = 293$ K and $T_2 = 323$ K

$$f'^{(n)}(r) = \frac{1}{b_1b_3 - b_2^2}\left[b_3h'^{(n-1)}_\omega - \frac{a_2b_3 - a_3b_2}{a_1}p'^{(n-1)}_\omega + \frac{b_2}{r}\left(\int (q_0H_0(b-r) - q'^{(n-1)}_\omega)rdr - C_1^{(n)}\right)\right]$$
$$= \frac{b_3}{b_1b_3 - b_2^2}\left[h'^{(n-1)}_\omega - \frac{a_2b_3 - a_3b_2}{a_1b_3}p'^{(n-1)}_\omega\right] + \frac{\gamma_1}{r}\left(\frac{q_0\left(r^2 - b^2\right)}{2}H_0(b-r) - \int q'^{(n-1)}_\omega)rdr - C_1^{(n)}\right),$$

where

$$\gamma_1 = \frac{b_2}{(b_1b_3 - b_2^2)}.$$

Introducing this function into the integrals included in the partial solution (9.13), we obtain

$$\int I_1(\beta r)f'^{(n)}rdr = \frac{\gamma_1 q_0}{2\beta}H_0(b-r)\left[\left(b^2 - r^2\right)I_0(\beta r) + \frac{2}{\beta}(bI_1(\beta b) - rI_1(\beta r))\right] + \frac{C_1^{(n)}\gamma_1}{\beta}I_0(\beta r) + \frac{\gamma_1 q_0}{2\beta}H_0(b-r)\left[\left(r^2 - b^2\right)I_0(\beta r) + \frac{b_3}{b_1b_3 - b_2^2}\left[\int I_1(\beta r)h'^{(n-1)}_\omega rdr - \frac{a_2b_3 - a_3b_2}{a_1b_3}\int I_1(\beta r)p'^{(n-1)}_\omega rdr\right] - \gamma_1\int I_1(\beta r)\left(\int q'^{(n-1)}_\omega)rdr\right)dr$$
$$\int K_1(\beta r)f'^{(n)}rdr = \frac{\gamma_1 q_0}{2\beta}H_0(b-r)\left[\left(b^2 - r^2\right)K_0(\beta r) + \frac{2}{\beta}(bK_1(\beta b) - rK_1(\beta r))\right] + \frac{C_1^{(n)}\gamma_1}{\beta}K_0(\beta r) + \frac{b_3}{b_1b_3 - b_2^2}\left[\int K_1(\beta r)h'^{(n-1)}_\omega rdr - \frac{a_2b_3 - a_3b_2}{a_1b_3}\int K_1(\beta r)p'^{(n-1)}_\omega rdr\right] - \gamma_1\int K_1(\beta r)\left(\int q'^{(n-1)}_\omega)rdr\right)dr. \quad (9.17)$$

As a result, the partial solution (9.13), taking into account integrals (9.17), takes the form:

$$\psi{\prime}_r^{(n)} = \frac{\gamma_1 q_0}{2\beta^2} H_0(b-r)\left[\frac{b^2}{r} - r + 2b(K_1(\beta b)I_1(\beta r) - I_1(\beta b)K_1(\beta r))\right] + \frac{C_1^{(n)}\gamma_1}{\beta^2 r} -$$
$$-K_1(\beta r)\left[\frac{b_3}{b_1b_3-b_2^2}\left[\int I_1(\beta r)h{\prime}_\omega^{(n-1)}rdr - \frac{a_2b_3-a_3b_2}{a_1b_3}\int I_1(\beta r)p{\prime}_\omega^{(n-1)}rdr\right] -\right.$$
$$\left. -\gamma_1\int I_1(\beta r)\left(\int q{\prime}_\omega^{(n-1)})rdr\right)dr\right] +$$
$$+I_1(\beta r)\left[\frac{b_3}{b_1b_3-b_2^2}\left[\int K_1(\beta r)h{\prime}_\omega^{(n-1)}rdr - \frac{a_2b_3-a_3b_2}{a_1b_3}\int K_1(\beta r)p{\prime}_\omega^{(n-1)}rdr\right] -\right.$$
$$\left. -\gamma_1\int K_1(\beta r)\left(\int q{\prime}_\omega^{(n-1)})rdr\right)dr\right].$$

The required displacements in the plate under the load (9.16) will be:

$$\begin{aligned}
\psi{\prime}^{(n)} &= C_2^{(n)}I_1(\beta r) + C_3^{(n)}K_1(\beta r) + \psi{\prime}_r^{(n)},\\
w{\prime}^{(n)} &= \frac{b_2}{b_3}\int \psi{\prime}^{(n)}dr + \frac{a_3}{b_3a_1}\int L_2^{-1}(p{\prime}_\omega^{(n-1)})dr + \frac{1}{b_3}\int L_3^{-1}(q{\prime})dr - \frac{1}{b_3}\int L_3^{-1}(q{\prime}_\omega^{(n-1)})dr -\\
&-\frac{C_1^{(n)}}{4b_3}r^2(\ln r - 1) + \frac{C_5^{(n)}r^2}{4b_3} + C_6^{(n)}\ln r + C_4^{(n)},\\
u{\prime}^{(n)} &= \frac{a_3}{a_1}w{\prime}^{(n)}{,}_r - \frac{a_2}{a_1}\psi{\prime}^{(n)} + \frac{1}{a_1}L_2^{-1}(p{\prime}_\omega^{(n-1)}) + \frac{C_7^{(n)}r}{2} + \frac{C_8^{(n)}}{r},
\end{aligned} \tag{9.18}$$

where

$$\int L_3^{-1}(q{\prime})dr = q_0\left(\frac{r^4-5b^4}{64} - \frac{b^4}{16}\ln\left(\frac{r}{b}\right) - \frac{b^2r^2}{8}\ln\left(\frac{r}{b}\right) + \frac{b^2r^2}{16}\right)H_0(b-r),$$
$$\int \psi dr = \frac{C_2^{(n)}I_0(\beta r)}{\beta} - \frac{C_3^{(n)}K_0(\beta r)}{\beta} + \frac{C_1^{(n)}\gamma_1}{\beta^2}\ln(r) + \frac{\gamma_1 q_0}{2\beta^2}H_0(b-r)\times$$
$$\left[\frac{b^2-r^2}{2} + b^2\ln\left(\frac{r}{b}\right) + \frac{2b}{\beta}(K_1(\beta b)I_0(\beta r) + I_1(\beta b)K_0(\beta r)) - \frac{2}{\beta^2}\right] -$$
$$\int\left(K_1(\beta r)\left[\frac{b_3}{b_1b_3-b_2^2}\left[\int I_1(\beta r)h{\prime}_\omega^{(n-1)}rdr - \frac{a_2b_3-a_3b_2}{a_1b_3}\int I_1(\beta r)p{\prime}_\omega^{(n-1)}rdr\right] -\right.\right.$$
$$\left.\gamma_1\int I_1(\beta r)\left(\int q{\prime}_\omega^{(n-1)})rdr\right)dr\right] +$$
$$I_1(\beta r)\left[\frac{b_3}{b_1b_3-b_2^2}\left[\int K_1(\beta r)h{\prime}_\omega^{(n-1)}rdr - \frac{a_2b_3-a_3b_2}{a_1b_3}\int K_1(\beta r)p{\prime}_\omega^{(n-1)}rdr\right] -\right.$$
$$\left.\left.\gamma_1\int K_1(\beta r)\left(\int q{\prime}_\omega^{(n-1)})rdr\right)dr\right]\right)dr.$$

In case of embedding or hinged support of the plate contour, from the part of boundary conditions ($\psi\prime^{(n)} = w\prime^{(n)} = 0 \; at \quad r = 1\cdot$) and conditions of boundedness of the solution in the centre, follow formulas for integration constants:

$$
\begin{aligned}
&C_1^{(n)} = -\frac{q_0 b^2}{2}, \quad C_6^{(n)} = -\frac{q_0 b^4}{16 b_3}, \quad C_8^{(n)} = 0, \quad C_3^{(n)} = \frac{\gamma_1 q_0}{\beta^2} b I_1(\beta r) + \\
&\left[\frac{b_3}{b_1 b_3 - b_2^2}\left(\int I_1(\beta r) h\prime_\omega^{(n-1)} r dr - \frac{a_2 b_3 - a_3 b_2}{a_1 b_3} \int I_1(\beta r) p\prime_\omega^{(n-1)} r dr\right) - \right.\\
&\left.\gamma_1 \int I_1(\beta r) \int q\prime_\omega^{(n-1)} r dr dr\right]\Bigg|_{r=0}, \\
&C_2^{(n)} = -\frac{1}{I_1(\beta)}\left(C_3^{(n)} K_1(\beta) + \frac{C_1^{(n)} \gamma_1}{\beta^2}\right) - \frac{1}{I_1(\beta)} \times \\
&\left(-K_1(\beta)\left[\frac{b_3}{b_1 b_3 - b_2^2}\left[\int I_1(\beta r) h\prime_\omega^{(n-1)} r dr - \frac{a_2 b_3 - a_3 b_2}{a_1 b_3} \int I_1(\beta r) p\prime_\omega^{(n-1)} r dr\right] - \right.\right.\\
&\left.\gamma_1 \int I_1(\beta r)\left(\int q\prime_\omega^{(n-1)}) r dr\right) dr\right]\Bigg|_{r=1} + \\
&I_1(\beta)\left[\frac{b_3}{b_1 b_3 - b_2^2}\left[\int K_1(\beta r) h\prime_\omega^{(n-1)} r dr - \frac{a_2 b_3 - a_3 b_2}{a_1 b_3} \int K_1(\beta r) p\prime_\omega^{(n-1)} r dr\right] - \right.\\
&\left.\left.(u\prime^{(n)} = 0, \;\; M' = M\prime_r^{(n-1)}, npur = 1) \;\; -\gamma_1 \int K_1(\beta r)\left(\int q\prime_\omega^{(n-1)}) r dr\right) dr\right]\Bigg|_{r=1}\right) \\
&C_4^{(n)} = -\frac{C_1^{(n)}}{4 b_3} - \frac{C_5^{(n)}}{4 b_3} - \\
&\frac{b_2}{b_3} \int \psi\prime^{(n)} dr\Big|_{r=1} - \frac{a_3}{b_3 a_1} \int L_2^{-1}(p\prime_\omega^{(n-1)}) dr\Big|_{r=1} + \frac{1}{b_3} \int L_3^{-1}(q\prime_\omega^{(n-1)}) dr\Big|_{r=1}.
\end{aligned}
\tag{9.19}
$$

Integration constants $C_5^{(n)}$, $C_7^{(n)}$ at hinged support of the plate contour ($u\prime^{(n)} = 0, M\prime = M\prime_r^{(n-1)}, \; at \; r = 1$) are equal to:

$$
\begin{aligned}
&C_5^{(n)} = 2\frac{a_7 a_1 + a_3^2}{a_1(a_6 + a_7)} L_3^{-1}(q\prime_\omega^{(n-1)})\Big|_{r=1} - \frac{2 a_3}{a_1} L_2^{-1}(p\prime_\omega^{(n-1)})\Big|_{r=1} + \\
&\frac{b_3}{a_6 + a_7} 2\left(L_3^{-1}(q\prime_\omega^{(n-1)})\right)_{,r}\Big|_{r=1} - \frac{2 b_3}{a_6 + a_7}\left(3 \sum_{k=1}^{3} \alpha_{0k} \int_{h_k} K_k T_k z dz + M\prime_r^{(n-1)}\right) + \\
&\left(b_3 - a_7 - \frac{a_3^2}{a_1}\right) \frac{b_3}{a_6 + a_7}\left(\frac{C_1^{(n)}}{2 b_3} + 2 C_6^{(n)}\right) \\
&C_7^{(n)} = -2\frac{a_6}{b_3 a_1} L_2^{-1}(p\prime_\omega^{(n-1)})\Big|_{r=1} + 2\frac{a_3}{b_3 a_1} L_3^{-1}(q\prime_\omega^{(n-1)})\Big|_{r=1} - \\
&\frac{a_3}{2 b_3 a_1} C_1^{(n)} - 2\frac{a_3}{a_1} C_6^{(n)} - \frac{a_3}{a_1 b_3} C_5^{(n)}.
\end{aligned}
\tag{9.20}
$$

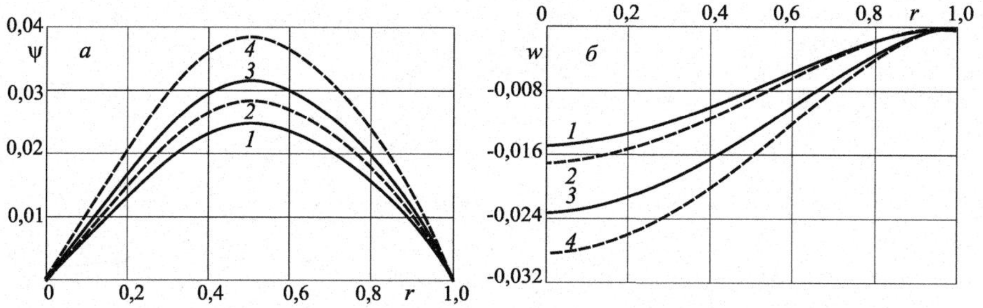

Fig. 9.5 Elastic and physically non-linear displacements at the radius of the load spot $b = 0{,}5$ ($T_1 = 293$ K, $T_2 = 323$ K)

Similar constants at the clamped contour ($u'^{(n)} = w'^{(n)} = 0$, at $r = 1$) are:

$$C_5^{(n)} = -2b_3C_6^{(n)} - \frac{C_1^{(n)}}{2} - \frac{2a_3}{a_1}L_2^{-1}(p_\omega^{(n-1)})\big|_{r=1} + 2L_3^{-1}(q_\omega^{(n-1)})\big|_{r=1},$$
$$C_7^{(n)} = -\frac{2}{a_1}L_2^{-1}(p_\omega^{(n-1)})\big|_{r=1}. \qquad (9.21)$$

Thus, the analytical solution of the problem on deformation of an elastoplastic circular three-layer plate under the action of a local circular load is given by recurrence formulae (9.12) with integration constants (9.19)–(9.21).

Numerical results are obtained for a hinge-supported plate with layers D16-T–fluoroplastic-4–D16-T at a circular load intensity $q_0 = 8$ MPa, and the following relative layer thicknesses: $h_1 = 0{,}02$; $h_2 = 0{,}06$; $h_3 = 0{,}4$.

Figure 9.5a, b shows the variation of relative shear in the fill and deflection of the plate along its radius. The curves are plotted at the load spot radius value $b = 0.5$ at different temperatures: *1*, *2*—elastic and physically nonlinear displacements at $T_1 = 293$ K; *3*, *4*—elastic and physically nonlinear displacements at $T_2 = 323$ K. Under isothermal loading, taking into account the physical nonlinearity of layer materials increases the calculated elastic displacements by 16%, at elevated temperature—by 22%.

Figure 9.6 illustrates the dependence of the maximum deflection of the considered plate on the radius of the load spot b at different temperatures (the numbering of the curves is the same). As the radius increases, the deflection grows non-linearly, reaching a maximum at the load acting on the entire outer surface of the top layer of the plate.

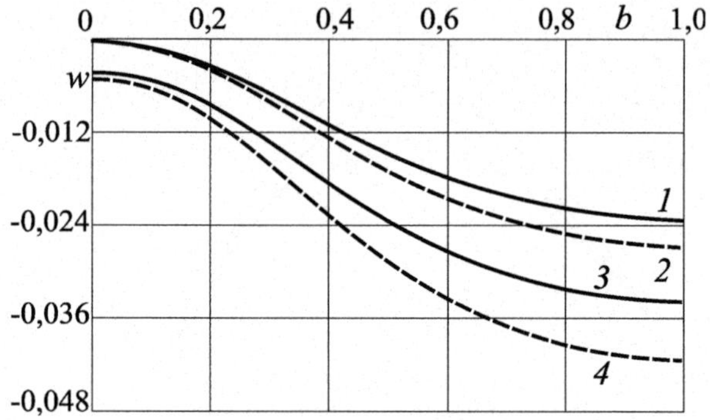

Fig. 9.6 Elastic and physically nonlinear deflections in the centre of the plate as a function of the radius of the load spot ($T_1 =$ 293 K, $T_2 = 323$ K)

Local uniformly distributed ring load. Consider the case of a local load uniformly distributed over an ring of relative radius $a \le r \le b$ (Fig. 8.14). Then it can be written as the difference of two loads (9.16):

$$q(r) = q_0(H_0(b - r) - H_0(a - r)). \tag{9.22}$$

The function $f\prime^{(n)}(r)$, included in (9.13), at load (9.22) takes the form:

$$f\prime^{(n)}(r) = \frac{b_3}{b_1 b_3 - b_2^2}\left[h\prime_\omega^{(n-1)} - \frac{a_2 b_3 - a_3 b_2}{a_1 b_3} p\prime_\omega(n-1)\right] + \\ \frac{q_0 \gamma_1}{2r}\left((r^2 - b^2)(H_0(b - r) - (r^2 - a^2)(H_0(a - r) - \int q\prime_\omega r dr - C_1^{(n)}\right). \tag{9.23}$$

The integral operator from the load (9.22) will be:

$$\int L_3^{-1}(q)dr = q_0\left(\frac{r^4 - 5b^4}{64} + \frac{b^2 r^2}{16} - \left(\frac{b^4}{16} + \frac{b^2 r^2}{8}\right) ln\left(\frac{r}{b}\right)\right) H_0(b - r) \\ -q_0\left(\frac{r^4 - 5a^4}{64} + \frac{a^2 r^2}{16} - \left(\frac{a^4}{16} + \frac{a^2 r^2}{8}\right) \ln\left(\frac{r}{a}\right)\right) H_0(a - r). \tag{9.24}$$

The required solution is obtained as the difference of two solutions (9.18) corresponding to the load components (9.22). As a result, the partial solution for the shear in the filler and the integral of the shear at the nth step will be:

$$\psi{\prime}_r^{(n)}(r) = \frac{\gamma_1 q_0}{2\beta^2}\left\{H_0(b-r)\left[\frac{b^2}{r} - r + 2b(K_1(\beta b)I_1(\beta r) - I_1(\beta b)K_1(\beta r))\right]\right.$$
$$\left. -H_0(a-r)\left[\frac{a^2}{r} - r + 2a(K_1(\beta a)I_1(\beta r) - I_1(\beta a)K_1(\beta r))\right]\right\} + \frac{C_1^{(n)}\gamma_1}{\beta^2 r}$$
$$-K_1(\beta r)\left[\frac{b_3}{b_1 b_3 - b_2^2}\left[\int I_1(\beta r)h{\prime}_\omega^{(n-1)}rdr - \frac{a_2 b_3 - a_3 b_2}{a_1 b_3}\int I_1(\beta r)p{\prime}_\omega^{(n-1)}rdr\right]\right.$$
$$\left. -\gamma_1\int I_1(\beta r)\left(\int q{\prime}_\omega^{(n-1)}rdr\right)dr\right]$$
$$+I_1(\beta r)\left[\frac{b_3}{b_1 b_3 - b_2^2}\left[\int K_1(\beta r)h{\prime}_\omega^{(n-1)}rdr - \frac{a_2 b_3 - a_3 b_2}{a_1 b_3}\int K_1(\beta r)p{\prime}_\omega^{(n-1)}rdr\right] - \right.$$
$$\left. \gamma_1\int K_1(\beta r)\left(\int q{\prime}_\omega^{(n-1)}rdr\right)dr\right],$$
$$\int \psi{\prime}^{(n)}dr = \frac{C_2^{(n)}I_0(\beta r)}{\beta} - \frac{C_3^{(n)}K_0(\beta r)}{\beta} + \frac{C_1^{(n)}\gamma_1}{\beta^2}\ln(r) +$$
$$\frac{\gamma_1 q_0}{2\beta^2}\left\{H_0(b-r)\left[\frac{b^2-r^2}{2} + b^2\ln\left(\frac{r}{b}\right) + \frac{2b}{\beta}(K_1(\beta b)I_0(\beta r) + I_1(\beta b)K_0(\beta r)) - \frac{2}{\beta^2}\right] - \right.$$
$$\left. H_0(a-r)\left[\frac{a^2-r^2}{2} + a^2\ln\left(\frac{r}{a}\right) + \frac{2a}{\beta}(K_1(\beta a)I_0(\beta r) + I_1(\beta a)K_0(\beta r)) - \frac{2}{\beta^2}\right]\right\} -$$
$$\int\left(K_1(\beta r)\left[\frac{b_3}{b_1 b_3 - b_2^2}\left[\int I_1(\beta r)h{\prime}_\omega^{(n-1)}rdr - \frac{a_2 b_3 - a_3 b_2}{a_1 b_3}\int I_1(\beta r)p{\prime}_\omega^{(n-1)}rdr\right] - \right.\right.$$
$$\left. \gamma_1\int I_1(\beta r)\left(\int q{\prime}_\omega^{(n-1)}rdr\right)dr\right]$$
$$+I_1(\beta r)\left[\frac{b_3}{b_1 b_3 - b_2^2}\left[\int K_1(\beta r)h{\prime}_\omega^{(n-1)}rdr - \frac{a_2 b_3 - a_3 b_2}{a_1 b_3}\int K_1(\beta r)p{\prime}_\omega^{(n-1)}rdr\right] - \right.$$
$$\left.\left. \gamma_1\int K_1(\beta r)\left(\int q{\prime}_\omega^{(n-1)}rdr\right)dr\right]\right)dr. \tag{9.25}$$

The deflection $w{\prime}^{(n)}(r)$, radial displacement $u{\prime}^{(n)}(r)$ and shear in the filler $\psi{\prime}^{(n)}(r)$ of the plate under consideration formally retain the form (9.12), but the integrals included in them will be defined by formulas (9.23)–(9.25).

Only those constants of integration will change, which are determined from the condition of boundedness of the solution in the centre of the plate (9.19):

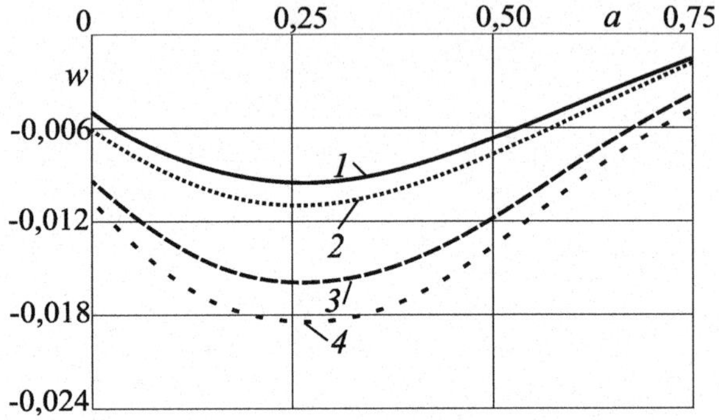

Fig. 9.7 Elastic and physically nonlinear deflections at the centre of the plate depending on the location of the load ring (T_1 = 293 K, T_2 = 323 K)

$$C_1^{(n)} = -\frac{q_0(b^2-a^2)}{2}, \quad C_6^{(n)} = \frac{q_0(b^4-a^4)}{16b_3}, \quad C_8^{(n)} = 0,$$
$$C_3^{(n)} = \frac{\gamma_1 q_0}{\beta^2}(bI_1(\beta b) - aI_1(\beta a)) +$$
$$+\left[\frac{b_3}{b_1b_3 - b_2^2}\left(\int I_1(\beta r)h{\prime}_\omega^{(n-1)}rdr - \frac{a_2b_3 - a_3b_2}{a_1b_3}\int I_1(\beta r)p{\prime}_\omega^{(n-1)}rdr\right) - \right. \tag{9.26}$$
$$\left.\gamma_1\int I_1(\beta r)\left(\int q{\prime}_\omega^{(n-1)}rdr\right)dr\right]\Bigg|_{r=0}.$$

Thus, the analytical solution of the problem of small elastic–plastic deformations for a circular three-layer plate under the action of an annular load is given by recurrence formulae (9.12) with integration constants (9.19)–(9.21), (9.26).

Figure 9.7 shows the variation of deflection at the centre of the plate as a function of the position of a load ring of constant width b–a = 0.25, as it advances towards the contour: *1, 2*—elastic and physically nonlinear at T_1 = 293 K; *3, 4*—elastic and physically nonlinear at T_2 = 323 K.

At $a = 0$ the load is distributed over a circle of radius b, at a = 0,75 the load ring is adjacent to the plate contour. The maximum deflection is observed at a load with inner radius a = 0,25.

Linear transverse load. Let a linear transverse load $Q(r)$ of constant intensity Q_0, applied along a circle of radius $r = a$, act on the considered three-layer plate (Fig. 8.16). The iteration solution of the system (9.7) in this case is constructed using the results (8.40) obtained for an elastic plate:

$$\psi\prime^{(n)} = C_2^{(n)} I_1(\beta r) + C_3^{(n)} K_1(\beta r) + \psi\prime_r^{(n)},$$
$$u\prime^{(n)} = \frac{a_3}{a_1} w\prime^{(n)},_r - \frac{a_2}{a_1}\psi\prime^{(n)} + \frac{1}{a_1} L_2^{-1}(p\prime_\omega^{(n-1)}) + \frac{C_7^{(n)} r}{2} + \frac{C_8^{(n)}}{r},$$
$$w\prime^{(n)} = \frac{b_2}{b_3}\int \psi\prime^{(n)} dr + \frac{a_3}{b_3 a_1}\int L_2^{-1}(p\prime_\omega^{(n-1)})dr + \frac{1}{b_3}\int L_3^{-1}(Q\prime)dr - \frac{1}{b_3}\int L_3^{-1}(q\prime_\omega^{(n-1)})dr -$$
$$\frac{C_1^{(n)}}{4b_3} r^2(\ln r - 1) + \frac{C_5^{(n)} r^2}{4b_3} + C_6^{(n)} \ln r + C_4^{(n)}, \tag{9.27}$$

where

$$\psi\prime_r^{(n)} = \frac{\gamma_1 Q_0 a}{\beta} H_0(a-r)\left(\frac{1}{\beta r} - K_1(\beta r) I_0(\beta a) - I_1(\beta r) K_0(\beta a)\right) + \frac{C_1^{(n)}\gamma_1}{\beta^2 r} -$$
$$K_1(\beta r)\left[\frac{b_3}{b_1 b_3 - b_2^2}\left[\int I_1(\beta r) h\prime_\omega^{(n-1)} r dr - \frac{a_2 b_3 - a_3 b_2}{a_1 b_3}\int I_1(\beta r) p\prime_\omega^{(n-1)} r dr\right] -\right.$$
$$\left.\gamma_1 \int I_1(\beta r)\left(\int q\prime_\omega^{(n-1)}) r dr\right) dr\right]$$
$$+I_1(\beta r)\left[\frac{b_3}{b_1 b_3 - b_2^2}\left[\int K_1(\beta r) h\prime_\omega^{(n-1)} r dr - \frac{a_2 b_3 - a_3 b_2}{a_1 b_3}\int K_1(\beta r) p\prime_\omega^{(n-1)} r dr\right] -\right.$$
$$\left.\gamma_1 \int K_1(\beta r)\left(\int q\prime_\omega^{(n-1)}) r dr\right) dr\right].$$

$$\int L_3^{-1}(Q\prime)dr = \frac{Q_0 a}{4}\left(r^2 - a^2 - a^2 \ln\left(\frac{r}{a}\right) - r^2 \ln\left(\frac{r}{a}\right)\right) H_0(a-r),$$
$$\int \psi\prime^{(n)} dr = \frac{C_2^{(n)} I_0(\beta r)}{\beta} - \frac{C_3^{(n)} K_0(\beta r)}{\beta} + \frac{C_1^{(n)}\gamma_1}{\beta^2}\ln(r) + \frac{\gamma_1 Q_0 a}{\beta^2} H_0(a-r)\times$$
$$\left(\ln\left(\frac{r}{a}\right) + I_0(\beta a) K_0(\beta r) - K_0(\beta a) I_0(\beta r)\right) -$$
$$\int\left(K_1(\beta r)\left[\frac{b_3}{b_1 b_3 - b_2^2}\left[\int I_1(\beta r) h\prime_\omega^{(n-1)} r dr - \frac{a_2 b_3 - a_3 b_2}{a_1 b_3}\int I_1(\beta r) p\prime_\omega^{(n-1)} r dr\right] -\right.\right.$$
$$\left.\gamma_1 \int I_1(\beta r)\left(\int q\prime_\omega^{(n-1)}) r dr\right) dr\right]$$
$$+I_1(\beta r)\left[\frac{b_3}{b_1 b_3 - b_2^2}\left[\int K_1(\beta r) h\prime_\omega^{(n-1)} r dr - \frac{a_2 b_3 - a_3 b_2}{a_1 b_3}\int K_1(\beta r) p\prime_\omega^{(n-1)} r dr\right] -\right.$$
$$\left.\left.\gamma_1 \int K_1(\beta r)\left(\int q\prime_\omega^{(n-1)}) r dr\right) dr\right]\right) dr.$$

Numerical results are obtained under the previously introduced assumptions at the linear load intensity $Q_0 = 4$ MN/m.

Figure 9.8 *a*, *b* shows the variation of shear in the filler and deflection of the plate under consideration along its radius. The curves are plotted in accordance with formulas (9.27) at the radius of the force circle $a = 0{,}75$: *1*, *2*—elastic and physically nonlinear at $T_1 = 293$ K; *3, 4*—elastic and physically nonlinear at $T_2 = 323$ K. Under isothermal loading, taking into account the physical nonlinearity of layer materials increases the calculated elastic deflections by 17%. A 30^0 increase in temperature causes them to rise by 24%. As you move along the radial coordinate, the relative shear extremum shift to the plate contour.

The graphs in Fig. 9.9 illustrate the variation of deflection at the centre of the plate as a function of the radius of the force circle: *1, 2*—elastic and physically nonlinear at $T_1 = 293$ K; *3, 4*—elastic and physically nonlinear at $T_2 = 323$ K. It should be noted, that in this case, for a degenerate linear load at the centre of the plate and its action on the contour, the temperature increase leads to a non-zero deflection (curves 3, 4).

The formulas (9.27) do not describe the deformation of the plate under the action of a concentrated force at its centre, since at $b = 0$ the circle degenerates and the resultant force of linear load goes to zero. To avoid this, we assume that this resultant force $Q = 2\pi a Q_0$ remains constant as the radius of the circle along which it is applied changes. This is possible if the intensity Q_0 is variable, compensating for changes in the radius a. To obtain the solution, we make the following substitution in (9.27): $Q_0 = Q/2\pi a$. Then we have the following displacements:

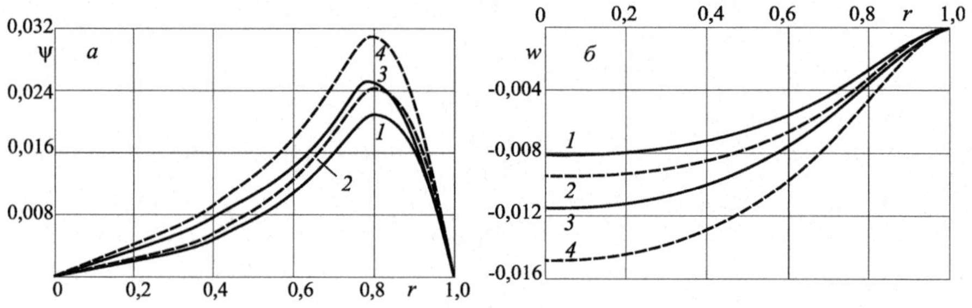

Fig. 9.8 Elastic and physically non-linear displacements at the radius of the force circle $a = 0,75$ ($T_1 = 293$ K, $T_2 = 323$ K)

Fig. 9.9 Elastic and physically nonlinear deflections at the centre of the plate as a function of the radius of the force circle ($T_1 = 293$ K, $T_2 = 323$ K)

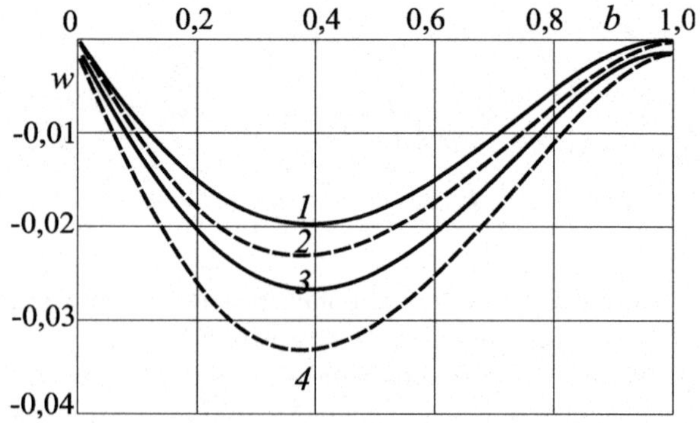

$$\begin{aligned}
&\psi'^{(n)} = C_2^{(n)} I_1(\beta r) + C_3^{(n)} K_1(\beta r) + \psi'^{(n)}_r, \\
&u'^{(n)} = \frac{a_3}{a_1} w'^{(n)},_r - \frac{a_2}{a_1}\psi'^{(n)} + \frac{1}{a_1} L_2^{-1}(p'^{(n-1)}_\omega) + \frac{C_7^{(n)} r}{2} + \frac{C_8^{(n)}}{r}, \\
&w'^{(n)} = \frac{b_2}{b_3}\int \psi'^{(n)} dr + \frac{a_3}{b_3 a_1}\int L_2^{-1}(p'^{(n-1)}_\omega) dr + \frac{1}{b_3}\int L_3^{-1}(Q')dr - \frac{1}{b_3}\int L_3^{-1}(q'^{(n-1)}_\omega)dr - \\
&-\frac{C_1^{(n)}}{4b_3} r^2(\ln r - 1) + \frac{C_5^{(n)} r^2}{4b_3} + C_6^{(n)} \ln r + C_4^{(n)},
\end{aligned} \tag{9.28}$$

where

$$\begin{aligned}
&\psi'^{(n)}_r = \frac{\gamma_1 Q}{2\pi\beta} H_0(a-r)\Big(\frac{1}{\beta r} - K_1(\beta r) I_0(\beta a) - I_1(\beta r) K_0(\beta a)\Big) + \frac{C_1^{(n)}\gamma_1}{\beta^2 r} - \\
&-K_1(\beta r)\Big[\frac{b_3}{b_1 b_3 - b_2^2}\Big[\int I_1(\beta r) h'^{(n-1)}_\omega r dr - \frac{a_2 b_3 - a_3 b_2}{a_1 b_3}\int I_1(\beta r) p'^{(n-1)}_\omega r dr\Big] - \\
&-\gamma_1 \int I_1(\beta r)\Big(\int q'^{(n-1)}_\omega r dr\Big) dr\Big] \\
&+I_1(\beta r)\Big[\frac{b_3}{b_1 b_3 - b_2^2}\Big[\int K_1(\beta r) h'^{(n-1)}_\omega r dr - \frac{a_2 b_3 - a_3 b_2}{a_1 b_3}\int K_1(\beta r) p'^{(n-1)}_\omega r dr\Big] - \\
&-\gamma_1 \int K_1(\beta r)\Big(\int q'^{(n-1)}_\omega r dr\Big) dr\Big], \\
&\int L_3^{-1}(Q') dr = \frac{Q}{8\pi}\Big(r^2 - a^2 - a^2 \ln\Big(\frac{r}{a}\Big) - r^2 \ln\Big(\frac{r}{a}\Big)\Big) H_0(a-r), \\
&\int \psi'^{(n)} dr = \frac{C_2^{(n)} I_0(\beta r)}{\beta} - \frac{C_3^{(n)} K_0(\beta r)}{\beta} + \frac{C_1^{(n)}\gamma_1}{\beta^2}\ln(r) + \frac{\gamma_1 Q}{2\pi\beta^2} H_0(a-r)\times \\
&\times\Big(\ln\Big(\frac{r}{a}\Big) + I_0(\beta a) K_0(\beta r) - K_0(\beta a) I_0(\beta r)\Big) - \\
&-\int\Big(K_1(\beta r)\Big[\frac{b_3}{b_1 b_3 - b_2^2}\Big[\int I_1(\beta r) h'^{(n-1)}_\omega r dr - \frac{a_2 b_3 - a_3 b_2}{a_1 b_3}\int I_1(\beta r) p'^{(n-1)}_\omega r dr\Big] - \\
&-\gamma_1 \int I_1(\beta r)\Big(\int q'^{(n-1)}_\omega r dr\Big) dr\Big] \\
&+I_1(\beta r)\Big[\frac{b_3}{b_1 b_3 - b_2^2}\Big[\int K_1(\beta r) h'^{(n-1)}_\omega r dr - \frac{a_2 b_3 - a_3 b_2}{a_1 b_3}\int K_1(\beta r) p'^{(n-1)}_\omega r dr\Big] - \\
&-\gamma_1 \int K_1(\beta r)\Big(\int q'^{(n-1)}_\omega r dr\Big) dr\Big]\Big) dr.
\end{aligned}$$

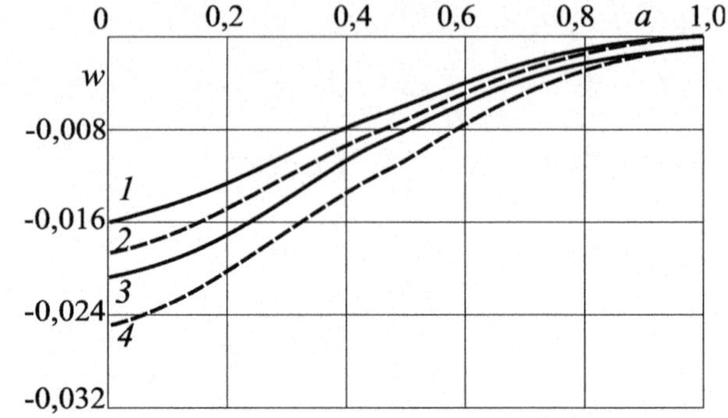

Fig. 9.10 Elastic and physically nonlinear deflections at the centre of the plate as a function of the radius of the force circle with a constant resultant ($T_1 = 293$ K, $T_2 = 323$ K)

The variation of the maximum deflection of the plate (9.28) as a function of the radius of the circle along which a distributed transverse load with a constant resultant force is applied is shown in Fig. 9.10: *1, 2*—elastic and physically nonlinear deflections at $T_1 = 293$ K; *3, 4*—elastic and physically nonlinear deflections at $T_2 = 323$ K. The maximum here is observed at $b = 0$, which corresponds to the concentrated force applied at the centre of the plate. As the load advances towards the contour, the deflection decreases to zero. On heating, the deflection increases proportionally and is different from zero on the contour.

Linear moments. Let the plate under study be subjected to linear moments of intensity $M_0 = \text{const}$, distributed on a circle of radius $r = a$ (Fig. 8.20). Let us solve the problem using the difference of solutions (9.27) for two identical linear transverse forces acting on closely spaced circles of radiuses $r = a–\xi$ and $r = a + \xi$ and directed in opposite directions. In this difference, we substitute $Q_0 = M_0/(2\xi)$ and set the parameter ξ to zero. Then

$$
\begin{aligned}
\psi'^{(n)} &= C_2^{(n)} I_1(\beta r) + C_3^{(n)} K_1(\beta r) + \psi'^{(n)}_r,\\
w'^{(n)} &= \frac{b_2}{b_3}\int \psi'^{(n)} dr + \frac{a_3}{b_3 a_1}\int L_2^{-1}(p'^{(n-1)}_{\omega})dr + \frac{1}{b_3}\int L_3^{-1}(M')dr - \frac{1}{b_3}\int L_3^{-1}(q'^{(n-1)}_{\omega})dr\\
&\quad - \frac{C_1^{(n)}}{4b_3} r^2(\ln r - 1) + \frac{C_5^{(n)} r^2}{4b_3} + C_6^{(n)} \ln r + C_4^{(n)},\\
u'^{(n)} &= \frac{a_3}{a_1} w'^{(n)}{}_{,r} - \frac{a_2}{a_1}\psi'^{(n)} + \frac{1}{a_1} L_2^{-1}(p'^{(n-1)}_{\omega}) + \frac{C_7^{(n)} r}{2} + \frac{C_8^{(n)}}{r},
\end{aligned}
\tag{9.29}
$$

where

$$\psi'^{(n)}_r = \frac{\gamma_1 M_0}{\beta} H_0(a-r)\left(\frac{1}{\beta r} - K_1(\beta r)(I_0(\beta a) + I_1(\beta a)\beta a) -\right.$$

$$-I_1(\beta r)(K_0(\beta a) - K_1(\beta a)\beta a)) + \frac{C_1^{(n)}\gamma_1}{\beta^2 r} -$$

$$-K_1(\beta r)\left[\frac{b_3}{b_1 b_3 - b_2^2}\left[\int I_1(\beta r) h'^{(n-1)}_\omega r dr - \frac{a_2 b_3 - a_3 b_2}{a_1 b_3}\int I_1(\beta r) p'^{(n-1)}_\omega r dr\right] -\right.$$

$$\left. -\gamma_1 \int I_1(\beta r)\left(\int q'^{(n-1)}_\omega) r dr\right) dr\right]$$

$$+I_1(\beta r)\left[\frac{b_3}{b_1 b_3 - b_2^2}\left[\int K_1(\beta r) h'^{(n-1)}_\omega r dr - \frac{a_2 b_3 - a_3 b_2}{a_1 b_3}\int K_1(\beta r) p'^{(n-1)}_\omega r dr\right] -\right.$$

$$\left. -\gamma_1 \int K_1(\beta r)\left(\int q'^{(n-1)}_\omega) r dr\right) dr\right],$$

$$\int L_3^{-1}(M_0 \prime) dr = \frac{M_0}{4}\left(2r^2 - 2a^2 - 3a^2 \ln\left(\frac{r}{a}\right) - r^2 \ln\left(\frac{r}{a}\right) H_0(a-r)\right).$$

The functions (9.29) formally describe the case of a concentrated moment applied at the centre of the plate, since at $a = 0$ the solution does not degenerate.

Numerical results are obtained under the previously introduced assumptions and at the intensity of linear moments $M_0 = 0{,}3$ MN.

Figure. 9.11, *a* and *b* show the variation of shear in the filler and deflection of a circular three-layer plate along its radius. The curves are plotted at the radius of the moment circle $a = 0{,}75$: *1, 2*—elastic and physically nonlinear deflections at $T_1 = 293$ K; *3, 4*—elastic and physically nonlinear deflections at $T_2 = 323$ K. Under isothermal loading, accounting for the physical nonlinearity of the layer materials changes the elastic deflection by 14%, and increasing the temperature by 30° leads to its increase by 21%.

In the case of shear, the maxima in the form of peaks are observed along the radius of the moment circle. Their magnitude decreases when the nonlinearity of the layer materials is taken into account.

The variation of the deflection *w* at the center of the plate as a function of the radius of the moment circle *a* is illustrated by Fig. 9.12: *1, 2*—elastic and physically nonlinear deflections at $T_1 = 293$ K; *3, 4*—elastic and physically nonlinear deflections at $T_2 = 323$ K. The maximum in this case is observed at the load applied at the center of the plate ($a = 0$). As the moment circle advances to the contour, the

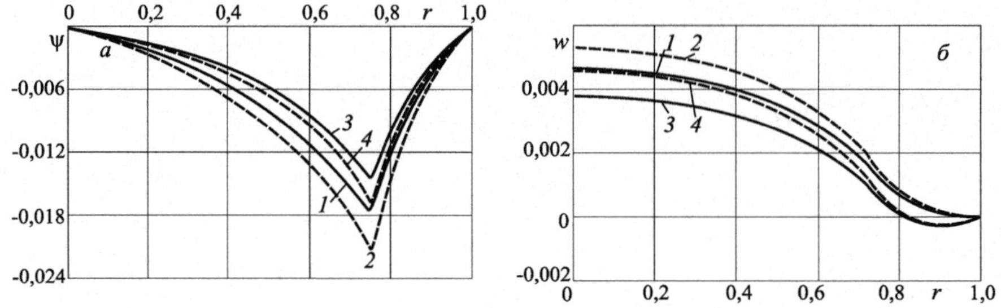

Fig. 9.11 Elastic and physically nonlinear displacements at the radius of the force circle $a = 0{,}75$ ($T_1 = 293$ K, $T_2 = 323$ K)

deflection decreases in magnitude, reaches zero in the vicinity of the radial coordinate $a = 0.38\ldots0.43$, changes sign and becomes non-zero on the contour when the temperature increases.

The deflections and shears in the last two cases (see Figs. 9.11, 9.12) increase proportionally with heating.

Integration constants at local loads were obtained numerically from the boundary conditions (9.19) and the condition of boundedness of the solution in the center of the plate.

Thus, taking into account the physical nonlinearity of the materials of the layers and the effect of the temperature field leads to a significant refinement of the stress–strain state of a three-layer circular plate under loading from the natural state.

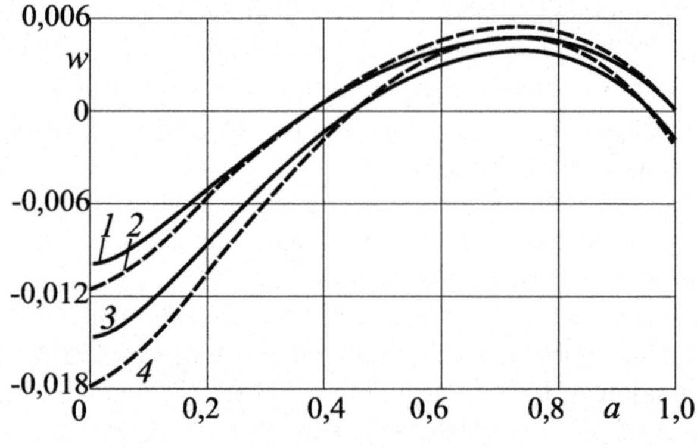

Fig. 9.12 Elastic and physically nonlinear deflections in the center of the plate as a function of the radius of the moment circle ($T_1 = 293$ K, $T_2 = 323$ K)

9.3 Repeated Bending of the Plate in the Temperature Field

Suppose that, starting from the moment $t = t_1$, instantaneous unloading and reloading with forces of the opposite sign p", q", varying according to the same law as at loading from the natural state, are carried out. These forces will create displacement u'', ψ'', w'', strain $\varepsilon\prime\prime_{\alpha}^{(k)}$ and stress $\sigma\prime\prime_{\alpha}^{(k)}$, $w(h_1^*) = 0$ ($\alpha = r$, φ) fields in the plate. In this case, we will assume that during unloading and subsequent alternating loading the temperature at all points of the plate remains unchanged, coinciding with the temperature field at the time of the beginning of unloading (i.e. $T_1(z) = T(z, t_1)$) and elasticity modules took constant values $G_k(z) \equiv G_k(T_1(z))$, $K_k(z) \equiv K_k(T_1(z))$, , and the corresponding loading trajectories still belong to the class of simple ones (according to Ilyushin).

We introduce differences for all values of stress–strain state and load, where the values with one stroke are stresses, strains and displacements in the plate before unloading, two strokes mark similar parameters during the second half cycle:

$$\begin{aligned}
&\sigma_{\alpha}^{(k)*} = \sigma\prime_{\alpha}^{(k)} - \prime\prime_{\alpha}^{(k)}, \sigma_{rz}^{(3)*} = \sigma\prime_{rz}^{(3)} - \sigma\prime\prime_{rz}^{(3)},\\
&s_{\alpha}^{(k)*} = s\prime_{\alpha}^{(k)} - s\prime\prime_{\alpha}^{(k)}, \sigma^{(k)*} = \sigma\prime^{(k)} - \sigma\prime\prime^{(k)},\\
&\varepsilon_{\alpha}^{(k)*} = \varepsilon\prime_{\alpha}^{(k)} - \varepsilon\prime\prime_{\alpha}^{(k)}, \varepsilon_{rz}^{(3)*} = \varepsilon\prime_{rz}^{(3)} - \varepsilon\prime\prime_{rz}^{(3)}, \ni_{\alpha}^{(k)*} = \ni\prime_{\alpha}^{(k)} - \ni\prime\prime_{\alpha}^{(k)},\\
&\varepsilon^{(k)*} = \varepsilon\prime^{(k)} - \varepsilon\prime\prime^{(k)}, q^* = q\prime - q\prime\prime, p^* = p\prime - p\prime\prime,\\
&u^* = u\prime - u\prime\prime, \psi^* = \psi\prime - \psi\prime\prime, w^* = w\prime - w\prime\prime, \alpha = r, \varphi\cdot
\end{aligned} \tag{9.30}$$

Due to the linear relationship between strains and displacements (8.2), similar relations will be valid for the corresponding values with asterisks in the layers of the plate:

$$\begin{aligned}
&\varepsilon_r^{(1)*} = u,_r^* + c\psi,_r^* - zw,_{rr}^*, \quad \varepsilon_{\varphi}^{(1)*} = \frac{1}{r}(u^* + c\psi^* - zw,_r^*), \quad \varepsilon_{rz}^{(1)*} = 0, \quad (c \le z \le c + h_1),\\
&\varepsilon_r^{(2)*} = u,_r^* - c\psi,_r^* - zw,_{rr}^*, \quad \varepsilon_{\varphi}^{(2)*} = \frac{1}{r}(u^* - c\psi^* - zw,_r^*), \quad \varepsilon_{rz}^{(2)*} = 0, \quad (-c \le z \le -c - h_2),\\
&\varepsilon_r^{(3)*} = u,_r^* + z\psi,_r^* - zw,_{rr}^*, \quad \varepsilon_{\varphi}^{(3)*} = \frac{1}{r}(u^* + z\psi^* - zw,_r^*), \quad \varepsilon_{rz}^{(3)*} = \frac{1}{2}\psi, \quad (-c \le z \le c),.
\end{aligned} \tag{9.31}$$

For stresses and strains (9.30), marked with an asterisk, we assume physical equations of state of the type (9.1):

$$\begin{aligned}
&w = \frac{b_2}{b_3}\left(\frac{C_2}{\beta} I_0(\beta r) - \frac{b_2 q_0}{8cb_3G_3} r^2\right) + \frac{q_0}{64b_3} r^4 + \frac{C_5 r^2}{4b_3} + C_4^{(n)}, \sigma^{(k)*} = 3K_k \varepsilon^{(k)*} (k = 1, 2, 3),\\
&s_{rz}^{(3)} = 2G_{\alpha}^{(3)} \ni_{rz}^{(3)*} (1 - \omega^{(3)*}), (\alpha = r, \varphi)\cdot
\end{aligned} \tag{9.32}$$

Unlike relations (9.1), the temperature is not explicitly contained here. According to Moskvitin's theory [1], we assume that the corresponding universal nonlinearity functions $C_5 = -\frac{2b_3}{a_6+a_7}\left\{\left(3+\frac{a_3^2}{b_3a_1}\right)\frac{q_0}{16}r_0^2+\right.$ in the bearing layers are expressed through the nonlinearity functions under loading from the natural state:

$$\omega^{(k)*} = \omega\prime^{(k)}(\varepsilon_u^{(k)*}, \varepsilon_{y1}^{(k)*}, T_1, A_1^*, \alpha_1^*). \tag{9.33}$$

The physical nonlinearity of the filler under repeated alternating loading is still described by relations (9.1). Its nonlinear characteristics do not change under cyclic deformation.

We express the stress components with asterisks in layers through strains with the help of physical relations (9.32):

$$\begin{gathered}
\sigma_\alpha^{(k)*} = \sigma_{\alpha e}^{(k)*} - \sigma_{\alpha\omega}^{(k)*}, \sigma_{rz}^{(3)*} = \sigma_{rze}^{(3)*} - \sigma_{rz\omega}^{(3)*}, \\
\sigma_{\alpha e}^{(k)*} = 2G_k\, э_\alpha^{(k)*} + 3K_k\varepsilon^{(k)*}, \sigma_{\omega\omega}^{(k)*} = 2G_k\omega_k(\varepsilon_u^{(k)*}, T_k)\, э_\alpha^{(k)*}, \\
\sigma_{rze}^{(3)*} = 2G_3\, э_{rz}^{(3)*}, \sigma_{rz\omega}^{(3)*} = 2G_3\omega_3(\varepsilon_u^{(3)*}, T_k)\, э_{rz}^{(3)*}\ (k = 1, 2, 3; \alpha = r, \varphi)\cdot
\end{gathered} \tag{9.34}$$

Performing a similar operation with the values of the internal force type (8.3), we obtain:

$$\begin{gathered}
T_\alpha^{(k)*} = T_{\alpha e}^{(k)*} - T_{\alpha\omega}^{(k)*}, \quad M_\alpha^{(k)*} = M_{\alpha e}^{(k)*} - M_{\alpha\omega}^{(k)*}, \\
H_\alpha^* = H_{\alpha e}^{(k)*} - H_{\alpha\omega}^{(k)*}, \quad Q^* = Q_e^* - Q_\omega^*.
\end{gathered} \tag{9.35}$$

Linear (index "e") and nonlinear (index "ω") components in (9.35) are calculated through the values of (9.34) by the formulas:

$$\begin{gathered}
T_e^{(k)*} = \int_{h_k} \sigma_{xe}^{(k)*}dz, \quad T_\omega^{(k)*} = dz, \quad M_e^{(k)*} = \int_{h_k} \sigma_{xe}^{(k)*}zdz, \\
M_\omega^{(k)*} = \int_{h_k} \sigma_{x\omega}^{(k)*}zdz, \quad Q_e^* = \int_{-c}^{c} \sigma_{xze}^{(3)*}dz, \quad Q_\omega^* = \int_{-c}^{c} \sigma_{xz\omega}^{(3)*}dz.
\end{gathered} \tag{9.36}$$

Then the generalised internal forces with asterisks will be:

$$\begin{gathered}
T_\alpha^* = T_{\alpha e}^* - T_{\alpha\omega}^* = \sum_{k=1}^{3} T_{\alpha e}^{(k)*} - \sum_{k=1}^{3} T_{\alpha\omega}^{(k)*}, \\
M_\alpha^* = M_{\alpha e}^* - M_{\alpha\omega}^* = \sum_{k=1}^{3} M_{\alpha e}^{(k)*} - \sum_{k=1}^{3} M_{\alpha\omega}^{(k)*}, \\
H_{\alpha e}^* = M_{\alpha e}^{(3)*} + c\left(T_{\alpha e}^{(1)*} - T_{\alpha e}^{(2)*}\right), \quad H_{\alpha\omega}^* = M_{\alpha\omega}^{(3)*} + c\left(T_{\alpha\omega}^{(1)*} - T_{\alpha\omega}^{(2)*}\right).
\end{gathered} \tag{9.37}$$

Formally, repeating the procedure of obtaining the equations of equilibrium in the internal forces with asterisks, we obtain:

$$
\begin{aligned}
& T_r^*,_r + \frac{1}{r}(T_r^* - T_\varphi^*) = -p^* + p_\omega^*, \\
& H_r^*, r + \frac{1}{r}(H_r^* - H_\varphi^*) - Q_r^* = h_\omega^*, \\
& M_r^*,_{rr} + \frac{1}{r}(2M_r^*,_r - M_\varphi^*,_r) = -q^* + q_\omega^*
\end{aligned}
\tag{9.38}
$$

Here nonlinear additives in right-hand sides are calculated through nonlinear components of generalised forces in (9.34):

$$
\begin{aligned}
p_\omega^* &= T_{r\omega}^*,_r + \frac{1}{r}(T_{r\omega}^* - T_{\varphi\omega}^*), \\
q_\omega^* &= M_{r\omega}^*,_{rr} + \frac{1}{r}(2M_{r\omega}^*,_r - M_{\varphi\omega}^*,_r), \\
q_\omega^* &= M_{r\omega}^*,_{rr} + \frac{1}{r}(2M_{r\omega}^*,_r - M_{\varphi\omega}^*,_r).
\end{aligned}
\tag{9.39}
$$

Linear components of generalised internal forces with asterisks are still expressed through displacements with asterisks by formulas of the type (8.12)–(8.14), therefore the system of differential Eqs. (9.38) with respect to displacements with asterisks is reduced to the following form:

$$
\begin{aligned}
& L_2(a_1 u^* + a_2 \psi^* - a_3 w,_r^*) = -p^* + p_\omega^*, \\
& L_2(a_2 u^* + a_4 \psi^* - a_5 w,_r^*) - 2cG_3 \psi^* = h_\omega^*, \\
& L_3(a_3 u^* + a_5 \psi^* - a_6 w,_r^*) = -q^* + q_\omega^*.
\end{aligned}
\tag{9.40}
$$

To solve the obtained system of nonlinear differential Eqs. (9.40), we apply the elastic solution method. The corresponding iterative equations are written in the form:

$$
\begin{aligned}
& L_2(a_1 u^{(n)*} + a_2 \psi^{(n)*} - a_3 w,_r^{(n)*}) = -p^{(n)*} + p_\omega^{(n-1)*}, \\
& L_2(a_2 u^{(n)*} + a_4 \psi^{(n)*} - a_5 w,_r^{(n)*}) - 2cG_3 \psi^{(n)*} = h_\omega^{(n-1)*}, \\
& L_3(a_3 u^{(n)*} + a_5 \psi^{(n)*} - a_6 w,_r^{(n)*}) = -q^{(n)*} + q_\omega^{(n-1)*}.
\end{aligned}
\tag{9.41}
$$

Additional "external" loads $p_\omega^{(n-1)*}$, $h_\omega^{(n-1)*}$, $q_\omega^{(n-1)*}$ in (9.41) at the first step ($n = 1$) are assumed to be equal to zero, and further are calculated from the results of the previous approximation using equations of the type (9.39):

$$
\begin{aligned}
p_{\omega}^{(n-1)*} &= T_{r\omega}^{(n-1)*},_r + \frac{1}{r}(T_{r\omega}^{(n-1)*} - T_{\varphi\omega}^{(n-1)*}),\\
h_{\omega}^{(n-1)*} &= H_{r\omega}^{(n-1)*},_r + \frac{1}{r}(H_{r\omega}^{(n-1)*} - H_{\varphi\omega}^{(n-1)*}) - Q_{\omega}^{(n-1)*},\\
q_{\omega}^{(n-1)*} &= M_{r\omega}^{(n-1)*},_{rr} + \frac{1}{r}(2M_{r\omega}^{(n-1)*},_r - M_{\varphi\omega}^{(n-1)*},_r),
\end{aligned}
\tag{9.42}
$$

where

$$
\begin{aligned}
T_{\alpha\omega}^{(n-1)*} &\equiv \sum_{k=1}^{3}\int_{hk} \sigma_{\alpha\omega}^{(k,n-1)*} dz = \sum_{k=1}^{3}\int_{hk} 2G_k\omega_k(\varepsilon_{\alpha}^{(k,n-1)*}, T_k)\ \text{э}_{\alpha}^{(k,n-1)*}\, dz,\\
M_{\alpha\omega}^{(n-1)*} &\equiv \sum_{k=1}^{3}\int_{hk} \sigma_{\alpha\omega}^{(k,n-1)*} z dz = \sum_{k=1}^{3}\int_{hk} 2G_k\omega_k(\varepsilon_{\alpha}^{(k,n-1)*}, T_k)\ \text{э}_{\alpha}^{(k,n-1)*}\, z dz,\\
H_{\alpha\omega}^{(n-1)*} &= M_{\alpha\omega}^{(3,\,n-1)*} + c\big(T_{\alpha\omega}^{(1,\,n-1)*} - T_{\alpha\omega}^{(2,\,n-1)*}\big),\\
Q_{\omega}^{(n-1)*} &= \int_{-c}^{c} 2G_k\omega_k(\varepsilon_{\alpha}^{(3,n-1)*}, T_k)\ \text{э}_{rz}^{(3,n-1)*}\, dz \quad (\alpha = r, \phi).
\end{aligned}
$$

The boundary value problem for determining the functions $u^{(n)*}(r)$, $\psi^{(n)*}(r)$, $w^{(n)*}(r)$ is closed by joining to the system of Eqs. (9.41), taking into account (9.42), the force or kinematic boundary conditions of the type (9.10), (9.11), i.e. on the contour of the plate ($r = 1$) the following conditions must be fulfilled:

- when rigidly embedded

$$
u^{(n)*} = \psi^{(n)*} = w^{(n)*} = w^{(n)*},_r = 0;
$$

- when hinged

$$
u^{(n)*} = \psi^{(n)*} = w^{(n)*} = M_r^{(n)*} = 0. \tag{9.43}
$$

In this case, the boundary value problem with asterisks (9.41)–(9.43), taking into account (9.36), (9.37), coincides with the boundary value problem for some fictitious three-layer elastoplastic plate which is isothermally loaded from the natural state by external forces p^*, q^*.

The fictitious plate geometrically coincides with the plate under consideration. Its elastic–plastic properties of materials are characterised by coordinate-variable shear modulus $G_k(z)$, volume modulus $K_k(z)$, universal nonlinearity functions (9.33). The system of differential equations for quantities with asterisks (9.41) coincides with (9.7) with the accuracy of notation and differs only by the absence of temperature terms in (9.34).

The analytical solution of the system (9.41) for continuum plates under *continuous loads* follows from (9.12), in which it is necessary to make appropriate substitutions of nonlinearity functions, loads and nonlinearity coefficients included in the approximation formulas:

$$\psi^{(n)*} = C_2^{(n)*} I_1(\beta r) + \psi_r^{(n)*},$$
$$u^{(n)*} = \frac{a_3}{a_1 a_2 - a_3^2}\left[L_3^{-1}(q^* - q_\omega^{(n-1)}) - \frac{a_6}{a_3}L_2^{-1}(p^* - p_\omega^{(n-1)*}) + \left(a_5 - \frac{a_2 a_6}{a_3}\right)\psi^{(n)*} + C_7^{(n)*} r\right],$$
$$w^{(n)*} = \frac{1}{b_3}\left[b_2\left(\frac{C_2^{(n)*}}{\beta} I_0(\beta r) + \int \psi_r^{(n)*} dr\right) - \right.$$
$$\left. - \int\left(\frac{a_3}{a_1}L_2^{-1}(p^* - p_\omega^{(n-1)*}) - L_3^{-1}(q^* - q_\omega^{(n-1)*})\right)dr + \frac{C_5^{(n)*} r^2}{4} + C_4^{(n)*}\right], \tag{9.44}$$

where the partial solution of the inhomogeneous modified Bessel equation is

$$\psi_r^{(n)*} = -K_1(\beta r)\int I_1(\beta r) f^{(n)*} r dr + I_1(\beta r)\int K_1(\beta r) f^{(n)*} r dr,$$

with such integrand function

$$f^{(n)*}(r) = \frac{b_3}{b_1 b_3 - b_2^2}\left[h_\omega^{(n-1)*} + \frac{a_2 b_3 - a_3 b_2}{a_1 b_3}(p^* - p_\omega^{(n-1)}) + \frac{b_2}{b_3 r}\int (q^* - q_\omega^{(n-1)*}) r dr\right].$$

At rigid embedding of the plate contour from (9.43) we obtain the following recurrence formulas for the integration constants in solution (9.44):

$$C_2^{(n)*} = \frac{\psi_r^{(n)*}}{I_1(\beta)}, \quad C_4^{(n)*} = -b_2\left(\frac{C_2^{(n)*} I_0(\beta)}{\beta} + \int \psi_r^{(n)*} dr\big|_{r=1}\right) +$$
$$+ \int\left(\frac{a_3}{a_1}L_2^{-1}(p^* - p_\omega^{(n-1)*}) - L_3^{-1}(q^* - q_\omega^{(n-1)*})\right)dr\Big|_{r=1} - \frac{C_5^{(n)*}}{4}, \tag{9.45}$$
$$C_5^{(n)*} = 2\left(\frac{a_3}{a_1}L_2^{-1}(p^* - p_\omega^{(n-1)*}) - L_3^{-1}(q^* - q_\omega^{(n-1)*})\right)\Big|_{r=1},$$
$$C_7^{(n)*} = \left(\frac{a_6}{a_3}L_2^{-1}(p * - p_\omega^{(n-1)*}) - L_3^{-1}(q^* - q_\omega^{(n-1)*})\right)_{r=1}.$$

At hinged support of the contour the integration constants $C_2^{(n)*}$, $C_4^{(n)*}$, $C_7^{(n)*}$ are determined by the formulas (9.45), and $C_5^{(n)*}$ is:

$$C_5^{(n)*} = \frac{2b_3}{a_6 + a_7}\Big[a_3 u^{(n)*},_r(1) + \psi^{(n)*},_r(1)\left(a_5 - \frac{a_6 a_2}{b_3}\right) +$$
$$\left(\frac{a_3}{a_1}\mathrm{L}_2^{-1}(p^* - p_\omega^{(n-1)*}) - \mathrm{L}_3^{-1}(q^* - q_\omega^{(n-1)*})\right),r\Big|_{r=1} +$$
$$\left(\frac{a_3}{a_1}\mathrm{L}_2^{-1}(p^* - p_\omega^{(n-1)*}) - \mathrm{L}_3^{-1}(q^* - q_\omega^{(n-1)*})\right)\Big|_{r=1}\Big].$$

After determining the values with asterisks, the required stress–strain state parameters of a three-layer elastic–plastic plate under repeated alternating loading follow from relations (9.30). For example, displacements and strains are:,

$$\varepsilon''^{(k)}_\alpha = \varepsilon'^{(k)}_\alpha - \varepsilon_\alpha^{(k)*}, \cdot\cdot \varepsilon''^{(3)}_{rz} = \varepsilon'^{(3)}_{rz} - \varepsilon_{rz}^{(3)*} \cdot \text{э}''^{(k)}_\alpha = \text{э}'^{(k)}_\alpha - \text{э}_\alpha^{(k)*}, \cdot\varepsilon''^{(k)} = \varepsilon'^{(k)} - \varepsilon^{(k)*},$$
$$u'' = u' - u^*, \psi'' = \psi' - \psi^*, w'' = w' - w^*, \tag{9.46}$$

where the values with one stroke correspond to the solution (9.12), with asterisks—to the solution (9.45).

Numerical results for repeated alternating loading with a uniformly distributed load are obtained under the same conditions as in the case of loading the plate from the natural state. Layer materials: D16T–Fluoroplastic–D16T. Load quantity is $q_0 = 5$ MPa. The boundary conditions correspond to the hinged support of the plate contour. The geometrical parameters of the plate are referred to its radius r_0, the relative thicknesses of the layers are: $h_1 = 0{,}02$; $h_2 = 0{,}06$; $h_3 = 0{,}4$.

Under reverse loading, the temperature remained the same T_1 and the nonlinearity functions $\omega^{(k)^*}\left(\varepsilon_u^*, T_1\right)$ $(k = 1, 2)$ (9.33), according to Moskvitin's hypothesis [1], in the bearing layers were taken in a form similar to (2.38):

$$\omega^{(k)^*}\left(\varepsilon_u^{(k)^*}, T_1\right) = \begin{cases} 0, & \\ A_2^*\left(1 - \dfrac{\varepsilon_{y1}^{(k)}}{\varepsilon_u^{(k)^*}}\right)^{\alpha_2^*}, & \varepsilon_u^{(k)^*} \le \varepsilon_{y1}^{(k)*}, \\ \varepsilon_u^{(k)^*} > \varepsilon_{y1}^{(k)^*} & \varepsilon_u^{(k)^*} > \varepsilon_{y1}^{(k)*}. \end{cases} \tag{9.47}$$

where for duralumin $A_2^* = 1$; $\alpha_2^* = 0,6$; $\varepsilon_{y1}^*(T_1) = \beta_2 \varepsilon_{y1}(T_1)$; $\beta_2 = 2,02$.

From the experimental data given in the monograph [1], it follows that deformation in the plastic region under loading from the natural state hardens the D16-T alloy, since $\beta_2 > 2{,}0$.

Figure 9.13 shows deflection (a) and relative shear (b) in the filler of a three-layer plate under alternating loading (one stroke—forward loading, two strokes—reverse loading): *1*—elasticity, *2*—isothermal elastic-plasticity, *3*—thermoelastic-plasticity. Since for duraluminium the cyclic hardening coefficient $\beta_2 = 2.02$ and the material becomes less plasticy at the second half-cycle, at isothermal ($T_1 = 293$ K) repeated loading the maximum deflection of the plate 3' is slightly less in modulus. When the plate was heated to temperature $T_2 = 323$ K, the deflection of the plate under direct loading (*3*') increased compared to the isothermal case. After changing the

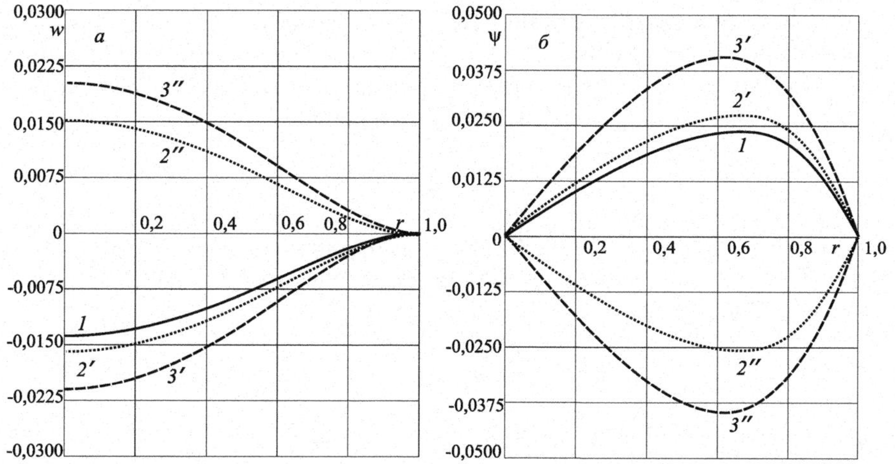

Fig. 9.13 Displacements under alternating isothermal and thermal-force loading(T_1 = 293 K, T_2 = 323 K)

sign of the load its maximum value slightly decreased (*3*") due to strain hardening of the materials. The deflection and relative shear under repeated loading decreased by 2–3% for both "cold" and "hot" plasticity conditions.

Figure 9.14 illustrates the change in radial stress σ_r at the centre of the plate when the sign of the external load changes: *1*—elastic stresses, *2*—elastic–plastic stresses (at T_1 = 293 K), *3*—elastoplastic thermo-force stresses (at T_2 = 323 K). Graphs with one stroke correspond to loading from natural state, with two strokes—repeated bending with alternating load. At the second half-cycle the stresses change sign, their extremums change insignificantly.

Let us consider several cases of plate bending by repeated local alternating loads.

Fig. 9.14 Radial stresses in the centre of the plate under alternating isothermal and thermo-force loading (T_1 = 293 K, T_2 = 323 K)

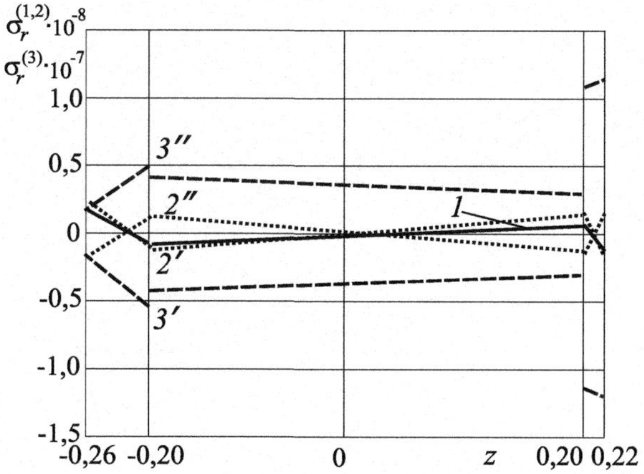

Load evenly distributed over the circle. Let a local vertical surface load $q'(r)$ uniformly distributed over a circle $r \le b$ act on the considered three-layer plate (Fig. 8.11). Starting from the moment $t = t_1$, instantaneous unloading and reloading with efforts $q''(r)$ of the opposite sign is carried out:

$$q''(r) = -q_0 H_0(b - r). \tag{9.48}$$

Introducing the differences (9.30) and taking the physical equations of state (9.32), we obtain a system of differential Eqs. (9.41) for the quantities with asterisks. Its iterative solution at a load of

$$q^*(r) = q''(r) - q'(r)$$

and constant temperature has the form:

$$\begin{aligned}
&\psi^{(n)*} = C_2^{(n)*} I_1(\beta r) + C_3^{(n)*} K_1(\beta r) + \psi_r^{(n)*}\\
&w^{(n)*} = \frac{b_2}{b_3}\int \psi^{(n)*} dr + \frac{a_3}{b_3 a_1}\int L_2^{-1}\left(p_\omega^{(n-)*}\right)dr + \frac{1}{b_3}\int L_3^{-1}(q^*)dr - \frac{1}{b_3}\int L_3^{-1}\left(q_\omega^{(n-1)*}\right)dr -\\
&-\frac{C_1^{(n)*}}{4b_3} r^2(\ln r - 1) + \frac{C_5^{(n)*} r^2}{4b_3} + C_6^{(n)*}\ln r + C_4^{(n)*}\\
&u^{(n)*} = \frac{a_3}{a_1} w_r^{(n)*} - \frac{a_2}{a_1}\psi^{(n)*} + \frac{1}{a_1} L_2^{-1}\left(p_\omega^{(n-1)*}\right) + \frac{C_7^{(n)*} r}{2} + \frac{C_8^{(n)*}}{r},
\end{aligned} \tag{9.49}$$

where

$$\begin{aligned}
\psi_r^{(n)*} &= \frac{\gamma_1 q_0}{2\beta^2} H_0(b - r)\left[\frac{b^2}{r} - r + 2b(K_1(\beta b) I_1(\beta r) - I_1(\beta b) K_1(\beta r))\right] + \frac{C_1^{(n)*}\gamma_1}{\beta^2 r}\\
&-K_1(\beta r)\left[\frac{b_3}{b_1 b_3 - b_2^2}\left[\int I_1(\beta r) h_\omega^{(n-1)*} r dr - \frac{a_2 b_3 - a_3 b_2}{a_1 b_3}\int I_1(\beta r) p_\omega^{(n-1)*} r dr\right]\right.\\
&\left.-\gamma_1\int I_1(\beta r)\left(\int q_\omega^{(n-1)*}) r dr\right) dr\right]\\
&+I_1(\beta r)\left[\frac{b_3}{b_1 b_3 - b_2^2}\left[\int K_1(\beta r) h_\omega^{(n-1)*} r dr - \frac{a_2 b_3 - a_3 b_2}{a_1 b_3}\int K_1(\beta r) p_\omega^{(n-1)*} r dr\right] -\right.\\
&\left.-\gamma_1\int K_1(\beta r)\left(\int q_\omega^{(n-1)*}) r dr\right) dr\right]
\end{aligned}$$

$$\int L_3^{-1}(q^*)dr = q_0\left(\frac{r^4-5b^4}{64}-\frac{b^4}{16}\ln\left(\frac{r}{b}\right)-\frac{b^2r^2}{8}\ln\left(\frac{r}{b}\right)+\frac{b^2r^2}{16}\right)H_0(b-r),$$

$$\int \psi^{(n)*}dr = \frac{C_2^{(n)*}I_0(\beta r)}{\beta}-\frac{C_3^{(n)*}K_0(\beta r)}{\beta}+\frac{C_1^{(n)*}\gamma_1}{\beta^2}\ln(r)+\frac{\gamma_1 q_0}{2\beta^2}H_0(b-r)\times$$

$$\times\left[\frac{b^2-r^2}{2}+b^2\ln\left(\frac{r}{b}\right)+\frac{2b}{\beta}(K_1(\beta b)I_0(\beta r)+I_1(\beta b)K_0(\beta r))-\frac{2}{\beta^2}\right]-$$

$$-\int\left(K_1(\beta r)\left[\frac{b_3}{b_1b_3-b_2^2}\left[\int I_1(\beta r)h_\omega^{(n-1)*}rdr-\frac{a_2b_3-a_3b_2}{a_1b_3}\int I_1(\beta r)p_\omega^{(n-1)*}rdr\right]-\right.\right.$$

$$\left.-\gamma_1\int I_1(\beta r)\left(\int q_\omega^{(n-1)*})rdr\right)dr\right]$$

$$+I_1(\beta r)\left[\frac{b_3}{b_1b_3-b_2^2}\left[\int K_1(\beta r)h_\omega^{(n-1)*}rdr-\frac{a_2b_3-a_3b_2}{a_1b_3}\int K_1(\beta r)p_\omega^{(n-1)*}rdr\right]-\right.$$

$$\left.\left.-\gamma_1\int K_1(\beta r)\left(\int q_\omega^{(n-1)*}\right)rdr\right)dr\right]\Bigg)dr.$$

The displacements of a three-layer elastoplastic plate under repeated alternating loading by a local circular load (9.48) can be obtained from relations (9.46) taking into account (9.18), (9.49).

Numerical results for repeated loading of the plate with local loads are obtained under the same conditions as for similar loading from the natural state.

Figure 9.15 shows the variation of deflection (*a*) and relative shear in the filler (*b*) along the plate radius under repeated alternating bending by a local circular load: *1* –elastic plate, *2*—isothermal elasticplasticity, *3*—thermoelasticplasticity. Here, as well as at continuous loading, after changing the sign of loading, the maximum value of displacements slightly decreased (curve *3*") by 3%, both at "cold" and "hot" plasticity, which is explained by cyclic hardening of the material.

Variable linear shear load. Let the plate be subjected to a linear shear force of constant intensity Q_0 applied along a circle of radius $r = a$ (Fig. 8.16). Starting from the moment $t = t_1$, instantaneous unloading and reloading with efforts of the opposite sign $Q''(r) = -Q_0$ is carried out. The corresponding iterative solution for displacements with asterisks in this case takes the following form:

$$\psi^{(n)*} = C_2^{(n)*}I_1(\beta r)+C_3^{(n)*}K_1(\beta r)+\psi_r^{(n)*},$$

$$w^{(n)*} = \frac{b_2}{b_3}\int\psi^{(n)*}dr+\frac{a_3}{b_3a_1}\int L_2^{-1}(p_\omega^{(n-1)*})dr+\frac{1}{b_3}\int L_3^{-1}(Q^*)dr-\frac{1}{b_3}\int L_3^{-1}(q_\omega^{(n-1)*})dr-$$

$$-\frac{C_1^{(n)*}}{4b_3}r^2(\ln r-1)+\frac{C_5^{(n)*}r^2}{4b_3}+C_6^{(n)*}\ln r+C_4^{(n)*}$$

$$u^{(n)*} = \frac{a_3}{a_1}w_r^{(n)*}-\frac{a_2}{a_1}\psi^{(n)*}+\frac{1}{a_1}L_2^{-1}\left(p_\omega^{(n-1)*}\right)+\frac{C_7^{(n)*}r}{2}+\frac{C_8^{(n)*}}{r},\qquad (9.50)$$

where

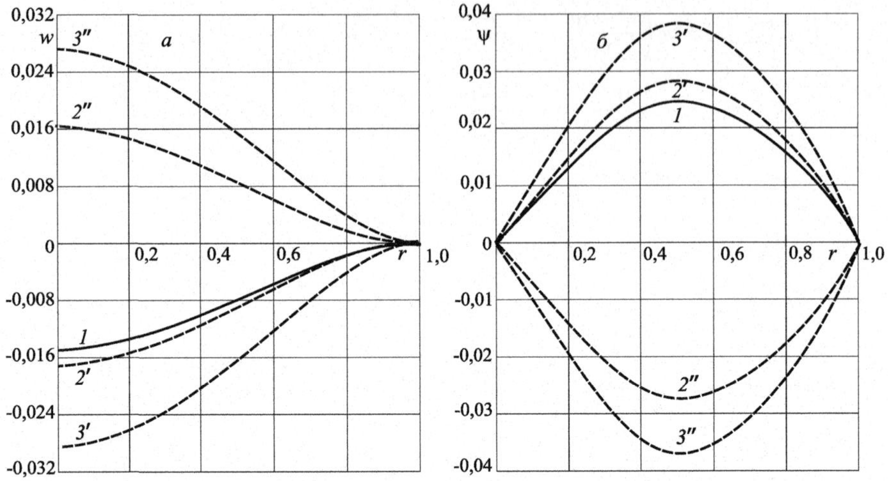

Fig. 9.15 Displacements under alternating isothermal and thermo-force loading by circular local load ($T_1 = 293$ K, $T_2 = 323$ K)

$$\psi_r^{(n)*} = \frac{\gamma_1 Q_0 a}{\beta} H_0(a-r)\left(\frac{1}{\beta r} - K_1(\beta r)I_0(\beta a) - I_1(\beta r)K_0(\beta a)\right) + \frac{C_1^{(n)*}\gamma_1}{\beta^2 r} -$$

$$K_1(\beta r)\left[\frac{b_3}{b_1b_3 - b_2^2}\left[\int I_1(\beta r)h_\omega^{(n-1)*} rdr - \frac{a_2b_3 - a_3b_2}{a_1b_3}\int I_1(\beta r)p_\omega^{(n-1)*} rdr\right] -\right.$$

$$\left.\gamma_1 \int I_1(\beta r)\left(\left(\int q_\omega^{(n-1)*}\right) rdr\right) dr\right] +$$

$$I_1(\beta r)\left[\frac{b_3}{b_1b_3 - b_2^2}\left[\int K_1(\beta r)h_\omega^{(n-1)*} rdr - \frac{a_2b_3 - a_3b_2}{a_1b_3}\int K_1(\beta r)p_\omega^{(n-1)*} rdr\right] -\right.$$

$$\left.-\gamma_1 \int K_1(\beta r)\left(\left(\int q_\omega^{(n-1)*}\right) rdr\right) dr\right]$$

$$\int L_3^{-1}(Q^*)dr = \frac{Q_0 a}{4}\left(r^2 - a^2 - a^2 \ln\left(\frac{r}{a}\right) - r^2 \ln\left(\frac{r}{a}\right)\right) H_0(a-r),$$

$$\int \psi^{(n)*} dr = \frac{1}{\beta}\left(C_2^{(n)*} I_0(\beta r) - C_3^{(n)*} K_0(\beta r)\right) + \frac{C_1^{(n)*}\gamma_1}{\beta^2}\ln(r) + \frac{\gamma_1 Q_0 a}{2\beta^2} H_0(b-r)\times$$

$$\times\left(\ln\left(\frac{r}{a}\right) + I_0(\beta a)K_0(\beta r) - K_0(\beta a)I_0(\beta r)\right)$$

$$-\int\left(K_1(\beta r)\left[\frac{b_3}{b_1b_3 - b_2^2}\left[\int I_1(\beta r)h_\omega^{(n-1)*} rdr - \frac{a_2b_3 - a_3b_2}{a_1b_3}\int I_1(\beta r)p_\omega^{(n-1)*} rdr\right] -\right.\right.$$

$$\left.-\gamma_1 \int I_1(\beta r)\left(\left(\int q_\omega^{(n-1)*}\right) rdr\right) dr\right]$$

$$+I_1(\beta r)\left[\frac{b_3}{b_1b_3 - b_2^2}\left[\int K_1(\beta r)h_\omega^{(n-1)*} rdr - \frac{a_2b_3 - a_3b_2}{a_1b_3}\int K_1(\beta r)p_\omega^{(n-1)*} rdr\right] -\right.$$

$$\left.\left.-\gamma_1 \int K_1(\beta r)\left(\left(\int q_\omega^{(n-1)*}\right) rdr\right) dr\right]\right) dr.$$

The displacements of a three-layer elastoplastic plate under repeated alternating loading by a local linear force load are obtained from relations (9.46) taking into account (9.18), (9.50).

Figure 9.16 shows the variation of deflection (*a*) and relative shear in the filler (*b*) along the plate radius under repeated alternating bending by a local circular load: *1*—elastic plate, *2*—isothermal elasticplasticity, *3*—thermoelasticplasticity. In this case, as before, after changing the sign of the load, the maximum displacement decreased by approximately the same amount.

Sign-variable linear moments. The plate under consideration is acted on by linear moments of intensity $M_0 = \text{const}$, distributed on a circle of radius $r = a$ (Fig. 8.20). The solution of the problem is obtained by using the difference of solutions (9.50) for two identical linear shear forces acting on closely spaced circles of radius $r = a\text{–}\xi$ and $r = a + \xi$. In this difference the substitution $Q_0 = M_0/(2\xi)$ is made and ξ tends to zero. Then:

$$\psi^{(n)*} = C_2^{(n)*} I_1(\beta r) + C_3^{(n)*} K_1(\beta r) + \psi_r^{(n)*},$$

$$w^{(n)*} = \frac{b_2}{b_3}\int \psi^{(n)*} dr + \frac{a_3}{b_3 a_1}\int L_2^{-1}\left(p_\omega^{(n-1)*}\right)dr + \frac{1}{b_3}\int L_3^{-1}(M^*)dr - \frac{1}{b_3}\int L_3^{-1}\left(q_\omega^{(n-1)*}\right)dr$$
$$- \frac{C_1^{(n)*}}{4b_3} r^2(\ln r - 1) + \frac{C_5^{(n)*} r^2}{4b_3} + C_6^{(n)*}\ln r + C_4^{(n)*},$$

$$u^{(n)*} = \frac{a_3}{a_1} w^{(n)*},_r - \frac{a_2}{a_1}\psi^{(n)*} + \frac{1}{a_1} L_2^{-1}(p_\omega^{(n-1)*}) + \frac{C_7^{(n)*} r}{2} + \frac{C_8^{(n)*}}{r},$$
(9.51)

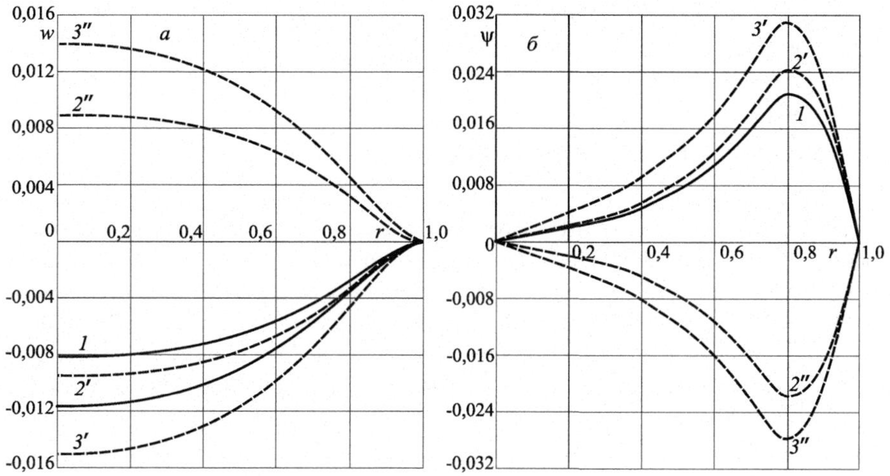

Fig. 9.16 Displacements under alternating isothermal and thermo-force loading by linear transverse force loading ($T_1 = 293$ K, $T_2 = 323$ K)

where

$$\psi_r^{(n)*} = \frac{\gamma_1 M_0}{\beta} H_0(a-r)\left(\frac{1}{\beta r} - K_1(\beta r)(I_0(\beta a) + I_1(\beta a)\beta a) -\right.$$

$$\left. - I_1(\beta r)(K_0(\beta a) - K_1(\beta a)\beta a)\right) + \frac{C_1^{(n)*}\gamma_1}{\beta^2 r} -$$

$$K_1(\beta r)\left[\frac{b_3}{b_1 b_3 - b_2^2}\left[\int I_1(\beta r) h_\omega^{(n-1)*} r dr - \frac{a_2 b_3 - a_3 b_2}{a_1 b_3}\int I_1(\beta r) p_\omega^{(n-1)*} r dr\right] -\right.$$

$$\left.\gamma_1 \int I_1(\beta r)\left(\int q_\omega^{(n-1)*}\right) r dr\right) dr\Big] +$$

$$I_1(\beta r)\left[\frac{b_3}{b_1 b_3 - b_2^2}\left[\int K_1(\beta r) h_\omega^{(n-1)*} r dr - \frac{a_2 b_3 - a_3 b_2}{a_1 b_3}\int K_1(\beta r) p_\omega^{(n-1)*} r dr\right] -\right.$$

$$\left.\gamma_1 \int K_1(\beta r)\left(\int q_\omega^{(n-1)*}) r dr\right) dr\right],$$

$$\int L_3^{-1}(M^*) dr = \frac{M_0}{4}\left(2r^2 - 2a^2 - 3a^2 \ln\left(\frac{r}{a}\right) - r^2 \ln\left(\frac{r}{a}\right)\right) H_0(a-r),$$

$$\int \psi r^{(n)} dr = \frac{C_2^{(n)*} I_0(\beta r)}{\beta} - \frac{C_3^{(n)*} K_0(\beta r)}{\beta} + \frac{C_1^{(n)*}\gamma_1}{\beta^2}\ln(r) + \frac{\gamma_1 M_0}{\beta^2} H_0(a-r)\times$$

$$\left(\ln\left(\frac{r}{a}\right) - 1 + K_0(\beta r)(I_0(\beta a) + I_1(\beta a)\beta a) - I_0(\beta r)(K_0(\beta a) - K_1(\beta a)\beta a)\right)$$

$$-\int\left(K_1(\beta r)\left[\frac{b_3}{b_1 b_3 - b_2^2}\left[\int I_1(\beta r) h_\omega^{(n-1)*} r dr - \frac{a_2 b_3 - a_3 b_2}{a_1 b_3}\int I_1(\beta r) p_\omega^{(n-1)*} r dr\right]\right.\right.$$

$$\left. - \gamma_1 \int I_1(\beta r)\left(\int q_\omega^{(n-1)*}) r dr\right) dr\right] +$$

$$I_1(\beta r)\left[\frac{b_3}{b_1 b_3 - b_2^2}\left[\int K_1(\beta r) h_\omega^{(n-1)*} r dr - \frac{a_2 b_3 - a_3 b_2}{a_1 b_3}\int K_1(\beta r) p_\omega^{(n-1)*} r dr\right] -\right.$$

$$\left.\left. - \gamma_1 \int K_1(\beta r)\left(\int q_\omega^{(n-1)*}) r dr\right) dr\right]\right) dr.$$

The displacements of a three-layer elastoplastic plate under repeated alternating loading by a linear moment load are obtained from relations (9.46) taking into account (9.18), (9.51).

Figure 9.17 shows the change of deflection (*a*) and relative shear in the filler (*b*) along the radius of the plate under repeated alternating bending by linear moments $M_0 = 0.3$ MN: *1*—elastic plate, 2—isothermal elasticplasticity, 3—thermoelastic-plasticity. In this case, after changing the sign of the load, the maximum value of displacements decreases as before due to cyclic hardening of materials of bearing layers.

Integration constants at local alternating loads follow from the boundary conditions (9.43) and the condition of boundedness of the solution in the center of the plate.

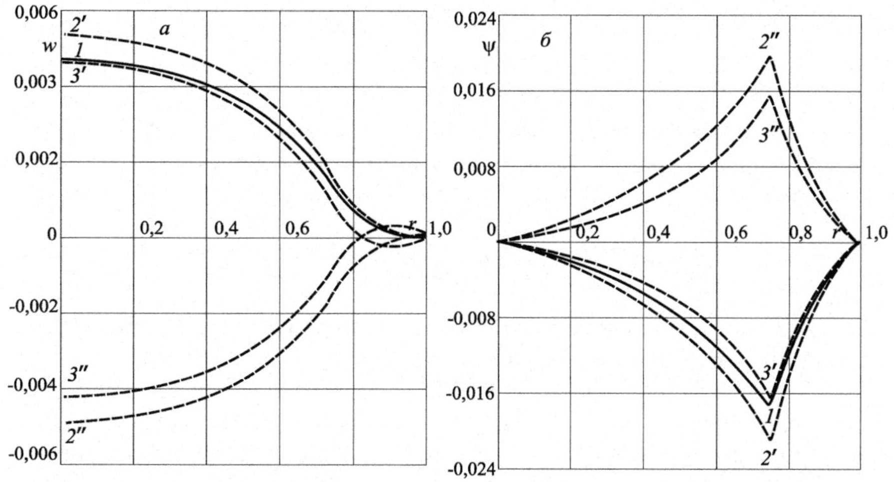

Fig. 9.17 Displacements under alternating isothermal and thermal-force loading by linear shear transverse moment load ($T_1 = 293$ K, $T_2 = 323$ K)

9.4 Cyclic Bending of a Plate in a Temperature Field

In a completely analogous way we obtain the solution for any mth variable loading. Let, after (m –1) alternating half-cycles of loading, starting from the moment $t_{m\text{-}1}$, there is an instantaneous unloading and a new loading by forces of the opposite sign $p^{(m)}$, $q^{(m)}$ (upper index "m" is load number). These loads will create fields of displacements $u^{(m)}$, $\psi^{(m)}$, $w^{(m)}$, strains $\varepsilon_\alpha^{(m,\,k)}$ and stresses $\sigma_\alpha^{(m,\,k)}$, $\sigma_{rz}^{(m,\,3)}$ ($\alpha = r, \varphi$) in the kth layer of the plate. In this case we will assume that during (m –1) alternating half-cycles the temperature in all points of the body remains unchanged after the first unloading, i.e. $T_m(z) = T_1(z, t_1)$ and the elasticity modules have retained the values $G_k(z) \equiv G_k(T_1(z))$, $K_k(z) \equiv K_k(T_1(z))$, and the corresponding loading trajectories belong to the class of simple (according to Ilyushin).

We keep for all values of stress–strain state and load differences, in which the values with index (m–1) are stresses, strains, displacements and loads in the plate before unloading, index m marks similar parameters in the process of mth half-cycle:

$$\begin{aligned}
\sigma_{\alpha\beta}^{(m,k)*} &= (-1)^n(\sigma_{\alpha\beta}^{(m-1,k)} - \sigma_{\alpha\beta}^{(m,k)}), \quad \varepsilon_{\alpha\beta}^{(m,k)*} = (-1)^n(\varepsilon_{\alpha\beta}^{(m-1,k)} - \varepsilon_{\alpha\beta}^{(m,k)}),\\
u^{(m)*} &= (-1)^n(u^{(m-1)} - u^{(m)}), \quad \psi^{(m)*} = (-1)^n(\psi^{(m-1)} - \psi^{(m)}),\\
w^{(m)*} &= (-1)^n(w^{(m-1)} - w^{(m)}),\\
q^{(m)*} &= (-1)^n(q^{(m-1)} - q^{(m)}),\ p^{(m)*} = (-1)^n(p^{(m-1)} - p^{(m)}).
\end{aligned} \tag{9.52}$$

The relations with asterisks (9.31), relating strains to displacements included in (9.52), will be valid in this case as well. For the stresses and strains marked with an asterisk, we assume physical equations of state of the type (9.32):

$$s_{\alpha}^{(m,k)*} = 2G_{\alpha}^{(k)}\ э_{\alpha}^{(m,k)*}\ (1 - \omega^{(m,k)*}),\ \sigma^{(m,k)*} = 3K_{k}\varepsilon^{(m,k)*}(k = 1, 2, 3),$$
$$s_{rz}^{(m,3)*} = 2G_{\alpha}^{(3)}\ э_{rz}^{(m,3)*}\ (1 - \omega^{(m,3)*}),\ (\alpha = r, \varphi) \tag{9.53}$$

We assume that the corresponding universal nonlinearity functions $\omega^{(m,k)*}$ in the bearing layers are expressed through the nonlinearity functions under loading from the natural state:

$$\omega^{(m,k)*} \equiv \omega_k(\varepsilon_u^{(m,\,k)*}, T_k) = \omega\prime^{(k)}(\varepsilon_u^{(m,k)*}, \varepsilon_y^{(m,k)*}, T_1, d_n^{(m)*}), \tag{9.54}$$

where $d_n^{(m)*}$ are experimentally determined material parameters included in the corresponding approximation formula.

The physical nonlinearity of the filler under mth alternating loading is still described by relations (9.1). Using the physical relations (9.53) in the components of stresses and strains with asterisks, as before, let us identify nonlinear components:

$$\begin{aligned}
&\sigma_{\alpha}^{(m,k)*} = \sigma_{\alpha e}^{(m,k)*} - \sigma_{\alpha\omega}^{(m,k)*}, \quad \sigma_{rz}^{(m,3)*} = \sigma_{rze}^{(m,3)*} - \sigma_{rz\omega}^{(m,3)*}\\
&\sigma_{\alpha e}^{(m,k)*} = 2G_k\ э_{\alpha}^{(m,k)*} + 3K_k\varepsilon^{(m,k)*}, \quad \sigma_{\alpha\omega}^{(m,\,k)*} = 2G_k\omega_k(\varepsilon_u^{(m,\,k)*}, T_k)\ э_{\alpha}^{(m,\,k)*},\\
&\sigma_{rze}^{(m,3)*} = 2G_3\ э_{rz}^{(m,3)*}, \quad \sigma_{rz\omega}^{(m,3)*} = 2G_3\omega_3(\varepsilon_u^{(m,3)*}, T_k)\ э_{rz}^{(m,3)*}, \quad (\alpha = r, \varphi),
\end{aligned} \tag{9.55}$$

Here the nonlinearity function $\omega_k(\varepsilon_u^{(m,k)*}, T_k)$, introduced in (9.54), is calculated according to Moskvitin's hypothesis [1] by a formula of the type (9.47):

$$\omega^{(m,k)*}(\varepsilon_u^{(m,k)*}) = \begin{cases} 0, & \varepsilon_u^{(m,k)*} \le \varepsilon_{y1}^{(m,k)*}, \\ A_2^{(m,k)*}\left(1 - \dfrac{\varepsilon_y^{(m,k)*}}{\varepsilon_u^{(m,k)*}}\right)^{\alpha_2^{(m,k)*}}, & \varepsilon_u^{(m,k)*} > \varepsilon_{ym}^{(m,k)*}. \end{cases} \tag{9.56}$$

where $A_2^{(m,k)*}$, $\alpha_2^{(m,k)*}$, $\varepsilon_{y1}^{(m,k)*}(T_1) = \beta_m\varepsilon_{y1}^{(m,k)*}(T_1)$ are nonlinearity parameters.

Performing a similar operation with the values of the internal force type (8.3), we obtain:

$$\begin{aligned}
&T_{\alpha}^{(m,k)*} = T_{\alpha e}^{(m,k)*} - T_{\alpha\omega}^{(m,k)*}, \quad M_{\alpha}^{(m,k)*} = M_{\alpha e}^{(m,k)*} - M_{\alpha\omega}^{(m,k)*}\\
&H_{\alpha}^{(m,k)*} = H_{\alpha e}^{(m,k)*} - H_{\alpha\omega}^{(m,k)*}, \quad Q^{(m,3)*} = Q_e^{(m,3)*} - Q_{\omega}^{(m,3)*}
\end{aligned} \tag{9.57}$$

The elastic (index "e") and inelastic (index "ω") components in (9.57) are calculated through the stress components (9.55) by the following formulae:

$$\begin{aligned}
&T_{\alpha e}^{(m,k)*} = \int\limits_{h_k} \sigma_{\alpha e}^{(m,k)*}dz, \quad T_{\alpha\omega}^{(m,k)*} = \int\limits_{h_k} \sigma_{\alpha\omega}^{(m,k)*}dz, \quad M_{\alpha}^{(m,k)^*} = \int_{h<} \sigma_{\alpha e}^{(k)^*} zdz,\\
&M_{\omega}^{(m,k)^*} = \int_{h_k} \sigma_{\alpha\omega}^{(k)^*} zdz, \quad Q_e^{(m,3)^*} = \int_{-c}^{c} \sigma_{rze}^{(m,3)^*} dz, \quad Q_{\omega}^{(m,k)^*} = \int_{-c}^{c} \sigma_{r\tau\omega}^{(m,3)^*} dz.
\end{aligned} \tag{9.58}$$

After that, the generalised internal forces will be:

$$T_{\alpha}^{(m)*} = T_{\alpha e}^{(m)*} - T_{\alpha\omega}^{(m)*} = \sum_{k=1}^{3} T_{\alpha e}^{(m,k)*} - \sum_{k=1}^{3} T_{\alpha\omega}^{(m,k)*},$$

$$M_{\alpha}^{(m)*} = M_{\alpha e}^{(m)*} - M_{\alpha\omega}^{(m)*} = \sum_{k=1}^{3} M_{\alpha e}^{(m,k)*} - \sum_{k=1}^{3} M_{\alpha\omega}^{(m,k)*},$$

$$H_{\alpha e}^{(m)*} = M_{\alpha e}^{(m,3)*} + c\left(T_{\alpha e}^{(m,1)*} - T_{\alpha e}^{(m,2)*}\right),$$

$$H_{\alpha\omega}^{(m)*} = M_{\alpha\omega}^{(m,3)*} + c\left(T_{\alpha\omega}^{(m,1)*} - T_{\alpha\omega}^{(m,2)*}\right). \tag{9.59}$$

Iterative equilibrium equations of the type (9.41) at the nth step for the values with asterisks at the mth half-cycle are written in the form:

$$\begin{aligned}
&L_2\left(a_1 u^{(m,n)*} + a_2\psi^{(m,n)*} - a_3 w_r^{(m,n)*}\right) = -p^{(m,n)*} + p_{\infty}^{(m,n-1)*},\\
&L_2(a_2 u^{(m,n)*} + a_4\psi^{(m,n)*} - a_5 w,_r^{(m,n)*}) - 2cG_3\psi^{(m,n)*} = h_{\omega}^{(m,n-1)*},\\
&L_3\left(a_3 u^{(m,n)*} + a_5\psi^{(m,n)*} - a_6 w_r^{(m,n)*}\right) = -q^{(m,n)*} + q_{\omega}^{(m,n-1)*}.
\end{aligned} \tag{9.60}$$

Additional "external" loads $p_{\omega}^{(m,n-1)^*}$, $h_{\omega}^{(m,n-1)^*}$, $q_{\omega}^{(m,n-1)^*}$ corresponding to inelastic components in (9.60), at the first step ($n = 1$) are assumed to be equal to zero, and further are calculated according to the results of the previous approximation:

$$\begin{aligned}
&p_{\omega}^{(m,n-1)*} = T_{r\omega}^{(m,n-1)*},_r + \frac{1}{r}\left(T_{r\omega}^{(m,n-1)*} - T_{\varphi\omega}^{(m,n-1)*}\right),\\
&h_{\omega}^{(m,n-1)*} = H_{r\omega}^{(m,n-1)*},_r + \frac{1}{r}(H_{r\omega}^{(m,n-1)*} - H_{\varphi\omega}^{(m,n-1)*}) - Q_{\omega}^{(m,n-1)*},\\
&q_{\omega}^{(m,n-1)*} = M_{r\omega}^{(m,n-1)*},_{rr} + \frac{1}{r}(2M_{r\omega}^{(m,n-1)*},_r - M_{\varphi\omega}^{(m,n-1)*},_r),
\end{aligned} \tag{9.61}$$

where

$$T_{\alpha\omega}^{(m,n-1)*} \equiv \sum_{k=1}^{3} \int_{h_k} \sigma_{\alpha\omega}^{(m,k,n-n-1)*} dz = \sum_{k=1}^{3} \int_{h_k} 2G_k\omega_k\left(\varepsilon_{\alpha}^{(m,k,n-1)*}, T_k\right) э_{\alpha}^{(m,k,k-1},$$

$$M_{\alpha\omega}^{(m,n-1)*} \equiv \sum_{k=1}^{3} \int_{h_k} \sigma_{\alpha\omega}^{(m,k,n-1)*} z dz = \sum_{k=1}^{3} \int_{h_k} 2G_k\omega_k\left(\varepsilon_{\alpha}^{(m,k,n-1)*}, T_k\right) э_{\alpha}^{(m,k,n-1)*} z dz,$$

$$H_{\alpha\omega}^{(m,n-1)*} = M_{\alpha\omega}^{(m,3,n-1)*} + c\left(T_{\alpha\omega}^{(m,1,n-1)*} - T_{\alpha\omega}^{(m,2,n-1)*}\right),$$

$$Q_{\omega}^{(m,\,n-1)*} = \int_{-c}^{c} 2G_k\omega_k(\varepsilon_{\alpha}^{(m,\,3,\,n-1)*}, T_k)\; э_{rz}^{(m,\,3,\,n-1)*}\, dz, \quad (\alpha = r, \phi).$$

The boundary value problem for determining the functions $u^{(m,n)*}(r)$, $\psi^{(m,n)*}(r)$, $w^{(m,n)*}(r)$ is closed by joining to the system of Eqs. (9.60) the kinematic boundary conditions of the type (9.43):

- when rigidly embedded

$$(u^{(m,n)*} = \psi^{(m,n)*} = w^{(m,n)*} = w^{(m,n)*}{}_{,r} = 0, \; at \quad r = 1;$$

- when hinged

$$(u^{(m,n)*} = \psi^{(m,n)*} = w^{(m,n)*} = M_r^{(m,n)*} = 0 \quad at \quad r = 1. \tag{9.62}$$

In this case, the boundary value problem with asterisks (9.60)–(9.62) taking into account (9.58), (9.59) coincides with the boundary value problem for some fictitious three-layer elastoplastic plate which is isothermally loaded from the natural state by external loads $q^{(m,n)*}$, $p^{(m,n)*}$. The iterative system of differential equations for quantities with asterisks (9.60) coincides with (9.41) with exactness of notation. Therefore, its analytical solution for a solid plate follows from the solution (9.44), in which it is necessary to make the substitutions of loads, nonlinearity functions and corresponding coefficients corresponding to the mth half-cycle:

$$\begin{aligned}
\psi^{(m,n)*} &= C_2^{(m,\,n)*} I_1(\beta r) + \psi_r^{(m,n)*},\\
u^{(m,n)*} &= \frac{a_3}{a_1 a_6 - a_3^2}\Bigg[L_3^{-1}\left(q^{(m)*} - q_\omega^{(m,n)*}\right) - \frac{a_6}{a_3} L_2^{-1}\left(p^{(m)*} - p_\omega^{(m,n)*}\right) + \\
&+ \left(a_5 - \frac{a_2 a_6}{a_3}\right)\psi^{(n)*} + C_7^{(m,n)*} r\Bigg],\\
w^{(m,n)*} &= \frac{1}{b_3}\Bigg[b_2\left(\frac{C_2^{(m,n)*}}{\beta} I_0(\beta r) + \int \psi_r^{(m,n)*} dr\right) - \\
&- \int\left(\frac{a_3}{a_1} L_2^{-1}\left(p^{(m)*} - p_\omega^{(m,n)*}\right) - L_3^{-1}\left(q^{(m)*} - q_\omega^{(m,n)*}\right)\right) dr + \frac{C_5^{(m,n)*} r^2}{4} + C_4^{(m,n)*}\Bigg].
\end{aligned} \tag{9.63}$$

Here the partial solution of the inhomogeneous modified Bessel equation and the integrand function are equal to:

$$\begin{aligned}
\psi_r^{(m,n)*} &= -K_1(\beta r)\int I_1(\beta r) f^{(m,n)*} r dr + I_1(\beta r)\int K_1(\beta r) f^{(m,n)*} r dr,\\
f^{(m,n)*}(r) &= \frac{b_3}{b_1 b_3 - b_2^2}\Bigg[h_\omega^{(m,n-1)*} + \frac{a_2 b_3 - a_3 b_2}{a_1 b_3}\left(p^{(m)*} - p_\infty^{(m,n-1)*}\right) + \\
&+ \frac{b_2}{b_3 r}\int\left(q^{(m)*} - q_\omega^{(m,n-1)*}\right) r dr.
\end{aligned}$$

It should be noted that the integration constants in the solution (9.63) depend on the half-cycle number m. When the plate contour is esmbedded, the following recurrence formulas are obtained for them from (9.62):

$$C_2^{(m,n)*} = -\frac{\psi_r^{(m,n)*}\big|_{r=1}}{I_1(\beta)}, \quad C_4^{(m,n)*} = -b_2\left(\frac{C_2^{(n)} I_0(\beta)}{\beta} + \int \psi_r^{(m,n)*}\, dr\big|_{r=1}\right) +$$

$$+ \int \left(\frac{a_3}{a_1} L_2^{-1}(p^{(m)*} - p_\omega^{(m,n-1)*}) - L_3^{-1}(q^{(m)*} - q_\omega^{(m,n-1)*})\right) dr\bigg|_{r=1} - \frac{C_5^{(m,n)*}}{4},$$

$$C_5^{(m,n)*} = 2\left(\frac{a_3}{a_1} L_2^{-1}\left(p^{(m)*} - p_\omega^{(m,n-1)*}\right) - L_3^{-1}\left(q^{(m)*} - q_\omega^{(m,n-1)*}\right)\right)\bigg|_{r=1},$$

$$C_7^{(m,n)*} = \left(\frac{a_6}{a_3} L_2^{-1}\left(p^{(m)*} - p_\omega^{(m,n-1)*}\right) - L_3^{-1}\left(q^{(m)*} - q_\omega^{(m,n-1)*}\right)\right)\bigg|_{r=1}. \tag{9.64}$$

When the contour is hinged, the integration constants $C_2^{(m,n)*}$, $C_4^{(m,n)*}$, $C_7^{(m,n)*}$ will be the same, only the constant $C_5^{(m,n)*}$ changes:

$$C_5^{(m,n)*} = \frac{2b_3}{a_6 + a_7}\Bigg[a_3 u^{(n)*},_r(1) + \psi^{(n)*},_r(1)\left(a_5 - \frac{a_6 b_2}{b_3}\right) -$$

$$\left(\frac{a_3}{a_1} L_2^{-1}\left(p^* - p_\omega^{(n-1)*}\right) - L_3^{-1}\left(q^* - q_\omega^{(n-1)*}\right)\right), r\big|_{r=1} \tag{9.65}$$

$$- \left(\frac{a_3}{a_1} L_2^{-1}(p^* - p_\omega^{(n-1)*}) - L_3^{-1}(q^* - q_\omega^{(n-1)*})\right)\bigg|_{r=1}\Bigg].$$

Thus, the obtained formulas (9.63) –(9.65) describe in iterations the analytical solution of the problem in displacements for the values with asterisks at an arbitrary mth half-cycle of alternating loading in the case of various types of boundary conditions.

The required solution of the problem is obtained from relations (9.52) taking into account solutions (9.12) and (9.64):

$$\sigma_x^{(m,k)} = \sigma{\prime}_x^k - \sum_{l=2}^{m} (-1)^k \sigma_x^{(l,k)*}, \quad \varepsilon_x^{(m,k)} = \varepsilon{\prime}_{ij}^k - \sum_{l=2}^{m} (-1)^k \varepsilon_x^{(l,k)*},$$

$$u^{(m)} = u\prime - \sum_{l=2}^{m} (-1)^l u^{(l)*}, \quad \psi^{(m)} = \psi\prime - \sum_{l=2}^{m} (-1)^l \psi^{(l)*},$$

$$w^{(m)} = w' - \sum_{l=2}^{m} (-1)^l w^{(l)*} \tag{9.66}$$

where values with one stroke are stresses, strains and displacements in the plate before the first unloading.

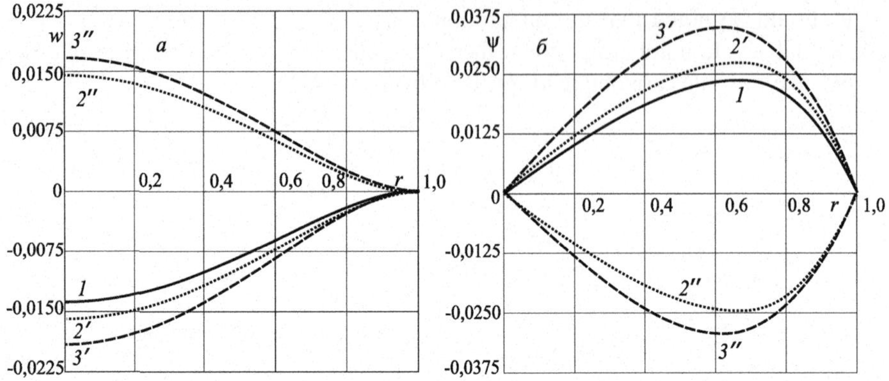

Fig. 9.18 Displacements under cyclic isothermal and thermal-force loading ($T_1 = 293$ K, $T_2 = 323$ K)

Similarly, the solutions at *m*t^{h} variable local loadings of the considered three-layer circular plate are constructed.

Numerical results were obtained for a three-layer plate (D16T—fluoroplastic-4—D16T) according to the solution (9.63) and integration constants (9.66). At eight half-cycles for duraluminium, the cyclic hardening coefficient was assumed to be $\beta_8 = 2.02$, and the parameters of plasticity functions (9.56) were assumed to be the same as for repeated alternating loading (9.47).

Figure 9.18 shows: deflection (*a*), relative shear in the filler (*b*) of an asymmetric in thickness three-layer plate under loading from the natural state: *1*—linear elasticity, *2*—elastic-plasticity, ($T_1 = 293$ K); *3*—at the eighth half-cycle ($T_2 = 323$ K). One stroke—direct loading, two strokes—reverse loading. Deflection and shear at the eighth half-cycle are reduced by 5–6% compared to the initial ones, due to cyclic strengthening of the material of bearing layers during the whole loading period.

Conclusions. Thus, the proposed methodology allows to obtain solutions of boundary value problems at any cycle of alternating loading by continuous and local loads of a three-layer elastoplastic circular plate according to the known solution of the corresponding problem on loading of a plate from a natural state in a temperature field. It should be emphasised that the above results are valid only in the region of small elastoplastic deformations and under simple variable loading or close to it.

Reference

1. Moskvitin, V.V. 1981. *Cyclic loading of structural elements*. Moscow, Nauka, 344 p. (in Russian).

Chapter 10
Circular Three-Layer Plate in a Neutron Flux

The axisymmetric bending of asymmetric in thickness circular elastic and elastic–plastic three-layer plates with a rigid incompressible in thickness filler in a neutron flux is investigated. Kinematic assumptions are based on the "broken" line hypothesis. Loads are continuous and local. Strains are small. The work of tangential stresses in the filler is taken into account. A numerical analysis of the dependence of the obtained solutions on the neutron flux intensity and the boundary conditions was performed.

10.1 Elastic Three-Layer Plate

It is assumed that at the initial moment of time the external surface $z = c + h_1$ of the considered elastic three-layer plate is subjected to an axisymmetric distributed force load $p(r),\ q(r)$ and neutron flux with density $\varphi_0 = \text{const}$ (Fig. 10.1).

Because the load is axisymmetric, there are no tangential displacements, and the plate deflection w, relative shear in the filler ψ, and radial displacement of the coordinate surface u are independent of the coordinate φ. Hereafter, these functions $w(r, t)$, $u(r, t)$ and $\psi(r, t)$ are considered to be unknown. Their change in time t is determined by the exposure time of the integral neutron flux $I_k(z, t)$. A rigid diaphragm on the contour is assumed to prevent the relative displacement of the layers.

It is assumed that the materials of the plate layers exhibit elastic properties during deformation. To describe the deformation of the layers, the physical equations of Hooke's law (2.33) are used, taking into account the radiation volume change:

$$\begin{aligned} s_\alpha^{(k)} &= 2G_k \text{э}_\alpha^{(k)}, \quad s_{rz}^{(3)} = 2G_k \text{э}_{rz}^{(3)}, \\ \sigma^{(k)} &= K_k\left(3\varepsilon^{(k)} - B_k I_k\right) \quad (k = 1,\ 2, 3; \quad \alpha = r, \varphi) \end{aligned} \tag{10.1}$$

S. Eduard et al., *Deformation of Three-layer Structural Elements in Thermal Radiation Fields*, https://doi.org/10.1007/978-981-97-7217-9_10

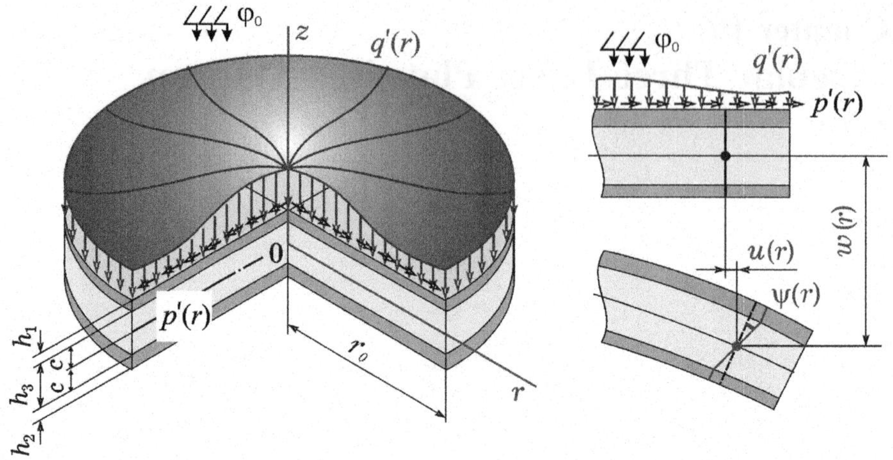

Fig. 10.1 Three-layer plate in neutron flux

where $s_{\alpha}^{(k)}$, $э_{\alpha}^{(k)}$ are deviator parts and $\sigma^{(k)}$, $\varepsilon^{(k)}$ are spherical parts of the stress and strain tensors; G_k, K_k—are shear and volume deformation modules; $s_{rz}^{(3)}$, $э_{rz}^{(3)}$ are shear stress and shear strain in the filler; $I_k(z, t)$ is integral neutron flux (2.56), accumulated by time t; B_k is material constant obtained from experience; k is layer number.

With the accepted kinematic hypotheses, the radial displacements in the layers $u_r^{(k)}$ will be defined through the required functions by formula (8.1), deformations by formula (8.2), and generalized internal forces by formula (8.3).

The equilibrium equations of the three-layer plate in generalized internal forces are obtained using the Lagrange principle of possible displacements in the form (8. 8), the force boundary conditions are (8.9).

The expression of generalized forces through the three required functions $w(r, t)$, $u(r, t)$, $\psi(r, t)$ is obtained by analogy with the Sect. 8.1, by substituting (10.1) into (8. 3) and considering strains (8.2). As a result, the generalized internal forces T_r, T_{φ}, Q are still defined by formulas (8.12–8.14). As a result, the formulas for generalized internal moments include the integral neutron flux $I_k(z, t)$:

$$
\begin{aligned}
M_r = &\left[K_1^+ h_1\left(c+\frac{h_1}{2}\right) - K_2^+ h_2\left(c+\frac{h_2}{2}\right)\right]u,_r + \left[K_1^- h_1\left(c+\frac{h_1}{2}\right) - K_2^- h_2\left(c+\frac{h_2}{2}\right)\right]\frac{u}{r} +\\
&+\left[cK_1^+ h_1\left(c+\frac{h_1}{2}\right) + cK_2^+ h_2\left(c+\frac{h_2}{2}\right) + \frac{2}{3}c^3K_3^+\right]\psi,_r +\\
&+\left[cK_1^- h_1\left(c+\frac{h_1}{2}\right) + cK_2^- h_2\left(c+\frac{h_2}{2}\right) + \frac{2}{3}c^3K_3^-\right]\frac{\psi}{r} -\\
&-\left[K_1^+ h_1\left(c^2+ch_1+\frac{h_1^2}{3}\right) + K_2^+ h_2\left(c^2+ch_2+\frac{h_2^2}{3}\right) + \frac{2}{3}c^3K_3^+\right]w,_{rr} -\\
&-\left[K_1^- h_1\left(c^2+ch_1+\frac{h_1^2}{3}\right) + K_2^- h_2\left(c^2+ch_2+\frac{h_2^2}{3}\right) + \frac{2}{3}c^3K_3^-\right]\frac{w,_r}{r} - \sum_{k=1}^{3} K_k B_k \int_{h_k} I_k z\,\mathrm{d}z,\\
M_\varphi = &\left[K_1^- h_1\left(c+\frac{h_1}{2}\right) - K_2^- h_2\left(c+\frac{h_2}{2}\right)\right]u,_r + \left[K_1^+ h_1\left(c+\frac{h_1}{2}\right) - K_2^+ h_2\left(c+\frac{h_2}{2}\right)\right]\frac{u}{r} +\\
&+\left[cK_1^- h_1\left(c+\frac{h_1}{2}\right) + cK_2^- h_2\left(c+\frac{h_2}{2}\right) + \frac{2}{3}c^3K_3^-\right]\psi,_r +\\
&+\left[cK_1^+ h_1\left(c+\frac{h_1}{2}\right) + cK_2^+ h_2\left(c+\frac{h_2}{2}\right) + \frac{2}{3}c^3K_3^+\right]\frac{\psi}{r} -\\
&-\left[K_1^- h_1\left(c^2+ch_1+\frac{h_1^2}{3}\right) + K_2^- h_2\left(c^2+ch_2+\frac{h_2^2}{3}\right) + \frac{2}{3}c^3K_3^-\right]w,_{rr} -\\
&-\left[K_1^+ h_1\left(c^2+ch_1+\frac{h_1^2}{3}\right) + K_2^+ h_2\left(c^2+ch_2+\frac{h_2^2}{3}\right) + \frac{2}{3}c^3K_3^+\right]\frac{w,_r}{r} - \sum_{k=1}^{3} K_k B_k \int_{h_k} I_k z\,\mathrm{d}z,\\
H_r = &\,c\left(K_1^+ h_1 - K_2^+ h_2\right)u,_r + c\left(K_1^- h_1 - K_2^- h_2\right)\frac{u}{r} +\\
&+\left[c^2\left(K_1^+ h_1 - K_2^+ h_2\right) + \frac{2}{3}c^3K_3^+\right]\psi,_r + \left[c^2\left(K_1^- h_1 - K_2^- h_2\right) + \frac{2}{3}c^3K_3^-\right]\frac{\psi}{r} -\\
&-\left[c\left(K_1^+ h_1(c+\frac{h_1}{2}) + K_2^+ h_2\left(c+\frac{h_2}{2}\right)\right) + \frac{2}{3}c^3K_3^+\right]w,_{rr} -\\
&-\left[c\left(K_1^- h_1(c+\frac{h_1}{2}) + K_2^- h_2\left(c+\frac{h_2}{2}\right)\right) + \frac{2}{3}c^3K_3^-\right]\frac{w,_r}{r} -\\
&-K_3B_3\int_{h_3} I_3 z\,\mathrm{d}z - \left(K_1B_1\int_{h_1} I_1\,\mathrm{d}z - K_2B_2\int_{h_2} I_2\,\mathrm{d}z\right),\\
H_\varphi = &\,c\left(K_1^- h_1 - K_2^- h_2\right)u,_r + c\left(K_1^+ h_1 - K_2^+ h_2\right)\frac{u}{r} +\\
&+\left[c^2\left(K_1^- h_1 - K_2^- h_2\right) + \frac{2}{3}c^3K_3^-\right]\psi,_r + \left[c^2\left(K_1^+ h_1 + K_2^+ h_2\right) + \frac{2}{3}c^3K_3^+\right]\frac{\psi}{r} -\\
&-\left[c\left(K_1^- h_1(c+\frac{h_1}{2}) + K_2^- h_2\left(c+\frac{h_2}{2}\right)\right) + \frac{2}{3}c^3K_3^-\right]w,_{rr} -\\
&-\left[c\left(K_1^+ h_1(c+\frac{h_1}{2}) + K_2^+ h_2\left(c+\frac{h_2}{2}\right)\right) + \frac{2}{3}c^3K_3^+\right]\frac{w,_r}{r} -\\
&-K_3B_3\int_{h_3} I_3 z\,\mathrm{d}z - \left(K_1B_1\int_{h_1} I_1\,\mathrm{d}z - K_2B_2\int_{h_2} I_2\,\mathrm{d}z\right),
\end{aligned}
\tag{10.2}
$$

where $K_k + \frac{4}{3}G_k \equiv K_k^+, \ \ K_k - \frac{2}{3}G_k \equiv K_k^-$.

After substituting the expressions for the generalized internal forces, including (10.2), into the equilibrium Eq. (8.8) and performing the necessary transformations, we obtain a system of differential equilibrium equations to determine the required functions $w(r, t)$, $u(r, t)$, $\psi(r, t)$ in the considered problem of deformation of a three-layer circular plate in a neutron flow:

$$\begin{aligned} &\mathrm{L}_2(a_1u + a_2\psi - a_3w,_r) = p,\\ &\mathrm{L}_2(a_2u + a_4\psi - a_5w,_r) - 2cG_3\psi = 0,\\ &\mathrm{L}_3(a_3u + a_5\psi - a_6w,_r) = -q. \end{aligned} \tag{10.3}$$

Here the linear differential operators L_2 (Bessel operator) and L_3 are defined by relations (8.16), the coefficients a_i—by formula (8.18).

It should be noted that the neutron flux parameters are not explicitly included in the system of Eq. (10.3) because they do not depend on the radial coordinate and their derivatives for this coordinate are zero. In addition, in the first two equations of the system (8.8), the terms containing the differences between generalized forces are reduced.

10.1.1 General Solution of the Boundary Value Problem of Elasticity Theory

The resulting system of equilibrium Eq. (10.3) coincides in form with a similar one for a three-layer unirradiated plate, for which the general analytical solution can be taken as (8.26):

$$\begin{aligned} &\psi = C_2I_1(\beta r) + C_3K_1(\beta r) + \psi_r,\\ &w = \frac{b_2}{b_3}\int \psi\,\mathrm{d}r - \frac{a_3}{b_3a_1}\int \mathrm{L}_2^{-1}(p)\mathrm{d}r + \frac{1}{b_3}\int \mathrm{L}_3^{-1}(q)\mathrm{d}r -\\ &-\frac{1}{4b_3}C_1r^2(\ln r - 1) + \frac{C_5r^2}{4b_3} + C_6\ln r + C_4,\\ &u = \frac{a_3}{a_1}w,_r - \frac{a_2}{a_1}\psi - \frac{1}{a_1}\mathrm{L}_2^{-1}(p) + \frac{C_7r}{2} + \frac{C_8}{r}, \end{aligned} \tag{10.4}$$

where $C_1, C_1, \ldots, C_1$ are integration constants, which may depend on the value of the neutron flux; $I_1(\beta r)$ is modified first-order Bessel function, $K_1(\beta r)$ is McDonald's function; ψ_r—is partial solution (8.25); $\mathrm{L}_2^{-1}, \mathrm{L}_3^{-1}$ are linear integral operators (8.26), the other parameters are introduced in (8.21).

The integral neutron flux enters the solution of the boundary value problem through the integration constants. Therefore, of the boundary conditions of interest are those that impose restrictions on the internal forces. For example, if the plate contour is hinged, the generalized radial moment on the contour is zero and the conditions must be fulfilled:

$$u = \psi = w = M_r = 0 \quad at\ r = r_0. \tag{10.5}$$

By substituting the solution (10.4) into the boundary conditions (10.5) and considering the boundedness of displacements in the center of the plate, we obtain the following integration constants in general form:

$$\begin{aligned}
&C_1 = C_3 = C_6 = C_8 = 0,\ C_2 = -\frac{\psi_r}{I_1(\beta r_0)}\bigg|_{r=r_0},\\
&C_4 = -\frac{b_2}{b_3}\left(\frac{C_2}{\beta} I_0(\beta r_0) + \int \psi_r \mathrm{d}r\bigg|_{r=r_0}\right) + \frac{1}{b_3}\int\left(\frac{a_3}{a_1}\mathrm{L}_2^{-1}(p) - \mathrm{L}_3^{-1}(q)\right)\mathrm{d}r\bigg|_{r=r_0} - \frac{C_5 r_0^2}{4b_3},\\
&C_5 = \frac{2b_3}{a_6 + a_7}\Bigg\{-\mathrm{L}_3^{-1}(q)_{,r}\Big|_{r=r_0} + \frac{a_3^2}{b_3 a_1 r_0}\left(\frac{a_6}{a_3}\mathrm{L}_2^{-1}(p) - \mathrm{L}_3^{-1}(q)\right)\bigg|_{r=r_0} +\\
&+\frac{a_7}{b_3 r_0}\left(\frac{a_3}{a_1}\mathrm{L}_2^{-1}(p) - \mathrm{L}_3^{-1}(q)\right)\bigg|_{r=r_0} - t\Bigg[\frac{\varphi_0 K_1 B_1}{\mu_1}\left(h_1 + c - \frac{1}{\mu_1} - \left(c - \frac{1}{\mu_1}\right)\exp(-\mu_1 h_1)\right) +\\
&+\frac{\varphi_{32} K_2 B_2}{\mu_2}\left(\left(c + h_2 + \frac{1}{\mu_2}\right)\exp(-\mu_2 h_2) - c - \frac{1}{\mu_2}\right) +\\
&+\frac{\varphi_{13} K_3 B_3}{\mu_3}\left(c - \frac{1}{\mu_3} + \left(c + \frac{1}{\mu_3}\right)\exp(-2\mu_3 c)\right)\Bigg]\Bigg\},\\
&C_7 = \frac{2a_3}{a_1 b_3 r_0}\left(\frac{a_6}{a_3}\mathrm{L}_2^{-1}(p) - \mathrm{L}_3^{-1}(q)\right)\bigg|_{r=r_0} - \frac{a_3}{a_1 b_3\, 5}.
\end{aligned} \tag{10.6}$$

Thus, the obtained relations (10.4) with the integration constants (10.6) make it possible to construct analytical solutions of the boundary value problems of deformation in the neutron flow of circular three-layer plates with an articulated contour fixed at different external loads, after finding appropriate partial solutions ψ_r of the nonuniform Bessel equation.

10.1.2 Plate Bending by Neutron Flux

Consider the deformation of a three-layer circular plate due to the incident neutron flux alone. If there is no force load in the general analytical solution (10.4): it is necessary to zero out the terms containing the values p and q; zero out the integrand function as well; take the partial solution as equal to zero ($f(r) = 0$, $\psi_r = 0$). As a result, we have:

$$\begin{aligned}
&\psi = C_2 I_1(\beta r) + C_3 K_1(\beta r),\\
&w = \frac{b_2}{b_3}\int \psi \mathrm{d}r - \frac{1}{4b_3} C_1 r^2(\ln r - 1) + \frac{C_5 r^2}{4b_3} + C_6 \ln r + C_4,\\
&u = \frac{a_3}{a_1} w,_r - \frac{a_2}{a_1}\psi + \frac{C_7 r}{2} + \frac{C_8}{r}.
\end{aligned} \tag{10.7}$$

To the zero integration constants (10.6) will be added C_2, i.e.:

$$\begin{aligned}
&C_1 = C_2 = C_3 = C_6 = C_8 = 0,\\
&C_4 = -\frac{C_5 r_0^2}{4b_3}; \quad C_7 = -\frac{a_3}{a_1 b_3} C_5,\\
&C_5 = \frac{2b_3}{a_6 + a_7} - t\left[\frac{\varphi_0 K_1 B_1}{\mu_1}\left(h_1 + c - \frac{1}{\mu_1} - \left(c - \frac{1}{\mu_1}\right)\exp(-\mu_1 h_1)\right) + \right.\\
&+\frac{\varphi_{32} K_2 B_2}{\mu_2}\left(\left(c + h_2 + \frac{1}{\mu_2}\right)\exp(-\mu_2 h_2) - c - \frac{1}{\mu_2}\right) +\\
&\left.+\frac{\varphi_{13} K_3 B_3}{\mu_3}\left(c - \frac{1}{\mu_3} + \left(c + \frac{1}{\mu_3}\right)\exp(-2\mu_3 c)\right)\right].
\end{aligned} \tag{10.8}$$

The solution (10.7) with integration constants (10.8) takes the form:

$$\begin{aligned}
&\psi = 0,\\
&w = \frac{C_5}{4b_3}(r^2 - r_0^2),\\
&u = \frac{C_5 a_3}{2a_1 b_3} r,
\end{aligned} \tag{10.9}$$

where a_1, a_3, b_3 are defined in (8.18, 8.21).

Thus, when a neutron flux irradiates a three-layer circular plate with an articulated contour, the relative shear in the filler is zero, the resulting radial displacements vary linearly along the radius, and the deflection varies quadratically.

Numerical results here and hereafter are obtained for a plate composed of D16T-fluoroplastic-D16T materials. The elastic characteristics of the layer materials are the same as before: physical constants $B_1 = B_2 = B_3 = 10^{23}$ m^2/neutron, which provides volume deformation in the layers in the order of 3–3.5%; for duralumin $\mu_1 = 1.26$ m^{-1}; for fluoroplastic $\mu_3 = 3.21$ m^{-1}; neutron flux action time is $t_0 = 60$ min; plate radius $r_0 = 1$ m. Deflection and thicknesses of layers are referred to the radius of the plate.

Figure 10.2 shows the change along the radius of deflection w (*a*) and the distribution along the height of the cross section on the plate contour ($r = r_0$) of linear strains ε_r^k (*b*) during irradiation with neutron fluxes of different intensities (neutron/(m^{2}s)): 1—$\varphi_1 = 10^{17}$, 2—$\varphi_2 = 10^{18}$. Layer thicknesses: $h_1 = 0.02$; $h_2 = 0.06$; $h_3 = 0.4$. Increasing the flow intensity by an order of magnitude causes approximately the same increase in deflection and radial strain.

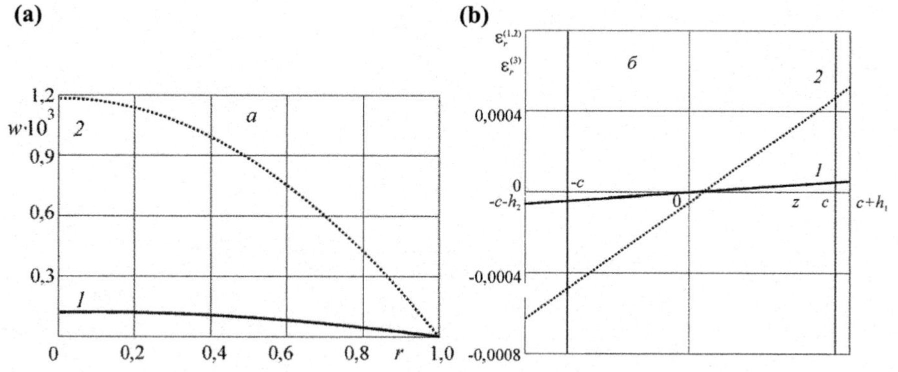

Fig. 10.2 Deflection w (a) and radial strain ε_r^k (b) during neutron irradiation

It should be noted that, when exposed to neutron irradiation, the volume strain in each layer increases, and the upper layer is irradiated with a flux of greater intensity, which causes its swelling and deflection, in some cases directed upward. The radial strain in the top layer and the adjoining part of the filler is positive, indicating their tensile strength. Neutron irradiation does not cause a relative shift in the filler.

The following numerical results for the deflections are obtained by irradiation with a neutron flux with an intensity of $\varphi_1 = 10^{18}$ neutrons/(m^{2}s), unless otherwise specified.

Figure 10.3 illustrates the dependence of plate deflection w on its asymmetry in thickness: 1—$h_1 = 0.01$, $h_2 = 0.07$; 2—$h_1 = 0.02$, $h_2 = 0.06$; 3—$h_1 = 0.04$, $h_2 = 0.04$; 4—$h_1 = 0.06$, $h_2 = 0.02$ at a constant filler thickness $h_3 = 0.4$. In this case, with a relatively thin upper carrier layer, the deflection is negative, and is directed downward, in the direction of the neutron flux fall. As this layer grows and the lower load-bearing layer weakens, the deflection changes its direction and increases in magnitude.

Fig. 10.3 Dependence of deflection on asymmetry of the three-layer plate by thickness

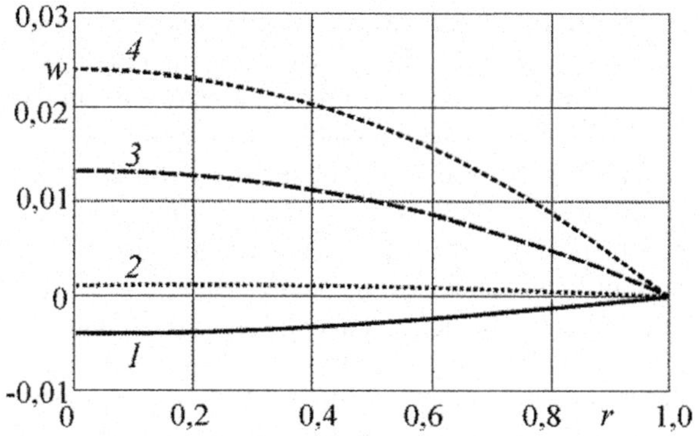

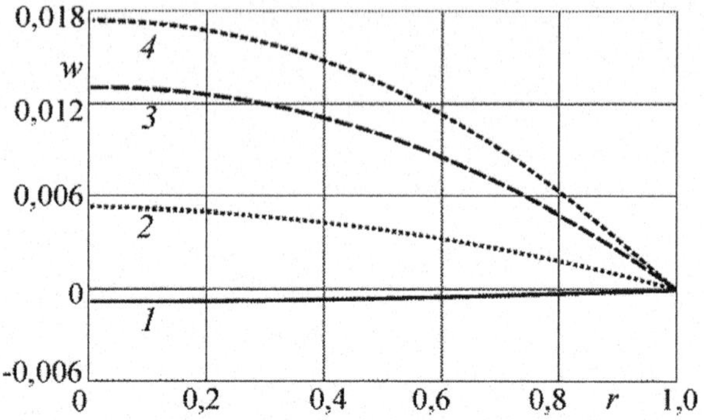

Fig. 10.4 Plate deflection at different thicknesses of the top layer

The change in deflection w along the radius of the plate at different thicknesses of the top layer h_1 is shown in Fig. 10.4: 1—$h_1 = 0.01$; 2—$h_1 = 0.02$; 3—$h_1 = 0.04$; 4—$h_1 = 0.06$ at a constant filler thickness $h_3 = 0.4$ and the bottom layer $h_2 = 0.04$.

As the thickness of the upper layer increases, the deflection increases, due to the growth of volumetric strain as its material swells in a larger neutron flux in terms of intensity.

As the thickness of the first layer increases, the deflection changes sign, so there is a critical value $h_1 = h_1^*$ at which it turns to zero. To determine this value in the solution (10.9) it is necessary to put: $w(h_1^*) = 0$. This is possible, if $C_5 = 0$, which gives the equation:

$$\begin{aligned}
&-\frac{\varphi_0 K_1 B_1}{\mu_1}\left(h_1 + c - \frac{1}{\mu_1} - \left(c - \frac{1}{\mu_1}\right)\exp(-\mu_1 h_1)\right)+ \\
&+\frac{\varphi_{32} K_2 B_2}{\mu_2}\left(\left(c + h_2 + \frac{1}{\mu_2}\right)\exp(-\mu_2 h_2) - c - \frac{1}{\mu_2}\right)+ \\
&+\frac{\varphi_{13} K_3 B_3}{\mu_3}\left(c - \frac{1}{\mu_3} + \left(c + \frac{1}{\mu_3}\right)\exp(-2\mu_3 c)\right) = 0
\end{aligned} \tag{10.10}$$

Equation (10.10) was examined numerically. It was obtained that for the chosen plate parameters and the value of neutron flux, the critical thickness is $h_1^* \approx 0.012$. If $h_1 < h_1^*$, the deflection is directed downwards, in the case of $h_1 = h_1^*$, the deflection is zero, with $h_1 > h_1^*$ the plate convex upwards.

The dependence of the deflection w on the thickness of the bottom layer h_2 is shown in Fig. 10.5: 1—$h_2 = 0.01$; 2—$h_2 = 0.04$; 3—$h_2 = 0.08$; 4—$h_2 = 0.16$ at constant thicknesses $h_1 = 0.04$ m and $h_3 = 0.4$ m. Here, as the thickness of the bottom layer increases, the plate deflection decreases consistently. If the layer is thick enough, the deflection changes sign. A numerical study of Eq. (10.10) showed that for the chosen plate parameters and neutron flux value, the critical thickness of the bottom layer is $h_2^* = 0.134$. If $h_2 < h_2^*$, the plate is convex upwards, in case $h_2 = h_2^*$, the deflection is approximately zero, in case $h_2 > h_2^*$, the deflection is directed downwards.

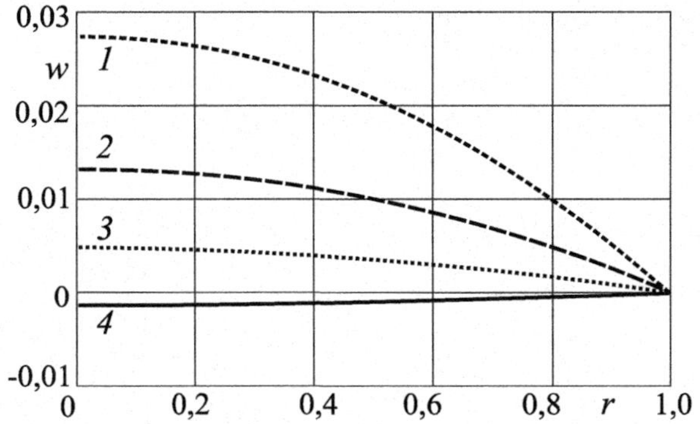

Fig. 10.5 Plate deflection at different thicknesses of the bottom layer

The plate deflection at different thicknesses of the filler plate at constant thicknesses of the bearing layers $h_1 = h_2 = 0.04$ is shown in Fig. 10.6: 1—$h_3 = 0.2$; 2—$h_3 = 0.03$; 3—$h_3 = 0.4$; 4—$h_3 = 0.5$. In this case, an increase in the thickness of the filler causes a decrease in the deflection of the plate.

Variation of radial strains ε_r^k and stresses σ_r^k (Pa) along the thickness of a three-layer circular plate on its contour at neutron flux intensity $\varphi_1 = 10^{17}$ (neutron/(m^{2}s)) and different thicknesses of the upper layer h_1 are shown in Figs. 10.7, 10.8, respectively ($h_3 = 0.4$, $h_2 = 0.04$): 1—$h_1 = 0.01$; 2—$h_1 = 0.02$.

In the example under consideration the strain also changes the sign, the choice of which is significantly influenced by the thickness of the upper load-bearing layer. Stresses in the layers under these conditions do not change the sign. In Fig. 10.7, the stress in the filler is increased by a factor of 10, for better visual perception.

Thus, due to neutron irradiation of a circular three-layer plate, volumetric strain occurs in its layers, caused by swelling of materials, due to which deflection and radial displacements arise in the plate. The direction of deflection depends on the properties of the materials and the thicknesses of the load-bearing layers.

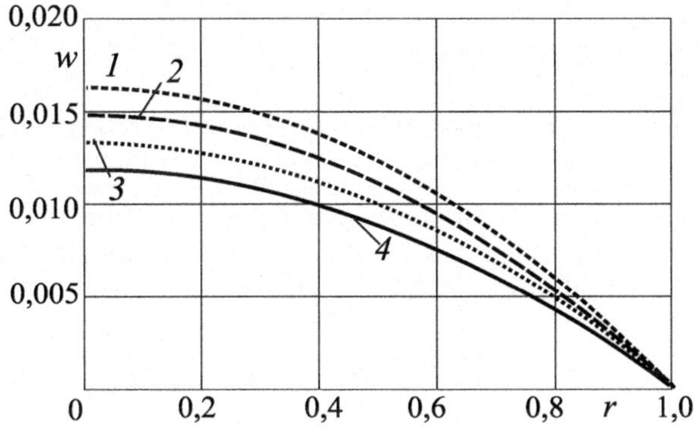

Fig. 10.6 Plate deflection at different filler thicknesses

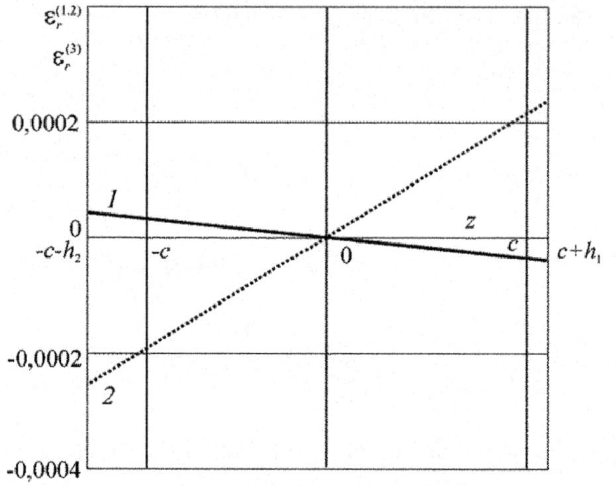

Fig. 10.7 Radial strains on the plate contour

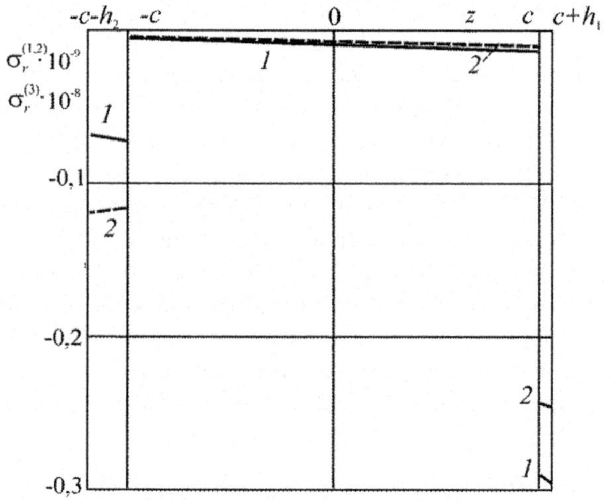

Fig. 10.8 Radial stresses on the plate contour

10.1.3 Bending by Uniformly Distributed Load

Consider the transverse bending of a circular three-layer plate under an evenly distributed load: $p(r) = 0$, $q(r) = q_0 = \text{const}$ in a neutron flux. In this case the partial solution ψ_r in (10.4) is

$$\psi_r = -\frac{b_2 q_0}{4cb_3 G_3} r.$$

The displacements follow from (10.4) and, taking into account the boundedness at the origin, take the form

$$\psi = C_2 I_1(\beta r) - \frac{b_2 q_0}{4cb_3G_3}r,$$
$$w = \frac{b_2}{b_3}\left(\frac{C_2}{\beta}I_0(\beta r) - \frac{b_2 q_0}{8cb_3G_3}r^2\right) + \frac{q_0}{64b_3}r^4 + \frac{C_5 r^2}{4b_3} + C_4^{(n)}, \quad (10.11)$$
$$u = \frac{a_3}{a_1a_6 - a_3^2}\left[\frac{q_0}{16}r_0^3 + \left(a_5 - \frac{a_2a_6}{a_3}\right)\psi + C_7 r\right],$$

where C_2, C_4, C_5, C_7 are non-zero integration constants.

By substituting the solution (10.11) into the boundary conditions for hinged support of the plate contour (10.5), and considering the expression for the generalized internal moment M_r (10.2), we obtain the following integration constants:

$$C_2 = \frac{b_2 q_0 r_0}{4cb_3G_3I_1(\beta r_0)},$$
$$C_4^{(n)} = \frac{b_2}{b_3}\left(\frac{C_2}{\beta}I_0(\beta r_0) - \frac{b_2 q_0}{8cb_3G_3}r_0^2\right) - \frac{q_0}{64b_3}r_0^4 - \frac{C_5 r_0^2}{4b_3},$$
$$C_5 = -\frac{2b_3}{a_6 + a_7}\left\{\left(3 + \frac{a_3^2}{b_3a_1}\right)\frac{q_0}{16}r_0^2 + \right.$$
$$+t\left[\frac{\varphi_0K_1B_1}{\mu_1}\left(h_1 + c - \frac{1}{\mu_1} - \left(c - \frac{1}{\mu_1}\right)\exp(-\mu_1h_1)\right) + \right. \quad (10.12)$$
$$+\frac{\varphi_{32}K_2B_2}{\mu_2}\left(\left(c + h_2 + \frac{1}{\mu_2}\right)\exp(-\mu_2h_2) - c - \frac{1}{\mu_2}\right) +$$
$$\left.\left.+\frac{\varphi_{13}K_3B_3}{\mu_3}\left(c - \frac{1}{\mu_3} + \left(c + \frac{1}{\mu_3}\right)\exp(-2\mu_3c)\right)\right]\right\},$$
$$C_7 = -\frac{a_3q_0}{8a_1b_3}r_0^2 - \frac{a_3}{a_1b_{3\,5}}C_5.$$

Thus, the solution (10.11) with integration constants (10.12) describes the displacements in an articulated three-layer circular plate when it is bent by an evenly distributed load in a neutron flux of intensity φ_0.

Numerical results were obtained for the previously considered three-layer plate at neutron flux rate $\varphi = 10^{18}$ (neutron/(m^{2}s)) and load value $q_0 = -1$ MПa. The deflection and thicknesses of the layers, as before, are referred to the radius of the plate.

The change in deflection w along the radius of the plate at different thicknesses of the upper load-bearing layer is shown in the Fig. 10.9: 1—$h_1 = 0.01$; 2—$h_1 = 0.02$. Thickness of the filler $h_3 = 0.4$; thickness of the bottom layer $h_2 = 0.04$.

For a plate with a thickness of $h_1 < h_1^* \approx 0.012$ m, the deflection from neutron irradiation is directed downward in the direction of the flux action and increases in modulus with the addition of a load directed in the same direction. In the case of $h_1 > h_1^*$, the deflection from the neutron irradiation is directed upwards and decreases in modulus under the action of a uniformly distributed load.

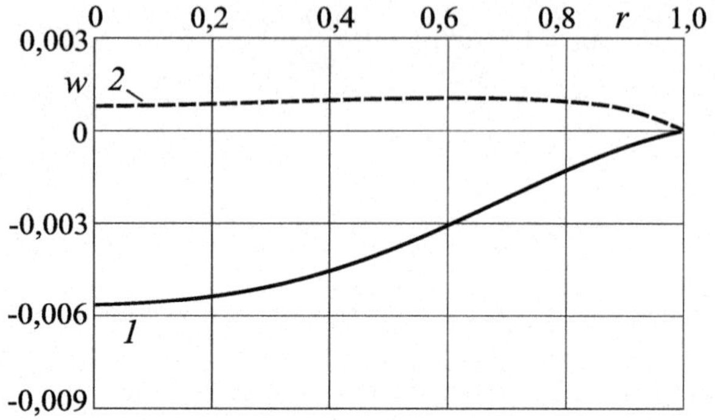

Fig. 10.9 Change of deflection w along the plate radius

10.1.4 Bending by a Load Distributed Over a Circle

Let the three-layer plate be subjected to a local surface load $q(r)$ uniformly distributed inside a circle of relative radius $b \leq r_0$ (Fig. 10.1). Then it can be written in the following analytical form (Fig. 10.10):

$$q(r) = q_0 H_0(b - r) \ (q_0 = \text{const}), \tag{10.13}$$

where $H_0(r)$ is Heaviside function (8.31).

To transform the general solution (10.4) for the load case (10.13), it is necessary to first write out some functions and calculate the integrals containing the Heaviside function.

The integral operator L_3^{-1}, introduced in (8.26), taken from the load (10.13), and the integral of it will be:

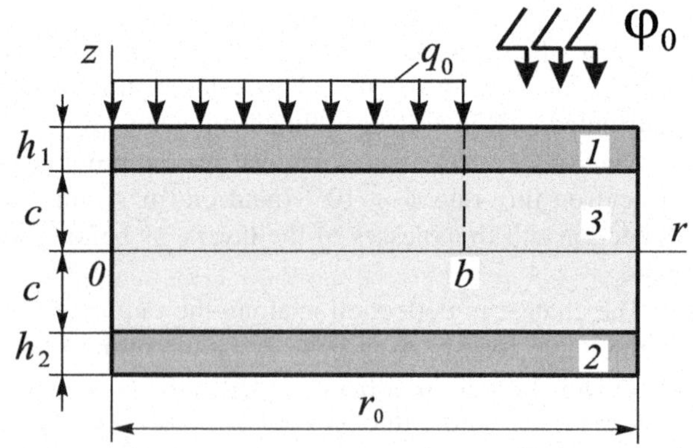

Fig. 10.10 Scheme of plate loading inside the circle

$$\mathrm{L}_3^{-1}(q(r)) = q_0\left(\frac{r^3}{16} - \frac{b^4}{16r} - \frac{b^2 r}{4}\ln\left(\frac{r}{b}\right)\right)H_0(b-r),$$
$$\int \mathrm{L}_3^{-1}(q(r))\mathrm{d}r = q_0\left(\frac{r^4 - 5b^4}{64} + \frac{b^2r^2}{16} - \left(\frac{b^4}{16} + \frac{b^2r^2}{8}\right)\ln\left(\frac{r}{b}\right)\right)H_0(b-r). \tag{10.14}$$

The function $f(r)$, included in the partial solution (8.25), at the load (10.13) takes the form:

$$f(r) = \frac{\gamma_1}{r}\left(\frac{q_0\left(r^2 - b^2\right)}{2}H_0(b-r) - C_1\right),\ \gamma_1 = \frac{b_2}{\left(b_1b_3 - b_2^2\right)}.$$

The integrals from its product on the Bessel functions included in the partial solution $\psi_r(r)$ in (10.4) are:

$$\int I_1(\beta r)f(r)r\mathrm{d}r = \frac{\gamma_1 q_0}{2\beta}H_0(b-r)\left[\left(r^2 - b^2\right)I_0(\beta r) + \right.$$
$$\left. + \frac{2}{\beta}(bI_1(\beta b) - rI_1(\beta r))\right] - \frac{C_1\gamma_1}{\beta}I_0(\beta r),$$
$$\int K_1(\beta r)f(r)r\mathrm{d}r = \frac{\gamma_1 q_0}{2\beta}H_0(b-r)\times \tag{10.15}$$
$$\times\left[\left(b^2 - r^2\right)K_0(\beta r) + \frac{2}{\beta}(bK_1(\beta b) - rK_1(\beta r))\right] + \frac{C_1\gamma_1}{\beta}K_0(\beta r).$$

The partial solution $\psi_r(r)$ and the integral of it, taking into account the results given in (10.15), will be:

$$\psi_r(r) = \frac{\gamma_1 q_0}{2\beta^2}H_0(b-r)\left[\frac{b^2}{r} - r + 2b(K_1(\beta b)I_1(\beta r) - I_1(\beta b)K_1(\beta r))\right] + \frac{C_1\gamma_1}{\beta^2 r},$$
$$\int \psi\mathrm{d}r = \frac{C_2I_0(\beta r)}{\beta} - \frac{C_3K_0(\beta r)}{\beta} + \frac{C_1\gamma_1}{\beta^2}\ln r + \frac{\gamma_1 q_0}{2\beta^2}H_0(b-r)\times$$
$$\times\left[\frac{b^2 - r^2}{2} + b^2\ln\left(\frac{r}{b}\right) + \frac{2b}{\beta}(K_1(\beta b)I_0(\beta r) + I_1(\beta b)K_0(\beta r)) - \frac{2}{\beta^2}\right]. \tag{10.16}$$

As a result, from the general solution (10.4), using expressions (10.14–10.16), formulas for displacements in an elastic circular three-layer plate deformed by a local circular load (10.13) in a neutron flux follow:

$$\psi = C_2 I_1(\beta r) + C_3 K_1(\beta r) +$$
$$+\frac{\gamma_1 q_0}{2\beta^2} H_0(b-r)\left[\frac{b^2}{r} - r + 2b(K_1(\beta b) I_1(\beta r) - I_1(\beta b) K_1(\beta r))\right] + \frac{C_1 \gamma_1}{\beta^2 r},$$
$$w = \frac{b_2}{b_3}\left[\frac{C_2 I_0(\beta r)}{\beta} - \frac{C_3 K_0(\beta r)}{\beta} + \frac{C_1 \gamma_1}{\beta^2}\ln r + \frac{\gamma_1 q_0}{2\beta^2} H_0(b-r)\times\right.$$
$$\left.\times\left[\frac{b^2 - r^2}{2} + b^2 \ln\left(\frac{r}{b}\right) + \frac{2b}{\beta}(K_1(\beta b) I_0(\beta r) + I_1(\beta b) K_0(\beta r)) - \frac{2}{\beta^2}\right]\right] +$$
$$+\frac{q_0}{b_3}\left(\frac{r^4 - 5b^4}{64} + \frac{b^2 r^2}{16} - \left(\frac{b^4}{16} + \frac{b^2 r^2}{8}\right)\ln\left(\frac{r}{b}\right)\right) H_0(b-r) -$$
$$-\frac{1}{4b_3} C_1 r^2 (\ln r - 1) + \frac{C_5 r^2}{4b_3} + C_6 \ln r + C_4,$$
$$u = \frac{a_3}{a_1} w_{,r} - \frac{a_2}{a_1}\psi + \frac{C_7 r}{2} + \frac{C_8}{r}. \tag{10.17}$$

The integration constants follow from the boundary conditions on the contour and the boundedness requirements of the solution (10.17) in the center of the plate. From the boundedness of the solution we have:

$$C_1 = -\frac{q_0 b^2}{2}, \quad C_3 = \frac{q_0 \gamma_1 b I_1(\beta b)}{\beta^2}, \quad C_6 = \frac{q_0 b^4}{16 b_3}, \quad C_8 = 0. \tag{10.18}$$

In the case of a hinged support of the contour from the conditions (10.5) we get:

$$C_2 = \frac{\gamma_1 q_0}{\beta^2 I_1(\beta r_0)}\left(\frac{b^2}{2} - b K_1(\beta r_0) I_1(\beta b)\right),$$
$$C_4 = -\frac{C_5}{4b_3} - \frac{b_2}{b_3}\int \psi \, dr\Big|_{r=r_0}, \quad C_5 = \left(\frac{C_1}{2} + 2b_3 C_6\right)\frac{a_1 b_3 - a_7 a_1 - a_3^2}{a_1(a_6 + a_7)} -$$
$$-\frac{2b_3 t}{a_6 + a_7}\left[\frac{\varphi_0 K_1 B_1}{\mu_1}\left(h_1 + c - \frac{1}{\mu_1} - \left(c - \frac{1}{\mu_1}\right)\exp(-\mu_1 h_1)\right) +\right.$$
$$+\frac{\varphi_{32} K_2 B_2}{\mu_2}\left(c\left(c + h_2 + \frac{1}{\mu_2}\right)\exp(-\mu_2 h_2) - c - \frac{1}{\mu_2}\right) +$$
$$\left.+\frac{\varphi_{13} K_3 B_3}{\mu_3}\left(c - \frac{1}{\mu_3} + \left(c + \frac{1}{\mu_3}\right)\exp(-2\mu_3 c)\right)\right], \quad C_7 = -\frac{2a_3}{a_1}\left(\frac{C_6}{r_0^2} + \frac{C_5}{2b_3}\right) \tag{10.19}$$

Thus, the required displacements in the plate under local circular load (10.13) are defined by formula (10.17) with integration constants (10.18, 10.19).

Numerical results for the analyzed plate of unit radius were obtained under the above conditions in the case of intensity of distributed circular load $q_0 = -1$ MPa and layer thicknesses $h_2 = 0.04$ and $h_3 = 0.4$. Neutron flux time is 60 min.

Figure 10.11 illustrates the dependence of the maximum deflection w in the center of the plate on the radius b of the circular load spot at different values of the neutron flux intensity (neutron/(m^{2}s)) and the thickness of the upper layer $h_1 = 0.04$: 1—$\varphi_1 = 0$, 2—$\varphi_2 = 10^{17}$, 3—$\varphi_3 = 10^{18}$. As the radius of the force spot increases, the deflection grows nonlinearly, reaching a maximum when the load is distributed over the entire outer surface of the upper layer. Neutron irradiation reduces the maximum deflection by modulus, first by 6.5%, and with increasing irradiation intensity—up to 34%.

Similar graphs of changes in the maximum deflection w in the center of the plate at other thicknesses of the upper load-bearing layer are shown in Fig. 10.12 ($h_1 =$ 0.01 m) and Fig. 10.13 ($h_1 = 0.02$ m). The numbering of the curves remains the same. If the thickness of the upper layer is less than the critical thickness $h_1 < h_1^* \approx 0.012$ (see Fig. 10.12), then an increase in the intensity of the neutron flux leads to an increase in the deflection modulus in the direction of the local load. In the case $h_1 > h_1^*$ (see Fig. 10.13), the deflection curve shifts upwards with increasing neutron flux intensity, i.e. the deflection from neutron irradiation prevails over the deflection from the force load.

The change in deflection w along the radius of the plate at neutron flux intensity $\varphi = 10^{18}$ (neutron/(m^{2}s)) and different radius b of the load spot with intensity $q_0 =$

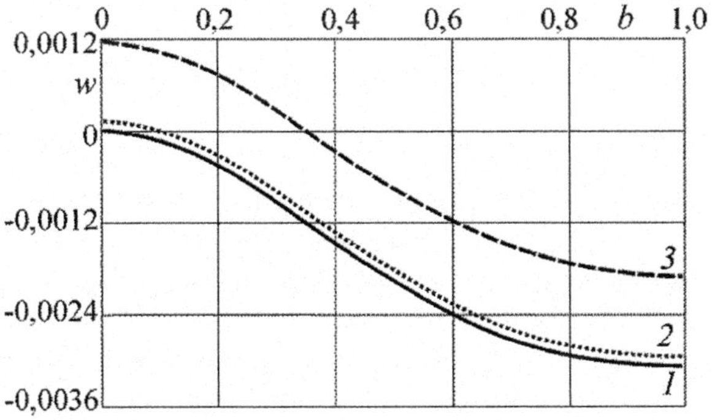

Fig. 10.11 Change of maximum deflection depending on the radius b of the load spot

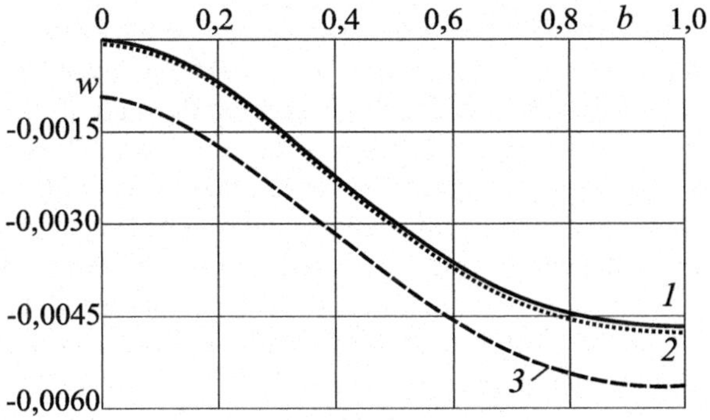

Fig. 10.12 Maximum deflection as a function of the radius b of the load spot ($h_1 = 0.01$)

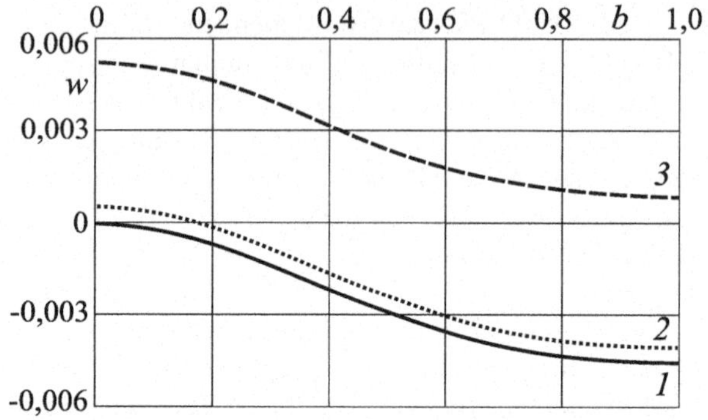

Fig. 10.13 Maximum deflection as a function of the radius b of the load spot ($h_1 = 0.02$)

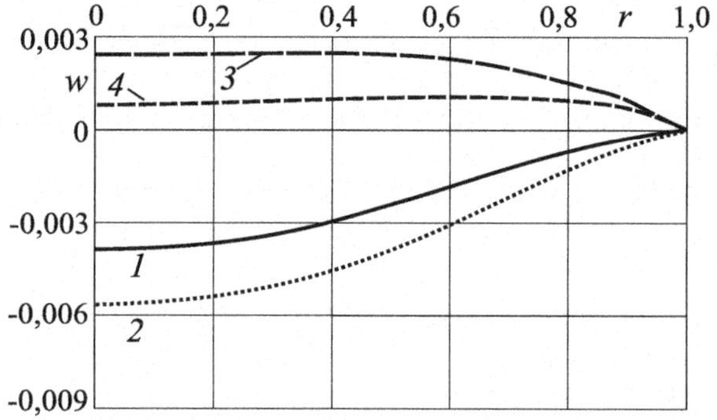

Fig. 10.14 Deflection w at different load spot radius b

−1 MPa is shown in Fig. 10.14: 1—$b = 0.5$, 2—$b = 1$ (in both cases $h_1 = 0{,}01$); 3—$b = 0.5$, 4—$b = 1$ (in both cases $h_1 = 0.02$). Thickness of the filler $h_3 = 0.4$, the bottom layer—$h_2 = 0.04$. At thickness $h_1 < h_1^* \approx 0.012$ the deflection is directed downward in the direction of the load and increases in module as the radius of its spot increases. In the case $h_1 > h_1^*$ the deflection is directed upwards and decreases in module as the radius of the load spot increases.

The variations of radial strains and stresses (MPa) along the plate thickness on the contour are shown in the Figs. 10.15, 10.16: 1—$b = 0.5$, 2—$b = 1$ (in both cases $h_1 = 0{,}01$); 3—$b = 0.5$, 4—$b = 1$ (in both cases $h_1 = 0.02$). Thickness of the filler $h_3 = 0.4$, the bottom layer—$h_2 = 0.04$, top layer—$h_1 = 0.02$.

In this case, the main influence on the sign of the strain has the thickness of the upper bearing layer. If it is less than the critical thickness $h_1 < h_1^*$ (curves 1, 2), then the strains in the upper layer are negative, and in the lower layer—positive, partially changing the sign with increasing radius of the circular load spot. In the case $h_1 > h_1^*$ the strains in the upper layer are positive, in the lower layer the strains are negative. The stresses under these conditions do not change the sign in the layers, remaining negative, which is caused by swelling of the materials of the layers and tightness in the hinged contour.

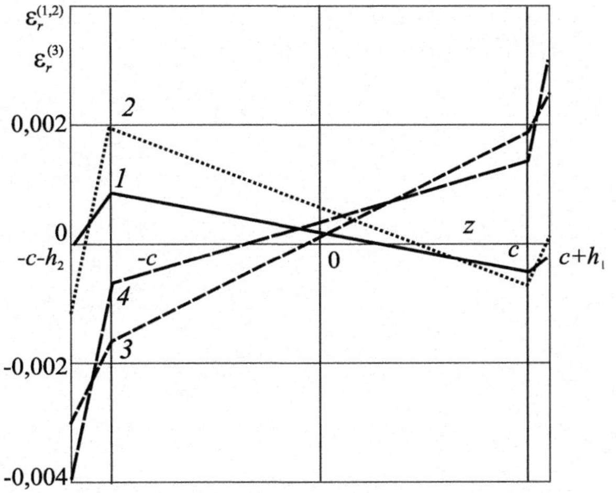

Fig. 10.15 Radial strains on the plate contour under the action of a load in a neutron flux

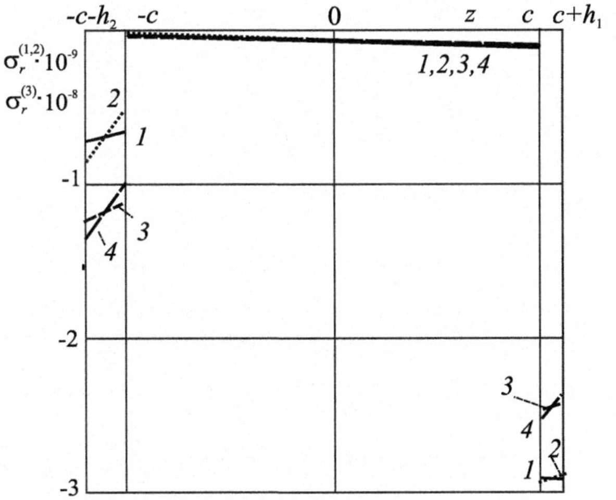

Fig. 10.16 Radial stresses on the plate contour under the action of a load in neutron flux

10.1.5 Bending of the Plate by a Ring Load

Consider the effect on the plate of a local load uniformly distributed over a ring of radius $a \leq r \leq b$ (Fig. 10.17).

$$q(r) = q_0(H_0(b - r) - H_0(a - r)). \tag{10.20}$$

To transform the general solution (10.4) in the case of a load of the type (10.20), let us first write out the necessary operators and integrals containing the Heaviside functions.

The operator L_3^{-1} from the load (10.20) and the integral of it are:

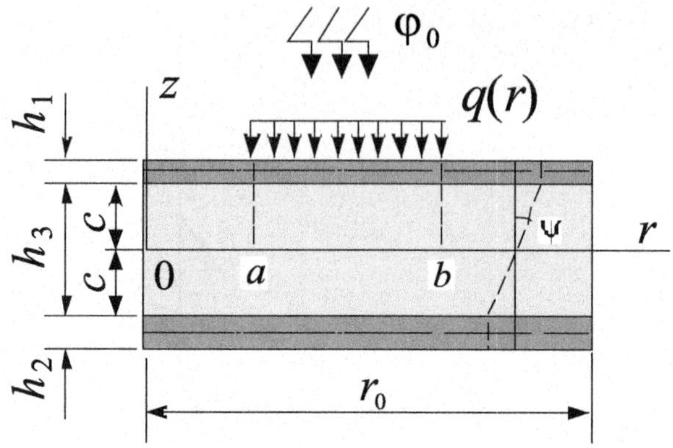

Fig. 10.17 Calculated scheme at ring load

$$
\begin{aligned}
&\mathrm{L}_3^{-1}(q(r)) = q_0\left[\left(\frac{r^3}{16} - \frac{b^4}{16r} - \frac{b^2 r}{4}\ln\left(\frac{r}{b}\right)\right)H_0(b-r) - \right.\\
&\left.-\left(\frac{r^3}{16} - \frac{a^4}{16r} - \frac{a^2 r}{4}\ln\left(\frac{r}{a}\right)\right)H_0(a-r)\right],\\
&\int \mathrm{L}_3^{-1}(q)\mathrm{d}r = q_0\left\{\left[\frac{r^4-5b^4}{64} - \frac{b^4}{16}\ln\left(\frac{r}{b}\right) - \frac{b^2r^2}{8}\ln\left(\frac{r}{b}\right) + \frac{b^2r^2}{16}\right]H(b-r) - \right.\\
&\left.-\left[\frac{r^4-5a^4}{64} - \frac{a^4}{16}\ln\left(\frac{r}{a}\right) - \frac{a^2r^2}{8}\ln\left(\frac{r}{a}\right) + \frac{a^2r^2}{16}\right]H(a-r)\right\}.
\end{aligned}
\tag{10.21}
$$

The desired solution is obtained as the difference of two solutions (10.17) corresponding to the load components (10.20). The relative shear in the filler is:

$$
\begin{aligned}
\psi &= C_2 I_1(\beta r) + C_3 K_1(\beta r) + \\
&+\frac{\gamma_1 q_0}{2\beta^2}H_0(b-r)\left[\frac{b^2}{r} - r + 2b(K_1(\beta b)I_1(\beta r) - I_1(\beta b)K_1(\beta r))\right] - \\
&-\frac{\gamma_1 q_0}{2\beta^2}H_0(a-r)\left[\frac{a^2}{r} - r + 2a(K_1(\beta a)I_1(\beta r) - I_1(\beta a)K_1(\beta r))\right] + \frac{C_1\gamma_1}{\beta^2 r}.
\end{aligned}
\tag{10.22}
$$

The deflection $w(r)$ and radial displacement $u(r)$ in a circular three-layer plate formally retain the form (10.4), but the integral of the relative shear included in them is as follows:

$$\int \psi \mathrm{d}r = \frac{C_2 I_0(\beta r)}{\beta} - \frac{C_3 K_0(\beta r)}{\beta} + \frac{C_1\gamma_1}{\beta^2}\ln(r) + \frac{\gamma_1 q_0}{2\beta^2}H_0(b-r)\times$$
$$\times\left[\frac{b^2-r^2}{2} + b^2\ln\left(\frac{r}{b}\right) + \frac{2b}{\beta}(K_1(\beta b)I_0(\beta r) + I_1(\beta b)K_0(\beta r)) - \frac{2}{\beta^2}\right] - \frac{\gamma_1 q_0}{2\beta^2}\times$$
$$\times H_0(a-r)\left[\frac{a^2-r^2}{2} + a^2\ln\left(\frac{r}{a}\right) + \frac{2a}{\beta}(K_1(\beta a)I_0(\beta r) + I_1(\beta a)K_0(\beta r)) - \frac{2}{\beta^2}\right]. \tag{10.23}$$

As a result, the formulas for deflection and radial displacement in a circular three-layer plate under ring loading still follow from (10.4), in which expressions (10.21), relative displacement (10.22), and the integral (10.23) must be considered:

$$w = \frac{b_2}{b_3}\left[\frac{C_2 I_0(\beta r)}{\beta} - \frac{C_3 K_0(\beta r)}{\beta} + \frac{C_1\gamma_1}{\beta^2}\ln r + \frac{\gamma_1 q_0}{2\beta^2}H_0(b-r)\times\right.$$
$$\left.\times\left[\frac{b^2-r^2}{2} + b^2\ln\left(\frac{r}{b}\right) + \frac{2b}{\beta}(K_1(\beta b)I_0(\beta r) + I_1(\beta b)K_0(\beta r)) - \frac{2}{\beta^2}\right]\right]-$$
$$-\frac{\gamma_1 q_0}{2\beta^2}H_0(a-r)\left[\frac{a^2}{r} - r + 2a(K_1(\beta a)I_1(\beta r) - I_1(\beta a)K_1(\beta r))\right] + \frac{C_1\gamma_1}{\beta^2 r}+$$
$$+\frac{q_0}{b_3}\left(\frac{r^4-5b^4}{64} + \frac{b^2r^2}{16} - \left(\frac{b^4}{16} + \frac{b^2r^2}{8}\right)\ln\left(\frac{r}{b}\right)\right)H_0(b-r)-$$
$$-\frac{q_0}{b_3}\left(\frac{r^4-5a^4}{64} + \frac{a^2r^2}{16} - \left(\frac{a^4}{16} + \frac{a^2r^2}{8}\right)\ln\left(\frac{r}{a}\right)\right)H_0(a-r)-$$
$$-\frac{1}{4b_3}C_1 r^2(\ln r - 1) + \frac{C_5 r^2}{4b_3} + C_6\ln r + C_4,$$
$$u = \frac{a_3}{a_1}w_{,r} - \frac{a_2}{a_1}\psi + \frac{C_7 r}{2} + \frac{C_8}{r}. \tag{10.24}$$

Integration constants for the hinged support of the contour of the plate are defined by formulas corresponding to (10.18, 10.19):

$$C_1 = -\frac{q_0\left(b^2 - a^2\right)}{2},\ C_3 = \frac{q_0\gamma_1(bI_1(\beta b) - aI_1(\beta a))}{\beta^2},\ C_6 = \frac{q_0\left(b^4 - a^4\right)}{16b_3},\ C_8 = 0,$$

$$C_2 = \frac{\gamma_1 q_0}{\beta^2 I_1(\beta r_0)}\left[\left(\frac{b^2}{2} - bK_1(\beta r_0)I_1(\beta b)\right) - \left(\frac{a^2}{2} - aK_1(\beta r_0)I_1(\beta a)\right)\right],$$

$$C_4 = -\frac{C_5}{4b_3} - \frac{b_2}{b_3}\int \psi \mathrm{d}r\Big|_{r=r_0},\ C_5 = \left(\frac{C_1}{2} + 2b_3C_6\right)\frac{a_1b_3 - a_7a_1 - a_3^2}{a_1(a_6 + a_7)} -$$
$$-\frac{2b_3t}{a_6 + a_7}\left[\frac{\varphi_0 K_1 B_1}{\mu_1}\left(h_1 + c - \frac{1}{\mu_1} - \left(c - \frac{1}{\mu_1}\right)\exp(-\mu_1 h_1)\right) + \right.$$
$$+\frac{\varphi_{32}K_2B_2}{\mu_2}\left(c\left(c + h_2 + \frac{1}{\mu_2}\right)\exp(-\mu_2 h_2) - c - \frac{1}{\mu_2}\right) +$$
$$\left. +\frac{\varphi_{13}K_3B_3}{\mu_3}\left(c - \frac{1}{\mu_3} + \left(c + \frac{1}{\mu_3}\right)\exp(-2\mu_3 c)\right)\right],\ C_7 = -\frac{2a_3}{a_1}\left(\frac{C_6}{r_0^2} + \frac{C_5}{2b_3}\right). \tag{10.25}$$

Here the integral of the relative shift is given by expression (10.23).

Thus, the required displacements in the three-layer hinged plate under the local ring load (10.20) are defined by formulas (10.22, 10.24) with integration constants (10.25).

Numerical results were obtained with ring load intensity $q_0 = -2$ MPa and layer thicknesses $h_2 = 0.04$, $h_3 = 0.4$.

The graphs of the dependence of the maximum deflection w in the center of the three-layer plate on the position of the load ring of constant width $(b - a) = 0.25$ at different values of the intensity of neutron flux (neutron/(m^{2}s)) (1—$\varphi_1 = 0$; 2—$\varphi_2 = 10^{17}$; 3—$\varphi_3 = 10^{18}$) are shown in Fig. 10.18 ($h_1 = 0.01$ m) and Fig. 10.19 ($h_1 = 0.02$ m).

Here, as in the case of a circular local load, if the thickness of the upper bearing layer is less than the critical one $h_1 < h_1^* \approx 0.012$, then an increase in the intensity of the neutron flux leads to a shift of the deflection towards the local load. The maximum absolute value of deflection is reached at a circular load with inner radius a = 0.25.

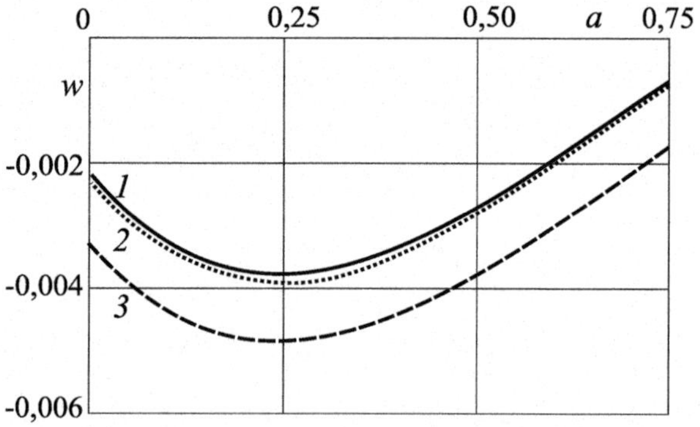

Fig. 10.18 Maximum deflection as a function of the coordinate of the inner radius of the load ring ($h_1 = 0.01$)

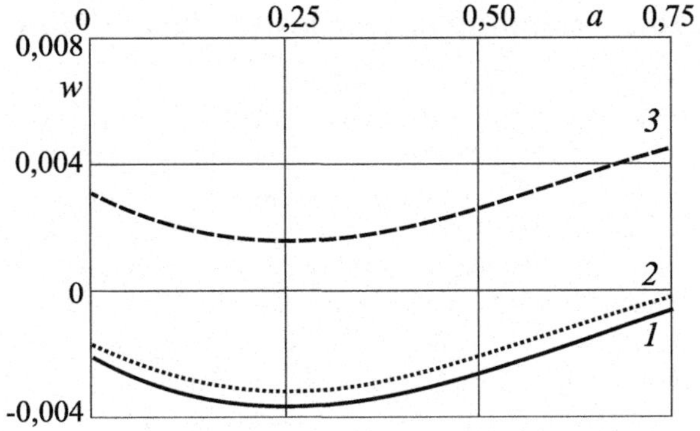

Fig. 10.19 Maximum deflection as a function of the coordinate of the inner radius of the load ring ($h_1 = 0.02$)

At the thickness of the upper layer $h_1 > h_1^*$, the deflection graph shifts upwards with increasing neutron flux intensity. The deflection reaches its maximum absolute value at a ring load with inner radius $a = 0.25$. In this case, at neutron flux intensity φ_3, the deflection from the neutron irradiation exceeds the deflection from the ring load and, as a result, is directed towards it.

The change in deflection w along the radius of the plate at the value of neutron flux intensity $\varphi = 10^{18}$ (neutron/(m^{2}s)) and different positions of the load ring (values of the inner radius a) is shown in Fig. 10.20: 1—$a = 0.25$, 2—$a = 0.75$ (in both cases $h_1 = 0.01$); 3—$a = 0.25$, 4—$a = 0.75$ (in both cases $h_1 = 0.02$).

If the thickness of the upper load-bearing layer is less than the critical $h_1 < h_1^* \approx 0.012$, then the deflection is directed downward toward the load and decreases in modulus as the radius of the power ring increases. In the case $h_1 > h_1^*$ the deflection is directed upwards and increases in modulus as the radius of the load ring increases.

Remark. Analytic solutions for the deformation in the neutron flux of a circular three-layer plate hinged to the contour under a linear force load can be obtained by a limit transition by setting the load ring thickness in the solutions (10.22–10.25) to zero and leaving constant its resultant force, as was done in Sect. 8.4.

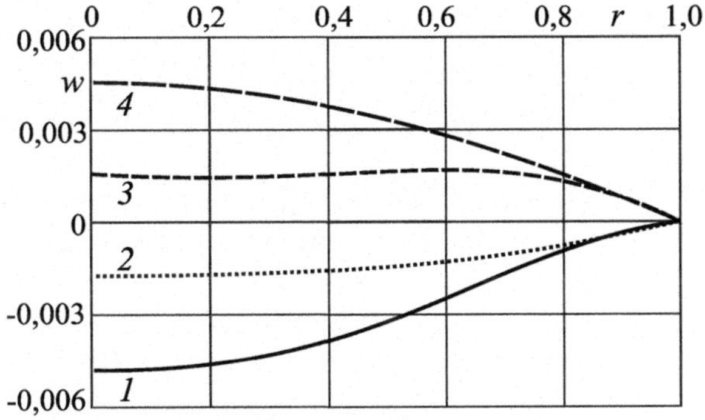

Fig. 10.20 Change of deflection w along the radius of the plate at different values of the inner radius of the load ring a

10.2 Bending of an Elastic–Plastic Plate in a Neutron Flux

It is assumed that the outer surface $z = c + h_1$ of a circular three-layer plate is subjected to an axisymmetric distributed force load $p'(r),\ q'(r)$ and a neutron flux of density φ_0 in the direction opposite to the external normal (see Fig. 10.1).

The materials of the plate layers can show elastic–plastic (load-bearing layers) and nonlinear-elastic (filler) properties in the deformation process. To describe the kinematics of the package, the previously accepted broken-line hypotheses are valid (see Sect. 8.1). Nonlinear physical equations of state of the type (2.33) are used to describe the deformation of the layer material:

$$\begin{aligned}
&s_{\alpha}^{\prime(k)} = 2G_k\left(1 - \omega^{\prime(k)}\left(\varepsilon_u^{(k)}, I\right)\right) Э_{\alpha}^{\prime(k)},\ s_{rz}^{\prime(3)} = 2G_3\left(1 - \omega^{\prime(3)}\left(\varepsilon_u^{(3)}, I\right)\right) Э_{rz}^{\prime(3)},\\
&\sigma^{\prime(k)} = K_k\left(3\varepsilon^{\prime(k)} - B_k I\right) \quad (k = 1, 2, 3;\ \alpha = r, \varphi),
\end{aligned} \tag{10.26}$$

where $\omega_k' \equiv \omega_k\left(\varepsilon_u^{\prime k}, I\right)$ is physical nonlinearity function of the material of the kth layer; $\varepsilon_u^{\prime(k)}$ is strain intensity; $I(z, t)$ is integral neutron flux, which by the moment t will pass through section z; B_k are material constants obtained experimentally; the slash above denotes loading from the natural state (at the first half-cycle).

We will further assume that the changes in time of external loads and boundary displacements occur in such a way that the corresponding loading trajectories do not belong to the class of complex loads, and the radiation growth of the plasticity limit does not exceed the growth of strain intensity in the irradiated points of the solid body, which would prevent the formation of plastic deformations.

The components of the stress tensor in the layers, using (10.26), will be represented through the deviator and the spherical part of the strain tensor:

$$\begin{aligned}
&\sigma_{\alpha}^{\prime(k)} = 2G_k Э_{\alpha}^{\prime(k)}\left(1 - \omega^{\prime(k)}\right) + K_k\left(3\varepsilon^{\prime(k)} - B_k I\right) \quad (k = 1, 2, 3),\\
&\sigma_{rz}^{\prime(3)} = s_{rz}^{\prime(3)} = 2G_3 Э_{rz}^{\prime(3)}\left(1 - \omega^{\prime(3)}\right) \quad (\alpha = r, \varphi).
\end{aligned}$$

Let us select the linear (index "e") and non-linear (index "ω") components in these stresses:

$$\begin{aligned}
&\sigma_{\alpha}^{\prime(k)} = \sigma_{\alpha e}^{\prime(k)} - \sigma_{\alpha\omega}^{\prime(k)},\quad \sigma_{rz}^{\prime(3)} = \sigma_{rze}^{\prime(3)} - \sigma_{rzw}^{\prime(3)},\\
&\sigma_{\alpha e}^{\prime(k)} = 2G_k Э_{\alpha}^{\prime(k)} + K_k\left(3\varepsilon^{\prime(k)} - B_k I\right),\quad \sigma_{\alpha\omega}^{\prime(k)} = 2G_k\omega_k Э_{\alpha}^{\prime(k)},\\
&\sigma_{rze}^{\prime(3)} = 2G_k Э_{rz}^{\prime(3)},\ \sigma_{rzw}^{\prime(3)} = 2G_3\omega_3 Э_{rz}^{\prime(3)} \quad (\alpha = r, \varphi).
\end{aligned} \tag{10.27}$$

The internal forces and moments in the layers of the plate will also be represented as the difference of linear and nonlinear parts:

$$\begin{aligned}
&T_{\alpha}^{\prime(k)} = T_{\alpha e}^{\prime(k)} - T_{\alpha\omega}^{\prime(k)},\ M_{\alpha}^{\prime(k)} = M_{\alpha e}^{\prime(k)} - M_{\alpha\omega}^{\prime(k)}\\
&H_{\alpha}^{\prime(k)} = H_{\alpha e}^{\prime(k)} - H_{\alpha\omega}^{\prime(k)},\ Q' = Q_e' - Q_\omega' \quad (\alpha = r, \phi).
\end{aligned}$$

The values $T_{\alpha e}^{\prime(k)}, T_{\alpha\omega}^{\prime(k)}, M_{\alpha e}^{\prime(k)}, M_{\alpha\omega}^{\prime(k)}, H_{\alpha e}^{\prime(k)}, H_{\alpha\omega}^{\prime(k)}, Q_e', Q_\omega'$ are calculated by formula (8.3), in which the stresses $\sigma_\alpha^{\prime(k)}$ must be replaced by linear and nonlinear components $\sigma_{\alpha e}^{\prime(k)}, \sigma_{\alpha\omega}^{\prime(k)}, \sigma_{rze}^{\prime(3)}, \sigma_{rz\omega}^{\prime(3)}$ from relations (10.27).

With generalized internal forces we do the same thing:

$$T_\alpha' = T_{\alpha e}' - T_{\alpha\omega}' = \sum_{k=1}^{3} T_{\alpha e}^{\prime(k)} - \sum_{k=1}^{3} T_{\alpha\omega}^{\prime(k)}, \quad M_\alpha' = M_{\alpha e}' - M_{\alpha\omega}' = \sum_{k=1}^{3} M_{\alpha e}^{\prime(k)} - \sum_{k=1}^{3} M_{\alpha\omega}^{\prime(k)},$$
$$H_{\alpha e}' = M_{\alpha e}^{\prime(3)} + c\left(T_{\alpha e}^{\prime(1)} - T_{\alpha e}^{\prime(2)}\right), \quad H_{\alpha\omega}' = M_{\alpha\omega}^{\prime(3)} + c\left(T_{\alpha\omega}^{\prime(1)} - T_{\alpha\omega}^{\prime(2)}\right). \tag{10.28}$$

The system of equilibrium equations of a circular three-layer plate in forces (8. 8) was obtained without using physical equations of state, so it remains valid in the case under consideration. Substituting into it the expressions for the internal forces (10.28), we obtain:

$$\begin{aligned} &T_r',_r + \frac{1}{r}(T_r' - T_\varphi') = -p' + p_\omega', \\ &H_r',_r + \frac{1}{r}(H_r' - H_\varphi') - Q' = h_\omega', \\ &M_r',_{rr} + \frac{1}{r}(2M_r',_r - M_\varphi',_r) = -q' + q_\omega'. \end{aligned} \tag{10.29}$$

Here, the linear components of the generalized internal forces are collected on the left side of the equations (the index "e" is omitted for convenience). The nonlinear additives are concentrated on the right and are included in the terms with the index "ω":

$$p_\omega' = T_{r\omega}',_r + \frac{1}{r}(T_{r\omega}' - T_{\varphi\omega}'), \quad q_\omega' = M_{r\omega}',_{rr} + \frac{1}{r}(2M_{r\omega}',_r - M_{\varphi\omega}',_r),$$
$$q_\omega' = M_{r\omega}',_{rr} + \frac{1}{r}(2M_{r\omega}',_r - M_{\varphi\omega}',_r). \tag{10.30}$$

The linear components of the generalized internal forces are still expressed through the displacements by formulas like (8.12–8.14), so the system of equilibrium Eq. (10.29) in displacements is reduced to the form:

$$\begin{aligned} &\mathrm{L}_2(a_1u' + a_2\psi' - a_3w',_r) = -p' + p_\omega', \\ &\mathrm{L}_2(a_2u' + a_4\psi' - a_5w',_r) - 2cG_3\psi' = h_\omega', \\ &\mathrm{L}_3(a_3u' + a_5\psi' - a_6w',_r) = -q' + q_\omega'. \end{aligned} \tag{10.31}$$

where the linear differential operators L_2, L_3 and the coefficients a_i are defined by relations (8.16, 8.18); the comma in the lower index denotes the differentiation operation by the following coordinate.

To close the boundary value problem of radiation elastic-plasticity in the case under consideration, the force (8.9) or kinematic (8.19) boundary conditions must be added to Eq. (10.31).

The resulting system of differential Eq. (10.31) is nonlinear. The method of elastic solutions is used to solve it. In this case, in iterative form this system will be:

$$\begin{aligned}
&\mathrm{L}_2\left(a_1 u'^{(n)} + a_2\psi'^{(n)} - a_3 w',_r^{(n)}\right) = -p' + p'^{(n-1)}_\omega,\\
&\mathrm{L}_2\left(a_2 u'^{(n)} + a_4\psi'^{(n)} - a_5 w',_r^{(n)}\right) - 2cG_3\psi'^{(n)} = h'^{(n-1)}_\omega,\\
&\mathrm{L}_3\left(a_3 u'^{(n)} + a_5\psi'^{(n)} - a_6 w',_r^{(n)}\right) = -q' + q'^{(n-1)}_\omega.
\end{aligned} \tag{10.32}$$

Here n is the number of approximation, $p'^{(n-1)}_\omega, h'^{(n-1)}_\omega, q'^{(n-1)}_\omega$ are additional "external" loads, which are assumed to be zero at the first step, and subsequently calculated from the results of the previous approximation by using formulas like (10.30):

$$\begin{aligned}
p'^{(n-1)}_\omega &= T'^{(n-1)}_{r\omega},_r + \frac{1}{r}\left(T'^{(n-1)}_{r\omega} - T'^{(n-1)}_{\varphi\omega}\right),\\
h'^{(n-1)}_\omega &= H'^{(n-1)}_{r\omega},_r + \frac{1}{r}\left(H'^{(n-1)}_{r\omega} - H'^{(n-1)}_{\varphi\omega}\right) - Q'^{(n-1)}_\omega,\\
q'^{(n-1)}_\omega &= M'^{(n-1)}_{r\omega},_{rr} + \frac{1}{r}\left(2M'^{(n-1)}_{r\omega},_r - M'^{(n-1)}_{\varphi\omega},_r\right).
\end{aligned} \tag{10.33}$$

The nonlinear additives in (10.33) are calculated by the expressions:

$$\begin{aligned}
T'^{(n-1)}_{r\omega} &\equiv \sum_{k=1}^{3}\int_{h_k}\sigma'^{(k,n-1)}_{\alpha\omega}\mathrm{d}z = \sum_{k=1}^{3}\int_{h_k} 2G_k\omega_k\left(\varepsilon'^{(k,n-1)}_\alpha, I\right)э'^{(k,n-1)}_\alpha \mathrm{d}z\ ,\\
M'^{(n-1)}_{r\omega} &\equiv \sum_{k=1}^{3}\int_{h_k}\sigma'^{(k,n-1)}_{\alpha\omega} z\ \mathrm{d}z = \sum_{k=1}^{3}\int_{h_k} 2G_k\omega_k\left(\varepsilon'^{(k,n-1)}_\alpha, I\right)э'^{(k,n-1)}_\alpha z\ \mathrm{d}z\ ,\\
H'^{(n-1)}_{r\omega} &= M'^{(3,n-1)}_{\alpha\omega} + c\left(T'^{(k,n-1)}_{\alpha\omega} - T'^{(2,n-1)}_{\alpha\omega}\right),\\
Q'^{(n-1)}_\omega &= \int_{-c}^{c} 2G_k\omega_k\left(\varepsilon'^{(k,n-1)}_\alpha, I\right)э'^{(k,n-1)}_\alpha \mathrm{d}z \quad (\alpha = r, \phi).
\end{aligned} \tag{10.34}$$

Thus, at each step of approximation we have a linear problem of elasticity theory for the plate under study with known additional "external" loads, which are calculated by formula (10.33, 10.34). The procedure for finding the solution of the system (10.32), taking into account its boundedness in the center of the plate, does not differ from that previously used for the corresponding thermoelastic-plasticity problem. As a result, we have the following recurrence solution to the radiation elastic-plasticity problem:

$$
\begin{aligned}
\psi'^{(n)} &= C_2^{(n)} I_1(\beta r) + \psi_r'^{(n)}, \\
u'^{(n)} &= \frac{a_3}{a_1 a_6 - a_3^2}\left[\mathrm{L}_3^{-1}\left(q' - q_\omega'^{(n-1)}\right) - \frac{a_6}{a_3}\mathrm{L}_2^{-1}\left(p' - p_\omega'^{(n-1)}\right) + \left(a_5 - \frac{a_2 a_6}{a_3}\right)\psi'^{(n)} + C_7^{(n)} r\right], \\
w'^{(n)} &= \frac{1}{b_3}\left[b_2\left(\frac{C_2^{(n)}}{\beta} I_0(\beta r) + \int \psi_r'^{(n)} \mathrm{d}\, r\right) - \right. \\
&\left. - \int\left(\frac{a_3}{a_1}\mathrm{L}_2^{-1}\left(p' - p_\omega'^{(n-1)}\right) - \mathrm{L}_3^{-1}\left(q' - q_\omega'^{(n-1)}\right)\right)\mathrm{d}\, r + \frac{C_5^{(n)} r^2}{4} + C_4^{(n)}\right],
\end{aligned}
\tag{10.35}
$$

where $\psi_r'^{(n)}$ is the partial solution of the inhomogeneous modified Bessel equation, which can be taken in the form:

$$
\psi_r'^{(n)} = -K_1(\beta r)\int I_1(\beta r) f'^{(n)} r\mathrm{d}r + I_1(\beta r)\int K_1(\beta r) f'^{(n)} r\mathrm{d}r,
$$

$f'^{(n)}(r)$ is integrating function

$$
f'^{(n)}(r) = \frac{b_3}{b_1 b_3 - b_2^2}\left[h_\omega'^{(n-1)} + \frac{a_2 b_3 - a_3 b_2}{a_1 b_3}\left(p' - p_\omega'^{(n-1)}\right) + \frac{b_2}{b_3 r}\int\left(q' - q_\omega'^{(n-1)}\right) r\mathrm{d}r\right].
$$

In the case of plate contour embedding boundary conditions, we obtain the following recurrence formulas for the integration constants:

$$
\begin{aligned}
C_2^{(n)} &= -\frac{\psi_r'^{(n)}\big|_{r=1}}{I_1(\beta)}, \quad C_4^{(n)} = -b_2\left(\frac{C_2^{(n)} I_0(\beta)}{\beta} + \int \psi_r'^{(n)} \mathrm{d}r\big|_{r=1}\right) + \\
&+ \int\left(\frac{a_3}{a_1}\mathrm{L}_2^{-1}\left(p' - p_\omega'^{(n-1)}\right) - \mathrm{L}_3^{-1}\left(q' - q_\omega'^{(n-1)}\right)\right)\mathrm{d}r\bigg|_{r=1} - \frac{C_5^{(n)}}{4}, \\
C_5^{(n)} &= 2\left(\frac{a_3}{a_1}\mathrm{L}_2^{-1}\left(p' - p_\omega'^{(n-1)}\right) - \mathrm{L}_3^{-1}\left(q' - q_\omega'^{(n-1)}\right)\right)\bigg|_{r=1}, \\
C_7^{(n)} &= \left(\frac{a_6}{a_3}\mathrm{L}_2^{-1}\left(p' - p_\omega'^{(n-1)}\right) - \mathrm{L}_3^{-1}\left(q' - q_\omega'^{(n-1)}\right)\right)\bigg|_{r=1}.
\end{aligned}
\tag{10.36}
$$

With a hinged support of the contour, the constants of integration $C_2^{(n)}$, $C_4^{(n)}$, $C_7^{(n)}$ will be the same as in (10.36), and the constant $C_5^{(n)}$ is defined by the expression

$$C_5^{(n)} = \frac{2b_3}{a_6 + a_7}\left[a_3 u'^{(n)},_r(1) + \psi'^{(n)},_r(1)\left(a_5 - \frac{a_6 b_2}{b_3}\right) - \right.$$
$$-\left(\frac{a_3}{a_1}\mathrm{L}_2^{-1}(p' - p_\omega'^{(n-1)}) - \mathrm{L}_3^{-1}(q' - q_\omega'^{(n-1)})\right),_r\Bigg|_{r=1} -$$
$$\left. -\left(\frac{a_3}{a_1}\mathrm{L}_2^{-1}(p' - p_\omega'^{(n-1)}) - \mathrm{L}_3^{-1}(q' - q_\omega'^{(n-1)})\right)\Bigg|_{r=1} - \sum_{k=1}^{3} K_k B_k \int_{h_k} I\, z\, \mathrm{d}z\right]. \quad (10.37)$$

The analytical form of the plasticity function in the kth bearing layer at some value of the integral neutron flux I in further numerical calculations is taken as follows:

$$\omega^{(k)}\left(\varepsilon_u^{(k)}, I\right) = \begin{cases} 0, & \varepsilon_u^{(k)} \le \varepsilon_y^{(k)}, \\ A_1^{(k)}\left(1 - \frac{\varepsilon_y^{(k)}}{\varepsilon_u^{(k)}}\right)^{\alpha_1^{(k)}}, & \varepsilon_u^{(k)} > \varepsilon_y^{(k)}. \end{cases} \quad (10.38)$$

where $\varepsilon_y'^{(k)} = \varepsilon_y'^{(k)}(I)$ is the strain yield strength, which is calculated by formula (2.40).

Thus, relations (10.35–10.37) provide an analytical recurrence solution to the boundary value problem describing the deformation of a three-layer elastic–plastic circular plate under neutron irradiation.

10.2.1 Repeated-Variable Bending of an Elastic–Plastic Plate

Let, starting from the moment $t = t_1$, instantaneous unloading and repeated loading with forces of the opposite sign p'', q'', changing by the same law as during loading from the natural state. These forces will create in the kth layer of the plate a field of displacements u'', ψ'', w'', strains $\varepsilon_\alpha''^{(k)}$ and stresses $\sigma_\alpha''^{(k)}$, $\sigma_{rz}''^{(3)}$ ($\alpha = r, \varphi$). We will assume that during the unloading and subsequent alternating loading, the effect of neutron flux has been stopped, and the integral neutron flux in all points of the body remains unchanged. It coincides with the level that was at the time of the start of unloading $I_1(z) = I(z, t_1)$.

As before, we introduce for all values of the stress–strain state and load differences of the type (9.30), in which the values with one stroke are the stresses, strains, and displacements in the plate before unloading at the integral neutron flux $I_1(z)$, with two strokes the similar parameters in the second half-cycle are marked:

$$
\begin{aligned}
&\sigma_{\alpha}^{(k)*} = \sigma_{\alpha}^{\prime(k)} - \sigma_{\alpha}^{\prime\prime(k)}, \quad \sigma_{rz}^{(3)*} = \sigma_{rz}^{\prime(3)} - \sigma_{rz}^{\prime\prime(3)}, \quad s_{\alpha}^{(k)*} = s_{\alpha}^{\prime(k)} - s_{\alpha}^{\prime\prime(k)}, \sigma^{(k)*} = \sigma^{\prime(k)} - \sigma^{\prime\prime(k)}, \\
&\varepsilon_{\alpha}^{(k)*} = \varepsilon_{\alpha}^{\prime(k)} - \varepsilon_{\alpha}^{\prime\prime(k)}, \quad \varepsilon_{rz}^{(3)*} = \varepsilon_{rz}^{\prime(3)} - \varepsilon_{rz}^{\prime\prime(3)}\ \text{э}_{\alpha}^{(k)*} = \text{э}_{\alpha}^{\prime(k)} - \text{э}_{\alpha}^{\prime\prime(k)}, \quad \varepsilon^{(k)*} = \varepsilon^{\prime(k)} - \varepsilon^{\prime\prime(k)}, \\
&q^{*} = q^{\prime} - q^{\prime\prime}, \quad p^{*} = p^{\prime} - p^{\prime\prime}, \\
&u^{*} = u^{\prime} - u^{\prime\prime}, \quad \psi^{*} = \psi^{\prime} - \psi^{\prime\prime}, \quad w^{*} = w^{\prime} - w^{\prime\prime}, \qquad (\alpha = r, \varphi).
\end{aligned}
\tag{10.39}
$$

Due to the linear connection of strains with displacements, the relations (8.2) will also be valid here for the corresponding magnitudes with asterisks. Formally, they will take the form (9.31).

For the stresses and strains with asterisks, introduced according to relations (10.39), we will accept physical equations of state of the type (10.26):

$$
\begin{aligned}
&s_{\alpha}^{(k)*} = 2G_{\alpha}^{(k)}\left(1 - \omega^{(k)*}\left(\varepsilon_{u}^{(k)*},\ I_1\right)\right)\text{э}_{\alpha}^{(k)*}, \ \mathrm{s}_{\mathrm{rz}}^{(3)*} = 2\mathrm{G}_{\alpha}^{(3)}\left(1 - \omega^{(3)*}\left(\varepsilon_{\mathrm{u}}^{(3)*},\ \mathrm{I}_1\right)\right)\text{э}_{\mathrm{rz}}^{(3)*}, \\
&\sigma^{(k)*} = 3K_k\varepsilon^{(k)*}, \qquad\qquad (k = 1, 2, 3; \quad \alpha = r, \varphi).
\end{aligned}
\tag{10.40}
$$

According to Moskvitin's theory [1], we assume that the universal nonlinearity functions $\omega^{(k)*}$ in the bearing layers can be expressed in terms of the nonlinearity functions under loading from the natural state

$$
\omega^{(k)*} = \omega^{\prime(k)}\left(\varepsilon_{u}^{(k)*},\ \varepsilon_{y}^{(k)*},\ I_1,\ A_1^{*}, \alpha_1^{*}\right),
\tag{10.41}
$$

i.e. for the materials of the bearing layers the formula (10.38) takes the form

$$
\omega^{(k)*}\left(\varepsilon_{u}^{(k)*}, I_1\right) = \begin{cases} 0, & \varepsilon_{u}^{(k)*} \le \varepsilon_{y1}^{(k)*}, \\ A_1^{(k)*}\left(1 - \dfrac{\varepsilon_{y1}^{(k)*}}{\varepsilon_{u}^{(k)*}}\right)^{\alpha_1^{(k)*}}, & \varepsilon_{u}^{(k)*} > \varepsilon_{y1}^{(k)*}. \end{cases}
$$

The physical nonlinearity of the filler under repeated alternating loading is still described by relations (10.26). Its nonlinear characteristics do not change during cyclic deformation. Using the physical relations (10.40), the expression of the stress components with asterisks through strains will be represented as:

$$
\begin{aligned}
&\sigma_{\alpha}^{(k)} = \sigma_{\alpha e}^{\prime(k)} - \sigma_{\alpha\omega}^{\prime(k)} \quad \sigma_{rz}^{\prime(3)} = \sigma_{rze}^{\prime(3)} - \sigma_{rzw}^{\prime(3)}, \\
&\sigma_{\alpha e}^{\prime(k)} = 2G\,\text{э}_{\alpha}^{\prime(k)} + K_k\left(3\varepsilon^{\prime(k)} - B_k I\right), \quad \sigma_{\alpha\omega}^{\prime(k)} = 2G_k\omega_k\,\text{э}_{\alpha}^{\prime(k)}, \\
&\sigma_{rze}^{\prime(3)} = 2G_3\,\text{э}_{rz}^{\prime(3)}, \sigma_{rzw}^{\prime(3)} = 2G_3\omega_3\,\text{э}_{rz}^{\prime(3)} \quad (\alpha = r, \varphi).
\end{aligned}
\tag{10.42}
$$

Here, unlike similar quantities with the stroke (10.27), the integral neutron flux is not contained explicitly, but is included in the nonlinearity functions.

By performing a similar operation with the values of internal forces introduced in (8.3), we obtain:

$$
\begin{aligned}
&T_{\alpha}^{(k)*} = T_{\alpha}^{(k)*} - T_{\alpha\omega}^{(k)*}, \quad M_{\alpha}^{(k)*} = M_{\alpha}^{(k)*} - M_{\alpha\omega}^{(k)*}, \\
&H_{\alpha}^{*} = H_{\alpha e}^{(k)*} - H_{\alpha\omega}^{(k)*}, \quad Q^{*} = Q^{*} - Q_{\omega}^{*}.
\end{aligned} \tag{10.43}
$$

The linear (index "*e*") and nonlinear (index "ω") terms in (10.43) are calculated through the stress components (10.42) by the expressions:

$$
\begin{aligned}
&T_{e}^{(k)*} = \int_{h_k} \sigma_{x}^{(k)*} dz, \quad T_{\omega}^{(k)*} = \int_{h_k} \sigma_{x\omega}^{(k)*} dz, \quad M_{e}^{(k)*} = \int_{h_k} \sigma_{x}^{(k)*} z dz, \\
&M_{\omega}^{(k)*} = \int_{h_k} \sigma_{x\omega}^{(k)*} z dz, \quad Q_{e}^{*} = \int_{-c}^{c} \sigma_{xz}^{(3)*} dz, \quad Q_{\omega}^{*} = \int_{-c}^{c} \sigma_{xz\omega}^{(3)*} dz.
\end{aligned} \tag{10.44}
$$

After that, the generalized internal forces with asterisks take the form:

$$
\begin{aligned}
&T_{\alpha}^{*} = T_{\alpha e}^{*} - T_{\alpha\omega}^{*} = \sum_{k=1}^{3} T_{\alpha e}^{(k)*} - \sum_{k=1}^{3} T_{\alpha\omega}^{(k)*}, \\
&M_{\alpha}^{*} = M_{\alpha e}^{*} - M_{\alpha\omega}^{*} = \sum_{k=1}^{3} M_{\alpha e}^{(k)*} - \sum_{k=1}^{3} M_{\alpha\omega}^{(k)*}, \\
&H_{\alpha e}^{*} = M_{\alpha e}^{(3)*} + c\left(T_{\alpha e}^{(1)*} - T_{\alpha e}^{(2)*}\right), \quad H_{\alpha\omega}^{*} = M_{\alpha\omega}^{(3)*} + c\left(T_{\alpha\omega}^{(1)*} - T_{\alpha\omega}^{(2)*}\right).
\end{aligned} \tag{10.45}
$$

After performing the appropriate procedures for constructing the equilibrium equations for the internal forces with asterisks, we obtain a system of differential equations in a form similar to (10.29):

$$
\begin{aligned}
&T_{r}^{*},_{r} + \frac{1}{r}\left(T_{r}^{*} - T_{\varphi}^{*}\right) = -p^{*} + p_{\omega}^{*}, \\
&H_{r}^{*},_{r} + \frac{1}{r}\left(H_{r}^{*} - H_{\varphi}^{*}\right) - Q_{r}^{*} = h_{\omega}^{*}, \\
&M_{r}^{*},_{rr} + \frac{1}{r}\left(2M_{r}^{*},_{r} - M_{\varphi}^{*},_{r}\right) = -q^{*} + q_{\omega}^{*}.
\end{aligned} \tag{10.46}
$$

Here the nonlinear additions in the right-hand parts are calculated through the nonlinear components of the generalized forces in (10.45):

$$
p_{\omega}^{*} = T_{r\omega}^{*},_{r} + \frac{1}{r}\left(T_{r\omega}^{*} - T_{\varphi\omega}^{*}\right), \quad q_{\omega}^{*} = M_{r\omega}^{*},_{rr} + \frac{1}{r}\left(2M_{r\omega}^{*},_{r} - M_{\varphi\omega}^{*},_{r}\right)
$$

$$
q_{\omega}^{*} = M_{r\omega}^{*},_{rr} + \frac{1}{r}\left(2M_{r\omega}^{*},_{r} - M_{\varphi\omega}^{*},_{r}\right). \tag{10.47}
$$

The linear components of the generalized internal forces with asterisks are expressed through the displacements with asterisks by formulas like (8.12–8.14), so the system of equilibrium Eq. (10.46) in displacements is reduced to this form:

$$
\begin{aligned}
&\mathrm{L}_2\left(a_1 u^* + a_2\psi^* - a_3 w,_r^*\right) = -p^* + p_\omega^*,\\
&\mathrm{L}_2\left(a_2 u^* + a_4\psi^* - a_5 w,_r^*\right) - 2cG_3\psi^* = h_\omega^*,\\
&\mathrm{L}_3\left(a_3 u^* + a_5\psi^* - a_6 w,_r^*\right) = -q^* + q_\omega^*.
\end{aligned}
\tag{10.48}
$$

To solve the resulting system of nonlinear differential Eq. (10.48) we apply the method of elastic solutions. The corresponding iteration equations will be written in the form:

$$
\begin{aligned}
&\mathrm{L}_2\left(a_1 u^{(n)*} + a_2\psi^{(n)*} - a_3 w,_r^{(n)*}\right) = -p^{(n)*} + p_\omega^{(n-1)*},\\
&\mathrm{L}_2\left(a_2 u^{(n)*} + a_4\psi^{(n)*} - a_5 w,_r^{(n)*}\right) - 2cG_3\psi^{(n)*} = h_\omega^{(n-1)*},\\
&\mathrm{L}_3\left(a_3 u^{(n)*} + a_5\psi^{(n)*} - a_6 w,_r^{(n)*}\right) = -q^{(n)*} + q_\omega^{(n-1)*}.
\end{aligned}
\tag{10.49}
$$

Additional "external" loads $p_\omega^{(n-1)*},\ h_\omega^{(n-1)*},\ q_\omega^{(n-1)*}$ in (10.49), at the first step ($n =$ 1) are taken as equal to zero, and then are calculated from the results of the previous approximation according to the formulas following from (10.47):

$$
\begin{aligned}
&p_\omega^{(n-1)*} = T_{r\omega}^{(n-1)*},_r + \frac{1}{r}\left(T_{r\omega}^{(n-1)*} - T_{\varphi\omega}^{(n-1)*}\right),\\
&h_\omega^{(n-1)*} = H_{r\omega}^{(n-1)*},_r + \frac{1}{r}\left(H_{r\omega}^{(n-1)*} - H_{\varphi\omega}^{(n-1)*}\right) - Q_\omega^{(n-1)*},\\
&q_\omega^{(n-1)*} = M_{r\omega}^{(n-1)*},_{rr} + \frac{1}{r}\left(2M_{r\omega}^{(n-1)*},_r - M_{\varphi\omega}^{(n-1)*},_r\right),
\end{aligned}
\tag{10.50}
$$

where

$$
\begin{aligned}
&T_{\alpha\omega}^{(n-1)*} \equiv \sum_{k=1}^{3}\int_{h_k} \sigma_{\alpha\omega}^{(k,\,n-1)*}\mathrm{d}z = \sum_{k=1}^{3}\int_{h_k} 2G_k\omega_k\left(\varepsilon_\alpha^{(k,\,n-1)*}, I_1\right)\mathfrak{z}_\alpha^{(k,\,n-1)*}\mathrm{d}z,\\
&M_{\alpha\omega}^{(n-1)*} \equiv \sum_{k=1}^{3}\int_{h_k} \sigma_{\alpha\omega}^{(k,\,n-1)*}z\mathrm{d}z = \sum_{k=1}^{3}\int_{h_k} 2G_k\omega_k\left(\varepsilon_\alpha^{(k,\,n-1)*}, I_1\right)\mathfrak{z}_\alpha^{(k,\,n-1)*}z\mathrm{d}z,\\
&H_{\alpha\omega}^{(n-1)*} = M_{\alpha\omega}^{(3,\,n-1)*} + c\left(T_{\alpha\omega}^{(1,\,n-1)*} - T_{\alpha\omega}^{(2,\,n-1)*}\right),\\
&Q_\omega^{(n-1)*} = \int_{-c}^{c} 2G_k\omega_k\left(\varepsilon_\alpha^{(3,\,n-1)*}, I_1\right)_{rz}^{(3,\,n-1)*}\mathrm{d}z\quad(\alpha = r, \phi).
\end{aligned}
$$

The boundary value problem for determination of functions $u^{(n)*}(r)$, $\psi^{(n)*}(r)$, $w^{(n)*}(r)$ is closed by adding to the system of Eqs. (10.49, 10.50) force (similar to 8.9) or kinematic boundary conditions:

- when embedding the contour of the plate

$$u^{(n)*} = \psi^{(n)*} = w^{(n)*} = w^{(n)*}, \ r = 0 \quad at \ r = 1$$

- with hinged support

$$u^{(n)*} = \psi^{(n)*} = w^{(n)*} = M_r^{(n)*} = 0 \quad at \ r = 1 \tag{10.51}$$

In this case, the boundary value problem (10.49–10.51) for the values with asterisks, taking into account (10.44, 10.45), coincides with the boundary value problem for some fictitious three-layer elastic–plastic plate, which experiences loading from the natural state by external forces p^*, q^* without regard to the neutron flux.

The fictitious plate is geometrically identical to the original plate under consideration. Its elastoplastic properties are characterized by the shear modulus G_k, the volume modulus K_k, and the universal nonlinearity functions (10.41). The system of differential equations for the values with asterisks (10.49) coincides with (10.32) and differs only in the absence of radiation terms in (10.42). Therefore, its analytical solution for solid plates follows from the solution (10.35), in which it is necessary to make appropriate substitutions for the nonlinearity functions, loads and coefficients d_m:

$$\begin{aligned}
&\psi^{(n)*} = C_2^{(n)*} I_1(\beta r) + \psi_r^{(n)*},\\
&u^{(n)*} = \frac{a_3}{a_1 a_6 - a_3^2}\left[\mathrm{L}_3^{-1}\left(q^* - q_\omega^{(n-1)*}\right) - \frac{a_6}{a_3}\mathrm{L}_2^{-1}\left(p^* - p_\omega^{(n-1)*}\right) + \left(a_5 - \frac{a_2 a_6}{a_3}\right)\psi^{(n)*} + C_7^{(n)*} r\right],\\
&w^{(n)*} = \frac{1}{b_3}\left[b_2\left(\frac{C_2^{(n)*}}{\beta} I_0(\beta r) + \int \psi_r^{(n)*} \mathrm{d}r\right) - \right.\\
&\left. - \int\left(\frac{a_3}{a_1}\mathrm{L}_2^{-1}\left(p^* - p_\omega^{(n-1)*}\right) - \mathrm{L}_3^{-1}\left(q^* - q_\omega^{(n-1)*}\right)\right)\mathrm{d}r + \frac{C_5^{(n)*} r^2}{4} + C_4^{(n)*}\right],
\end{aligned} \tag{10.52}$$

where the partial solution of the modified Bessel equation

$$\begin{aligned}
&\psi_r^{(n)*} = -K_1(\beta r)\int I_1(\beta r) f^{(n)*} r\mathrm{d}r + I_1(\beta r)\int K_1(\beta r) f^{(n)*} r\mathrm{d}r,\\
&f^{(n)*}(r) = \frac{b_3}{b_1 b_3 - b_2^2}\left[h_\omega^{(n-1)*} + \frac{a_2 b_3 - a_3 b_2}{a_1 b_3}\left(p^* - p_\omega^{(n-1)*}\right) + \frac{b_2}{b_3 r}\int\left(q^* - q_\omega^{(n-1)*}\right) r\mathrm{d}r\right].
\end{aligned}$$

If the plate contour is embedded, from (10.51) we obtain the following recurrence formulas for the integration constants in the solution (10.52):

$$C_2^{(n)*} = -\frac{\psi_r^{(n)*}\big|_{r=1}}{I_1(\beta)},\ C_4^{(n)*} = -b_2\left(\frac{C_2^{(n)}I_0(\beta)}{\beta} + \int \psi_r^{(n)*}dr\bigg|_{r=1}\right) +$$
$$+\int\left(\frac{a_3}{a_1}L_2^{-1}(p^* - p_\omega^{(n-1)*}) - L_3^{-1}(q^* - q_\omega^{(n-1)*})\right)dr\bigg|_{r=1} - \frac{C_5^{(n)*}}{4},$$
$$C_5^{(n)*} = 2\left(\frac{a_3}{a_1}L_2^{-1}(p^* - p_\omega^{(n-1)*}) - L_3^{-1}(q^* - q_\omega^{(n-1)*})\right)\bigg|_{r=1},$$
$$C_7^{(n)*} = \left(\frac{a_6}{a_3}L_2^{-1}(p^* - p_\omega^{(n-1)*}) - L_3^{-1}(q^* - q_\omega^{(n-1)*})\right)\bigg|_{r=1}. \tag{10.53}$$

With a hinged support of the contour, the constants of integration $C_2^{(n)*}$, $C_4^{(n)*}$, $C_7^{(n)*}$ will be determined by formula (10.53), and for $C_5^{(n)*}$ we obtain the following expression:

$$C_5^{(n)*} = \frac{2b_3}{a_6 + a_7}\left[a_3 u^{(n)*},_r(1) + \psi^{(n)*},_r(1)\left(a_5 - \frac{a_6 b_2}{b_3}\right) -\right.$$
$$-\left(\frac{a_3}{a_1}L_2^{-1}(p^* - p_\omega^{(n-1)*}) - L_3^{-1}(q^* - q_\omega^{(n-1)*})\right),_r\bigg|_{r=1} -$$
$$\left.-\left(\frac{a_3}{a_1}L_2^{-1}(p^* - p_\omega^{(n-1)*}) - L_3^{-1}(q^* - q_\omega^{(n-1)*})\right)\bigg|_{r=1}\right]. \tag{10.54}$$

We will find the required parameters of the stress–strain state of a three-layer elastic–plastic plate under repeated sign-variable loading from relations of the type (10.14). For example, the displacements are:

$u'' = u' - u^*,\quad \psi'' = \psi' - \psi^*,\quad w'' = w' - w^*,$

where the values with the stroke are given by formulas (10.35–10.42), and with the asterisk—by (10.52–10.54).

10.2.2 Cyclic Bending of the Plate in a Neutron Flux

In a completely likewise we obtain the solution for any mth variable loading. Let, after $(m - 1)$ sign-variable semicycles of loading, starting from the moment $t_{m\text{-}1}$, there is instantaneous unloading and new loading with forces of opposite sign $p^{(m)}$, $q^{(m)}$ (the upper index "m" is the number of loading). These forces will create displacement $u^{(m)}$, $\psi^{(m)}$, $w^{(m)}$, strain $\varepsilon_\alpha^{(m,\,k)}$ and stress $\sigma_\alpha^{(m,\,k)}$, $\sigma_{rz}^{(m,\,3)}$ ($\alpha = r, \varphi$) fields in the kth load-bearing layer of the plate. In this case we will assume that during the $(m - 1)$ sign-variable half-cycles the integral neutron flux in all points of the body has remained the same as at the beginning of the first unloading, i.e. $I'(z) = I_1(z, t_1)$.

We define for all values of the stress–strain state and load differences, in which the values with index $(m - 1)$ are stresses, strains, displacements and loads in the plate before unloading, index m denotes similar parameters during the mth half-cycle:

$$
\begin{aligned}
&\sigma_{\alpha\beta}^{(m,\,k)*} = (-1)^n\Big(\sigma_{\alpha\beta}^{(m-1,\,k)} - \sigma_{\alpha\beta}^{(m,\,k)}\Big), \quad \varepsilon_{\alpha\beta}^{(m,\,k)*} = (-1)^n\Big(\varepsilon_{\alpha\beta}^{(m-1,\,k)} - \varepsilon_{\alpha\beta}^{(m,\,k)}\Big),\\
&u^{(m)*} = (-1)^n\big(u^{(m-1)} - u^{(m)}\big), \quad \psi^{(m)*} = (-1)^n\big(\psi^{(m-1)} - \psi^{(m)}\big),\\
&w^{(m)*} = (-1)^n\big(w^{(m-1)} - w^{(m)}\big),\\
&q^{(m)*} = (-1)^n\big(q^{(m-1)} - q^{(m)}\big), \quad p^{(m)*} = (-1)^n\big(p^{(m-1)} - p^{(m)}\big).
\end{aligned}
\tag{10.55}
$$

The relations (8.2) linking the strains and displacements will also be valid here for the magnitudes with asterisks, which are introduced in (10.55). For the stresses and strains marked with an asterisk, accept physical equations of state of the type (10.40):

$$
\begin{aligned}
&s_\alpha^{(m,k)*} = 2G_1^{(k)}\,\mathfrak{z}_\alpha^{(m,k)*}\big(1-\omega^{(m,k)*}\big), \qquad \sigma^{(m,k)*} = 3K_k\varepsilon^{(m,k)*},\\
&s_{rz}^{(m,\,3*)} = 2G_1^{(3)}\,\mathfrak{z}_{rz}^{(m,\,3)*}\big(1-\omega^{(m,\,3)*}\big) \qquad (k=1,\ 2,\ 3; \quad \alpha = r, \varphi).
\end{aligned}
\tag{10.56}
$$

Applying Moskvitin's theory [1], we assume that in Eq. (10.56) the universal nonlinearity functions $\omega^{(m,\,k)*}$ in the bearing layers can be expressed through the corresponding nonlinearity functions when loaded from the natural state:

$$
\omega^{(m,k)*} = \omega'^{(k)}\Big(\varepsilon_u^{(m,\,k)*},\ \varepsilon_{y1}^{(m,\,k)*},\ I_1,\ A_1^{(m,\,k)*},\ \alpha_1^{(m,\,k)*}\Big),
\tag{10.57}
$$

i.e. for the materials of the bearing layers the formula (10.38) takes the form

$$
\omega^{(m,k)*}\big(\varepsilon_u^{(m,k)*}, I_1\big) = \begin{cases} 0, & \varepsilon_u^{(m,k)*} \le \varepsilon_{y1}^{(m,k)*},\\ A_1^{(m,k)*}\left(1 - \dfrac{\varepsilon_{y1}^{(m,k)*}}{\varepsilon_u^{(m,k)*}}\right)^{\alpha_1(m,k)*}, & \varepsilon_u^{(m,k)*} > \varepsilon_{y1}^{(m,k)*}, \end{cases}
$$

where $A_1^{(m,k)*}$, $\alpha_1^{(m,k)*}$, $\varepsilon_{y1}^{(m,k)*} = \varepsilon_y^{(m,k)*}(I_1)$ are material nonlinearity parameters obtained experimentally at the mth half-cycle of loading.

The physical nonlinearity of the filler is still described by relations (10.26).

We repeat the procedure of obtaining equilibrium equations for the values with asterisks introduced in (10.55). As a result, we obtain a system of the type (10.46), which in displacements takes the form coinciding with (10.48) with the accuracy of notation. To solve the resulting system of nonlinear differential equations we apply the method of elastic solutions. The corresponding iterative equations of equilibrium at the nth approximation step for the values with asterisks at the mth load half-cycle are:

$$
\begin{aligned}
&\mathrm{L}_2\big(a_1u^{(m,n)*} + a_2\psi^{(m,n)*} - a_3w,_r^{(m,n)*}\big) = -p^{(m,n)*} + p_\omega^{(m,n-1)*},\\
&\mathrm{L}_2\big(a_2u^{(m,n)*} + a_4\psi^{(m,n)*} - a_5w,_r^{(m,n)*}\big) - 2cG_3\psi^{(m,n)*} = h_\omega^{(m,n-1)*},\\
&\mathrm{L}_3\big(a_3u^{(m,n)*} + a_5\psi^{(m,n)*} - a_6w,_r^{(m,n)*}\big) = -q^{(m,n)*} + q_\omega^{(m,n-1)*}.
\end{aligned}
\tag{10.58}
$$

The additional "external" loads $p_{\omega}^{(m,\,n-1)*}$, $h_{\omega}^{(m,\,n-1)*}$, $q_{\omega}^{(m,\,n-1)*}$ corresponding to the inelastic components in (10.58) are taken to be zero at the first step ($n = 1$) and are subsequently calculated using the results of the previous approximation by formulas(10.50) with the number of half-cycle m added to the upper index of the terms:

$$
\begin{aligned}
p_{\omega}^{(m,\,n-1)*} &= T_{r\omega}^{(m,\,n-1)*}{}_{,r} + \frac{1}{r}\left(T_{r\omega}^{(m,\,n-1)*} - T_{\varphi\omega}^{(m,\,n-1)*}\right),\\
h_{\omega}^{(m,\,n-1)*} &= H_{r\omega}^{(m,\,n-1)*}{}_{,r} + \frac{1}{r}\left(H_{r\omega}^{(m,\,n-1)*} - H_{\varphi\omega}^{(m,\,n-1)*}\right) - Q_{\omega}^{(m,\,n-1)*},\\
q_{\omega}^{(m,\,n-1)*} &= M_{r\omega}^{(m,\,n-1)*}{}_{,rr} + \frac{1}{r}\left(2M_{r\omega}^{(m,\,n-1)*}{}_{,r} - M_{\varphi\omega}^{(m,\,n-1)*}{}_{,r}\right),
\end{aligned}
\tag{10.59}
$$

where

$$
\begin{aligned}
T_{\alpha\omega}^{(m,\,n-1)*} &\equiv \sum_{k=1}^{3}\int_{h_k}\sigma_{\alpha\omega}^{(k,\,m,\,n-1)*}\mathrm{d}z = \sum_{k=1}^{3}\int_{h_k} 2G_k\omega_k\left(\varepsilon_{\alpha}^{(k,\,m,\,n-1)*}, I_1\right)\mathfrak{z}_{\alpha}^{(k,\,m,\,n-1)*}\mathrm{d}z,\\
M_{\alpha\omega}^{(m,\,n-1)*} &\equiv \sum_{k=1}^{3}\int_{h_k}\sigma_{\alpha\omega}^{(k,\,m,\,n-1)*}z\mathrm{d}z = \sum_{k=1}^{3}\int_{h_k} 2G_k\omega_k\left(\varepsilon_{\alpha}^{(k,\,m,\,n-1)*}, I_1\right)\mathfrak{z}_{\alpha}^{(k,\,m,\,n-1)*}z\mathrm{d}z,\\
H_{\alpha\omega}^{(m,\,n-1)*} &= M_{\alpha\omega}^{(3,\,m,\,n-1)*} + c\left(T_{\alpha\omega}^{(1,\,m,\,n-1)*} - T_{\alpha\omega}^{(2,\,m,\,n-1)*}\right),\\
Q_{\omega}^{(m,\,n-1)*} &= \int_{-c}^{c} 2G_k\omega_k\left(\varepsilon_{\alpha}^{(3,\,m,\,n-1)*}, I_1\right)_{rz}^{(3,\,m,\,n-1)*}\mathrm{d}z\ (\alpha = r, \varphi).
\end{aligned}
$$

Values (10.59) serve as corrections for plasticity and physical nonlinearity of the layer materials, the nonlinearity functions in which depend on the value of the integral neutron flux, fixed before the first unloading.

The boundary value problem for $u^{(m,\,n)*}(r)$, $\psi^{(m,\,n)*}(r)$, $w^{(m,\,n)*}(r)$ is closed by adding to the system of Eq. (10.58) the force (similar to 8.9) or kinematic boundary conditions:

- when embedding the contour of the plate

$$
u^{(m,\,n)*} = \psi^{(m,\,n)*} = w^{(m,\,n)*} = w^{(m,\,n)*},\ r = 0\quad at\ r = 1;
$$

- with hinged support

$$
u^{(m,\,n)*} = \psi^{(m,\,n)*} = w^{(m,\,n)*} = M_{r}^{(m,\,n)*} = 0\quad at\ r = 1;
\tag{10.60}
$$

In this case, the boundary problem for values with asterisks on the mth half-cycle matches with (10.49, 10.51). Therefore, its analytical solution formally matches the solution of (10.52), in which it is necessary to make the appropriate half-cycle substitutions of loads, nonlinearity functions, and appropriate coefficients. As a result:

$$
\begin{aligned}
\psi^{(m,\,n)*} &= C_2^{(m,\,n)*} I_1(\beta r) + \psi_r^{(m,\,n)*},\\
u^{(m,\,n)*} &= \frac{a_3}{a_1 a_6 - a_3^2}\Big[L_3^{-1}\Big(q^{(m)*} - q_\omega^{(m,\,n)*}\Big) - \frac{a_6}{a_3} L_2^{-1}\Big(p^{(m)*} - p_\omega^{(m,\,n)*}\Big) +\\
&+ \Big(a_5 - \frac{a_2 a_6}{a_3}\Big)\psi^{(n)*} + C_7^{(m,\,n)*} r\Big],\\
w^{(m,\,n)*} &= \frac{1}{b_3}\Big[b_2\Big(\frac{C_2^{(m,\,n)*}}{\beta} I_0(\beta r) + \int \psi_r^{(m,\,n)*}\,\mathrm{d}r\Big) -\\
&- \int\Big(\frac{a_3}{a_1} L_2^{-1}\Big(p^{(m)*} - p_\omega^{(m,\,n)*}\Big) - L_3^{-1}\Big(q^{(m)*} - q_\omega^{(m,\,n)*}\Big)\Big)\mathrm{d}r + \frac{C_5^{(m,\,n)*} r^2}{4} + C_4^{(m,\,n)*}\Big],
\end{aligned}
\tag{10.61}
$$

where the partial solution of the inhomogeneous modified Bessel equation

$$
\begin{aligned}
\psi_r^{(m,\,n)*} &= -K_1(\beta r)\int I_1(\beta r) f^{(m,\,n)*} r\mathrm{d}r + I_1(\beta r)\int K_1(\beta r) f^{(m,\,n)*} r\mathrm{d}r,\\
f^{(m,\,n)*}(r) &= \frac{b_3}{b_1 b_3 - b_2^2}\Big[h_\omega^{(m,\,n-1)*} + \frac{a_2 b_3 - a_3 b_2}{a_1 b_3}\big(p^{(m)*} - p_\omega^{(m,\,n-1)*}\big) +\\
&+ \frac{b_2}{b_3 r}\int\big(q^{(m)*} - q_\omega^{(m,\,n-1)*}\big) r\mathrm{d}r\Big].
\end{aligned}
$$

It should be noted, that the integration constants in the solution (10.61) depend on the cycle number m and are determined from the boundary conditions (10.60). In the case of the plate contour embedding we obtain:

$$
\begin{aligned}
C_2^{(m,\,n)*} &= -\frac{\psi_r^{(m,\,n)*}\big|_{r=1}}{I_1(\beta)},\quad C_4^{(m,\,n)*} = -b_2\Big(\frac{C_2^{(n)} I_0(\beta)}{\beta} + \int \psi_r^{(m,\,n)*}\,\mathrm{d}r\big|_{r=1}\Big) +\\
&+ \int\Big(\frac{a_3}{a_1} L_2^{-1}\big(p^{(m)*} - p_\omega^{(m,\,n-1)*}\big) - L_3^{-1}\big(q^{(m)*} - q_\omega^{(m,\,n-1)*}\big)\Big)\mathrm{d}r\Big|_{r=1} - \frac{C_5^{(m,\,n)*}}{4},\\
C_5^{(m,\,n)*} &= 2\Big(\frac{a_3}{a_1} L_2^{-1}\big(p^{(m)*} - p_\omega^{(m,\,n-1)*}\big) - L_3^{-1}\big(q^{(m)*} - q_\omega^{(m,\,n-1)*}\big)\Big)\Big|_{r=1},\\
C_7^{(m,\,n)*} &= \Big(\frac{a_6}{a_3} L_2^{-1}\big(p^{(m)*} - p_\omega^{(m,\,n-1)*}\big) - L_3^{-1}\big(q^{(m)*} - q_\omega^{(m,\,n-1)*}\big)\Big)\Big|_{r=1}.
\end{aligned}
\tag{10.62}
$$

With a hinged support of the contour $C_2^{(m,\,n)*}$, $C_4^{(m,\,n)*}$, $C_7^{(m,\,n)*}$ will be the same, and the integration constant $C_5^{(m,\,n)*}$ is defined by the expression

$$C_5^{(m,\,n*)} = \frac{2b_3}{a_6 + a_7}\left[a_3 u^{(n)*},_r(1) + \psi^{(n)*},_r(1)\left(a_5 - \frac{a_6 b_2}{b_3}\right) - \right.$$
$$- \left(\frac{a_3}{a_1} L_2^{-1}\left(p^* - p_\omega^{(n-1)*}\right) - L_3^{-1}\left(q^* - q_\omega^{(n-1)*}\right)\right),_r\Bigg|_{r=1} - \qquad (10.63)$$
$$\left. - \left(\frac{a_3}{a_1} L_2^{-1}\left(p^* - p_\omega^{(n-1)*}\right) - L_3^{-1}\left(q^* - q_\omega^{(n-1)*}\right)\right)\Bigg|_{r=1}\right].$$

The obtained formulas (10.61–10.63) describe in iterations the analytical solution of the problem for the values with asterisks for an arbitrary *m*th half-cycle loading of a three-layer circular plate in a neutron flux in the case of different kinds of boundary conditions. The desired solution of the cyclic sign-variable loading of the analyzed plate follows from relations (10.55), taking into account solutions (10.35) and (10.61):

$$\sigma_x^{(m,\,k)} = \sigma_x'^k - \sum_{l=2}^{m} (-1)^k \sigma_x^{(l,\,k)*}, \quad \varepsilon_x^{(m,\,k)} = \varepsilon_{ij}'^k - \sum_{l=2}^{m} (-1)^k \varepsilon_x^{(l,\,k)*},$$
$$u^{(m)} = u' - \sum_{l=2}^{m} (-1)^l u^{(l)*}, \quad \psi^{(m)} = \psi' - \sum_{l=2}^{m} (-1)^l \psi^{(l)*}, \quad w^{(m)} = w' - \sum_{l=2}^{m} (-1)^l w^{(l)*},$$

where the values with one stroke are the stresses, strains and displacements in the plate before the first unloading.

10.2.3 Parametric Analysis of Solutions

The dependence of the stress–strain state of an elastic–plastic three-layer circular plate on the neutron flux under single, repeated, and cyclic loading is analyzed numerically.

For the plate (D16T-fluoroplastic-D16T) were taken: $h_1 = 0.02$; $h_2 = 0.06$; $h_3 = 0.4$; $B = 10$–23 m^2/neutron, which provides volume strain in layers 3–3.5%; for duralumin—$\mu = 1.26$ cm, for fluoroplastic—$\mu = 3.21$ cm; other required material parameters are given in Tables 2.1, 2.2 and 2.3. The dependence of the strain yield strength on the integral neutron flux *I* was taken into account by relations (2.40), the volume strain by relations (2.57). Neutron flux intensity—$\varphi_0 = 10^{18}$ neutrons/(m^{2}s), time of its effect—$t = 3600$ s, intensity of the distributed load $q' = -5$ MPa, the contour is hinged.

Figures 10.21, 10.22 and 10.23 show the results of the analysis of the three-layer plate under a single load in the neutron flux. Numerical analysis has demonstrated the acceleration of convergence of the method of elastic solutions in the calculation of the irradiated plate (see Fig. 10.21): *a*—deflection, *b*—relative shear.

This is explained by an increase in the yield strength and a decrease in the plastic properties of metals under neutron exposure. The maximum difference of the results

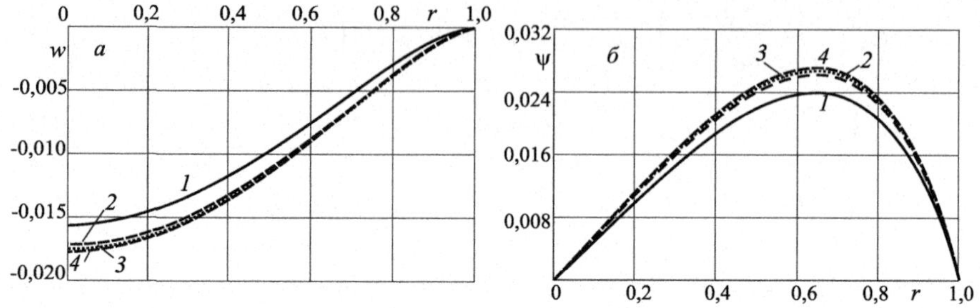

Fig. 10.21 Convergence of the method of elastic solutions

in the 4th approximation, which is taken as the expected solution, from the previous one is about 0.4%, and from the initial one is 14%.

Figure 10.22 compares deflections (*a*) and relative shear (*b*) in the unirradiated and irradiated plate: 1, 3—elastic and elastic–plastic displacements without irradiation; 2, 4—elastic and elastic–plastic displacements after irradiation. In both cases curves *4* are located between graphs 2 and 3. Consequently, neutron irradiation increases displacement due to volumetric deformation, but decreases plastic deformation by increasing the yield strength (at 2–3%).

The distribution of the areas of physical nonlinearity of the layer materials (gray filling) in the cross section of the three-layer plate, taking into account the neutron irradiation, is illustrated in Fig. 10.23. Here the nonlinearity areas occupy a slightly smaller volume than in the case of isothermal and thermoforce loading (see Fig. 8. 2).

Figure 10.24 shows the change in deflection w (*a*) and shear ψ (*b*) along the axis of the three-layer plate after 2 and 8 half cycles of loading. Curves with one stroke correspond to loading from the natural state, two strokes—repeated alternating loading: 1, 2—elastic and elastic–plastic plates without irradiation; 3—elastic–plastic irradiated plate; 3^8—bending on the 8th half-cycle of the irradiated plate. During repeated loading, the displacement decreases by 2–3%. At the 8th half-cycle, the decrease is 4–5% compared to the non-irradiated plate. This is explained by the cyclic hardening

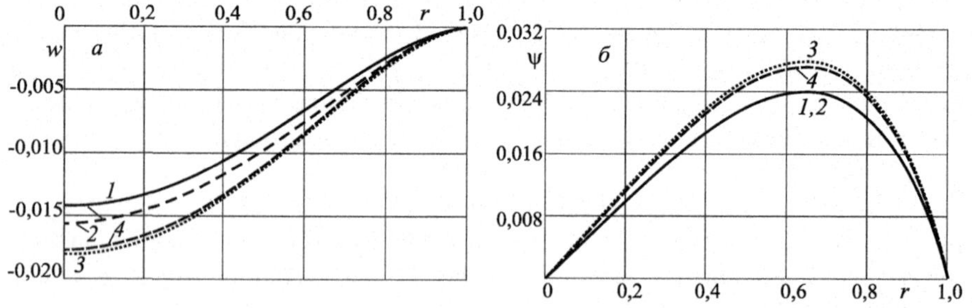

Fig. 10.22 The influence of neutron irradiation on elastic and elastic–plastic displacements in the plate

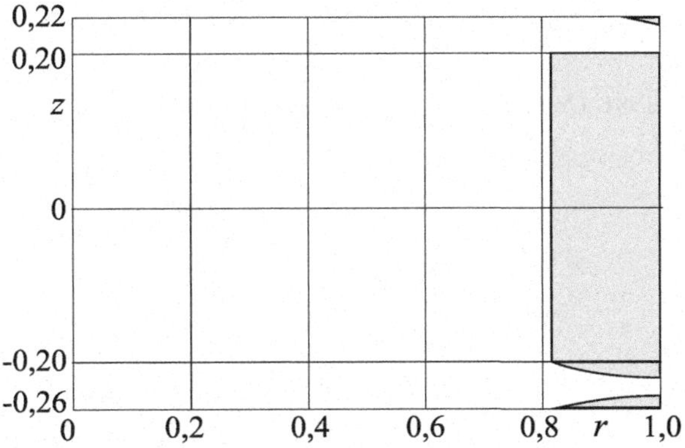

Fig. 10.23 Distribution of the areas of physical nonlinearity of the layer materials in the cross section of the three-layer plate

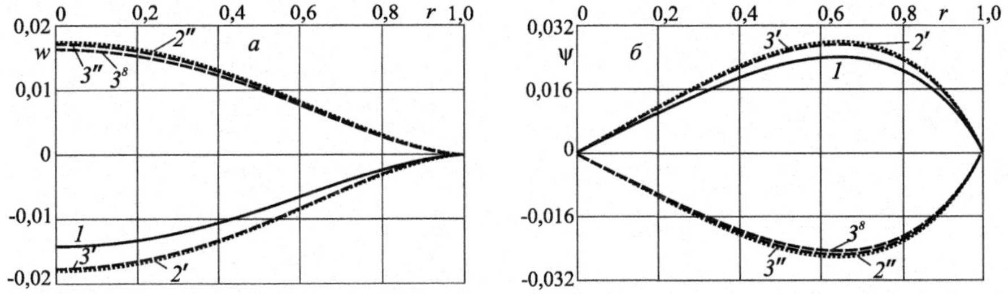

Fig. 10.24 Change of displacements along the axis of the three-layer plate after 2 and 8 half-cycles of loading

of the material; in the calculations, the parameters of the plasticity functions (10.57) were taken the same as those for repeated alternating loading (9.47).

10.3 Conclusions

The methodology proposed in this section and the obtained analytical solutions make it possible to perform a parametric study of the stress–strain state of an elastic–plastic three-layer circular plate under single, repeated and cyclic loads according to the known solution of the corresponding problem of radiation-force loading from the natural. It should be emphasized that these results are valid only in the area of small elastic–plastic deformations and under simple alternating loads.

The developed methods can be used in the design organizations of the construction, machine-building and energy profile when calculating the composite elements of structures.

Reference

1. Moskvitin V.V. 1981. *Cyclic loading of structural elements*, 344. Moscow, Nauka (in Russian)

The manufacturer's authorised representative in the EU is Springer Nature Customer Service Centre GmbH, Europaplatz 3, 69115 Heidelberg, Germany. If you have any concerns regarding our products, please contact ProductSafety@springernature.com

Printed and bound by CPI Group (UK) Ltd, Croydon, CR0 4YY
01/10/2026
02225853-0009